Springer Series in Advanced Manufacturing

The **Springer Series in Advanced Manufacturing** includes advanced textbooks, research monographs, edited works and conference proceedings covering all major subjects in the field of advanced manufacturing.
The following is a non-exclusive list of subjects relevant to the series:

1. Manufacturing processes and operations (material processing; assembly; test and inspection; packaging and shipping).
2. Manufacturing product and process design (product design; product data management; product development; manufacturing system planning).
3. Enterprise management (product life cycle management; production planning and control; quality management).

Emphasis will be placed on novel material of topical interest (for example, books on nanomanufacturing) as well as new treatments of more traditional areas.

As advanced manufacturing usually involves extensive use of information and communication technology (ICT), books dealing with advanced ICT tools for advanced manufacturing are also of interest to the Series.

Springer and Professor Pham welcome book ideas from authors. Potential authors who wish to submit a book proposal should contact Anthony Doyle, Executive Editor, Springer, e-mail: anthony.doyle@springer.com.

INDEXED IN SCOPUS

Tianyuan Liu · Jinsong Bao · Yu Zheng · Yuqian Lu

Industrial Intelligence: Methods and Applications

Springer

Tianyuan Liu
College of Mechanical Engineering
Donghua University
Shanghai, China

Jinsong Bao
College of Mechanical Engineering
Donghua University
Shanghai, China

Yu Zheng
School of Mechanical Engineering
Shanghai Jiao Tong University
Shanghai, China

Yuqian Lu
Department of Mechanical
and Mechatronics Engineering
University of Auckland
Auckland, New Zealand

ISSN 1860-5168 ISSN 2196-1735 (electronic)
Springer Series in Advanced Manufacturing
ISBN 978-3-031-81476-1 ISBN 978-3-031-81477-8 (eBook)
https://doi.org/10.1007/978-3-031-81477-8

This Springer imprint is published by the registered company Springer Nature Switzerland AG
The registered company address is: Gewerbestrasse 11, 6330 Cham, Switzerland

Preface

Intelligent manufacturing is a unifying goal for countries in the era of Industry 4.0.

With the rapid development of artificial intelligence, industrial Internet of things, cloud computing, edge computing, digital twins, etc., the realization of industrial intelligence has become possible. Intelligent technology is driving the transformation of manufacturing to intelligence at an unprecedented rate, and the impact of this transformation and upgrading is comparable to the driving force generated by mass production at the beginning of the 20th century, and the impact of intelligence is reflected in all aspects.

Industrial Intelligence in the Design Stage. Current industrial product design processes rely excessively on manual experience and process optimization processes rely on repetitive testing. This design mode causes a large amount of wasted human and material resources and fails to form quantitative design goals. The integration of intelligent technology can identify and optimize structural design features, conduct structured intelligent analysis of design documents, and carry out intelligent optimization of process design. The related research covers the processes of conceptual design, program design, technical design, construction design, and production design.

Industrial Intelligence in Manufacturing Stage. Manufacturing is an important process in industrial production. However, there are a lot of uncertainties and contingencies in the manufacturing process, which bring impacts on the quality of production. Therefore, this book addresses the typical scenarios in the manufacturing stage-quality inspection, personnel work behavior detection and manufacturing-oriented human–machine collaboration technology, and gives solutions from the perspective of industrial intelligence.

Industrial Intelligence in the Operation and Maintenance Stage. Stable operation of industrial processes as well as equipment maintenance is the guarantee for realizing intelligent manufacturing. The integration of artificial intelligence technology can strengthen the quality control of the production process and realize transparent production. Equipment is the specific executor of production activities, and the prediction of the health state of equipment based on intelligent technology is crucial for the improvement of equipment service life and production quality.

Industrial Intelligence Hot Spot Research. The current level of development of artificial intelligence technology in various manufacturing stages varies, which is due to the varying degree of digitization at different stages. Digital twin is an important enabling technology to realize the information physical system, based on the digital twin technology is of great significance for the digital transformation of industrial processes and the realization of intelligence. It is important to enhance the degree of human collaboration and insight into the manufacturing process based on assistive means such as AR.

Globally, the manufacturing industry still needs to seize the new opportunities for the development of intelligent manufacturing and strive to move forward. Taking digitalization, networking, intelligent transformation and upgrading as the main line, effectively promoting the process of industrial intelligence is an important grasp of the development of the manufacturing industry.

Shanghai, China Tianyuan Liu
Shanghai, China Jinsong Bao
Shanghai, China Yu Zheng
Auckland, New Zealand Yuqian Lu

Contents

Chapter 1
Overview of Industrial Intelligence

The pivotal role of Artificial Intelligence (AI) technology in smart manufacturing represents a prominent area of inquiry. Presently, an increasing array of devices, both large and small, deployed across factory floors are outfitted with sensors. These devices collect and share voluminous data while capturing myriad production actions. Utilizing AI's analytical prowess, this data is analyzed in real-time to enhance decision-making processes. Such advancements provide business users with critical insights into various operational improvements, ranging from reducing asset downtime and boosting manufacturing efficiency to automating production, forecasting demand, optimizing inventory levels, and enhancing risk management.

Concurrently, the evolution of Manufacturing Industry Intelligence (MII) is intrinsically linked with the progression of smart manufacturing, central to which is the integration of intelligence. MII fosters intelligent, resilient manufacturing systems that are fault-tolerant, on-demand, and self-organizing. Its foundational concept involves delivering on-demand manufacturing services to end users through the optimal coordination of distributed manufacturing resources, enhanced by AI methodologies. This book addresses Manufacturing Industrial Intelligence, specifically focusing on AI algorithms, techniques, and application methods within the manufacturing sector. Defined as a systems discipline, it is committed to the development, validation, and deployment of a broad spectrum of machine learning algorithms for industrial applications, aimed at achieving sustainable performance.

1.1 Overview of Intelligent Manufacturing

This section introduces the basic concepts and typical reference architectures for realizing intelligent manufacturing, as well as the four core technologies for virtual-real integration of intelligent manufacturing, respectively. Through the introduction of this section, it provides basic preparation for the subsequent contents.

T. Liu et al., *Industrial Intelligence: Methods and Applications*, Springer Series in Advanced Manufacturing, https://doi.org/10.1007/978-3-031-81477-8_1

1.1.1 Concepts and Reference Models

(1) **The Concept of Intelligent Manufacturing**

The concept of Intelligent Manufacturing is perceived as an imported notion, predominantly articulated in two English phrases: "Smart Manufacturing" and "Intelligent Manufacturing". Interestingly, Wikipedia only recognizes "Smart Manufacturing" thus far. An analysis of academic papers and enterprise technical reports reveals a regional distinction in terminology: in the United States, the term "Smart Manufacturing" is predominantly used by academia and industry, whereas both terms find usage internationally. The leading journal in this domain is titled "Journal of Intelligent Manufacturing". It is notable that academician Yang Shuzi was among the first to publish on the Intelligent Manufacturing System, translating it as "Intelligent Manufacturing System (IMS)". The choice between 'smart' and 'intelligent' largely reflects a preference; 'intelligent' suggests a slightly higher degree of cognition, whereas 'smart' connotes quickness and agility. When applied to manufacturing, however, the terms exhibit negligible difference, suggesting mere variations in contextual usage rather than a substantive distinction.

Intelligent manufacturing has undergone systematic development over the past 30 years. Despite the absence of a uniform definition, it is commonly accepted that intelligent manufacturing integrates cutting-edge information technologies—such as the Internet of Things, big data, cloud computing, deep learning—with advanced automation, sensing, control, and digital manufacturing technologies across the entire product lifecycle. This integration facilitates intelligent sensing, human–computer interaction, decision-making, and execution, thereby revolutionizing units, workshops, factories, and both intra- and inter-enterprise operation management and optimization within new manufacturing systems. Definitions of intelligent manufacturing encompass several key aspects: the product lifecycle, the integration of information and industrial technologies, intelligent methodologies, operation management, and the conceptualization of new manufacturing systems.

In China, the depth of theoretical research in intelligent manufacturing is still evolving, with definitions of its scope and essence remaining somewhat indistinct. Academician Zhou Ji delineates three developmental stages of intelligent manufacturing: digitization, networking, and intelligence. He emphasizes that these stages are interdependent and vary in focus. The transformation and upgrading of manufacturing intelligence fundamentally depend on digital transformation, followed by networked connectivity and intelligent decision-making. The essence of intelligent manufacturing, therefore, is to actualize the value of data within the manufacturing network, enabling data to generate value. Notably, VR/AR technology plays a crucial role in maximizing the value transfer of data.

(2) **Mainstream Intelligent Manufacturing Reference Model**

Since the United States introduced the concept of smart manufacturing in the 1980s, it has garnered significant attention and has been prioritized as a national initiative by numerous countries. Currently, Germany's "Industry 4.0" strategy, the U.S. industrial

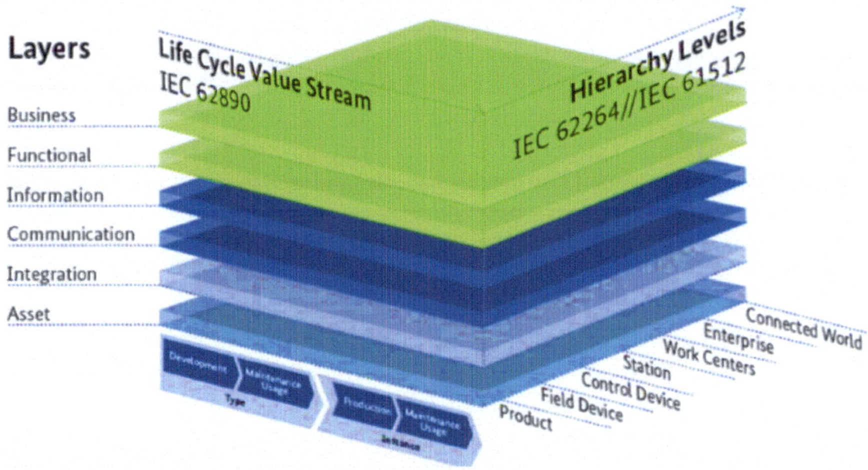

Fig. 1.1 German Industry 4.0 reference model (RAMI 4.0)

Internet strategy, and Made in China 2025 exert substantial global influence. These initiatives underscore the development of four notable smart manufacturing reference models.

1. **German Industry 4.0 Reference Model (RAMI 4.0)**

Initiated in April 2013 at the Hannover Messe, Germany's "Industry 4.0" strategy focuses on establishing a highly flexible, personalized, and digitized smart manufacturing paradigm. This paradigm is achieved through real-time connectivity, mutual identification, and effective communication among individuals, equipment, and products via cyber-physical systems (CPS). The RAMI 4.0 model, illustrated in Fig. 1.1, demonstrates three significant shifts.

Decentralization of Production: Moving from centralized to decentralized production methods, diminishing the importance of scale in industrial output.

Product Personalization: Transitioning from standardized products to those that are highly personalized, anticipating future scenarios where production is automated and tailored to individual preferences, potentially on a single-item basis.

User Engagement: Expanding from limited to comprehensive user involvement, where users engage not only at the beginning and end of the production process but also actively participate throughout the entire value creation cycle.".

The German Industry 4.0 reference model delineates three principal features, as depicted in Fig. 1.1.

Horizontal Integration of the Value Network: This involves the application of Cyber-Physical Systems (CPS) to enhance the synergy between research, development, and application. It fosters cooperation among enterprises in areas such as sustainable development, commercial confidentiality, standardization, and employee training, aligning with the international standard IEC 62890.

Vertical Integration of the Entire Value Chain: This feature pertains to the digitization of the complete value chain—from product design, R&D, and planning to production and service—through the implementation of CPS within the enterprise. It corresponds to the international standard IEC 62264 (or ISA95).

End-to-End System Engineering: At the factory production level, CPS facilitates the customization of special IT structural modules tailored to individual needs. This ensures the organic integration of data collected by sensors and controllers with the ERP management system, thereby establishing an intelligent factory.

2. **Industrial Internet Reference Model (IIRA)**

The concept of the 'industrial Internet' was first introduced by General Electric in 2012. It shares foundational principles with Industry 4.0, advocating for the integration of people, data, and machines to establish an open, global industrial network. This concept extends beyond mere manufacturing processes to encompass the entire value chain of the product life cycle. In contrast to Industry 4.0, the Industrial Internet places a greater emphasis on software, networks, and big data, aiming to foster the integration of physical and digital systems, facilitate the convergence of communication, control, and computation, and create a cohesive cyber-physical system environment.

The industrial Internet framework is built on three core components.

Intelligent Devices: These involve embedding information technology into equipment to transform it into smart, interconnected products. This includes providing digital instrumentation for industrial machines, which represents the initial phase of the Industrial Internet revolution. Enhancements in machine-to-machine interactions are driven by three factors: the reduced cost of instrumentation, the increased computational power of microprocessor chips, and advancements in big data analytics.

Intelligent Systems: These systems interconnect devices to form cohesive units. Beyond traditional networked systems, intelligent systems encompass machine instrumentation and software that are deployed and integrated extensively across different units and networks. As more devices are connected, the system not only expands but also evolves autonomously, becoming progressively more intelligent through the synergies among interconnected machines.

Intelligent Decision-Making: This involves real-time processing and judgment based on insights derived from big data and the Internet. Intelligent decision-making transpires when sufficient data from smart devices and systems enable data-driven learning, facilitating the shift of operational functions from human operators to digital security systems at the network layer.

In March 2014, major industry leaders in the United States, including General Electric, IBM, Cisco, Intel, and AT&T, established the Industrial Internet Consortium (IIC) and introduced the Industrial Internet Reference Architecture (IIRA), illustrated in Fig. 1.2. The IIRA seeks to dismantle technical barriers through the development of universal standards, facilitating data sharing across devices from diverse manufacturers and utilizing the Internet to enhance traditional industrial processes. This initiative aims to better integrate the physical and digital realms.

The Consortium has initiated the formulation of a unified reference architecture for the Industrial Internet. This framework will delineate the functional areas, technologies, and standards pertinent to the Industrial Internet of Things (IoT), steering the development of related standards. Furthermore, it aims to assist hardware and software developers in creating IoT-compatible products. The overarching objective is to achieve comprehensive integration of sensors, networks, computers, cloud computing systems, large-scale enterprises, vehicles, and numerous other entities, thereby enhancing efficiency across the entire industrial chain. Currently, the IIRA and the German RAMI 4.0 reference model are undergoing system integration, as depicted in Fig. 1.3.

3. **Japanese Industrial Innovation Value Chain Reference Model (IVRA)**

On December 8, 2016, the Industrial Value Chain Initiative (IVI), spearheaded by a consortium of manufacturing firms, equipment vendors, and system integration companies, unveiled the Industrial Value Chain Reference Architecture (IVRA). This architecture, designed for smart factories, builds upon the existing infrastructure of the Japanese manufacturing industry and is depicted in Fig. 1.4. The IVRA, fundamentally akin to the RAMI 4.0 model, adopts a three-dimensional approach. Each

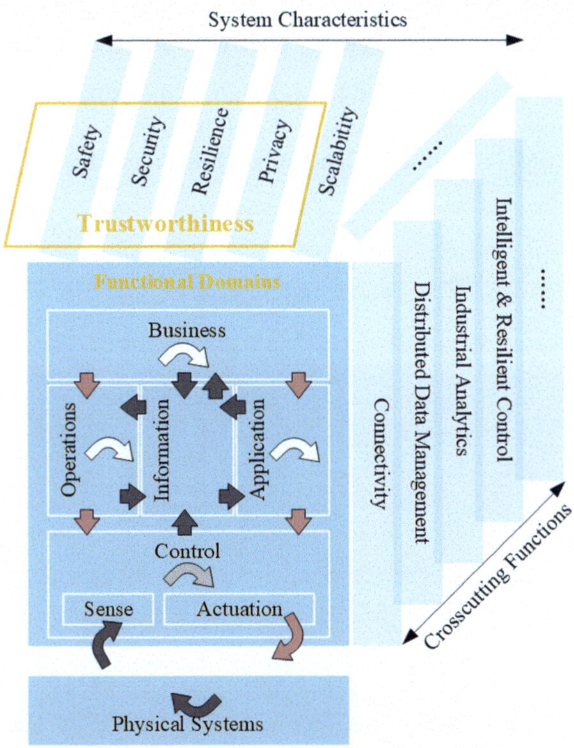

Fig. 1.2 Industrial Internet Reference Architecture (IIRA)

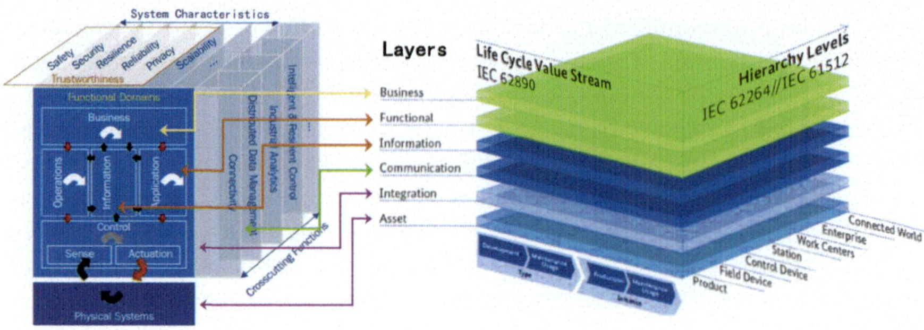

Fig. 1.3 IIRA and RAMI 4.0 convergence

segment is termed a 'Smart Manufacturing Unit (SMU)', with the manufacturing site conceptualized as a single unit assessed along three axes:

- Resource Axis (Vertical): This axis is segmented into layers of employees, processes, products, and equipment.
- Execution Axis (Horizontal): It follows the Plan, Do, Check, and Action sequence of the PDCA cycle.
- Management Axis (Inward): This axis focuses on Quality (Q), Cost (C), Delivery (D), and Environment (E), collectively referred to as QCDE activities.

In the Industrial Value Chain Reference Architecture (IVRA), the manufacturing industry and engineering chains are articulated through an array of Smart Manufacturing Units (SMUs), termed 'Generalized Functional Blocks (GFBs)'. Each

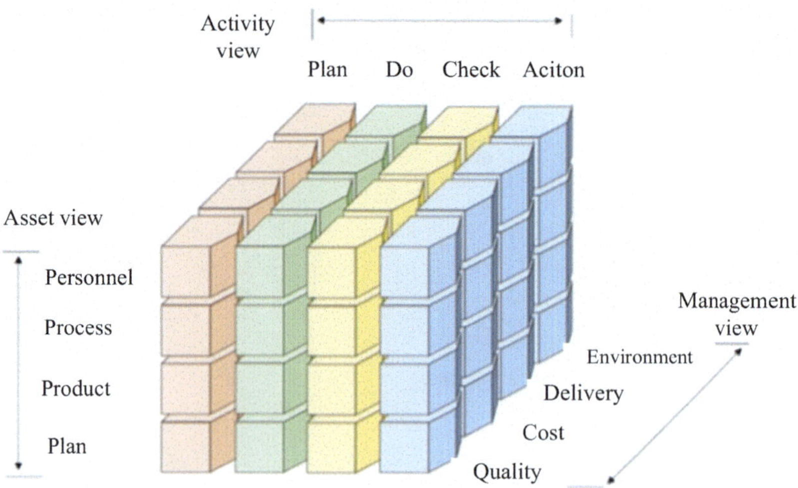

Fig. 1.4 IVCI architecture in Japan

GFB comprises multiple SMUs, representing the enterprise scale vertically across enterprise, department, plant, and equipment layers, and horizontally depicting the production process through five phases: market demand and design, architecture and realization, production, maintenance, and R&D. Inwardly, GFBs reflect the demand and supply process encompassing stages such as basic planning, raw material procurement, production execution, logistics and sales, and after-sales service.

IVRA further categorizes the interconnections between SMUs as 'Process Link Units (PLUs)', subdivided into four components: value, material, information, and data. Precision in the transfer of these components among SMUs enhances smart manufacturing efficiency.

Distinct from RAMI 4.0 or IIRA, IVRA's notable feature is its integration of a 'sense of presence', which includes the tangible involvement of specific employee operations within SMUs. Emblematic of the Japanese manufacturing ethos, particularly exemplified by the Toyota Production System, this approach emphasizes the optimal utilization of human resources to augment on-site production capabilities and efficiency. Consequently, IVI's innovative reference framework for smart factories is poised to set a new global standard, embedding the unique value orientation of Japanese manufacturing.

4. **China Intelligent Manufacturing System Architecture (IMSA)**

Drawing from the intelligent manufacturing advancements in Germany and the United States, China has streamlined its intelligent manufacturing system compared to the RAMI 4.0 framework by consolidating products and equipment into a single equipment hierarchy. As illustrated in Fig. 1.5, the Made in China 2025 reference model is structured across three dimensions: life cycle, system level, and intelligent function. This model is segmented into five distinct layers: the first layer encompasses the production base automation system; the second layer, the production execution system; the third layer, the product full life cycle management system; the fourth layer, the enterprise control and support system; and the fifth layer, the enterprise computing and data center (private cloud).

The life cycle refers to the integration of a series of interconnected value-creating activities, and different industries have different life cycles; the life cycle dimension is refined into design, production, logistics, sales and services, but ignores the difference between sample development and product production; the intelligent function dimension highlights system integration, data integration and information integration at all levels; the standard system architecture is uniquely proposed; and it focuses on solving the basic bottlenecks encountered in the current work of advancing the Intelligent manufacturing work encountered in the data integration, interconnection and other basic bottlenecks.

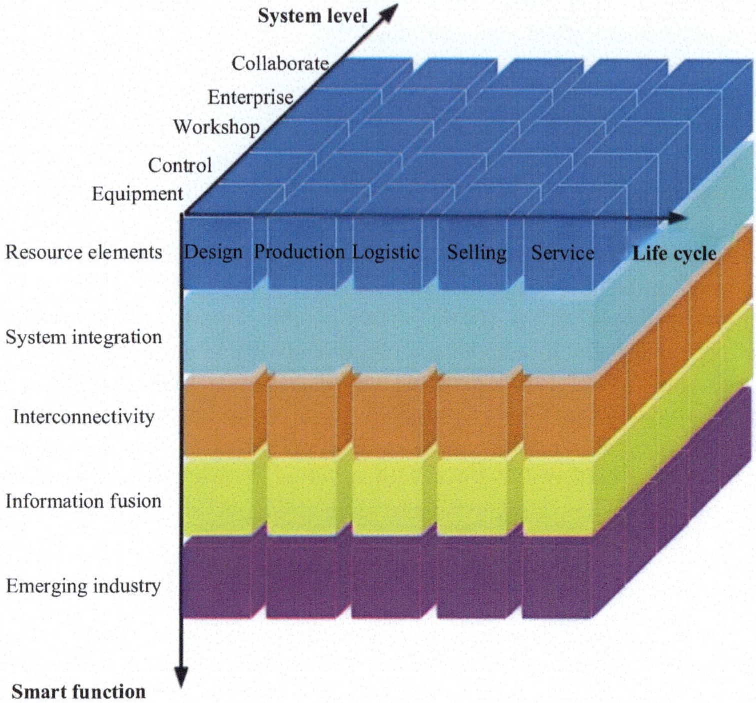

Fig. 1.5 Made in China 2025 reference model

1.1.2 The Core of Intelligent Manufacturing (CPS)

Intelligent manufacturing is characterized by the in-depth application of virtual-real integration technology, and there are four types of virtual-real integration from the point of view of virtual-real integration: mobile, cloud, edge and big data-driven, as shown in Fig. 1.6.

Intelligent manufacturing makes the traditional pyramid of multi-level management structure to a flat network structure (as shown in Fig. 1.7), reducing the levels and intermediate links, and this paradigm change makes the intelligent manufacturing information physics system play a central role.

(1) **Cyber Physical System (CPS)**

C The concept of Cyber Physical Systems (CPS) was first detailed by the U.S. National Science Foundation (NSF) in 2006 during the inaugural international symposium. Since then, the U.S. government, academia, and industry sectors have prioritized CPS research and its application, recognizing its strategic importance in capturing the vanguard of global industrial competition. By 2013, Germany's

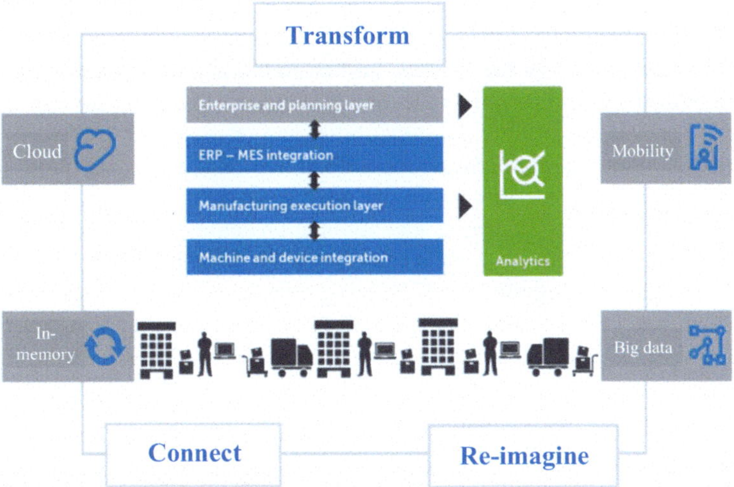

Fig. 1.6 Intelligent manufacturing virtual-real integration approach

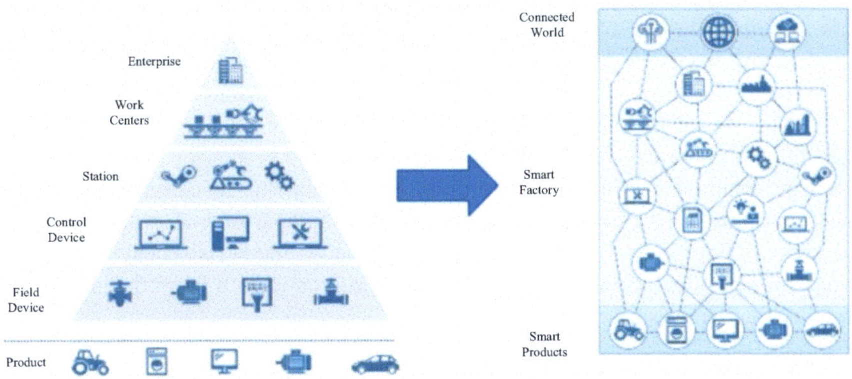

Fig. 1.7 Manufacturing management structure migration

'Industry 4.0 Implementation Recommendations' had designated CPS as the cornerstone technology, launching strategic initiatives in standardization, technology research and development, and the establishment of validation and testing platforms. Emerging from the confluence of control and information technologies, CPS is increasingly pivotal in integrating manufacturing with the internet, positioning itself as a foundational technology system for the new global industrial revolution.

"Made in China 2025" underscores the transformative role of intelligent manufacturing, such as intelligent equipment and smart factories, advocating for enhanced research and development in CPS-related domains including control systems, industrial software, and big data platforms. The State Council's guiding opinions on integrating manufacturing with the internet further emphasize the construction of CPS reference models and comprehensive technical standards, fostering environments for testing and verification to support compatibility and interoperability.

At present, with the accelerated implementation and deepening promotion of 'Made in China 2025', there is an urgent need to explore the origins, conceptual nuances, technological components, and application contexts of CPS. This research aims to build a national consensus and support the development of a robust manufacturing sector. Differing definitions of CPS by significant research entities like the U.S. National Institute of Standards and Technology, the German National Academy of Sciences and Engineering, and the European Union's Seventh Framework Program all converge on the core principle: CPS facilitates automated data flows between cyber and physical spaces, enabling state awareness, real-time analysis, scientific decision-making, and precise implementation. This closed-loop system addresses the complexities and uncertainties in production and application services, enhancing resource allocation efficiency and optimization.

CPS's impact spans four key areas: State perception through diverse sensors; Real-time analysis via industrial software; Scientific decision-making supported by big data platforms; Precise execution through hardware such as controllers and actuators. Together, these components—encompassing hardware, software, networking, and platform services—form the 'new four bases,' complementing the 'four bases' outlined by 'Made in China 2025' to fortify the national manufacturing foundation.

The Ministry of Industry and Information Technology (MIIT) white paper elucidates three hierarchical levels of CPS: unit, system, and system-of-systems, emphasizing the gradual evolution from individual components to extensive, interconnected systems. This progression extends from the departmental to the enterprise and industry levels, eventually reaching the broader industrial ecosystem, facilitating efficient integration of resources and driving innovation in products, business models, and market structures.

1. **Unit Level**

The unit level is the smallest unit of the information-physical system with indivisibility. It can be a component or a product, through "a hard" (such as with sensing, control functions of the robotic arm and transmission bearings, etc.) and "a soft" (such as embedded software) can be composed of "sensing-analysis-decision-making-execution "The data of the closed loop, with a sense, calculable, interactive, extendable, self-decision-making function. Each minimal unit is an information carrier that can be identified, located, accessed, and networked. By describing and modeling the identity information, geometry, functional information, and operation state of physical entities in the information space, a minimal digital unit can be mapped in the virtual space, and along with the processing, assembling, and integrating of physical entities, it can be continuously superimposed, expanded, and upgraded. This process is also the process of synchronous evolution of the smallest unit in both virtual and physical space to the system level and the system of system level.

2. **System Level**

The system level is an organic combination of "a hard, a soft, a network". Multiple smallest units of the cyber physical system (unit level) through the industrial network (such as industrial field bus, industrial Ethernet, etc., referred to as "a network"), to achieve a wider range, a wider range of automatic flow of data, you can constitute an intelligent production line, intelligent workshop, intelligent factory, to achieve a number of unit-level CPS interconnection, interoperability and interoperability, to further improve the optimal allocation of manufacturing resources, breadth, depth and precision. Interconnection, interoperability and interoperation of multiple unit-level CPSs are realized, which further improves the breadth, depth and precision of the optimal allocation of manufacturing resources. System-level CPS realizes self-organization, self-configuration, self-decision-making and self-optimization of local manufacturing resources based on state sensing, information interaction and real-time analysis of multiple unit-level minimum units. Distributed control system (DCS) and data acquisition and monitoring system (SCADA) composed of sensors, control terminals, configuration software, industrial network, etc. are system-level CPS, and intelligent production line composed of CNC machine tools, robots, AGV carts, conveyor belts, etc. are system-level CPS, and production scheduling, equipment management, material distribution, planning, planning and management are carried out for production elements such as human, machine, material, material, and environment through manufacturing execution system (MES). The intelligent workshop, which is composed of equipment management, material distribution, production planning and quality monitoring, is also a system-level CPS.

3. **System of Systems (i.e., SoS Level)**

System of Systems (i.e. SoS level) is an organic combination of multiple system-level CPSs, covering the four major elements of "one hard, one soft, one network and one platform". SoS level CPS realizes cross-system and cross-platform interconnectivity, interoperability and interoperability through the big data platform, and facilitates the integration, exchange and sharing of heterogeneous data from multiple sources. The closed-loop automatic flow realizes comprehensive information perception, in-depth analysis, scientific decision-making and precise execution within the scope of the whole bureau.

(2) **Cyber Physical Production System (CPPS)**

An information-physical system in a manufacturing scenario can be defined as a Cyber-Physical Production System (CPPS, Cyber-Physical Production System).

1. **Industrial IoT and Interconnection: Toward CPPS Interconnectivity**

The emergence of industrial Internet of Things (IoT) has largely promoted the development of intelligent manufacturing and industrial transformation and upgrading, but with the changes of the times and technological innovations, intelligent manufacturing has put forward higher and higher requirements for industrial IoT, so the research on IoT is mainly focused on the following three aspects:

Radio Frequency Identification (RFID). This technology can complete data input and output without direct contact with the recognized items through spatial coupling of radio frequency signals. Therefore, RFID is widely used in logistics, vehicle management, assembly line operation, military management, etc. However, the technology faces risks in data security, privacy, and replicability.

Network communication issues. There are various options for information transmission between machines, people and products connected by IoT, especially wireless transmission is particularly important for the realization of IoT interconnection. The wireless transmission and wireless sensor network (WSN) is the main, through the wireless communication method to form a multi-hop self-organizing system, sensing the object information sent to the observer, with a large network size, self-organization, good dynamic performance, high reliability and a series of advantages. However, the current IoT sensor network in intelligent sensing node technology, its own inspection and diagnosis technology and security encryption technology needs to be further researched and strengthened.

Data fusion and intelligent technology. Due to the existence of a large number of sensors on the Internet of things, in the process of information perception, due to the existence of redundant information, if each sensor individually transmits data to the convergence node, then it will cause a waste of bandwidth and resources, which will definitely affect the timeliness of information collection, so it is necessary to integrate the information to get the information that the user is satisfied with, so that the object can actively communicate with the user. It can be seen, to truly realize the intelligence of the Internet of Things, only a comprehensive study to solve the perception technology, communication technology, fusion technology, intelligent control technology and cloud computing technology, the Internet of Things can be recognized and accepted by the market.

Industrial Internet of Things has five major functions of information perception, transmission, processing, decision-making and implementation, which clearly reflect its several significant features, specifically analyzed as follows.

Ubiquity, interconnectivity (information perception) using RFID, sensors, positioning equipment and other perception devices, to build a ubiquitous network for the manufacturing plant, to realize the interconnectivity of personnel, equipment, materials, products, workshops, factories, information systems, and even all links of the industrial chain, as well as the comprehensive perception and collection of resource attributes, manufacturing status, production process and other data.

Reliability and real-time (information transmission) Connecting the physical entities of manufacturing workshops to the IOT information network, relying on a variety of information communication methods, realizing reliable, real-time exchange and transmission of multi-source information within the network coverage area, and opening up the end-to-end data chain of manufacturing enterprises.

Relevance, integration (information processing) through a variety of data processing methods, intelligent analysis and processing of massive sensory information, the formation of standard information that can be used for optimal decision-making, and support for the integration and control of multi-source manufacturing information from heterogeneous sensing devices.

Autonomous and adaptive (information decision-making) According to the real-time collection of multi-source manufacturing data, it independently analyzes and judges the execution process and the behavior of resources, realizes the dynamic response of the manufacturing process, and based on the relevant knowledge, data model and intelligent algorithms, it realizes the adaptive decision-making of the dynamic resource and energy allocation and production control for the manufacturing process.

Precision, synergistic (information efficiency) According to the decision-making scheme, adjust the manufacturing resources or manufacturing process, so that the object is in the expected state of execution, and through the integration and sharing of real-time data, to achieve synergistic optimization of all units, the whole process, and all the links and precise control.

Taking discrete manufacturing workshop as an example, the manufacturing process describes a series of operational processes by processing and assembling raw materials and transforming them into products, which involves a variety of production factors such as equipment, tooling, materials, personnel, distribution vehicles, etc. and a number of activities such as production, quality control, monitoring, management and control, etc. In view of the different application objectives such as workshop resource management, production scheduling, logistic optimization, quality control, etc., although the various manufacturing Internet of Things architectures proposed by experts For different application objectives such as workshop resource management, production scheduling, logistics optimization, quality control, etc., although experts and scholars have proposed various manufacturing IOT architectures at different levels and with different coverage, they can be described as a system structure based on the main line of "sensing-transmission-processing-application" of discrete workshop manufacturing data.

IOT perception layer discrete manufacturing workshop equipment, personnel, materials, tools, work in progress and other types of production elements and production activities generated by the state, operation, process and other real-time multi-source data is the basis of the optimization and control of the production process for the characteristics of the different production elements and data collection and application needs, through the workshop site configuration RFID, sensors, ultra-wideband (Ultra Wide Band, UWB) and other types of data. Band, UWB) and other types of sensing equipment, to realize the interconnection of all types of production elements and data collection, to ensure that the manufacturing workshop multi-source information of real-time reliable access.

Data transmission layer discrete manufacturing workshop production state, material flow, environmental parameters, equipment operation, quality testing and other data distribution, multi-source, for different sensing equipment has different data transmission characteristics and needs, selectively through the Internet, industrial Ethernet, field bus, industrial LAN, industrial sensor network to realize the effective transmission and exchange of sensing information, to ensure that the workshop site Stable transmission and application of production data.

Analysis and processing layer discrete manufacturing workshop with strong metal interference, blocking and covering and other complex environmental characteristics, as well as multi-species and variable batch mixed-line production, production conditions, production factors moving and other complex production characteristics, which leads to redundancy, disorder and strong uncertainty of manufacturing data, for the typical characteristics of the manufacturing data with large capacity, low value density, such as data calibration, smoothing, filtering, fusion, For manufacturing data with typical characteristics such as large volume and value density, data checking, smoothing, filtering, fusion, classification, correlation and other processing operations are carried out to transform it into effective data that can be applied in production and management, classified and stored, and the value-added application of massive data is realized through a variety of intelligent computation and analysis methods.

The application service layer uses the sensed data for manufacturing workshop management and production process control optimization, providing workshop holographic digital display, manufacturing resource visualization management, real-time monitoring of manufacturing process, dynamic distribution of materials, dynamic scheduling of production, quality diagnosis and traceability and other functional services, and realizes the cooperation with the information systems such as MES, ERP, PDM and other information systems through the unified data integration interface. Through the unified data integration interface, it realizes close integration with manufacturing execution system (MES), enterprise resource planning (ERP), product data management (PDM) and other information systems, and realizes transparent, real-time and accurate management, control and optimization of the manufacturing site on a variety of visualization terminals.

2. **Industrial Intelligence and Software: The Computing Kernel that Empowers CPPS**

Industrial software is the modeling, codification and instrumentation of the laws of industrial R&D and design, production and manufacturing, operation and management, service and other full life cycle aspects, which is the carrier of industrial knowledge, technology accumulation and experience system, and the core of realizing industrial digitalization, networking and intelligence. In short, industrial software is the codification of algorithms, algorithms are the abstract description of the solution to the real problem, the core of the simulation tool is a set of algorithms, the core of the scheduling plan is a set of algorithms, enterprise resource planning is also a set of algorithms. Industrial software defines the information-physical system, the essence of which is to create a "state perception-real-time analysis-scientific decision-making-accurate implementation" of the closed loop of data, to construct a system of rules for the automatic flow of data, to cope with the uncertainty of the manufacturing system, and to realize the efficient allocation of manufacturing resources. Industrial software is the thinking understanding of intelligent manufacturing, the worldview, values and methodology behind perception and control, information transmission and analysis and decision-making, and the brain of intelligent manufacturing. Industrial software supports and defines intelligent manufacturing and constructs the rule system of data flow.

The logic behind intelligent manufacturing lies in the establishment of a set of computer information space and physical space based on the automatic flow of data closed-loop empowerment system to solve the complexity and uncertainty of the production process.

The general term for the application technology of artificial intelligence in the industrial field can be considered industrial intelligence, also known as industrial artificial intelligence. The Chinese Academy of Engineering's "Artificial Intelligence 2.0 Consulting Report" defines new-generation artificial intelligence as "artificial intelligence based on new information environment, new technologies and new development goals", which include new fields of intelligence from macro to micro, smart cities, digital economy, smart manufacturing and so on. Macroscopically, the specific research content of industrial artificial intelligence includes intelligent learning capability (such as machine learning), language capability (such as natural language processing), perception capability (such as image recognition), reasoning capability (such as automated reasoning), memory capability (such as knowledge representation), planning capability (such as automated planning), and execution capability (such as robots), and so on.

As shown in Fig. 1.8, AI technology in the field of intelligent manufacturing is being continuously applied to various fields such as image recognition, voice recognition, intelligent robots, intelligent driving/autonomous driving, troubleshooting and predictive maintenance, quality monitoring, etc., covering a wide range of aspects from R&D and innovation, production management, quality control, troubleshooting, and so on. Artificial intelligence can provide intelligent guidance for complex processes. Taking product development and design as an example, industrial design software, after integrating the AI module, can understand the needs of designers, and can also interface with diversified data such as regional economy, social opinion, social media, and so on, and the resulting data model can intelligently recommend relevant product design and development programs to the designer, and even independently design multiple preliminary product programs for the designer to choose from. Artificial intelligence plays a role in production and manufacturing management, innovates the production model, and improves production efficiency and product quality. Artificial intelligence technology through the Internet of Things on the production process, equipment conditions, process parameters and other information for real-time collection; product quality, defects detection and statistics; in the offline state, the use of machine learning technology to mine the relationship between product defects and the Internet of Things historical data, the formation of control rules; in the online state, through the enhancement of learning technology and real-time feedback, control of the production process to reduce product defects; at the same time Integrating expert experience to continuously improve the learning results. In the maintenance service link, the system uses sensors to monitor the state of the equipment, establishes an analysis model of equipment failure through machine learning, and replaces workpieces that may fail before the failure occurs, so as to guarantee the continuous and trouble-free operation of the equipment. Taking CNC machine tools as an example, the machine learning algorithm model and intelligent sensors and other technical means to monitor the power, current, voltage and other

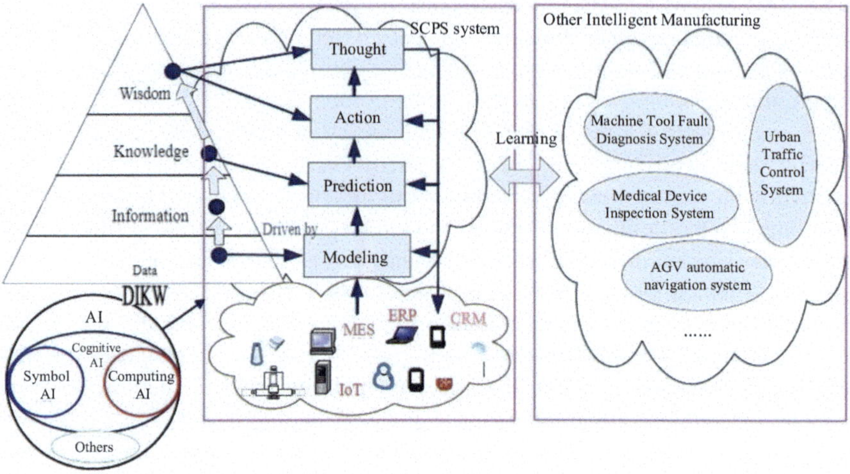

Fig. 1.8 Artificial intelligence-driven smart manufacturing

information of the cutting tool, spindle and feed motor in the machining process, to identify the force, wear and tear of the cutting tool, the state of breakage and the stability of the machine tool machining state, and according to these states, real-time adjustment of machining parameters (spindle speed, feed speed) and machining instructions, to prejudge when to change the tool in order to Improve machining accuracy, shorten production line downtime, and improve the safety of equipment operation.

3. **Industrial Cloud and Industrial Big Data Platform: Decision and Control Platform for CPPS**

The industrial cloud and intelligent service platform is a highly integrated, open and shared data service platform, a cross-system, cross-platform, cross-field data distribution center, data storage center, data analysis center and data sharing center, based on which the industrial cloud service platform promotes the development, integration and open sharing of basic data and tools such as professional software libraries, application model libraries, product knowledge libraries, test and evaluation libraries, case expert libraries, etc., to It realizes the optimization of resource allocation for the management of the whole factors, processes, industry chain and life cycle of production, so as to improve production efficiency, innovate mode of business and build a new industrial ecology. This will bring products, machines, people, and business from closed to open, from independent to system, will reorganize the relationship between customers, suppliers, vendors, and internal organizations, reconfigure the operation mode of information, product, and capital flows in the production system, and rebuild a new industrial value chain and competitive pattern. International giants are accelerating the construction of industrial cloud and intelligent service platform, downward integration of hardware resources, upward bearing software applications, accelerating the pace of integration of global strategic resources, to seize the right to formulate rules, the right to speak standards, the right to dominate the ecological dominance and the competitive high ground. Analogous to the human body, the industrial cloud and intelligent service platform constitutes a decision-making organ that can receive, store and analyze data and information like a brain, and analyze and form decisions.

Industrial big data refers to all kinds of data and related technologies and applications generated in the industrial field around the typical intelligent manufacturing model, from customer demand to sales, orders, planning, R&D, design, process, manufacturing, procurement, supply, inventory, shipment and delivery, after-sales service, operation and maintenance, end-of-life or recycling and remanufacturing of the entire product lifecycle. It is centered on product data, which greatly extends the scope of traditional industrial data, and also includes industrial big data-related technologies and applications. There are three main sources of industrial big data: the first is production and operation-related business data. The first type is production and operation-related business data, which mainly comes from the scope of traditional enterprise informatization and is collected and stored inside enterprise information systems, including traditional industrial design and manufacturing software, enterprise resource planning (ERP), product lifecycle management (PLM), supply chain management (SCM), customer relationship management (CRM), and environmental management system (EMS), etc. Through these enterprise information systems, a large amount of data has been accumulated and stored. These enterprise information systems have accumulated a large amount of product development data, production data, operational data, customer information data, logistics supply data, and environmental data. This type of data is a traditional data asset in the industrial field, and is gradually expanding its scope in the context of new technology applications such as mobile Internet. The second category is equipment IoT data. It mainly refers to industrial production equipment and target products in the IoT operation mode, real-time generation and collection of data covering operation and running conditions, working conditions, environmental parameters and other data reflecting the operation status of equipment and products. Such data is the new and fastest growing source of industrial big data. Narrowly defined industrial big data refers to this type of data, that is, industrial equipment and products quickly generated and the existence of a large number of time-series differences in the data. The third category is external data. It refers to data from external Internet sources of the enterprise related to the production activities and products of the industrial enterprise, for example, environmental regulations for evaluating the environmental performance of the enterprise, and macro-socio-economic data for predicting the product market. Industrial big data technology is a series of technologies and methods that enable the value contained in industrial big data to be mined and displayed, including data acquisition, pre-processing, storage, analysis and mining, visualization and intelligent control. Industrial big data application is the process of integrating and applying a series of industrial big data technologies and methods to a specific industrial big data set to obtain valuable information. The research and breakthrough of industrial big data technology, its essential goal is to discover new patterns and knowledge from complex data sets, mining to get valuable new information, so as to promote the product innovation of industrial enterprises, improve the level of operation and production efficiency and expand new business models. Industrial big data has four typical characteristics: real-time, accuracy, and closed-loop.

Real-time: industrial big data mainly comes from manufacturing and product operation and maintenance, production lines, equipment, industrial products, instruments, etc. are high-speed operation, from the frequency of data collection, data processing, data analysis, abnormality discovery and response, etc. all have high real-time requirements.

Accuracy: mainly refers to the authenticity, completeness and reliability of the data, and pays more attention to the quality of the data, as well as the reliability of the processing and analyzing techniques and methods.

Closed-loop: This includes the closure and association of the data chain in the horizontal process of the whole life cycle of the product, as well as the need to support the dynamic and continuous adjustment and optimization under the closed-loop scenarios of state perception, analysis, feedback and control in the process of longitudinal data collection and processing of intelligent manufacturing. In addition to the above four basic typical characteristics, the industry generally believes that industrial big data also has the characteristics of integration, transparency, predictability and so on.

Industrial big data is the key technology of intelligent manufacturing, and its main role is to open up the physical world and the information world, and promote the transformation of production-oriented manufacturing to service-oriented manufacturing. It has a wide range of application prospects in intelligent manufacturing. In the product lifecycle process of product market demand acquisition, product R&D, manufacturing, operation, service until end-of-life recycling, industrial big data plays a great role in intelligent design, intelligent production, networked collaborative manufacturing, intelligent service, personalized customization and other scenarios. In intelligent design, automated design and digital simulation optimization can be realized through product data analysis; in intelligent production, industrial big data technology can be applied in manufacturing, such as human–machine intelligent interaction, industrial robots, simulation optimization of manufacturing processes, digital control, condition monitoring, etc., to improve the accuracy of production failure prediction and comprehensively optimize production efficiency; in networked collaborative manufacturing, industrial big data technology can be realized in the process of product lifecycle, networked collaborative manufacturing, intelligent services, personalized customization and other scenarios. In networked collaborative manufacturing, industrial big data technology can realize the application of intelligent management, such as product lifecycle management, customer relationship management, supply chain management, production, supply and marketing as a whole, etc., and achieve process synergy and transparency through equipment networking and intelligent control; in intelligent services, industrial big data can realize product intelligence and remote maintenance through the collection, analysis and optimization of product operation and usage data, meanwhile, industrial big data can realize the application of intelligent inspection and supervision, and at the same time, industrial big data can realize the application of intelligent inspection and supervision, which can improve production failure prediction accuracy and comprehensively optimize production efficiency. data can realize the application of intelligent inspection and supervision.

1.2 Industrial Intelligence Definition

Industrial intelligence is a concrete manifestation of intelligent manufacturing. This section will introduce the basic concepts of industrial intelligence and the classification of typical industrial intelligence scenarios. Through this section, readers are provided with a perceptual understanding of industrial intelligence.

1.2.1 Industrial Intelligence Concept

In recent years, with the rapid development of a new generation of information technology, the wave of artificial intelligence is sweeping the world. Machine Learning and Deep Learning are currently the most advanced technologies in the field of Artificial Intelligence, and its the core technology of Artificial Intelligence. Machine Learning (ML) is a subfield of artificial intelligence, which encompasses almost all the methods that have the greatest impact on the world. Machine learning theory focuses on designing and analyzing algorithms that allow computers to learn automatically. Deep Learning (DL) is a subcategory of Machine Learning. It derives its ideas from the way the human brain works and is a learning process that utilizes deep neural networks to solve feature representations. If your first contact with machine learning is deep learning, you may find a deep learning "hammer" in your hand, and all machine learning problems look like "nails". Therefore, machine learning is a technology to achieve artificial intelligence and deep learning a technology to achieve machine learning. The relationship between the three is shown in Fig. 1.9.

As you can see from the above figure, artificial intelligence has a wide range of research areas, machine learning is just one of the branches, while deep learning is just a branch of machine learning. Artificial intelligence is mainly composed of machine learning, expert system, evolutionary computation, fuzzy logic, computer vision, natural language processing, recommender system and so on. And machine learning consists of supervised learning, semi-supervised learning, unsupervised learning and reinforcement learning. With the rapid development of science and technology, the research branches of artificial intelligence are getting richer and richer, promoting the rapid development of various industries, of which the manufacturing industry is a typical example.

The application of artificial intelligence in the manufacturing field is called industrial intelligence, which is an important content of the transformation and development of digitalization, networking and intelligence in manufacturing. With the rapid

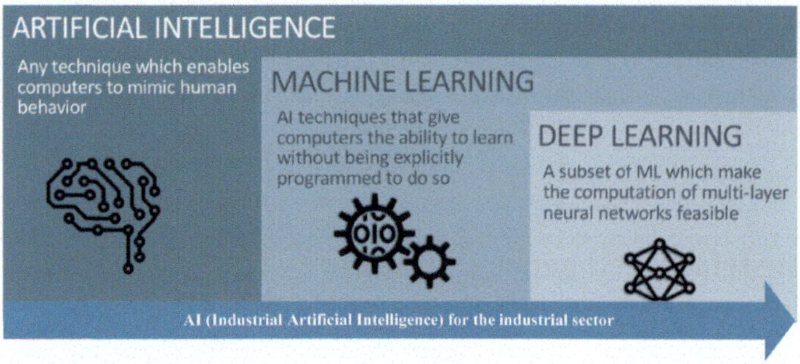

Fig. 1.9 Relationship between industrial artificial intelligence and artificial intelligence

development of artificial intelligence technology, combined with the mechanism model, engineering knowledge and industrial big data accumulation, the formation of artificial intelligence models in the manufacturing field, and integrated with industrial software and industrial Internet platform, will form a series of convergence and innovation of technology, products and models. Artificial intelligence empowered manufacturing field, will optimize to enhance production efficiency, improve product quality, reduce production costs, typical application scenarios are intelligent products and equipment, intelligent factories and production lines, intelligent management and services, intelligent supply chain and logistics, intelligent R & D and design, intelligent monitoring and decision-making. It will promote revolutionary changes in the industrial model, comprehensively reshape the manufacturing value chain, and greatly improve the innovation and competitiveness of the manufacturing industry [1].

In order to improve the core competitiveness of the manufacturing industry, the governments and organizations of the world's major developed countries have actively introduced relevant strategic policies to enhance the level of industrial intelligence has become a global consensus and trend. For example, Germany's "Industry 4.0" [2], the United States' "Industrial Internet" [3] and "Made in China 2025" [4]. With the introduction and promotion of these policies, artificial intelligence technology and manufacturing industry continue to integrate deeply so that the manufacturing industry presents a new manufacturing paradigm - intelligent manufacturing. Intelligent manufacturing is a kind of production method that allows enterprises to become smarter in research and development, production, management, service and other aspects, and its application enables manufacturing enterprises to realize production intelligence, management intelligence, service intelligence and product intelligence.

Artificial intelligence technology has gone through many waves of technological change and large-scale application, from the early expert system, machine learning gradually developed to the current continued hot deep learning and many other technological changes and, its theory and technology is becoming increasingly mature, and the field of application is also expanding. At the same time, with the rapid progress and continuous maturation of hardware computing power, software algorithms and solutions, industrial production has gradually become the key exploration direction of artificial intelligence, and industrial intelligence has emerged.

Industrial intelligence (also known as industrial artificial intelligence) is the fusion of artificial intelligence technology and industrial development, throughout the design, production, management, service and other aspects of the industrial field, to achieve the ability to mimic or exceed the human perception, analysis, decision-making and other technologies, methods, products and application systems. It can be considered that the essence of industrial intelligence is the combination of general artificial intelligence technology with industrial scenarios, mechanisms, and knowledge to realize innovative applications such as design mode innovation, intelligent decision-making in production, and optimal allocation of resources. It needs to have the ability of self-perception, self-learning, self-execution, self-decision-making, and self-adaptation to adapt to the variable industrial environment and accomplish diverse industrial tasks, and ultimately achieve the purpose of enhancing enterprise insight, improving production efficiency or equipment product performance [5].

The core enabling technologies of CPPS are shown in Fig. 1.10. Currently, the fourth scientific and technological revolution and industrial change is flourishing, the core content of which is the digitalization, networking and intelligent development of the industrial economy. As a strategic technology to help this round of scientific and technological revolution and industrial change, the new round of artificial intelligence technology represented by deep learning, knowledge graph, etc. shows an explosive trend, and industrial intelligence has ushered in a new stage of development. Through comprehensive real-time perception of massive data, end-to-end deep integration and intelligent modeling and analysis, industrial intelligence has raised the level of enterprise analysis and decision-making to a whole new level. With the deepening of the integration of industrial intelligence technology and domain knowledge, it is closer to the actual needs of the industry, and its development is becoming more and more mature in computer vision, natural language processing, knowledge mapping, etc.

(1) **Data Technology**

Data is the proverbial new oil that is driving the growth of M4.0. Therefore, it is crucial to understand that smart factories are primarily about data or rather actionable data that leads to information, knowledge and insights. The more data that is analyzed, the more informed the decisions will be. Prior to the Fourth Industrial Revolution, there was a heavy reliance on manual methods to collect machine data-very inefficient, highly susceptible to human error, and unable to provide real-time visibility into operations. As the competitive landscape of manufacturing changes, automating the data collection process for machines and applications is critical to future success. Inspection of the factory floor with sensors can help create a complete view of the capacity and performance of a company's assets, including manufacturing equipment, inventory, personnel, and more. Different types of signals can be captured from the equipment, such as cutting forces, vibrations, acoustic emissions,

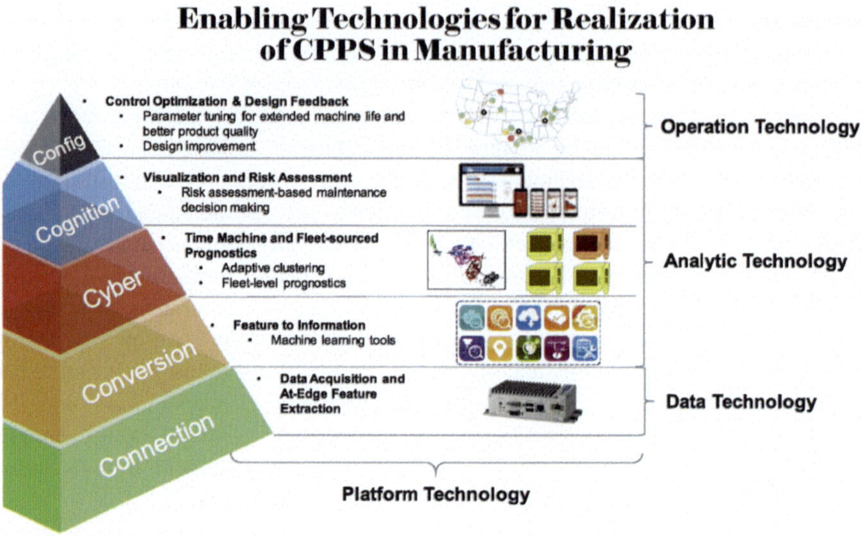

Fig. 1.10 Enabling techniques for implementing CPPS in the manufacturing industry

temperatures, currents, oils, pressures, and more. Data can be static (machine IDs, process parameters, job IDs, etc., i.e. primarily background information) or dynamic (machine health, inventory levels, asset utilization, etc.)). Data can be generated at component level, machine level or shop floor level and can be broadly categorized into structured and unstructured data. The role of data technology is to efficiently track, control and record large and diverse streams of manufacturing data in real time. Communication and management of data has also become an important part of industrial AI data technology. It is expected that the communication technology should be characterized by features such as high data transfer rate, low waiting time, high reliability, high security, accurate traceability and scalability.

(2) **Analytical Techniques**

Many companies today have invested heavily in data acquisition hardware and sensors to capture and store large amounts of process data. But they typically use them only for tracking purposes, not as a basis for improving operations. It is important to know what to do with the information collected. Analytics refers to the application of statistics and other mathematical tools to these data streams to evaluate and improve practices. Through analytics, manufacturers can investigate even the smallest changes in the production process and go beyond lean manufacturing programs (e.g., Six Sigma). It allows them to break down the production process into the most specific tasks or activities to identify specific processes/components that are underperforming or causing bottlenecks. Data-driven modeling allows manufacturing companies to discover hidden patterns, unknown correlations and other useful information in the manufacturing system and integrate the information gained with other technologies to improve productivity and innovation. In addition to data analysis, data visualization tools are an important part of analytics. The up-to-the-minute, real-time information provided by data analytics would be wasted if the results could not be communicated clearly and effectively to staff. Easy-to-interpret and easy-to-use graphs, charts and reports make it easier for manufacturers to understand the data being analyzed, track important metrics and assess distance or proximity to goals. Unknown correlations and other useful information from the manufacturing system and integrate the information gained with other technologies to improve productivity and innovation. In addition to data analysis, data visualization tools are an important part of analytics. The up-to-the-minute, real-time information provided by data analytics would be wasted if the results could not be communicated clearly and effectively to staff. Easy-to-interpret and easy-to-use graphs, charts and reports make it easier for manufacturers to understand the data being analyzed, track important metrics and assess distance or proximity to goals. Unknown correlations and other useful information from the manufacturing system and integrate the information gained with other technologies to improve productivity and innovation. In addition to data analysis, data visualization tools are an important part of analytics. The up-to-the-minute, real-time information provided by data analytics would be wasted if the results could not be communicated clearly and effectively to staff. Easy-to-interpret and easy-to-use graphs, charts and reports make it easier for manufacturers to understand the data being analyzed, track important metrics and assess distance or proximity to goals. Data visualization tools are an essential element of analytics. The up-to-the-minute, real-time information provided by data analysis is wasted if the results are not communicated clearly and effectively to staff. Easy-to-explain and easy-to-use graphs, charts and reports make it easier for manufacturers to understand the data being analyzed, track important metrics and assess the distance or proximity to goals. Data visualization tools are an essential element of analytics. The up-to-the-minute, real-time information provided by data analysis is wasted if the results are not communicated clearly and effectively to staff. Easy-to-explain and easy-to-use graphs, charts and reports make it easier for manufacturers to understand the data being analyzed, track important metrics and assess distance or distance to goals.

(3) **Platform Technology**

A platform is hardware or software that plays an application-supporting role in an industrial environment, such as connecting devices, processing data (collecting/extracting, storing, analyzing, and visualizing), and finally delivering it to the final application. Platforms are at the center of the Industrial AI concept, providing the tools and flexibility needed to develop application-centric functionality unique to each industry. Platform technology helps to orchestrate, integrate, deploy and support technologies such as digital twins, data storage, connectivity across the shop floor, edge intelligence, robotics integration and more.

(4) **Operations Technology**

Based on the information derived from the analysis, operations technology is combined with other technologies aimed at enterprise control and optimization through systems such as product lifecycle management, enterprise resource planning, manufacturing execution systems, customer relationship management, and customer relationship management. supply chain management. Finally, the results of analyzing the collected data can be fed back to equipment designers for closed-loop lifecycle redesign. Additionally, supervisory control and feedback of the physical space is managed through the OT level. Advanced OT integrated in manufacturing systems is used to form closed-loop management systems where tasks are generated and executed by intelligent agents operating in a distributed and autonomous manner. OT enables manufacturing ecosystems with features such as self-configuration, self-adjustment and self-optimization, which ultimately improves flexibility and resilience across the entire production system and results in higher efficiency and economic impact.

1.2.2 Industrial Intelligence Classification

(1) **Industrial Machine Vision**

Computer vision and machine vision are often thought of as one and the same; they are different terms for overlapping technologies. Computer vision broadly refers to the capture and automation of image analysis and focuses on image analysis functions in a wide range of theoretical and practical applications. Traditionally, machine vision refers to the use of computer vision in industrial or practical applications or processes that require the execution of certain functions or outcomes based on image analysis performed by a vision system. The vision system uses software to recognize pre-programmed functions. The system can trigger various set "actions" based on the findings. The connection and difference between the two are shown in Fig. 1.11.

1. **Signal Processing**

Signal processing involves processing electronic signals to clean them up (e.g., to eliminate noise), to extract information, to prepare them for output to a display or to

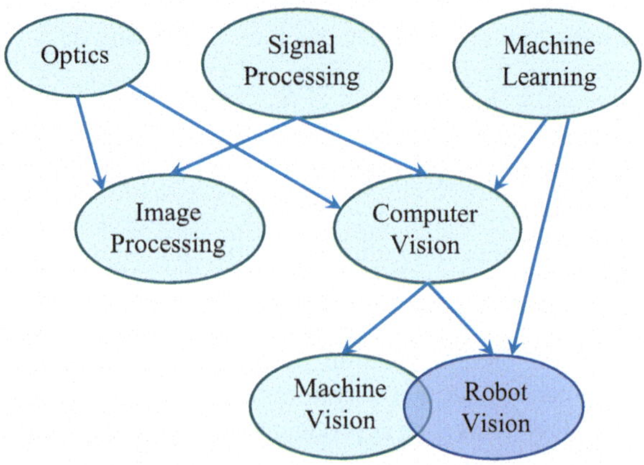

Fig. 1.11 Computational vision versus machine vision

prepare them for further processing. Anything can be more or less a signal. Various types of signals can be processed, such as analog electrical signals, digital electronic signals, frequency signals, etc. An image is basically just a two (or more) dimensional signal. For robot vision, we are interested in image processing, so we are talking about image processing.

2. **Image Processing and Computer Vision**

Computer vision and image processing are like cousins, but they have very different goals. Image processing techniques are mainly used to improve the quality of an image, convert it to another format (e.g. histogram) or alter it for further processing. On the other hand, computer vision is more about extracting information from an image to make sense of it. So you can use Image Processing to convert a color image to grayscale and then use Computer Vision to detect objects in that image. If we look further up the genealogy, we see that both fields have been heavily influenced by the field of physics (especially optics).

3. **Pattern Recognition and Machine Learning**

So far, so simple. It starts to get more complex when we include pattern recognition in the family tree or speak more broadly about machine learning. This branch of the family focuses on recognizing patterns in data, which is important for many of the more advanced functions required by Robot Vision. For example, in order to be able to recognize objects from their images, the software must be able to detect whether the objects it sees are similar to previous objects. Thus, machine learning is another parent of computer vision and signal processing.

However, not all computer vision techniques require machine learning. You can also use machine learning for signals that are not images. In fact, the two fields are often combined in such a way that "computer vision" detects features and information from an image, which is then used as input to a machine learning algorithm. For example, computer vision detects the size and color of parts on a conveyor belt, and then machine learning determines whether those parts are faulty based on knowledge of what a good part looks like.

4. **Machine Vision**

Now we get into machine vision and everything changes. This is because "machine vision" is a completely different term than all the previous ones. It is more about the specific application than the technology. Machine vision refers to the use of vision in industry for automated inspection, process control and robot guidance. The rest of the "family" belongs to the scientific field, while machine vision belongs to the engineering field.

In some ways, you can think of it as a child of computer vision because it uses techniques and algorithms from computer vision and image processing. However, although it is used to guide robots, it robot vision is not exactly the same.

5. **Robot Vision**

Finally, we arrive at robot vision. Robot vision incorporates techniques from all previous terms. In many cases, robot vision and machine vision are used interchangeably. However, there are some subtle differences. Some machine vision applications (e.g., part inspection) have nothing to do with robotics-the part is placed only in front of the vision sensor that is looking for faults.

Furthermore, robotic vision is not just an engineering field. It is a science with its own specific research area. Unlike pure computer vision research, robot vision must incorporate aspects of robotics into its techniques and algorithms, such as kinematics, reference system calibration, and the robot's ability to physically affect its environment. Visual serving is a perfect example of a technology that can only be called "robot vision" rather than "computer vision". It involves controlling the motion of a robot by using feedback from the robot's position as detected by a vision sensor.

The inputs and outputs for each of the technology categories described above are shown in Table 1.1.

Computer vision is the use of computers and related equipment to simulate the working mechanism of the biological visual system, so that computers can observe and understand the world through vision like humans. Its main task is to have the ability to adapt to the environment autonomously by processing the captured images or videos to obtain three-dimensional information about the corresponding scene.

The research of computer vision is largely directed to the content of images. It includes image processing and pattern recognition, in addition to the description of spatial shapes, geometric modeling, and recognition processes [6]. Realizing image understanding is the ultimate goal of computer vision [7].

Image processing techniques refer to the conversion of an input image into an image that is actually required. Image processing techniques generally include image compression, enhancement and restoration, matching, description and recognition in 3 parts. The main methods used are image transformation, image coding and compression, image enhancement and restoration, image segmentation, image description and image classification (recognition). In computer vision mainly preprocessing and feature extraction is done with the help of image processing techniques.

Pattern recognition techniques refer to the statistical characteristics or structural information extracted from an image to classify the image into certain categories [8]. Common pattern recognition in daily life is text recognition or fingerprint recognition. In computer vision pattern recognition techniques are mainly used to recognize and classify certain parts of an image, such as segmented regions.

Table 1.1 Machine vision and computer vision technologies

Technology	Input	Output
Signal processing	Electrical signals	Electrical signals
Image processing	Pictures	Picture
Computer vision	Pictures	Messages/functions
Pattern recognition/machine learning	Information/functions	Information
Machine vision	Pictures	Information
Robot vision	Pictures	Body movements

Image understanding is the semantic understanding of an image. It is a discipline that takes the image as the object and the knowledge as the core, and studies what targets are in the image, the interrelationships between the targets, what the image is the scene and how to apply the scene. Image understanding requires knowledge of the physical laws of scene imaging and knowledge of the content of the scene in addition to sophisticated image processing.

With the development of computer vision and artificial intelligence disciplines, the related research contents are constantly expanding and covering each other, image understanding is both an extension and expansion of computer vision research, as well as a new field of research on human intelligence, permeating the research process of artificial intelligence, and in recent years it has been used in the fields of industrial vision, human–computer interaction, visual navigation, virtual reality, specific image analysis and interpretation, and bio-visual research, etc. It has been widely used in industrial vision, human–computer interaction, visual navigation, virtual reality, specific image analysis and interpretation, and biological vision research.

Computer vision is the "eye" of artificial intelligence and the core technology for perceiving the objective world. Since entering the twenty-first century, the field of computer vision has been developing vigorously, with a large number of theories and methods emerging, and remarkable results have been achieved in many core issues. However, the current computational vision technology still has certain limitations, such as poor interpretability and high dependence on image data quality. However, with the rapid development of information technology, computer vision technology will be more mature, for example, the current 5G speed and capacity, which brings a broader prospect for computer vision AI.

(2) **Industrial Natural Language Processing**

We are currently in the era of big data, where the amount of data is multiplying exponentially, but most of the data in industry comes from a wide range of sources and is complex in structure, of which only 21% is presented in a structured form. Most of the data exists mainly in the form of text, which is highly unstructured and cannot be understood and applied by systems and programs in industry, so there is a need to develop a new technology that enables systems and applications to understand all types of data.

Natural Language Processing (NLP) is an important direction in the field of computer science and artificial intelligence, which is a science that integrates linguistics, computer science and mathematics. Natural Language Processing is the development of applications or services that can understand human language, and its use of computers to process information such as the shape, sound, and meaning of natural language, that is, the input, output, recognition, analysis, comprehension, and generation of words, phrases, sentences, and chapters of the operation and processing. It has many important implications for the way computers and humans interact. The research and application of natural language processing its architecture system is the core issue, and its architecture diagram is shown in Fig. 1.12.

The NLP architecture diagram mainly consists of NLP data, foundation, technology and application and is layered. Various types of data are collected from the

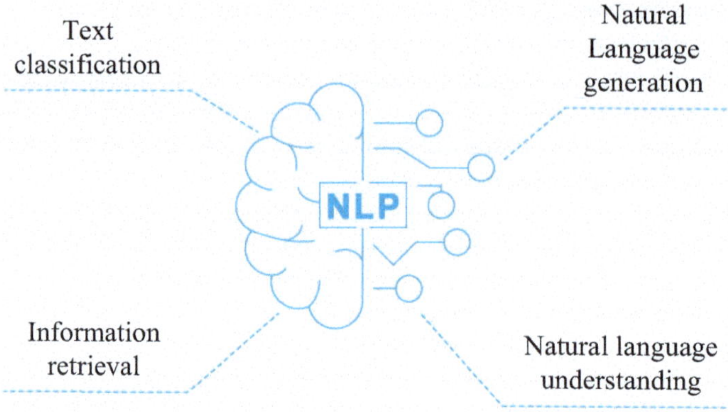

Fig. 1.12 NLP architecture

bottom of the system, and these data are processed through lexical analysis, syntactic analysis, semantic analysis and other processing using relevant search, recognition and comprehension techniques, so that the system software can understand and apply the data. From the architecture diagram, we can see that the research system of NLP is getting richer and richer, and the technology is getting more and more mature. But there are still some limitations of its own, such as over-reliance on data, syntactic ambiguity and so on.

Natural language processing research is gradually shifting from the semantic transfer of lexical semantic components to further, narrative understanding. However natural language processing at the human level is an artificial intelligence complete problem. It is the equivalent of solving the central AI problem of making computers as smart as people, or powerful AI. the future of natural language processing in general is therefore also closely integrated with AI development.

(3) **Industrial Knowledge Graph**

Most of the current research is in the state of weak artificial intelligence, for example, using machine learning and deep learning algorithms to train a model that is more like a machine with statistical knowledge, and is not very intelligent. Real artificial intelligence should be like a human being, with the ability to analyze and reasoning machine intelligence. The knowledge graph is currently working to achieve this goal.

Knowledge graph technology is an important part of artificial intelligence technology, which is essentially a large-scale semantic network that contains a variety of entities, concepts, and semantic relationships. The technical system of knowledge graph is shown in Fig. 1.13.

The technical system under discussion predominantly hinges on five crucial processes: knowledge modeling, knowledge extraction, knowledge fusion, knowledge storage, and knowledge computing. This system has evolved into a relatively mature framework. Knowledge modeling involves creating conceptual models that map knowledge, utilizing either top-down or bottom-up approaches. Knowledge

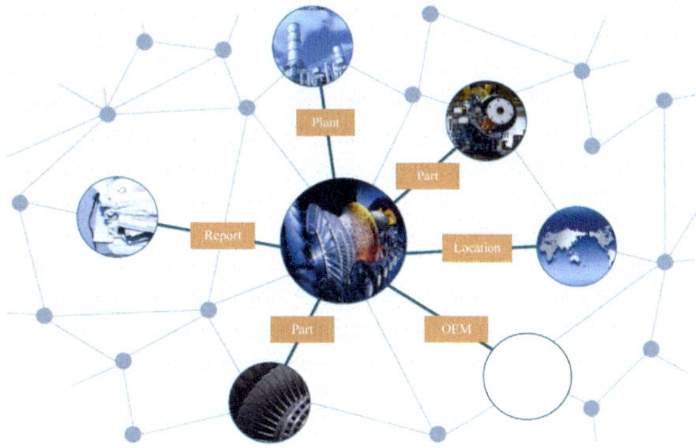

Fig. 1.13 Knowledge graph techniques

extraction refers to the process of retrieving data from various sources and types, and integrating this knowledge into a graph. Knowledge fusion entails aligning and merging knowledge from disparate sources to establish globally unified knowledge identifiers and associations, including both data schema and data layer fusions. Knowledge storage encompasses single and hybrid storage schemes, with storage media categorized into native platforms (e.g., Neo4j, AllegroGraph) and those based on existing databases (e.g., MySQL, MongoDB). The selection of a storage method is contingent on specific characteristics, indicating the absence of a unified storage approach. Knowledge computing, a pivotal aspect of industrial knowledge graph capability output, involves graph mining and knowledge reasoning. Graph mining utilizes graph theory to explore and mine the graph structure, while knowledge reasoning includes ontology-based and rule-based reasoning, tailored to business-specific rule definitions and structured on ontology and rule definitions to facilitate the reasoning process and yield results.

Currently, knowledge graph technology is well-established and extensively applied in the industrial sector. It significantly enhances the ability to organize, manage, and comprehend vast internet data, translating this information into formats that more closely mirror human cognitive processes. Thus, establishing a knowledge base equipped with semantic processing and open interconnection capabilities can substantially enhance intelligent information services such as intelligent search, question answering, and personalized recommendations. Envisioning a future where intelligent machines possess cognitive capabilities, knowledge graphs represent the foundational knowledge base of this 'brain.' This technology holds critical importance for big data intelligence and is poised to profoundly influence fields such as natural language processing, information retrieval, and artificial intelligence.

(4) **Industrial Unmanned Systems**

Advancements in science and technology have significantly expanded and enhanced human capabilities to understand, transform, and utilize the world. Mechanization and electrification have augmented physical capabilities, while informatization and intelligence have bolstered human intellect. Unmanned systems, which integrate these elements, are poised to revolutionize production modes and induce profound, disruptive changes.

Since the turn of the twenty-first century, unmanned systems have evolved rapidly across various domains—from air and space to land and sea, and from physical to information systems. A plethora of intelligent unmanned systems have proliferated, including unmanned aircraft, vehicles, ships, and submersible vehicles, as well as robots deployed in industries such as agriculture, logistics, transportation, education, healthcare, and the military. In manufacturing, unmanned factories serve as a prime example. These factories, equipped with numerous robots and high-end intelligent devices, increasingly operate with minimal human intervention. The integration of cyber-physical systems enhances these factories with capabilities for perception, cognition, and self-regulation. A notable instance is the Siemens unmanned factory in Chengdu, which autonomously handles 75% of its processes. Since its inception, this facility has maintained a consistent workforce and footprint, yet has augmented its production capacity eightfold, producing an item every second with a product quality pass rate of 99.9985%. This illustrates how unmanned factories significantly liberate human labor while ensuring high product quality.

Unmanned systems are emerging as a pivotal component of the new scientific and technological revolution and industrial transformation. Although still in the nascent stages of development and facing various challenges, continuous enhancements are expected to resolve existing issues, heralding a future where unmanned systems play a central role in advancing industrial capabilities.

References

1. Shuquan Z, Qian W, Zhixia W, Kan X (2019) Industrial intelligence technology and applications. Shanghai Science and Technology Press
2. Kagermann H, Helbig J, Hellinger A, et al (2013) Recommendations for implementing the strategicinitiative INDUSTRIE 4.0: securing the future of German manufacturing industry; final report of the Industrie 4.0 Working Group. Forschungsunion
3. Jianhong F (2014) Development status and trend of intelligent manufacturing equipment. J Mech Electr Eng 31(8)
4. Li L (2018) China's manufacturing locus in 2025: with a comparison of "Made-in-China 2025" and "Industry 4.0". Technol Forecast Soc Change 135:66–74
5. Industrial intelligence white paper 2020 edition. http://www.miit.gov.cn/n973401/n5993937/n5993968/c7887033/content.html
6. Vernon D (1991) Machine vision: automated visual inspection and robot vision. Printice-Hall.
7. Cootes TF, Taylor C J (2001) Statistical models of appearance for computer vision. Technical report, www.isbe.man.ac.uk/bim/refs.html
8. Xuegong Z (2010) Pattern recognition, 3rd edn. Tsinghua University Press, Beijing

Chapter 2
Foundations of Industrial Intelligence

In order to realize industrial intelligence, a number of fundamental tools are needed. These include the algorithmic foundations of industrial intelligence, the basic processes for realizing industrial intelligence, and the broader industrial intelligence application scenarios.

2.1 Fundamentals of Industrial Intelligence Algorithms

Intelligence algorithms are tools for establishing correlations between industrial data and industrial insights. This subsection analyzes the task of machine learning for industrial intelligence and then introduces classical algorithms for industrial intelligence.

2.1.1 Machine Learning Tasks for Industrial Intelligence

Machine learning is a multi-disciplinary cross-discipline involving probability theory, statistics, approximation theory, convex analysis, algorithmic complexity theory and other disciplines. It specializes in studying how computers simulate or implement human learning behaviors to acquire new knowledge or skills, and reorganize the existing knowledge structure so that it can continuously improve its own performance. With the rapid development of the new generation of information technology, machine learning technology system is becoming more and more mature. Next will introduce supervised learning, semi-supervised learning, unsupervised learning and reinforcement learning and several other typical machine learning methods [1].

T. Liu et al., *Industrial Intelligence: Methods and Applications*, Springer Series in Advanced Manufacturing, https://doi.org/10.1007/978-3-031-81477-8_2

(1) Supervised Learning

Supervised learning is the process of tuning the parameters of a classifier to achieve a required performance using a set of samples of a known class, also known as supervised training or instructor learning. Supervised learning infers a function from labeled data.Each example includes an input vector and a target output.

The process aims to deduce a model from these inputs and outputs. A supervised learning algorithm is one that analyzes that training data and produces an inferred function that it can use to map out new instances. An optimal scenario would allow the algorithm to correctly determine the class labels of the invisible instances. This requires that the learning algorithm be formed in a "reasonable" way from the training data to the unseen instances.

Supervised learning splits into regression and classification.

In regression, existing data points are analyzed to fit a model y = f(x), where y is the label. This model predicts y for new inputs x. Regression is mainly for continuous variables, such as predicting the price of a house. Classification is to analyze the input feature vectors and for a new vector get its label. Classification differs from regression by targeting discrete outputs, like tumor property estimation.

Classifiers encompass Neural Networks, SVMs, k-NN, Gaussian Mixtures, Bayesian, Decision Trees, and RBF. Supervised learning models are categorized as Probabilistic or Non-probabilistic, and as Discriminative or Generative based on observed variable distribution. Supervised learning, including algorithms like neural networks and decision trees, is extensively studied and applied. However, it requires predefined outputs and lacks environmental interaction, limiting its use in complex control scenarios.

(2) Semi-supervised/Unsupervised Learning

1. Semi-supervised Learning

Semi-Supervised Learning (SSL) is an important pattern recognition and machine learning approach that blends supervised and unsupervised techniques, utilizing both labeled and a vast array of unlabeled data for tasks. When semi-supervised learning is used, it will require as few people as possible to do the work, and at the same time, it can bring a relatively high level of accuracy, therefore, semi-supervised learning is getting more and more attention.

Semi-supervised learning can be further categorized into pure semi-supervised learning and direct inference learning. Pure semi-supervised learning treats unlabeled data as non-targets for prediction, whereas direct push learning considers these unlabeled samples as the actual targets, aiming for optimal generalization on them. In essence, pure semi-supervised learning operates under an "open-world" assumption, aspiring to generalize to unseen data, whereas direct push learning adopts a "closed-world" view, focusing on predicting the observed unlabeled data. The architecture of the two is shown in Fig. 2.1.

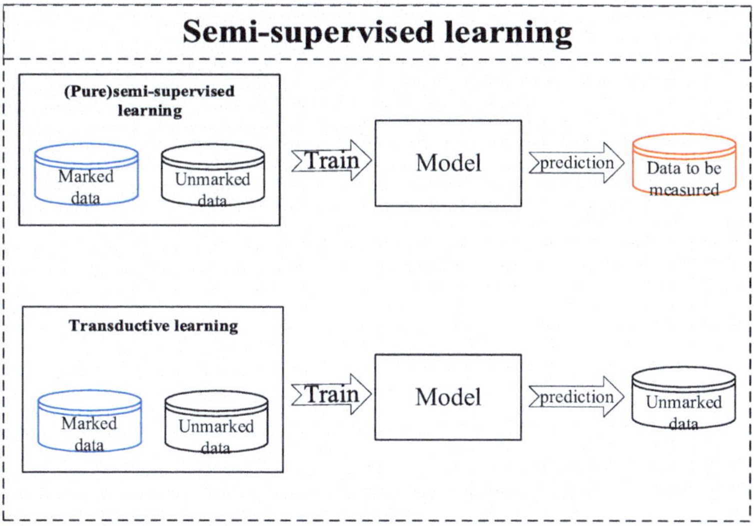

Fig. 2.1 Semi-supervised learning classification

Since its inception, semi-supervised learning has primarily dealt with synthetic lab data. Future research should identify tasks where it can markedly enhance performance and broaden its real-world applications. Additionally, it should establish a universally compelling protocol for employing these methods. In addition, there are many semi-supervised classification methods, while research on semi-supervised regression problems is more limited. In the future, the relationship between semi-supervised classification and semi-supervised regression has to be continued to be investigated, and other semi-supervised regression methods have to be proposed.

2. **Unsupervised Learning**

Unsupervised learning is a data processing method that realizes the classification of samples through data analysis of a large number of samples of the object under study without category information. Unsupervised learning mainly consists of direct methods based on the estimation of probability density function and indirect clustering methods based on the similarity between samples presented.

In unsupervised learning, the data is not specifically labeled and the learning model is designed to infer some intrinsic structure of the data. There are two general ideas for unsupervised learning:

The first idea is to instruct the Agent without assigning it an explicit classification, but rather to use some form of incentive system when successful. This training is typically framed as a decision-making problem, aiming for decisions that maximize rewards, not just classification, which is why it's known as reinforcement learning.

The second idea is called Clustering, this type of learning is not aimed at maximizing a utility function, but rather finding approximate points in the training data, and this section will focus on this type of unsupervised learning idea.

The second idea of unsupervised learning common application scenarios include the learning of association rules and clustering. Common algorithms include Apriori, K-Means, EM and so on.

The data passed to the algorithm by unsupervised learning is very rich in internal structure, while the objectives and rewards used for training are very sparse. Much of what an unsupervised learning algorithm learns must consist of understanding the data itself, rather than applying that understanding to a specific task. Such properties make unsupervised learning more promising for applications.

(3) **Reinforcement Learning**

Reinforcement Learning (RL), also termed as Evaluation Learning or Augmented Learning, is a machine learning paradigm that addresses how an agent learns strategies to maximize rewards or accomplish goals through environmental interactions. The biggest difference between reinforcement learning and supervised or unsupervised learning is that it doesn't require a lot of "data feeding". Instead, you learn certain skills by trying them out on your own.

Reinforcement learning is a machine learning method that allows an Agent to learn through experimentation in an interactive environment and based on its own actions and errors fed back from experience. Supervised learning and reinforcement learning both involve input–output mappings, but they differ: supervised learning gives direct feedback on correct behaviors, whereas reinforcement learning relies on rewards and penalties to signal appropriate actions. Reinforcement learning differs in terms of goals when compared to unsupervised learning. Unsupervised learning aims to discover data similarities and differences, while reinforcement learning seeks an action model to maximize the agent's total rewards, as illustrated in Fig. 2.2.

A common model for reinforcement learning is the standard Markov Decision Process (MDP). As given, reinforcement learning can be categorized into model-based RL and model-free RL, as well as active RL and passive RL. Variants of reinforcement learning include inverse reinforcement learning, hierarchical reinforcement learning, and reinforcement learning for partially observable systems. The algorithms used to solve reinforcement learning problems can be categorized into

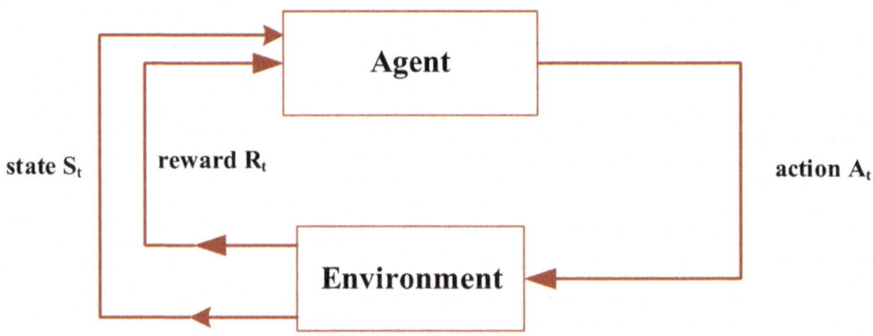

Fig. 2.2 Intensive learning schematic

policy search algorithms and value function algorithms [2]. Deep learning models can be used in reinforcement learning to form deep reinforcement learning.

As reinforcement learning's knowledge base expands, it increasingly integrates with other AI fields, like digital twins, computer vision, and knowledge representation, fueling rapid advancements in manufacturing.

The above content mainly introduces some techniques and methods involved in the current industrial intelligence, but the important thing is how to use these methods and techniques to realize industrial intelligence? Therefore, the realization process of industrial intelligence will be introduced in detail next.

2.1.2 Classical Algorithms for Common AI

The classical algorithms for industrial intelligence are probabilistic modeling methods, kernel methods, decision trees and random forests, and feedforward and backpropagation neural networks [2].

(1) **Probabilistic Modeling**

Probabilistic Modeling (PM) is the application of statistical principles to data analysis. It was one of the earliest forms of machine learning and is still widely used today. One of the best known algorithms is the Simple Bayes algorithm. Plain Bayes is a class of machine learning classifiers based on the application of Bayes' theorem, which assumes that the features of the input data are all independent. Another closely related model is logistic regression.

1. **Plain Bayes**

Bayesian classification encompasses algorithms grounded in Bayes' theorem, leveraging probabilistic statistics to categorize sample data by assessing the likelihood of membership in each class. The Plain Bayesian method is a classification method based on Bayes' theorem with the assumption of conditional independence of features.

In tackling classification, Bayesian classification aims to select the best category label by considering the probabilities in an ideal scenario with full knowledge of all relevant probabilities. When exact predictions are challenging, Bayesian methods use the probabilities of classification-related events. By combining prior and posterior probabilities, these methods mitigate the biases of subjective priors and overfitting from relying solely on sample data, effectively addressing the "inverse probability" issue.The core algorithm of Bayesian classification method is the Bayesian formula:

$$P(y_i|x) = \frac{P(x|y_i)P(y_i)}{P(x)} \tag{2.1}$$

In Eq. (2.1), $P(y_i)$ is the a priori probability (or marginal probability), $P(x|y_i)$ is the likelihood probability, i.e., the probability that attribute x occurs under a given category y_i; $P(y_i|x)$ is the a posteriori probability, which is the probability that a specific thing occurs under a particular condition, i.e., the probability that x belongs to category y_i; $P(x)$ is the a priori probability of x, which is the probability that x belongs to category y_i; $P(x)$ is the a posteriori probability of x. probability that is recorrected after more information is obtained, and is the probability that a specific thing occurs under a given condition, i.e., the probability that x belongs to the category y_i; $P(x)$ is the a priori probability (or marginal probability) of x, also called the standardized constant. In other words:

$$\text{Posterior probability} = \frac{\text{Likelihood} \times \text{prior probability}}{\text{Standardized Constants}} \tag{2.2}$$

The simple Bayesian classification algorithm is the most straightforward among Bayesian classifiers. It operates on the same principle: calculate the conditional probability of each category given the item in question, and assign the item to the category with the highest probability.

Based on the Bayesian classification algorithm, the simple Bayesian classification algorithm also needs to satisfy the assumption of "feature conditions are independent of each other", that is to say, "simple Bayes = Bayesian formula + assumption of conditional independence", and then through the given training set. Then, through the given training set, we learn the joint probability distribution from input to output, and then based on the learned model, we input X to find the output Y which maximizes the posterior probability, and then we judge which category the item to be classified belongs to according to the result of Y.

2. **Logistic Regression**

Logistic regression belongs to the supervised learning inside machine learning, which is a very classical binary classifier to solve the classification problem with the idea of regression. Because of the simplicity of the model of logistic regression, parallelizable, and the strong interpretability of the parameters after training, it has always been loved by the industry.

As depicted in Fig. 2.3, logistic regression is a variant of linear regression that incorporates a logistic function, typically the Sigmoid function. This nonlinear transformation enables it to effectively handle binary classification problems.

Logistic regression is particularly prevalent in epidemiology for identifying risk factors and predicting the likelihood of disease occurrence. For instance, to study gastric cancer risks, one could compare two cohorts with differing characteristics and lifestyles, with the dependent variable being the presence or absence of gastric cancer. The dependent variable here is whether or not gastric cancer is present, i.e., "yes" or "no", and the independent variables can include many things, such as age, gender, dietary habits, Helicobacter pylori infection, and so on. The independent variables can be either continuous or categorical.

(2) **Kernel Method-SVM**

Support Vector Machines (SVMs) and more generally Kernel Methods are representative of the statistical learning techniques that emerged in the mid-1990s. In

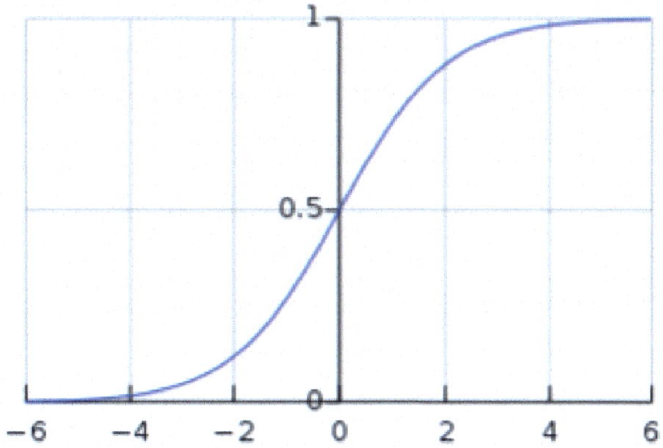

Fig. 2.3 Sigmoid function image

1963, V. N. Vapnik introduced the concept of "support vectors" and later developed the VC dimension and the principle of minimizing structural risk as part of statistical learning theory. By the mid-1990s, this approach became mainstream in machine learning, with the kernel method, exemplified by Support Vector Machines (SVMs), proving effective in text categorization applications. The kernel method, epitomized by SVMs, has shown its strength in text categorization. Subsequently, SVMs gained widespread acceptance, and the kernel trick became ubiquitous across machine learning domains.

As shown in Fig. 2.4, given a training sample set D, which contains samples of class A and class B, classification learning hopes to find a division hyperplane in the sample space based on the training set D to separate samples of different classes. In the sample space, the dividing hyperplane can be described by the following linear equation:

$$\omega^T + b = 0 \tag{2.3}$$

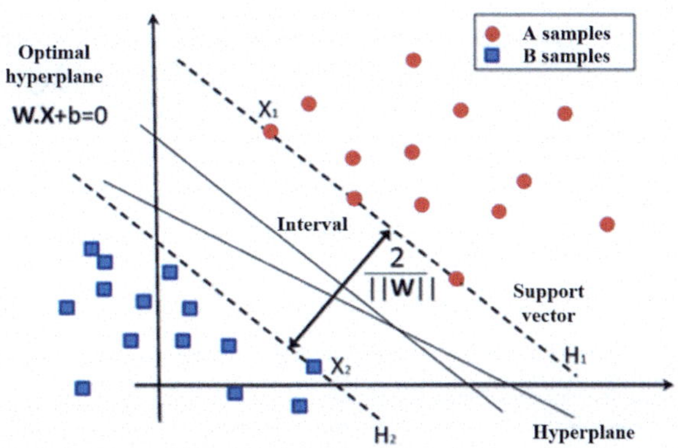

Fig. 2.4 Support vector machine

where w is the normal vector, which determines the direction of the hyperplane, and b is the displacement term, which determines the distance between the hyperplane and the origin. Then the distance from any point x in the sample space to the hyperplane is:

$$r = \frac{|\omega^T + b|}{\|\omega\|} \tag{2.4}$$

Assuming that the hyperplane (ω, b) correctly classifies the sample, i.e., for $(x_i, y_i) \in D$, if $y_i = +1$, there is $\omega^T x_i + b > 0$; if $y_i = -1$, there is $\omega^T x_i + b < 0$. Let:

$$\begin{cases} \omega^T x_i + b \geq +1 \\ \omega^T x_i + b \leq -1 \end{cases} \tag{2.5}$$

The few training sample points closest to the hyperplane that make the above equation hold are called support vectors (support vectors), and the sum of the distances (which also becomes the interval) of two dissimilar support vectors to the hyperplane is:

$$\Upsilon = \frac{2}{\|\omega\|} \tag{2.6}$$

To find the dividing hyperplane with maximum margin, that is, to find the constraint parameters ω and b that are satisfied when γ is maximized; that is, to maximize $\|\omega^{-1}\|$, which is equivalent to minimizing $\|\omega^2\|$.

(3) **Decision Trees and Random Forests**

1. **Decision Tree**

As shown in Fig. 2.5, Decision Tree (DT) is a basic classification and regression method, and this paper focuses on decision trees for classification. A decision tree model has a tree structure and represents the process of classifying instances based on features in a classification problem. It can be thought of as a collection of if–then rules or as a conditional probability distribution defined on the feature space and class space, and its main advantage is that the model is readable and fast for classification. Decision tree learning usually consists of three steps: feature selection, decision tree generation and decision tree pruning.

2. **Random Forest**

As shown in Fig. 2.6, Random Forest is a classifier composed of multiple decision trees, or more accurately, Random Forest is a strong classifier formed by combining multiple weak classifiers.

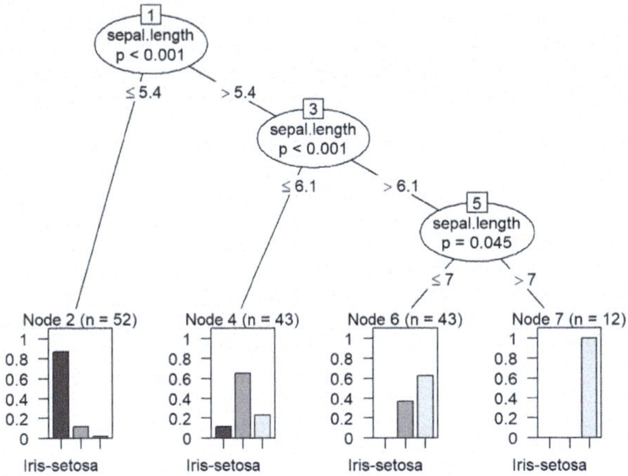

Fig. 2.5 Decision tree

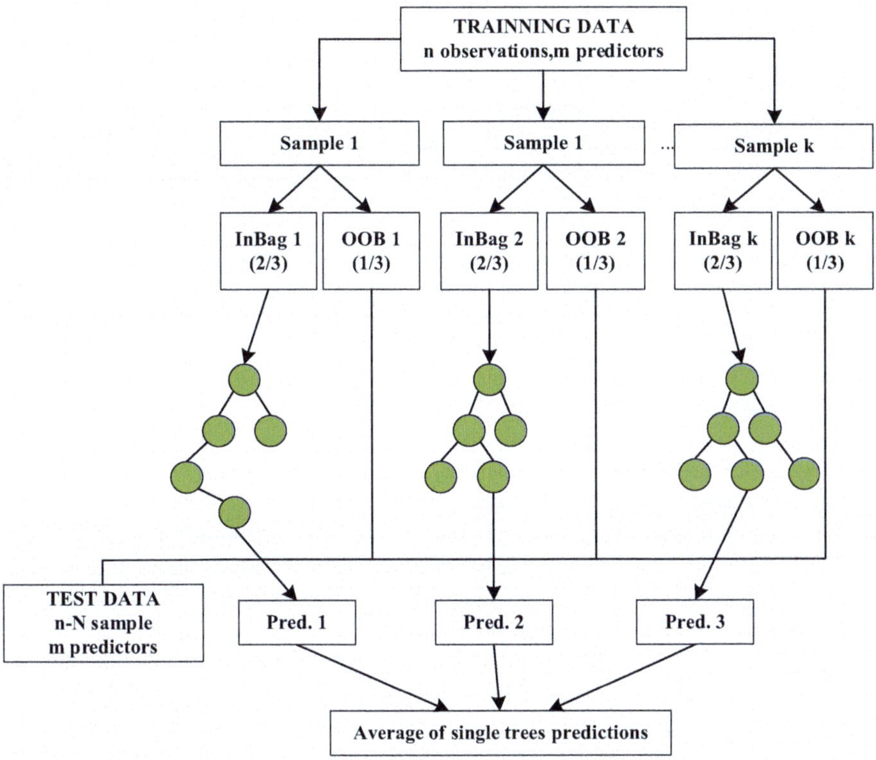

Fig. 2.6 Random forest

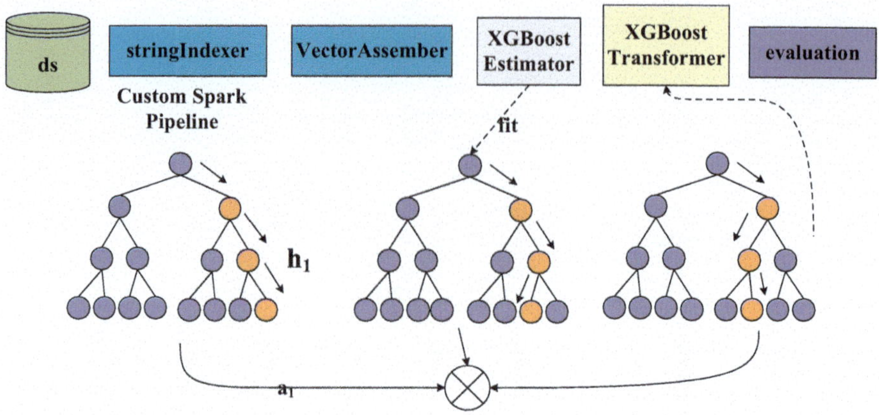

Fig. 2.7 XGBoost

3. **XGBoost**

XGBoost is one of the Boosting algorithms. The idea of Boosting algorithm is to integrate many weak classifiers together to form a strong classifier. As shown in Fig. 2.7, XGBoost is a type of Boosting tree model which is integrating many tree models together to form a very strong classifier, and the tree model used is the CART regression tree model.

(4) **Feedforward and Backpropagation Neural Networks**

In machine learning, a neural network is a model that mimics the human brain's complex information processing, drawing on biological neural network principles and network topology theories, after understanding and abstracting the brain's structure and external stimulus response mechanisms.

The most basic unit of neural network is neuron, as shown in the following figure on the right for the 1943 McCulloch and Pitts proposed "M-P neuron model", f for the activation function to sign function as an example (Fig. 2.8 on the left, respectively, for the activation function of the sign function and the Sigmoid function), f is the activation function, and f is the activation function of the sign function. The M-P model is a weighted summing and reactivation process, which can complete the linearly differentiable classification problem.

The Perceptron, on the other hand, consists of two layers of neurons. The Perceptron excels at logical AND, OR, and NOT operations due to their linear separability—different cases can be divided by a linear hyperplane. However, it struggles with linearly non-separable problems like the XOR gate. Instead, to realize the simulation of heteroscedastic gates, neurons that simulate three kinds of logic gates, namely, with, or, and without, need to be connected to form a two-layer perceptual machine. The concept of multi-layer perceptron (later renamed as neural network) was born, and except for the neurons in the input and output layers, the neurons in all other layers are called hidden layer, as shown in Fig. 2.9.

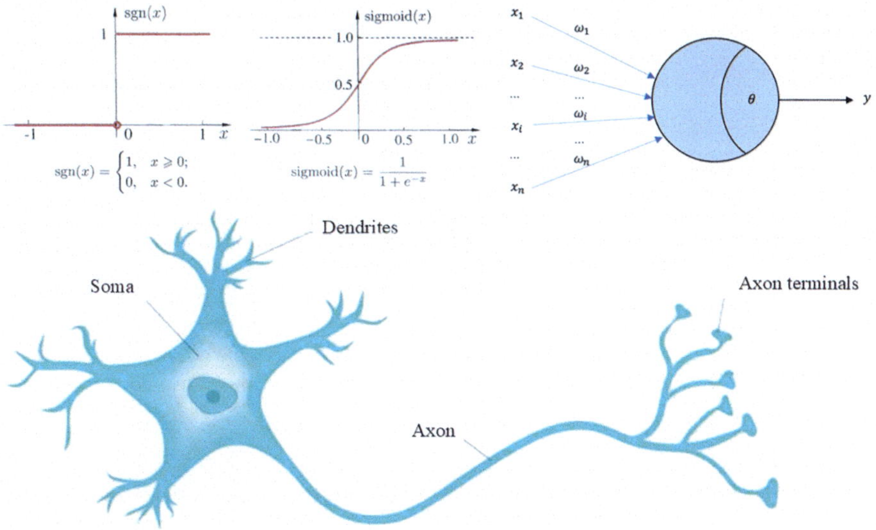

Fig. 2.8 Neural network biology and mathematical modeling

Fig. 2.9 Perceptron and multilayer perceptrons

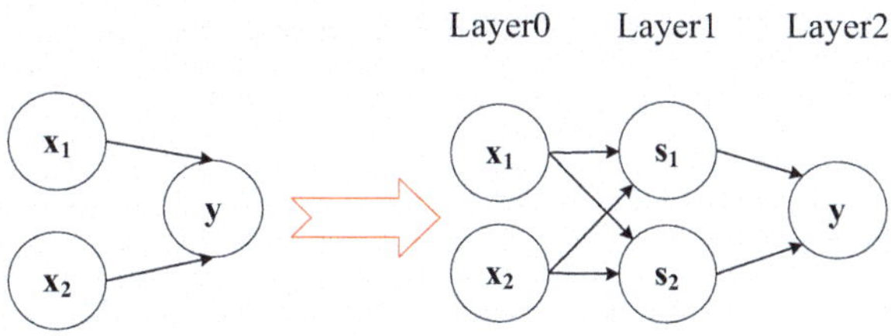

A single hidden layer feedforward neural network is shown in Fig. 2.10, where the circles represent the inputs to the neural network and the "+1" circles are called bias nodes. The input layer accepts external inputs and processes the signals through the hidden layer and the output layer, and finally outputs the results in the output layer neurons. The learning process of a neural network is actually learning how to process the signal better so that its predictions are as accurate as possible.

Multilayer networks have a stronger learning ability than single-layer perceptual machines, but the training of multilayer networks also requires more powerful learning algorithms to realize. At this time, the emergence of error backpropagation algorithm (BP algorithm for short) has brought light to multilayer networks. BP algorithm is based on the gradient descent method, which mainly relies on two links (excitation propagation, weight update) to optimize the model parameters repeatedly and iteratively. The excitation propagation mainly refers to the excitation response of the network to the inputs during the forward propagation process and the corresponding errors acquired by the hidden layer and the output layer during the backward propagation phase. Weight update, on the other hand, refers to the adjustment

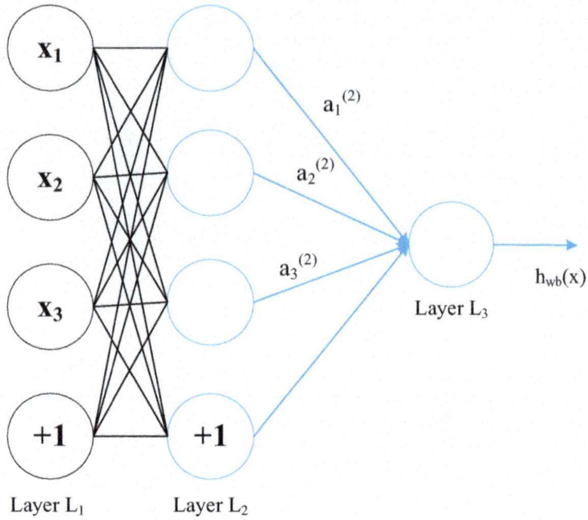

Fig. 2.10 Single hidden layer feedforward neural network

of connection weights and thresholds by the errors of the output layer in the back-propagation process, and the continuous optimization of the model in the iterative learning.

2.1.3 Deep Learning Algorithm

Deep neural networks, with their numerous hidden layers, automate the transformation of data across multiple representation spaces, unlike SVMs or decision trees which require human-designed layers and only transform data into one or two contiguous spaces.

Machine learning tasks that once relied heavily on manual feature extraction (such as target detection, machine translation, and speech recognition) are now being addressed by a variety of end-to-end deep learning paradigms. Common deep neural networks are convolutional neural networks for image processing, recurrent neural networks suitable for processing time-series data, and graph neural networks adept at processing non-Euclidean spatial data.

(1) **Convolutional Neural Network**

Convolutional Neural Networks (CNN) is a class of feed-forward neural networks containing convolutional computation with deep structure, which is one of the representative algorithms for deep learning. As shown in Fig. 2.11, the basic structure of CNN consists of an input layer, a convolutional layer, a pooling layer, a fully connected layer, and an output layer. The convolutional and pooling layers are generally taken several, using alternating settings of convolutional and pooling layers, i.e., a convolutional layer is connected to a pooling layer, and a convolutional layer

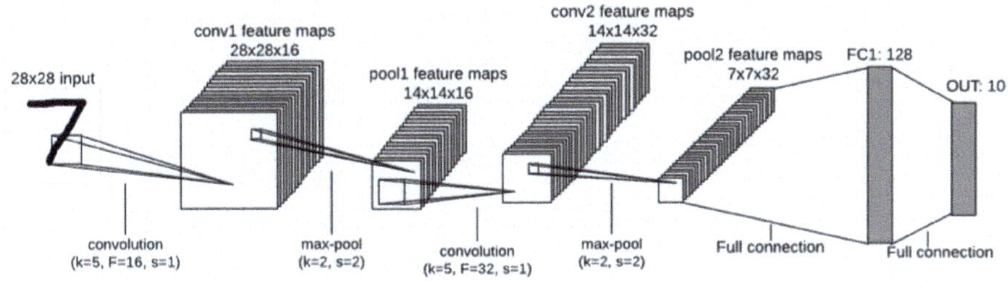

Fig. 2.11 Convolutional neural network

is connected to the pooling layer and then a convolutional layer, and so on. AIn convolutional neural networks (CNNs), each neuron in the output feature map is locally connected to the input. The neuron's input value is computed by applying weights to the local inputs, summing them, and adding a bias, a process analogous to convolution, hence the name CNN.

In the 1990s, Lecun et al. published a paper to design and train a CNN using the BP algorithm (the model was named LeNet-5), which is a classical CNN structure to classify handwritten numbers. However, when facing more complex problems, LeNet-5 did not have satisfactory results due to the lack of large-scale training data and sufficient computational power at that time. After that, many methods were designed to overcome the difficulty of training deep CNNs. Among them, the most famous one is the AlexNet structure proposed by Krizhevsky, which made a major breakthrough in the image recognition task. AlexNet is similar to LeNet-5, but the hierarchy is a bit deeper. The nonlinear activation function ReLu and the Dropout method were also used, achieving excellent results.

AlexNet was a big success and set off a research boom in convolutional neural networks. After this, researchers proposed other improvements, the most famous of which have to be the four ZFNet, VGGNet, GoogleNet and ResNet. Structurally, CNNs have evolved to have more layers. For instance, the ResNet, winner of ILSVRC 2015, is over 20 times deeper than AlexNet and more than 8 times deeper than VGGNet. Increasing depth allows the network to leverage enhanced nonlinearity to approximate the structure of the objective function and to better capture feature representations. However, this also increases the overall complexity of the network, making it difficult to optimize and easy to overfit [3].

(2) **Recurrent Neural Network**

As shown in Fig. 2.12, Recurrent Neural Network (RNN) is a class of recurrent neural networks that take sequence data as input, recursively in the evolutionary direction of the sequence and all the nodes (recurrent units) are connected in a chain. Recurrent neural network (RNN) research originated in the '80s and '90s and evolved into a prominent deep learning algorithm by the early twenty-first century. Variants like Bidirectional RNNs (Bi-RNNs) and Long Short-Term Memory (LSTM) networks are widely recognized types of RNNs.

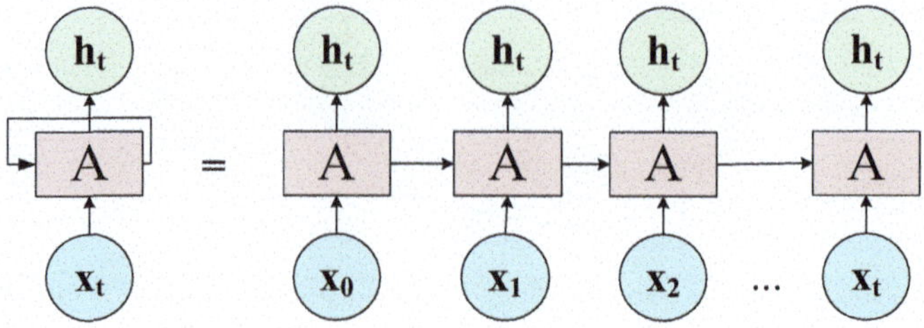

Fig. 2.12 Recurrent neural network

In 2017, Facebook AI introduced a seq2seq architecture using CNNs with gating units to replace RNNs, enhancing performance and significantly accelerating training. Soon after, Google introduced the Transformer, utilizing self-attention to replace both RNNs and CNNs, simplifying the model further. In 2018, the Allen Institute for Artificial Intelligence introduced ELMo, a context-dependent word representation method. It uses a bi-directional LSTM language model to learn context-specific vector representations for words, showing improvements across six NLP tasks. The OpenAI team built upon this foundation by proposing the GPT pre-training model, which replaces LSTMs with Transformers for training the language model. GPT attaches Softmax directly to the pre-trained model's last layer for task-specific outputs and fine-tunes it for enhanced task performance. Shortly after, Google proposed the BERT model, which extends the unidirectional language model in GPT to a bi-directional language model (Masked Language Model) and introduces the sentence prediction task in the pre-training.The BERT model achieved the best results in 11 tasks, which is another milestone work in the field of NLP [4].

(3) **Graph Neural Network Model**

As shown in Fig. 2.13, the concept of Graph Neural Networks (GNN) was first proposed by Scarselli et al. in 2009, which extends existing neural networks for processing data represented in a graph. In a graph, each node is defined by its characteristics and related nodes [5].

GNNs categorize into five types: GCNs, Graph Attention Networks, Graph Autoencoders, Graph Generative Networks, and Graph Spatial–temporal Networks.

Among them, Graph Convolutional Networks (GCNs) are the basis for many complex graph neural network models, including autoencoder-based models, generative models, and spatio-temporal networks. The following figure visualizes the steps involved in learning node representations for graph neural networks. GCN methods can be subdivided into two main categories, spectral-based and spatial-based [6]. Spectral-based methods define graph convolution by introducing filters from the perspective of graph signal processing, where the graph convolution operation is interpreted as the removal of noise from the graph signal. The spatial-based method in graph convolution networks aggregates neighbor features for node representation, and graph pooling can be integrated with convolution layers to simplify the graph into higher-level structures [7–9].

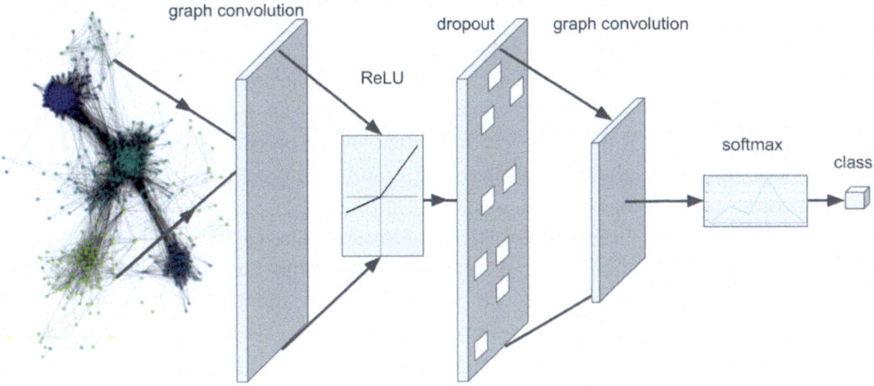

Fig. 2.13 Graph neural network

2.2 Industrial Intelligent Processes

The goal of industrial process intelligent manufacturing is to realize intelligent production, and its key is to realize manufacturing process intelligence. Optimal control in industrial processes involves using intelligent optimization and control integration to adapt to production changes, make decisions, and track set values, thereby optimizing operational metrics like quality, efficiency, and energy use [10]. The realization process of machine learning is shown in Fig. 2.14 and described in detail next.

2.2.1 *Data Collection*

The goal of machine learning is to make the learned model work well for "new samples", i.e., to make the model generalizable. A model that is too effective in learning might overemphasize the training data's specific traits, mistaking them for

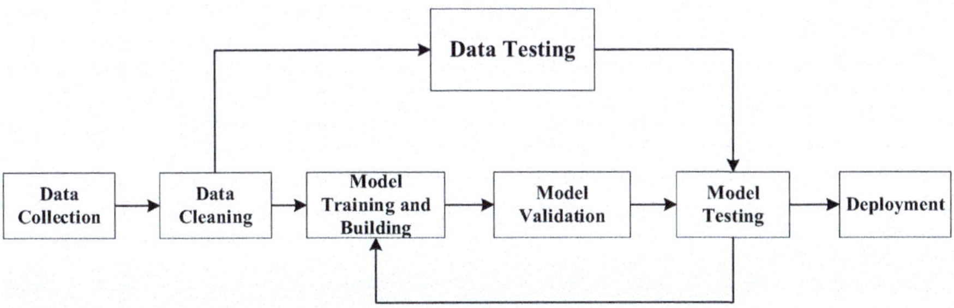

Fig. 2.14 Machine learning implementation process

universal, leading to overfitting and diminished generalization. "Underfitting" is due to the fact that the model does not learn the general properties from the training samples.

"There is no science without measurement" is a famous quote from the scientist Mendeleev. In the field of machine learning, the evaluation of models is also crucial. Only by choosing an evaluation method that matches the problem, can we quickly identify problems in the model selection or training process, and iteratively optimize the model. Model evaluation is mainly divided into two stages: offline evaluation and online evaluation. For different types of machine learning problems such as classification, ranking, regression, sequence prediction, etc., the selection of evaluation metrics varies. Knowing the precise definition of each evaluation metric, selecting appropriate evaluation metrics in a targeted manner, and making model adjustments based on the feedback of evaluation metrics are all key issues in the model evaluation stage. Training models for various types of machine learning to ensure good metrics is not possible without processing the system data, and how the data is divided for training models is extremely important to ensure the accuracy of these metrics.

Before using a dataset to train a model, we need to divide the entire dataset into a training set, a test set and a validation set. The training set is employed to develop models through various methods and ideas. The validation set guides the selection of the best model via cross-validation, refining its performance through iterations. Ultimately, the model's effectiveness is assessed using the test set. If the dataset is well partitioned, the speed of model application can be increased. If the data set is not well partitioned, the deployment of the model application will be greatly affected, or even the model cannot be applied. How to divide the dataset, different sizes of dataset training set, validation set, test set will be divided differently.

(1) **Delineation of Data for Small-Scale Datasets**

In traditional machine learning, the 70%/30% split is standard: 70% of the data is allocated for model training (the training set), and 30% for model testing (more accurately referred to as the validation or development set). Including a validation set, the dataset might be divided into 60%/20%/20% proportions: 60% for the training set, 20% for the validation set, and the remaining 20% for the test set. The above two division criteria are only applicable to small-scale datasets, if for millions, or even tens of millions of datasets need to modify the dataset division principle.

(2) **Delineation of Data for Large-Scale Datasets**

Currently we are in the era of big data, so the data size of many datasets is millions or even hundreds of millions, and if you want to get a good model is inseparable from big data. When facing this kind of large-scale dataset, the traditional data division principle is no longer applicable, the validation set and test set can be divided below 30% and 20%. For the million-level dataset, we can use the 98%/1%/1% rule to divide the dataset.

In summary, datasets are randomly split into training, validation, and test sets. Models are trained on the training set, validated on the validation set, and continuously refined based on performance. The best-performing model is then retrained with data from both the training and validation sets, and its final performance is evaluated using the test set.

2.2.2 Data Preprocessing, Feature Engineering and Feature Learning

The whole process of machine learning in industrial intelligent system mainly includes scene selection, data preprocessing, feature engineering, model training, model evaluation and optimization, and model fusion and other steps.

In machine learning, data quality sets the upper limit for results, with algorithms striving to reach that limit. Effective data preprocessing and feature engineering are crucial steps, closely intertwined in the machine learning process. Therefore, the next step will be a detailed introduction to data preprocessing and feature engineering.

(1) **Data Preprocessing**

The purpose of data preprocessing is to make the original data more suitable for neural network processing, including vectorization, standardization, processing of missing values and feature extraction.

1. **Vectorization**

All inputs and targets of a neural network must be floating-point tensors (which can be integer tensors in specific cases). Whatever data is being processed (sound, image or text), it must first be converted into a tensor, a step called data vectorization.

2. **Value Normalization**

There are various ranges of values for features when predicting house prices, some features are smaller floating point numbers and others are relatively large integers. Before entering this data into the network, you need to standardize each feature separately so that its mean is 0 and standard deviation is 1.

The input data has features that take most values in the range 0–1 and features that have features that should all take values in roughly the same range.

3. **Handling Missing Values**

In general, it is safe for neural networks to set missing values to 0, as long as 0 is not a meaningful value. The network can learn from the data that 0 means missing data and will ignore this value.

(2) **Feature Engineering**

Feature engineering (FE) is the process of using one's knowledge about the data and machine learning algorithms to perform hard-coded transformations on the data (not learned by the model) to improve the model before inputting the data. The essence of this is to formulate the problem in a simpler way, thus making it easier. Good features can use fewer resources, solve problems more elegantly and can solve problems using less data. Feature engineering mainly includes feature construction, feature extraction, feature selection and so on [11].

1. **Feature Construction**

Feature construction refers to the manual construction of new features from the original data [12]. It takes time to observe the original data, think about the potential form of the problem and the data structure, data sensitivity and machine learning experience can help feature construction. Feature construction requires strong insight and analytical skills, requiring us to be able to identify some physically meaningful features from the raw data. Assuming that the raw data is tabular data, you can generally create new features by using blended or combined attributes, or decompose or slice the original features to create new ones. Feature construction benefits from domain expertise or experience to create valuable new features. Unlike feature extraction, which uses established methods to transform data into features, construction involves manual creation, such as merging or splitting existing features to form new ones.

2. **Feature Extraction**

Generally before the feature selection, it extracts the object is the original data, the purpose is to automatically build new features, the original data is converted into a set of features with obvious physical significance (such as Gabor, geometric features, texture features) or statistical significance [13]. Generally commonly used methods include dimensionality reduction (PCA, ICA, LDA, etc.), SIFT, Gabor, HOG, etc. for images, bag-of-words model, word embedding model for text, etc.

3. **Feature Selection**

Feature selection refers to the process of selecting a subset of relevant features from a given set of features is called feature selection [14] Features can be categorized into two types: Attributes or features that are useful for the current learning task are called relevant features; Attributes or features that are not useful for the current learning task, called irrelevant features.

Feature selection may reduce the predictive ability of the model because the eliminated features may contain valid information, and discarding this information will somewhat reduce the performance of the model [15]. However, it is also a trade-off between computational complexity and model performance. If as many features as possible are retained, the performance of the model improves, but at the same time the model becomes more complex and the computational complexity improves as well. If as many features as possible are eliminated, the performance of the model decreases, but the model becomes simpler and the computational complexity is reduced.

Common feature selections are filtered, wrapped and embedded methods.

2.2.3 *Model Training and Reinforcement*

Different machine learning problems have different evaluation metrics, but they also have common evaluation metrics, some of which are highlighted below.

(1) **Accuracy**

Accuracy refers to the proportion of correctly classified samples to the total number of samples. The accuracy rate, while straightforward for classification problems, has limitations. It becomes less reliable when class distributions are highly imbalanced, as the larger classes can disproportionately influence the accuracy metric. For example, when the negative samples accounted for 99%, the classifier predicts all the samples as negative samples can also get 99% accuracy, in other words, the overall accuracy is high, does not mean that the proportion of small categories of high accuracy.

(2) **Precision and Recall**

Precision rate is the ratio of the number of correctly categorized positive samples to the number of samples that the classifier determines to be positive [16]. Recall measures the ratio of correctly identified positive instances to the actual positives. Precision and Recall are interrelated; enhancing Precision often means classifying samples as positive more conservatively, which can lead to missing out on true positives. In order to increase the Precison value, the classifier needs to try to predict the samples as positive samples only when it is "more confident", but then it will often be too conservative and miss a lot of positive samples that it is "not confident", resulting in a lower Recall value. In the sorting problem, there is usually no definite threshold to determine the results as positive or negative samples, but rather, it adopts the Precision value of the results returned by the TopN and the Recall value to measure the performance of the ranking model. performance, i.e., the result of TopN returned by the model is considered to be the positive sample determined by the model, and then Precision at N positions and Recall at the first N positions are calculated.

(3) **F1 Score**

The F1 score is the harmonic mean of Precision and Recall, where the normal mean treats all values equally and the harmonic mean back gives higher weight to lower values, so a classifier can only get a high F1 score if Recall and Precision are both high. The F1 score is more favorable for classifiers that have similar Precision and Recall. But this doesn't always meet your expectations; in some cases, you care more about precision, while in other cases, you may really care about recall. The trade-off between precision and recall will be well worth thinking about.

(4) **Mean Square Error, Root Mean Square Error, Absolute Percentage Error**

1. Mean square error:

$$MSE = \frac{1}{n}\sum_{i=1}^{n}(y_{pred} - y_i)^2$$

2. Root mean square error:

$$RMES = \sqrt{MSE}$$

3. Both the mean square error and the root mean square error can be affected by outliers, which can affect the final model assessment. Mean Absolute Percentage Error:

$$MAPE = \sum_{i=1}^{n} \left| \frac{y_{pred} - y_i}{y_i} \right| \times \frac{100}{n}$$

The average absolute percentage error improves the robustness of the outliers, which is equivalent to normalizing the error at each point and reduces the impact of absolute errors from individual outliers.

(5) **ROC Curve**

Binary classifiers are the most common and widely used classifiers in the field of machine learning. There are many indicators for evaluating binary classifiers, such as precision, recall, F1 score, P-R curve, etc. However, it is found that these indicators can more or less only reflect the performance of the model in a certain aspect, in contrast, the ROC curve has a lot of advantages, and it is often used as one of the most important indicators for evaluating binary classifiers. ROC curve is the abbreviation of Receiver Operating Characteristic Curve, the Chinese name is "Subject Operating Characteristic Curve", the horizontal coordinate of ROC curve is the False Positive Rate (FPR), and the vertical coordinate is the True Positive Rate (TPR), the FPR and TPR are calculated as follows, respectively:

$$FPR = \frac{FP}{N}$$

$$TPR = \frac{TP}{P}$$

P is the number of true positive samples, N is the number of true negative samples, TP is the number of positive samples predicted by the classifier among P positive samples, and FP is the number of positive samples predicted among N negative samples. AUC, or Area Under the ROC Curve, measures the model's performance through the ROC curve. A higher AUC indicates that the classifier is more effective at ranking true positives higher than false positives, signifying better classification performance. The ROC curve can minimize the interference caused by different test sets and measure the performance of the model more objectively.

A new round of information revolution and industrial change is flourishing, and the intelligent development of industry is the focus and trend of global concern. Major developed countries prioritize and implement strategic policies to foster AI applications in manufacturing and industry. The U.S. issued the National Strategic Plan for AI R&D and the Advanced Manufacturing Leadership Strategic Report in October

2016 and 2018, emphasizing product lifecycle optimization, advanced robotics, big data mining, and manufacturing system cybersecurity. Since 2015, Japan has issued four industrial intelligence policy documents, including the New Robotics Strategy, the 2015 Manufacturing White Paper, the Japan Advanced Integrated Intelligence Platform Plan (AIP), and the AI Industrialization Roadmap. These focus on advanced robotics, big data mining, and advancing intelligent predictive systems for equipment failure. The EU released its Digital Industrial Strategy in May 2016, focusing on the research and development of advanced robots and industrial autonomy systems. The reinforcing effect of this series of measures will all greatly promote the development and application of industrial intelligence.

According to Markets report, the AI manufacturing market size is expected to reach USD 17.2 billion by 2025, growing at a CAGR of 49.5% during the forecast period (2018–2025). Automation Technology expects that AI can account for up to 2.2% of the value added to the manufacturing industry by 2035, which is ranked as the top of the 16 major industries in the society. According to the McKinsey report, AI could increase productivity in Germany's industrial sector by 0.8% to 1.4% per year. Accenture's analysis predicts that by 2035, AI applications will boost the manufacturing industry's value-added growth rate by 2.0%, the highest increase among all sectors in China. It can be seen that industrial intelligence has a better application prospect.

2.3 Industrial Intelligence Scenarios

As shown in Fig. 2.15, IoT Analytics (Analytics) recently released its report on the Industrial AI Market 2020–2025. As part of the study, the Analytics team identified 33 different use cases that employ AI tools and techniques on (primarily) IoT-connected data sources and industrial enterprise assets. These 33 industrial AI use cases have been categorized into 10 broader use case categories that make up the majority of the 2019 industrial AI market.

2.3.1 Predictive Maintenance

The largest use case for industrial AI is "predictive maintenance" (expected to account for more than 24% of the overall market by 2018). Predictive maintenance utilizes advanced analytics (e.g., machine learning) to determine the condition of individual assets or entire sets of assets (e.g., plants) [17].

Predict when maintenance should be performed. As shown in Fig. 2.16, predictive maintenance typically combines various sensor readings (sometimes from external data sources) and performs predictive analytics on thousands of recorded events. Using supervised ML to predict the remaining useful life of an asset is the most common technique used in predictive maintenance.

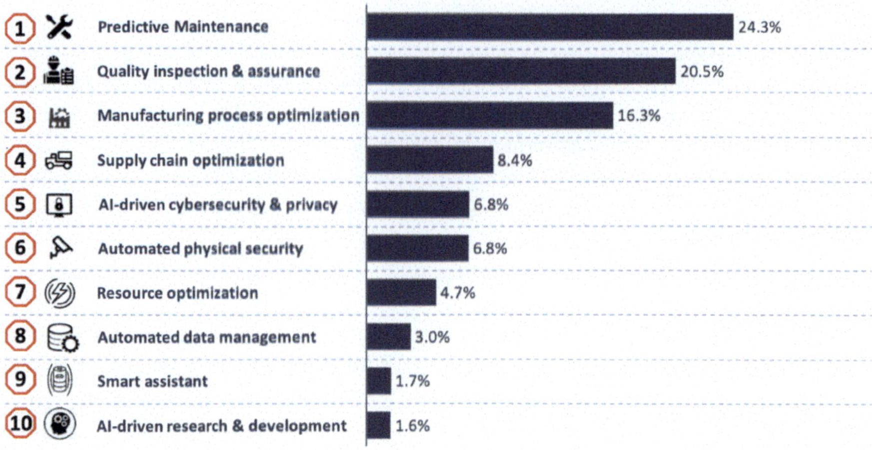

Fig. 2.15 AI market report

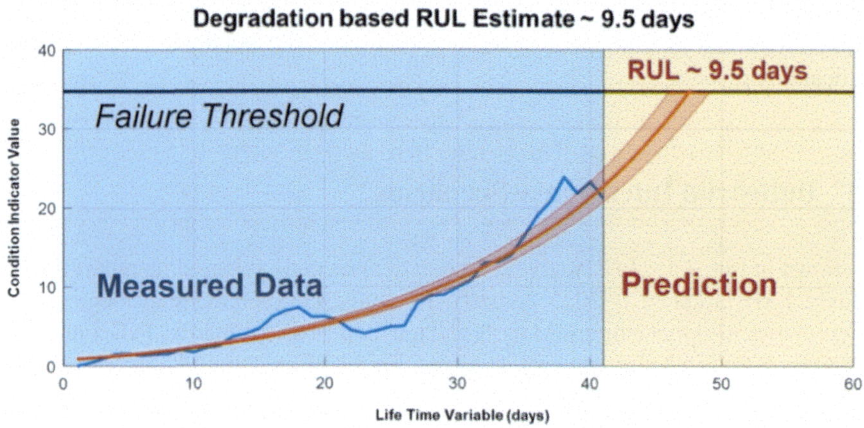

Fig. 2.16 Prediction of remaining useful life (RUL) by regression ML technique

At a given failure threshold, that equipment asset is likely to break within 9.5 days. German rail operator Deutsche Bahn, for example, uses data from railroad turnouts to predict failures, thereby reducing massive unplanned delays; Nissan runs the AI Predictive maintenance platform, which performs RUL prognostics on 7500 assets. The company claims a 50% reduction in unplanned downtime and a payback period of < 3 months. Nissan scaled the solution from 20 critical assets to thousands without increasing the workload of the on-site PDM team.

One of the biggest challenges of predictive maintenance is eliminating data imbalances, as there is often not enough failure data to store all assets. Data is said to be imbalanced when the share represented by failure events is less than some specific required share of the data set. In order to make accurate predictions about the data, this imbalance must first be removed. There are two primary methods for achieving balanced data: data sampling and cost-sensitive learning algorithms.

2.3.2 *Quality Inspection and Assurance*

Quality inspection and quality assurance is the second largest industrial AI use case category at 20.5%. While there are multiple ways to perform AI-based quality inspections, automated optical inspection is by far the largest subcategory [18].

Automated optical inspection is a technology in which a camera automatically scans the device under test for catastrophic failures (e.g., missing parts) and/or quality defects (e.g., rounded dimensions or shapes or tilted parts). Computer vision is the basis for optical inspection. Once the image is recognized, semi-supervised ML is the most efficient technique for classifying the image into fault categories. This use case primarily offers cost savings, especially beneficial for large-scale manufacturing where even minor reductions in waste or testing time can lead to substantial financial benefits. After implementing optical inspection at one of these facilities, BOSCH claimed savings of $1.3 million due to a 45% reduction in total test time.

A large German automotive company, for example, is introducing a deep learning-enabled system for quality control. The system is fully integrated into the final assembly process, eliminating the need for a separate test area in a controlled environment. It is expected that this solution will soon replace the traditional cameras currently in use, which will result in significant savings.

2.3.3 *Manufacturing Process Optimization*

Automated manufacturing process optimization is perhaps one of the most obvious but still hardest AI use cases to implement. One way that this optimization can be achieved is through autonomous machines or robots [19]. The idea behind these autonomous assets is that they save money by replicating monotonous manual tasks in the manufacturing process. Before being put into production, the autonomous machine/robot performs the same task over and over again, learning each time until it reaches a sufficient level of accuracy. Reinforcement learning techniques are commonly used to train robots and autonomous machines, as shown in Fig. 2.17. With this technique, a robot can teach itself to do something relatively quickly under human supervision. The "brain" of such robots/machines is usually a neural network.

The reinforcement learning system shown in Fig. 2.17 replicates the natural human learning process: an agent (e.g., a robot) takes a certain action that is observable by the environment. A certain reward is generated and sent back to the agent. After repeating the process several times, the robot's "brain" (i.e. neural network) optimizes its behavior to adapt to the environment.

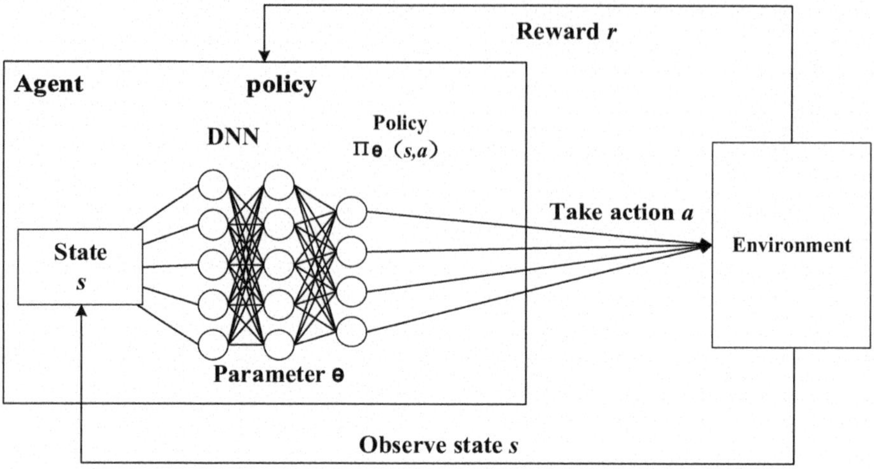

Fig. 2.17 Manufacturing process optimization

Fig. 2.18 Robot factory

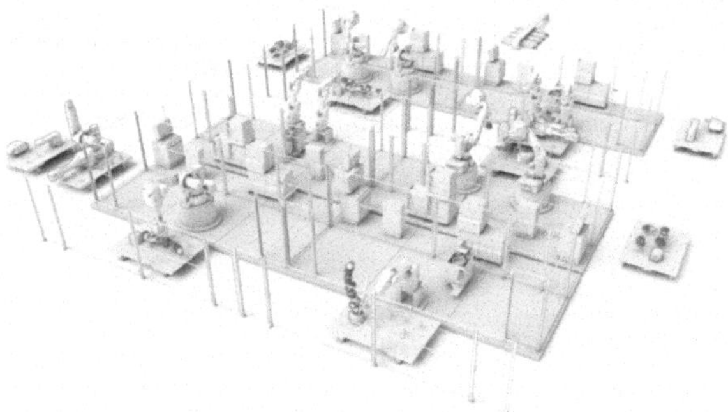

For example, ABB is investing $150 million in a "state-of-the-art, automated and flexible" robotics factory in Shanghai (Fig. 2.18). In this factory, ABB will use robots to build robots. These robots will have autonomous and collaborative elements. The autonomy of the robots is established through the use of AI and digital twins.

2.3.4 Supply Chain Optimization

Eight percent of all industrial AI implementations are improvements to the industrial supply chain. Using AI tools to improve inventory management is one of the key applications. Predictive inventory management uses predictive analytics for a variety of inventory-related tasks, including reducing inventory planning time, minimizing inventory costs, optimizing repairs and finding the best reorder point. For these tasks, the most commonly used techniques include time series analysis, probabilistic modeling (Markov and Bayesian models), and simulation (e.g., Monte Carlo simulation).

Continental Airlines, for example, has built software to predict the optimal point at which to change tires on its fleet. The underlying model predicts overall miles traveled and compares it to a baseline to generate operations. In this way, Continental has reduced its tire inventory and improved road safety.

2.3.5 Cybersecurity and Privacy Powered by Artificial Intelligence

Artificial intelligence-driven cybersecurity and privacy involves aspects such as cyber threat detection. It typically involves observing network infrastructure and detecting the threat of cyberattacks in real time. It also typically includes activities such as network traffic analysis, endpoint detection and response, and malware sandboxing. AI-based cyber threat detection is often part of a larger cyber security solution that also uses a variety of preventative measures (e.g., firewalls).

2.3.6 Automated Physical Security

Surveillance and physical threat detection requires real-time monitoring of production sites or workers in order to automatically detect physical security threats and/or potential safety hazards.

2.3.7 Automated Data Management

Because data is often stored on multiple systems and in multiple locations, it is difficult to access and analyze data quickly and comprehensively. As a result, several industrial companies have begun to adopt data management solutions to perform tasks such as data collection, data filtering, data cleansing and integration in real time [20].

2.3.8 Intelligent Assistant

Voice assistants are one example of intelligent assistants in a manufacturing setup. Integrating voice assistant technology into their real-time industrial monitoring systems allows staff to gain insights without having to write explicit commands or print long status reports.

2.3.9 Artificial Intelligence-Driven R&D

Automated component design is a major use case in AI-driven R&D. The goal: Given a set of predefined constraints, let the software independently develop dozens of different designs in a short time. The best design is selected afterwards. For this purpose, digital twins and simulations usually complement AI techniques.

2.3.10 Self-resource Exploration

Autonomous Resource Exploration (ARE) is a technology that analyzes and processes large amounts of imagery (e.g., radar, satellite, or drone imagery) to detect optimal points for resource extraction, and is particularly important in the mining and quarrying industry, as well as in the oil and gas industry. ai is especially useful in detecting the ability to capture hard-to-access areas (e.g., the seafloor or mountain ranges).

References

1. Chou C-H (2016) Machine learning, vol 1. Tsinghua University Press, Beijing, pp 1–425
2. Chollet F (2018) Python deep learning (trans., L. Zhang), vol 8. People's Post and Telecommunications Press, Beijing, pp 1–292
3. Zhou F, Jin L, Dong J (2017) A review of convolutional neural network research. J Comput 40(6):1229–1251
4. Yang L, Wu Y, Wang J, Liu Y (2018) A review of recurrent neural network research. Comput Appl 38(S2):1–6+26
5. Zhou J, Cui G, Zhang Z et al (2018) Graph neural networks: a review of methods and application. arXiv: Learning
6. Hinton GE, Osindero S, Teh Y (2006) A fast learning algorithm for deep belief nets. Neural Comput 18(7):1527–1554
7. Krizhevsky A, Sutskever I, Hinton GE (2012) Imagenet classification with deep convolutional neural networks. In: International conference on neural information processing systems, Lake Tahoe, Nevada, USA, pp 1097–1105
8. Lecun Y, Bengio Y, Hinton GE (2015) Deep learning. Nature 521(7553):436–444
9. Zhang WJ, Yang GS, Lin YZ et al (2018) On definition of deep Learning. In: IEEE World Automation Congress, Stevenson, Washington, USA, pp 232–236
10. Wang JJ, Ma YL, Zhang LB et al (2018) Deep learning for smart manufacturing: methods and applications. J Manuf Syst 48(C):144–156
11. Lecun Y, Bottou L, Bengio Y et al (1998) Gradient-based learning applied to document recognition. Proc IEEE 86(11):2278–2324

12. Simonyan K, Zisserman A (2014) Very deep convolutional networks for large-scale image recognition. arXiv preprint, arXiv: 1409.1556
13. Szegedy C, Liu W, Jia Y Q et al (2015) Going deeper with convolutions. In: IEEE Conference on computer vision and pattern recognition, Boston, Massachusetts, USA, pp 1–9
14. He KM, Zhang XY, Ren SQ et al (2016) Deep residual learning for image recognition. In: IEEE Conference on computer vision and pattern recognition, Las Vegas, Nevada, USA, pp 770–778
15. Howard AG, Zhu ML, Chen B et al (2017) Mobile nets: efficient convolutional neural networks for mobile vision applications. arXiv preprint, arXiv: 1704.04861
16. Hu J, Shen L, Albanie S et al (2020) Squeeze-and-excitation networks. IEEE Trans Pattern Anal Mach Intell 42(8):2011–2023
17. Cui D, Wang GQ, Lu YP et al (2020) Reliability design and optimization of the planetary gear by a GA based on the DEM and Kriging model. Reliab Eng Syst Safety 203:107074
18. Guan SY, Wang XK, Hua L et al (2021) Quantitative ultrasonic testing for near-surface defects of large ring forgings using feature extraction and GA-SVM. Appl Acoust 173:107714
19. Tian SS, Li YX, Kang JN et al (2021) Multi-robot path planning in wireless sensor networks based on jump mechanism PSO and safety gap obstacle avoidance. Futur Gener Comput Syst 118:37–47
20. Lu C, Gao L, Pan QK et al (2019) A multi-objective cellular grey wolf optimizer for hybrid flowshop scheduling problem considering noise pollution. Appl Soft Comput 75:728–749

Chapter 3
Intelligent Design and Structural Optimization

In the new era of manufacturing development characterized by knowledge and information, product design is receiving more and more universal attention from countries around the world. Product design is a complex creative activity involving human creativity. With the further development of market demand, the functional requirements of products are increasing, the complexity is increasing, and product design is becoming more difficult. How to realize the intelligentization of product design and quickly complete the development and improvement of products has become an important topic in manufacturing research.

In this context, the theory, method and technology of intelligent design have been applied more and more. Intelligent design, as a new advanced design technology, mainly researches how to apply artificial intelligence technology to product design to realize machine intelligence in order to build a product intelligent design system. Intelligent design system for product design is to realize the essence of product innovation design and rapid design, which is also a powerful weapon for enterprises to develop new products, win the market, and obtain the survival and development space. Product planning, program design, detailed design and improvement design are the main processes of product design. Program design, a key stage, determines both product properties and innovativeness. Since mechanical products are often composed of a series of parts and components with multilevel structure characteristics, an intelligent optimization algorithm can be sought to realize the scheme design of mechanical products. The genetic algorithm from artificial intelligence applies biological evolution mechanisms and simulates its characteristics to solve optimization problems and complete the overall design of mechanical products. As an important research direction in the field of artificial intelligence, instance-based reasoning provides technical support for the further design of parts after the generation of mechanical product solutions. Instance-based reasoning is based on rich instances, according to the search algorithm, find similar instances in the instance library, modify and store the similar instances according to certain rules, and realize the detailed design of mechanical product parts. Traditional CAD technology fails

T. Liu et al., *Industrial Intelligence: Methods and Applications*, Springer Series in Advanced Manufacturing, https://doi.org/10.1007/978-3-031-81477-8_3

to meet product design needs. Combining AI and CAD with databases and expert systems to create intelligent CAD is the main development direction. Genetic algorithm and example-based intelligent CAD system combine to build mechanical product intelligent design systems, shortening development cycle and improving efficiency.

Intelligent design mainly uses computers to replace human experts to process knowledge, data and information, so that computers can undertake complex design tasks at a higher level and realize the integration, intelligence and automation of design. Modern manufacturing industry puts forward the following three requirements for intelligent design.

1. Higher and higher design quality. With the development of the economy and the improvement of product processing accuracy, the quality of product design has also put forward higher requirements.
2. Shorter and shorter design cycle. Product competition is the efficiency of competition, intelligent design should reflect the characteristics of automation, intelligence, improve production efficiency, reduce labor intensity.
3. More and more complex design objects and environmental requirements. With the continuous expansion of market demand, mechanical products and equipment should have more functions and higher adaptability to complex environments.

3.1 State of the Art in Intelligent Design Research

In order to realize intelligent design, it is necessary to study its methods as well as tools. Therefore, this section analyzes the current state of research on intelligent design research methods and intelligent CAD design of products, respectively.

3.1.1 Intelligent Design Research Methods

Intelligent design is not the same as automated design. Design is an extremely complex activity, and it is currently impossible to use computers to fully simulate and replace humans in product design. Computer systems and designers are complementary in terms of design, and it is often necessary for designers to be involved in the intelligent design process. Intelligent design simulates designers' cognitive and behavioral characteristics to achieve product design customization, automation and intelligence, enabling the design system to use rich knowledge for reasoning, judgment and decision-making.

With the in-depth study of intelligent design technology, a series of mature intelligent design methods have been gradually formed. Typical intelligent design methods comprise object-oriented knowledge representation, constraint satisfaction-based method, prototype-based method, rule-based method and instance-based method.

These intelligent design methods simulate human design behaviors in different ways, providing theoretical and design frameworks for the development and application of the technology, and are significant for intelligent design. These intelligent design methods simulate human design behavior from different aspects, offering a theoretical foundation and design framework for the technology's development and application, and are significant for intelligent design.

(1) **Object-Oriented Representation of Knowledge**

Object-Oriented Knowledge Representation (OOKR) is a method of knowledge representation in which the attributes and characteristics of the objects described by the knowledge are represented as some kind of data structure so that they can be stored and utilized by computers [1].

OOKR overcomes the shortcomings of traditional knowledge representation methods such as generative rules in a single form. It can express both descriptive and process behavioral knowledge and has strong numerical computation processing ability. Object-oriented thinking in intelligent design conforms to people's thinking habits and has better code reusability and maintainability.

(2) **Intelligent Design Methods Based on Constraint Satisfaction**

Constraint-Satisfied Design (CSD, Constraint-Satisfied Design) is an intelligent design methodology based on constraint satisfaction, which refers to solving the design problem as a Constraint-Satisfied Problem (CSP, Constraint-Satisfied Problem) [2]. In AI technique, the CSP problem is solved by searching for the solution that satisfies all constraints in the solution space. However, when the design problem is relatively complex and often involves more variables, it is difficult to find the real solution in the huge solution space by searching method. Therefore, CSD is often used to determine whether the design solution can satisfy the constraints after other intelligent design methods have been generated.

(3) **Prototype-Based Intelligent Design Methodology**

Prototype-Based Design (PBD, Prototype-Based Design) is an intelligent design method that uses design prototypes as the structure of the attribute space of the design solution to solve the content of the attribute space [3]. Design prototypes are typical construction forms of design solutions summed up by humans for specific design issues. When encountering a new design problem, the designer will choose one of these prototypes as the structure of the solution and then utilize other design methods to determine the specific content of the solution.

(4) **Rule-Based Intelligent Design Methodology**

Rule-based design (RBD, Rule-Based Design), also often referred to as "expert system design method", is one of the most commonly used intelligent design methods [4]. It is used to simulate human design behavior by reasoning step by step with process, logical or empirical design rules. RBD expresses design knowledge as rules and obtains the solution of the design problem through the reasoning mechanism.

(5) **Example-Based Intelligent Design Methodology**

Case-Based Design (CBD, Case-Based Design) is a relatively advanced intelligent design methodology, which is used to simulate the design behavior of human beings using long-term accumulated design examples, design knowledge and experience to provide examples of reference and decision support for the current design problems, and is an application of CBD technology in intelligent design [5].

CBD draws successful experience and knowledge from the past successful examples and applies them in the new program. The reasoning process of expert systems is different from rule-based reasoning. Based on CBR reasoning with examples, in general, in CBR, the characteristics of examples and the problem-solving environment are stored in the database as a data model. When new problems arise, the most similar examples are retrieved from the database and modified to meet the requirements of current problem-solving. When encountering a new problem to be solved, the most similar instance to the current problem is retrieved from the database and modified to meet the requirements for solving the current problem.

3.1.2 *Current Status of Research on Intelligent CAD Design of Products*

The rapid development of intelligent design is prominently reflected in the development of intelligent mechanical CAD design. In the early stage of mechanical intelligent design, computing-related work was mainly handled by the traditional CAD system. With the gradual maturity of the computer integrated system CIMS, mechanical intelligent CAD design has gradually tended towards a more intelligent direction. It can not only handle conventional mechanical CAD design but also undertake more complex creative design of mechanical products, and has gradually been widely welcomed in the mechanical design industry.

As an instance reasoning technique in intelligent design, it is getting more and more popular in the research craze and wide attention. Application systems based on CBR technology are also increasing in various fields, such as software engineering, mechanical and electronic design, medical and legal, etc. There are shadows of CBR intelligent design.

Zhang et al. [6] from Beijing Institute of Technology proposed an intelligent design method for structural aspects of fixtures based on ontologies and knowledge components. Chen et al. [7] from Hunan University of Technology combined examples and rule-based reasoning to determine the main characteristic parameters in the whole system and their relationships with each other. Zhang [8] proposed an emergency decision-making mechanism for environmental mass events based on CBR technology and designed a three-layer architectural model for environmental mass event instances by analyzing their characteristics. Xu et al. [9] constructed an intelligent design framework for machine tool guideway based on CBR intelligent design, researched the key technology, and efficiently accomplished the variant and serialization design of the guideway.

Wang [10] applies CBR-based technology to the construction of e-commerce website, solves the problem of long construction cycle, and the effect is remarkable. Yang [11] conducted an in-depth research and discussion on the decision scheme correction algorithm and similar case retrieval algorithm, which provided new ideas for the design and development of intelligent decision support system.

Qiu et al. [12] proposed a two-level similarity measure, practice has proved that this method of similarity matching of the outer contour of the punch loaded parts has a better discriminatory quality effect and high computer operating efficiency. Nanjing University of Aeronautics and Astronautics, such as Liu Jinshan. To solve problems of ignoring workpiece geometric features in complex fixture design, a geometric method for fixture intelligent design and 3D models was proposed.

Liu [13] studied the theory of BP, RBF neural network. Combined with the secondary development of CATIA using the CAA platform, effectively combined neural network algorithms with CATIA software to improve surface reconstruction accuracy and achieve surface optimization.

3.2 Product Intelligent Design Framework

The establishment of a product intelligent design system aims to automate product design using knowledge and reduce human labor intensity. Knowledge is the main theme of the design, and the instance library provides support. Mechanical product intelligent design system generally contains hardware systems, artificial intelligence algorithms, multilevel instance library and so on. The mechanical product intelligent design framework is shown in Fig. 3.1.

The human–machine interface and problem description enter the system. It contains two levels of instance libraries. The first stores mechanical product part examples and generates the overall design via AI algorithms. The second stores functional and structural parameters of mechanical parts and realizes further design and model establishment through AI algorithms. The interactive human–machine interface is convenient for users to operate, and supports adding and deleting instance knowledge in the instance knowledge base, which is convenient for instance knowledge maintenance.

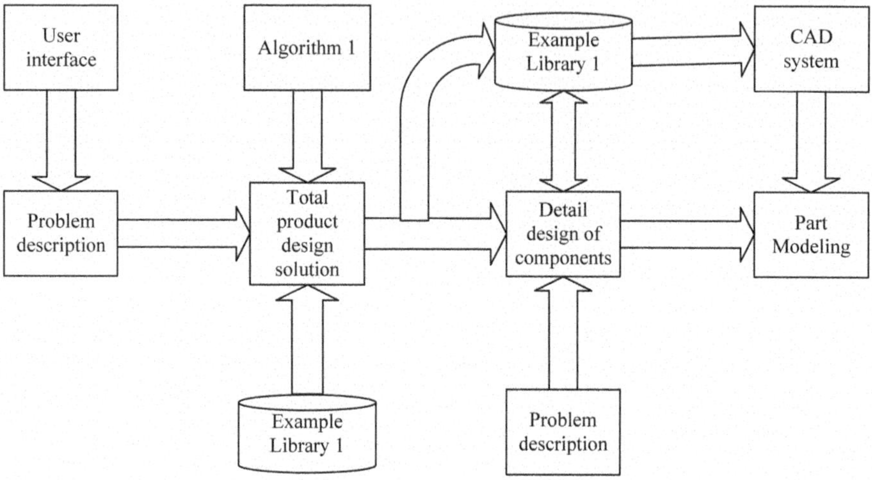

Fig. 3.1 Product intelligent design flowchart

3.3 Key Technologies in Intelligent Design

The key technologies in intelligent design include 3D model retrieval technology, automatic identification of processing features, and design solution optimization technology. This section will introduce them one by one to help readers understand.

3.3.1 3D Model Retrieval Techniques

View-based 3D model retrieval has become a research hotspot. The method firstly represents the 3D model as a collection of 2D views, and then adopts deep learning techniques for classification and retrieval, but the existing methods need to be improved in terms of accuracy and efficiency. Convolutional neural network can be applied to 3D model retrieval technology. The method involves indexing and model retrieval. In the indexing stage, select representative views and input them to the trained Convolutional Neural Network (CNN) to extract features and classify. Then, organize the features by their feature categories to build the index in the retrieval stage. In the retrieval stage, the representative views of the input model are classified into a category using CNN and voting algorithm. Only the features of this category are selected for similarity matching instead of all categories, reducing the search space. Moreover, as the number of retrieval views gradually increases, the retrieval process is prematurely terminated as soon as the 3D model can be identified.

(1) Overarching Framework

As shown in Fig. 3.2, the 3D model retrieval method is divided into three steps: (1) CNN training; (2) 2D representative view selection and indexing; (3) model retrieval In the CNN training stage, the 3D model is firstly converted to 2D view, and then the CNN is trained through the 2D view. In the 2D representative view selection and indexing stage, the 3D model is first converted to 2D view. Then, the K-means algorithm is used to select the 2D representative view. Finally, this view is used as the input of the trained CNN for feature extraction and indexing. K-means algorithm is used to select the 2D representative view, and finally the 2D representative view is used as the input of the trained CNN for feature extraction and index building. In the model retrieval stage, the input can be either an image or a model. If it's a picture, retrieval is direct. If it's a model, a 2D representative view is generated first, then retrieval is done through the view. It uses the way of classification first and retrieval later, with each view matched in only one category for similarity. A similarity matching method based on the increment of the view is proposed to effectively improve retrieval efficiency.

(2) Index Building

The number of 2D views and projection angles have a significant impact on the effective representation of a 3D model. If too many 2D views will increase the time complexity of retrieval. Therefore, it is possible to K-means to different numbers of 2D representative views as follows:

Step 1: Generate 40 2D projected views for each 3D model.

Step 2: Cluster the views of each model with K-means algorithm. When clustering 2D views with K-means, the value of category K must be determined first. Experiments show that 10–20 views are matched best. Therefore, the value of K is roughly in the range of 10–20.

Step 3: Select the view closest to the center of each class as the representative view of this 3D model. Divide the 2D views of the 3D model into K classes, calculate the distance between the view and the center of the class by Euclidean distance in each class, take the view with the smallest distance as the representative view of the class, and finally get K 2D views.

Step 4: Input the representative view into the trained ResNet50, finally the output features are indexed according to the category, the fully connected layer outputs the depth features of the representative view. The Softmax layer outputs the classification results. The index building process is shown in Fig. 3.3.

(3) Model Retrieval

The task of the model search is to find similar or identical models in the model library according to the input. The input can be an image or a 3D model. If the input

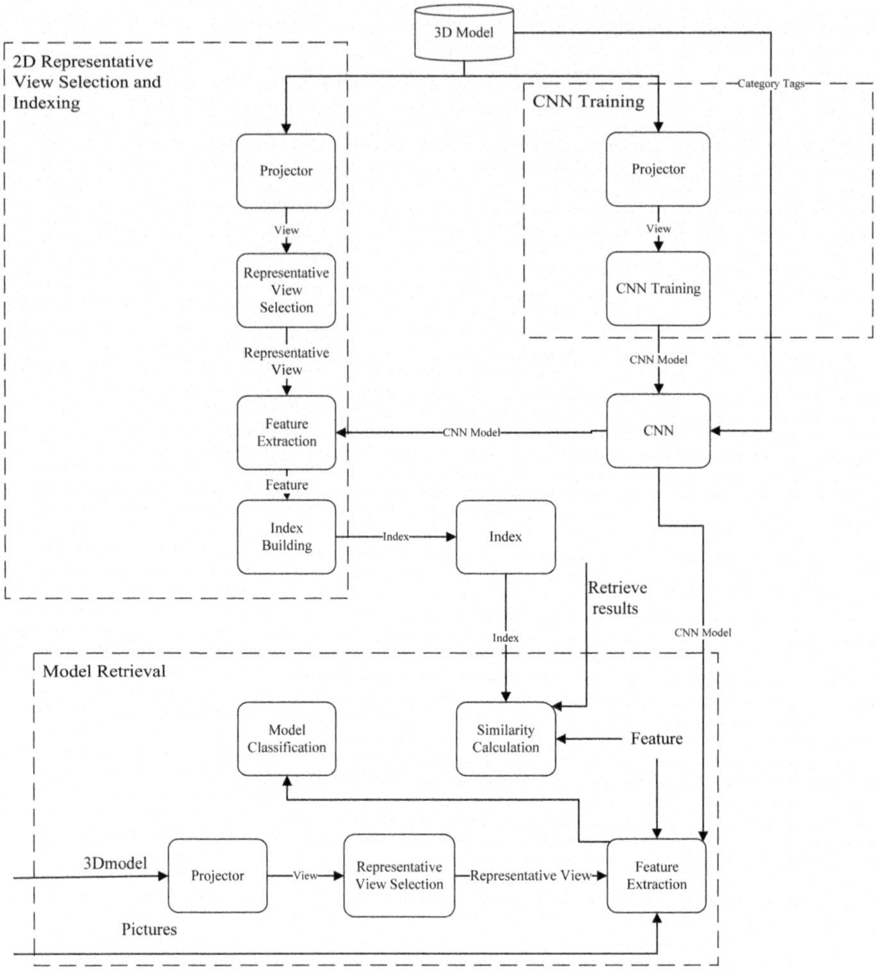

Fig. 3.2 Retrieval process

is an image, the trained CNN determines the category and extracts the features in a category, and the retrieval result is obtained by the following equation:

$$I = \arg\min_{i,j} dis\left(W, F_{i,j}\right)$$

I represents the retrieval result, $dis()$ represents the computation of Euclidean distance, W is the feature value of the input image, $F_{i,j}$ is the eigenvalue of the j-th view of the i-th model, $1 \leq i \leq m$, m is the number of models in a category, $1 \leq j \leq n_i$, n_i is the number of representative views of the i-th model, model i is the retrieved result.

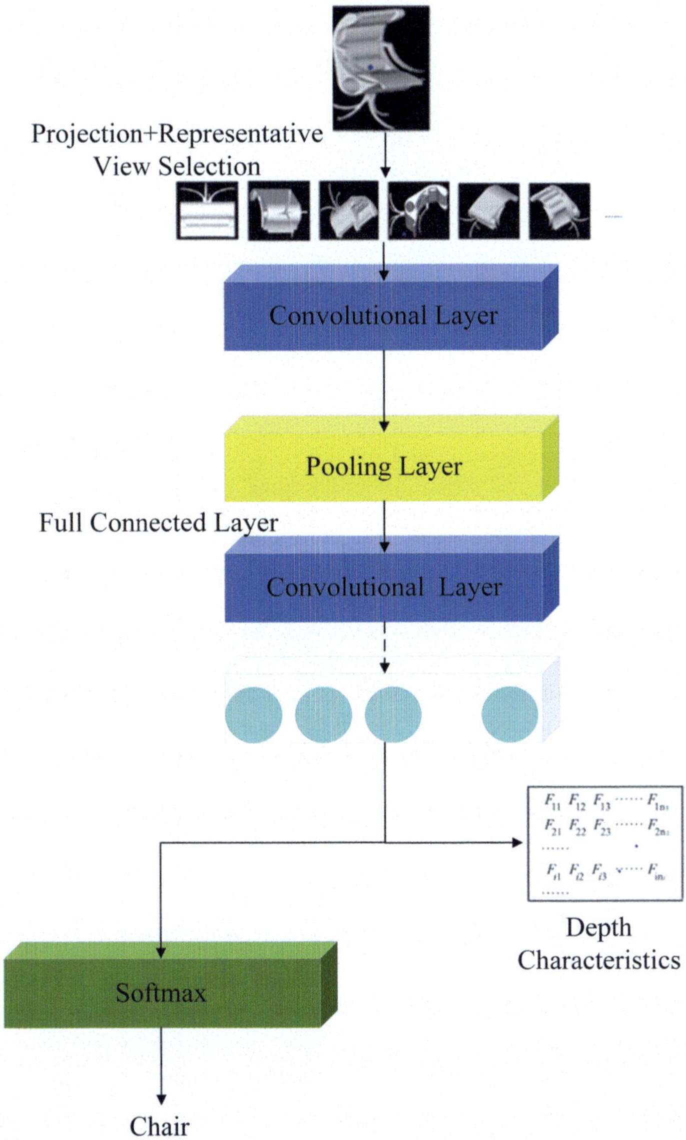

Fig. 3.3 Index building process

If the input is a 3D model, the retrieval is divided into 3 steps: (1) generating a 2D representative view; (2) inputting the view to the CNN to extract features and categorize. Considering misidentification in the classifier, causing individual views of the input model not to be fully in the same category, a voting strategy is used to determine the retrieval category of the model. (3) similarity matching is performed in the determined category.

Category_Vector represents the category vector, where the c-th element indicates the number of views that are categorized into the c-th category. *Category_Vector* initialize as follows:

$$Category\_Vector = [0, 0, \ldots, 0]$$

where *Category_Vector* is a c-dimensional vector representing the c-categories in the model library.

When a representative view is assigned to class c, this vector is updated as follows:

$$Category\_Vector[c] = Category\_Vector[c] + 1$$

Finally, the following categories were identified for the model:

$$C = argmax\ Category_c\_Vector[c]$$

C denotes the classification result. If C cannot be uniquely determined, randomly choose among the remaining views to continue voting until C is uniquely determined.

After the models are accurately categorized by the voting strategy, the retrieval strategy is as follows: (1) if the distance between an input view and a model view is small enough, i.e., dis < n, it can be determined that this model is the required model; (2) when the number of representative views selected in sequence reaches 5, and all 5 views are classified into the same model of the same class, it can be determined that this model is the required model (3) If the representative views are matched to different models in the same category, the cumulative distance value is calculated. If a model has the smallest cumulative distance value, it is the required model.

3.3.2 Automatic Recognition of Processing Characteristics

Reasonable selection of mechanical design parameters is a very important part of mechanical design and manufacturing. In mechanical design and manufacturing process design, there are many design and process parameters to be determined and selected. Many of them are not precise values from calculations but are chosen by design engineers and craftsmen based on experience within a certain range. Making full use of historical data and generating new designs and evaluating them through deep learning methods is of great significance to the design field.

Automatic machining feature recognition technology is a core technology and key support in the field of intelligent design and manufacturing, and it is a constituent unit of computer-aided design and manufacturing system. At present, there are many methods for machining feature recognition, and the traditional methods are graph-based methods, rule-based methods, volume decomposition-based methods, trace-based methods and hybrid methods. Each of these methods has its own advantages and disadvantages, but they generally have the following problems: (1) focusing on specific types of CAD representations without learning ability; (2) poor noise immunity to input CAD models.

In recent years, deep learning has achieved great success in the field of computer vision and pattern recognition, and has been widely applied to recognition tasks such as classification, segmentation, and target detection. This has led researchers to start applying deep learning to the field of processing feature recognition. FeatureNet [14] is a three-dimensional convolutional neural network that takes voxel models as input. This method can effectively identify single machining features and segment complex multi-feature parts into single ones by using specific algorithms, thus realizing the feature recognition of multi-feature models.

FeatureNet automatically generates 24 common machining feature data sets through the SolidWorks secondary development interface, as shown in Fig. 3.4. CAD modeling software-generated models are in boundary representation. As this can't be trained directly in the CNN, they are converted to binary 3D voxel models. 1 means inside the shape, 0 means outside. Using this approach, the entire model can be represented as a long string of binary numbers, which can be used as inputs for training the deep neural network.

The convolutional neural network architecture can use the spatial structure of voxels to learn simple structures like planes and surfaces. By stacking multiple convolutional layers, more complex structures can be encoded hierarchically, enabling the network to process complex features on the CAD model. FeatureNet, which is used to recognize individual features on a 3D voxel model, consists of eight layers: an input layer, four convolutional layers, a maximal pool, a fully connected layer, and a classified output layer, as shown in Figs. 3.3, 3.4 and 3.5.

The input layer is a $64 \times 64 \times 64$ 3D voxel model. After that, there are four convolutional layers. The three parameters of these layers are the number of convolutional kernels, the size of the kernels, and the spatial step size. The parameter values of the four layers are (32, 7, 2), (32, 5, 1), (64, 4, 1), and (64, 3, 1). The fourth convolutional layer is followed by a maximum pooling layer that downsamples the feature map along the spatial dimension. The two parameters of the pooling layer are the filter

Fig. 3.4 24 common machining features in the FeatureNet dataset

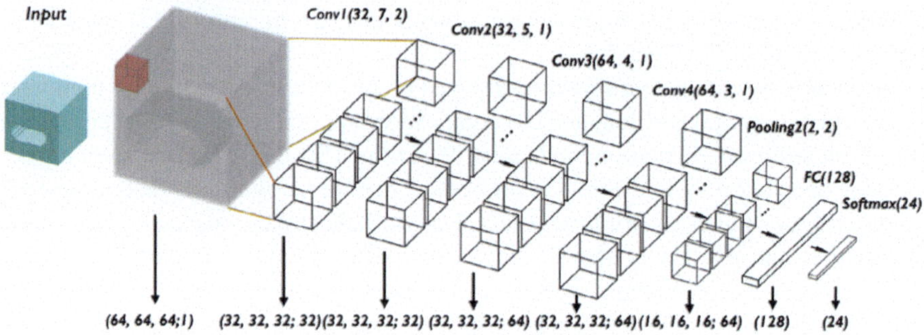

Fig. 3.5 FeatureNet: 3D convolutional neural network for processing feature recognition

size and the step size. The pooling layer is followed by a fully connected layer, and finally the predicted probabilities for each category are output through a Softmax classifier.

3.3.3 *Design Optimization*

(1) **Deep Learning Based 3D Modeling Method for Sketches**

Sketching and 3D modeling are the two main steps in industrial design. Sketching is usually done first as it allows the designer to express their vision quickly and close. The design sketch is then converted into a 3D model for downstream engineering and manufacturing, using CAD tools that provide a high level of accuracy and editability. However, design sketching and CAD modeling are often performed by different experts with different skills, making design iterations cumbersome, expensive, and time-consuming. Therefore, how to efficiently implement sketch-based 3D modeling becomes a very important research direction. In recent years, researchers tend to use data-driven approaches for computer graphics and sketch-based modeling techniques. Deep learning techniques, with their strong data representation learning capabilities, are widely used in various segments of computer graphics. The research related to the implementation of sketch-based 3D modeling using deep learning techniques is introduced and their own reflections are given.

Sketch2CAD [15], as a learning-based interactive modeling system, is similar to a translation software that converts sketching operations into corresponding CAD modeling operations. The user draws a sketch on the CAD model, and the system infers the type of operation and its corresponding parameters based on the sketch and the local CAD model.

1. **Sketch Classification Network**

As shown in Fig. 3.6, the first stage categorizes sketches into possible CAD operations (stretching, resection, rounding, scanning). In addition to the user sketch, the classifier also inputs a depth map and a normal map of the current 3D model, thus providing powerful contextual information about the intended operation. Using the sketch, depth map and normal map as inputs, both the classification task and the segmentation task are performed to obtain the type of modeling operation.

The second phase segments the sketch and contextual information into parts specific to the target CAD operation. For example, the sketch of the rounding operation is segmented into two contour curves, and the context mapping map is segmented to form the base surface for applying the chamfering operation to generate the parameters of the modeling operation.

The third phase instantiates the CAD operation by fitting parametric curves or shapes to the segmented strokes and projecting these strokes onto selected surfaces of the 3D model.

Figure 3.7 shows the structure of Sketch2CAD's classification network with the loss function shown below, where O is the true operation type of the input training samples, the weight W_0 = 1/N_O, is the number of training samples of type O, and P_O is the predicted probability of operations of type O. Figure 3.7 shows the structure of Sketch2CAD's classification network with the loss function below. Here, O is the true operation type of the input training samples. The weight is the number of training samples of type O, and is the predicted probability of operations of type O. We use the weights for different operation types to avoid potential statistical bias in training due to their frequency of comparison. We use weights for different operation types to avoid potential statistical bias settings in training due to the frequency of their comparison.

2. **Sketch Partition Network**

As shown in Fig. 3.8, each segmentation network is a U-Net structure, and Unet networks are typically characterized by a U-shaped symmetric structure with a convolutional layer on the left and an upsampling layer on the right. The original Unet structure contains 4 convolutional layers and the corresponding 4 upsampling layers. Therefore, in implementation, you can either implement the network from scratch, initialize the weights and train the model. Or you can borrow the convolutional layer structure and corresponding trained weight files of some existing networks like resnet, vgg, etc., along with the upsampling layer at the back for training. Because in deep learning model training, using existing weight model files can greatly accelerate the training speed.

Another feature is that the feature maps obtained from each convolutional layer of the Unet network are stacked to the corresponding upsampling layer, thus realizing that the feature maps of each layer can be effectively used in subsequent computations. That is, cross-layer connectivity. In this way, compared with other network structures like FCN, Unet avoids directly supervising and calculating the loss in

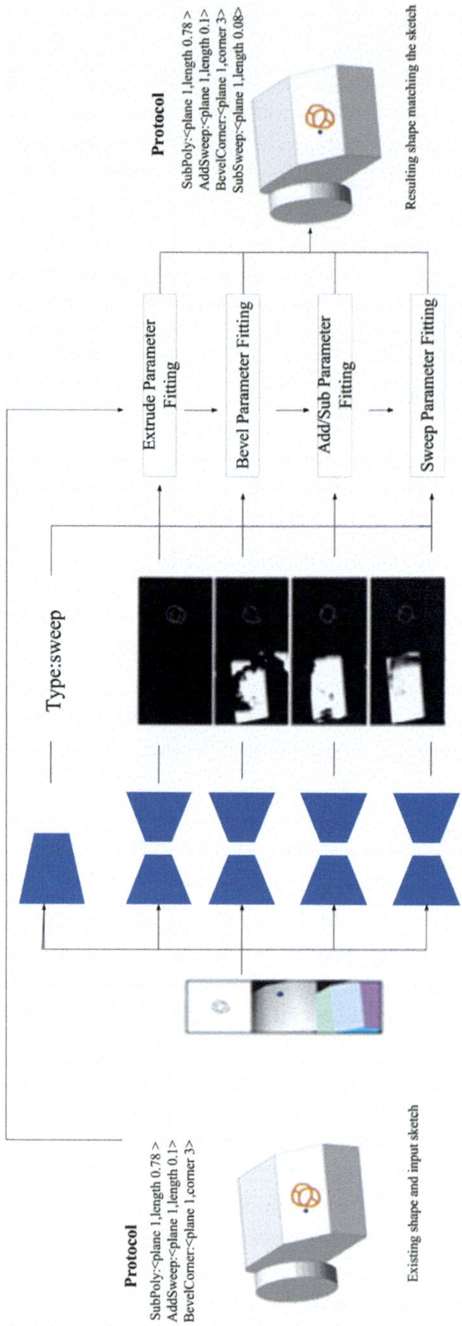

Fig. 3.6 Overall framework of the Sketch2CAD approach

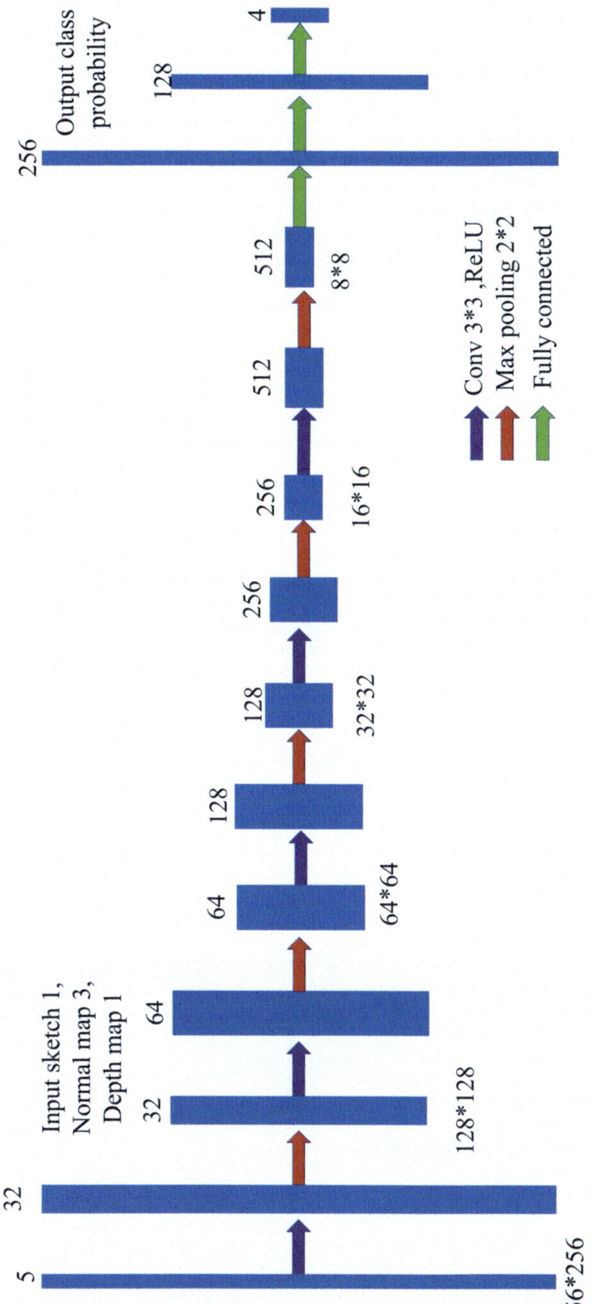

Fig. 3.7 Sketch2CAD's classification network

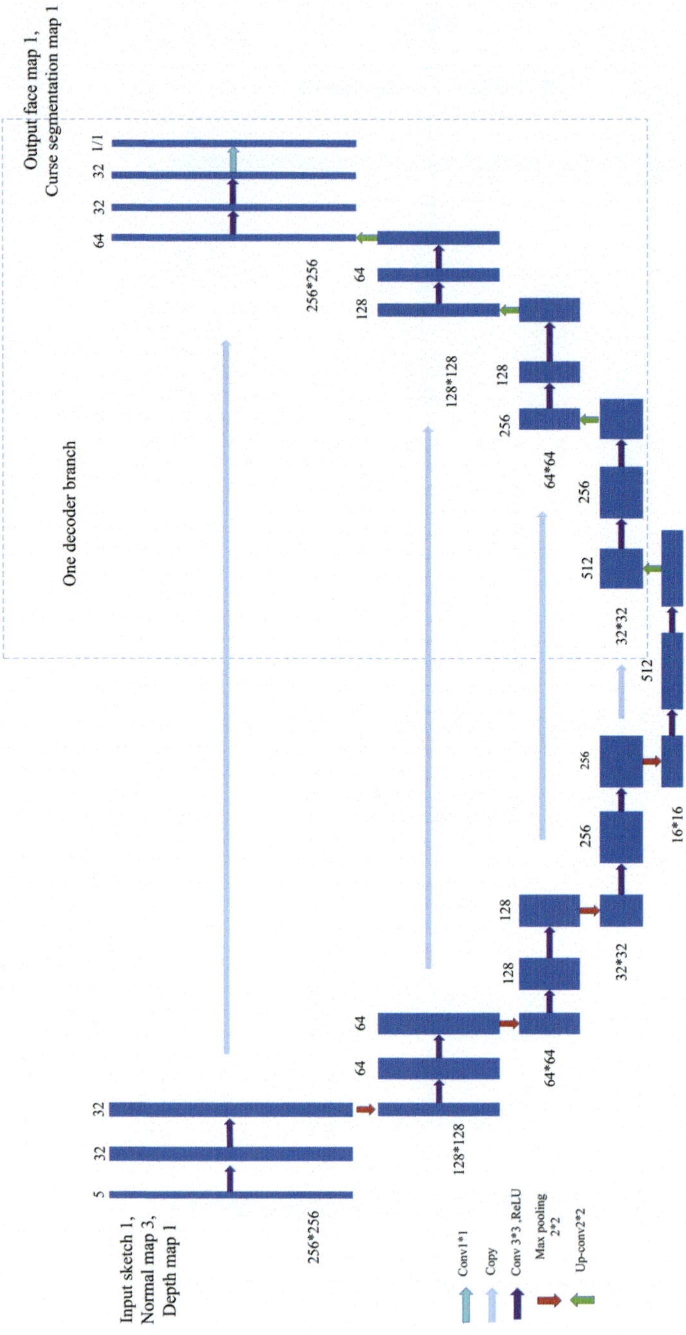

Fig. 3.8 Segmentation network for Sketch2CAD

high-level feature maps. Instead, it combines the features in low-level feature maps. This enables the final feature maps to contain both high-level and many low-level features, achieving the fusion of features at different scales and improving the model result accuracy. The result of the model is to improve the accuracy of the model.

The two decoder branches output two graphs: the probabilistic graph F of the base plane and the curvilinear segmentation graph C, both with a space size of 256 × 256 and a channel width of 1. The loss function is:

$$\mathcal{L}_{reg}(S, D, K) = \frac{1}{256^2}\left\|F - \tilde{F}^2\right\| + \frac{1}{|\tilde{M}|}\left\|\tilde{M} \odot (C - \tilde{C})\right\|^2$$

The parameters of its loss function are explained as follows:

1. $\tilde{F}$ is the true basemap; the foreground pixel has the value $\tilde{F}(x, y) = 1$, and the background pixel has the value 0;
2. $\tilde{C}$ is the real stroke map;
3. $\tilde{M}$ is the stroke pixel mask corresponding to the stroke map;
4. $\odot$ is the component product.

Given the predicted base map, the base map F is first binarized by thresholding 0.5 and then the original CAD model is rendered to find the corresponding faces. Different operations have their specific curve segmentation maps C. The principle of designing curve mapping is that the input and output is a learnable mapping. Corresponding to different operations with different regression details.

(2) **Manufacturability Analysis of Machined Features in 3D Models**

Design for manufacturability (DFM), which means starting from improving the manufacturability of parts, makes the parts and various processes easy to manufacture, with low manufacturing costs, high efficiency, and a low cost ratio. Design for manufacturability analysis means that the designer, with the help of a set of tools and a related knowledge base at the product design stage, assesses the manufacturability of the product. Under the premise of guaranteeing the function and quality of the product, the design is modified so that the product design meets the requirements of the manufacturing process and has good manufacturability.

Traditional DFM mainly uses manual methods to analyze the development efficiency, cost, and manufacturability of products, which can take a lot of time. At present, researchers have applied deep learning methods to the manufacturability analysis of CAD models. They enhance 3D voxel data through the surface normal of the object boundary and apply it to the 3D convolutional neural network to identify local machining features and analyze their manufacturability. They also use the 3D gradient class activation heat map (3D-GradCAM) to locate the machining features. This allows designers to identify non-manufacturable features and modify them.

Taking the manufacturability of drilled holes as an example, the important geometric parameters of holes are the diameter, depth and location of the holes. The DFM of a drilled hole has the following main requirements: (1) the ratio of the depth to the diameter of the hole should be less than a given value; (2) the hole should be at a certain distance from the edge of the material; and (3) different holes should be at a certain distance from each other. The manufacturability of the holes is determined based on their compliance with the above rules. These rules are used to construct hole feature manufacturable datasets and non-manufacturable datasets for use in 3D convolutional neural networks. Different categories of machining feature manufacturability datasets can be obtained by recognizing the historical data related to the industry by FeatureNet.

The deep learning based manufacturability analysis framework is shown in Fig. 3.9. The input to the network is the voxelized CAD model, and the results of the manufacturability analysis are obtained through convolutional, pooling, and fully connected layers. This is a dichotomous network, i.e., manufacturable and non-manufacturable. After obtaining the manufacturability analysis results, the location of machined features is located and visualized by 3D-GradCAM.

To obtain feature localization maps using 3D-GradCAM, the spatial importance of each feature map in the last convolutional layer of the 3D-CNN needs to be calculated. The spatial importance of each feature map can be seen as its weight, which is the global average of the gradients returned from a specific category. The visualization results are shown in Fig. 3.10, with green for manufacturable features and red and yellow for non-makeable features.

(3) **Automatic Generation and Evaluation of Design Solutions**

Currently, deep learning has been applied to various research areas in design optimization. Researchers have integrated topology optimization and generative models

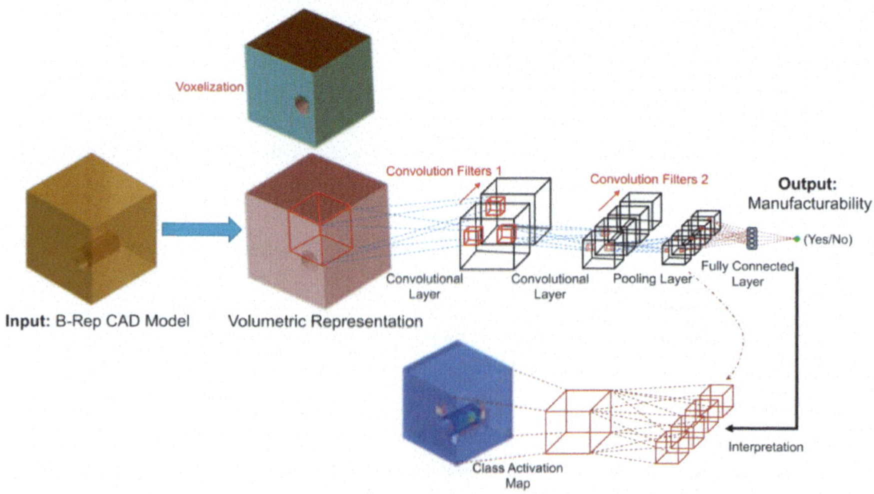

Fig. 3.9 Deep learning based manufacturability analysis framework

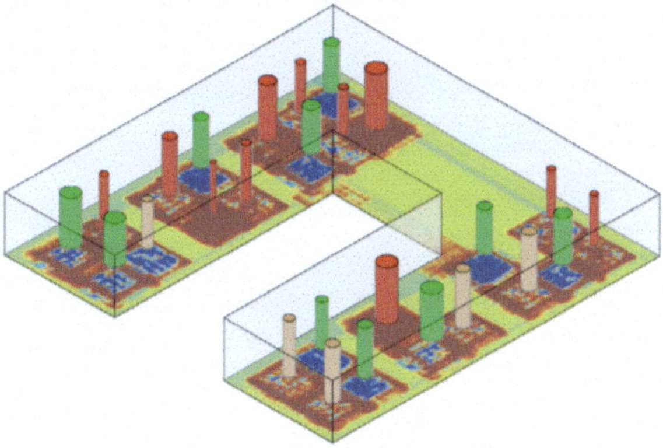

Fig. 3.10 Visualization of CAD model manufacturability analysis results

to construct a deep generative design framework to generate a large number of designs and filter out good new designs based on a limited number of historical designs.

Topology optimization (topology optimization) is a mathematical method for optimizing the distribution of materials in a given region based on given load cases, constraints, and performance metrics, and is a type of structural optimization.

Generative model is a module of Generative Adversarial Networks (GAN), which produces good output from mutual game learning with Discriminative Model. Boundary Equilibrium GAN (BEGAN) is a type of GAN that uses a self-encoder as a discriminator. The architecture of the generative model in this study is shown in Fig. 3.11. The encoder consists of five convolutional layers, four downsampling layers, and one fully-connected layer. The encoder uses the reconstruction error function as the loss function, which can effectively reconstruct the similarity data into training data. data reconstructed as training data. The structure of the decoder is similar to that of the encoder, except that the subsampling is replaced with upsampling for the reduction of the generated new design.

The deep generative design framework consists of two main components, iterative design exploration and design evaluation, as shown in Fig. 3.12. Iterative design exploration is the integration of topology optimization and generative modeling to produce new designs, while design evaluation quantifies and evaluates the novelty and key attributes of new designs. Each phase is explained below:

In stage 1, previous designs in the market and industry were collected and used as reference designs for phase 2. In this study, the reference design is defined as the baseline design to create the new design in topology optimization.

In stage 2, the new design is obtained by modeling based on the reference design. In this step, the topology optimization has multiple objective functions: (1) engineering performance; and (2) difference from the reference design (i.e., L1 distance in the pixel direction) aimed at improving aesthetics and diversity.

In stage 3, similar designs collected in the topology optimization are filtered by user-set guidelines and thresholds for similarity calculation. This process is repeated in stage 6 to filter out the designs generated by the generative model.

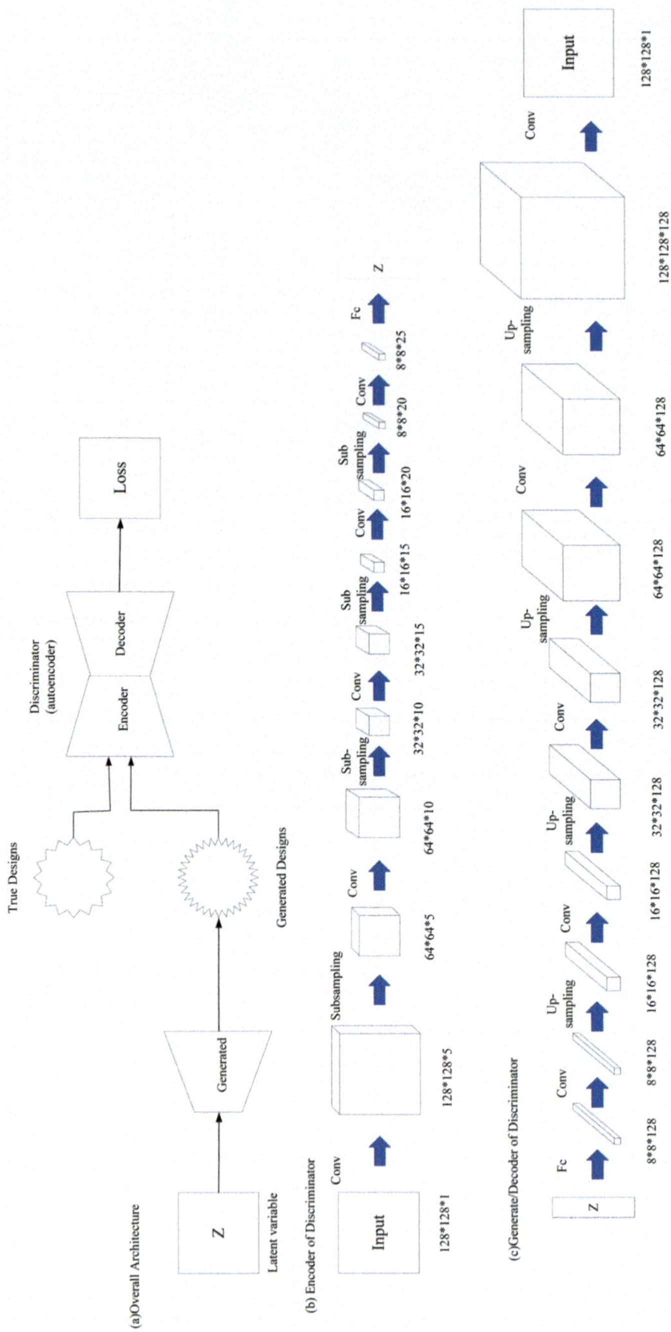

Fig. 3.11 Network architecture of BEGAN and autoencoder

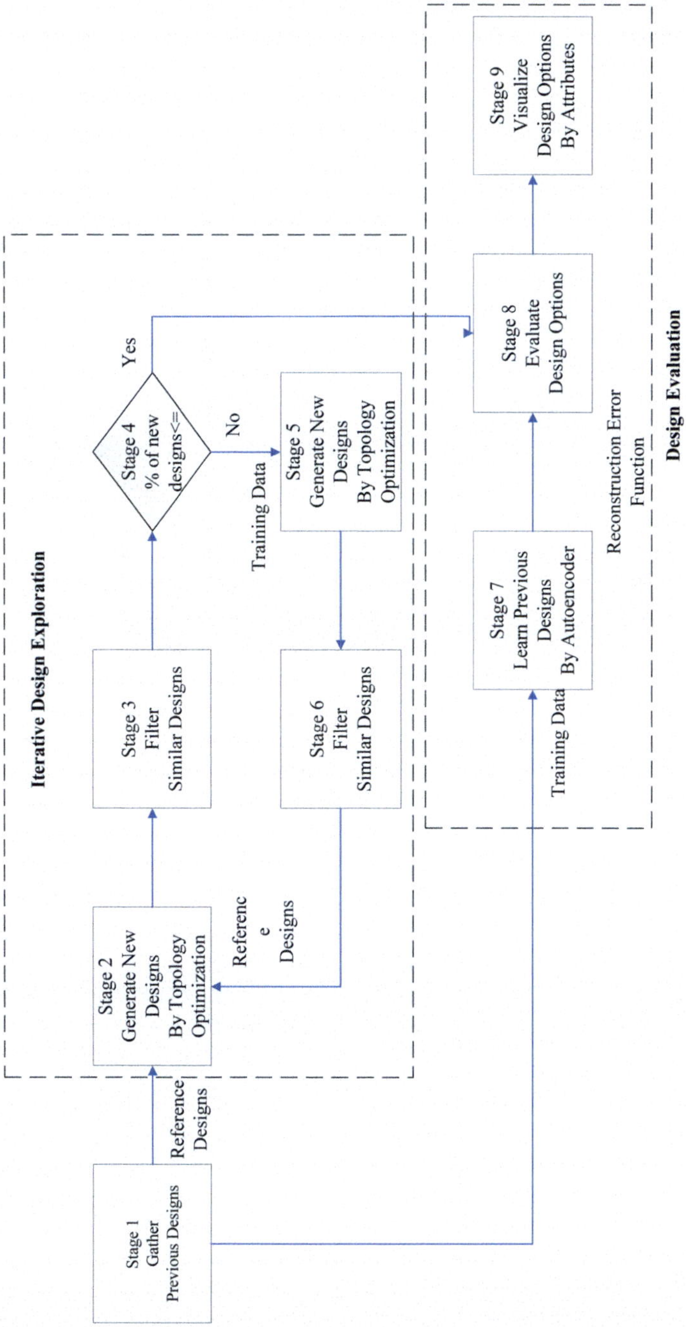

Fig. 3.12 Deep generative design framework

In stage 4, the ratio of the number of new designs in the current iteration to the total number of designs in the previous iteration is calculated. If it is less than a threshold set by the user, the iterative design exploration is exited and jumps to stage 8; otherwise, stage 5 is entered.

In stage 5, after learning the summarized designs in the current iteration, new designs are created by generative modeling, and they are used as reference designs in stage 2 after similar designs are filtered out in stage 6. The purpose of this iterative process is to create a large number of various designs starting from a small number of previous designs in stage 1.

In stage 7, the historical designs collected in stage 1 are used as training data, and the reconstruction error function is used as a loss function for the autoencoder. This function can be used to assess the novelty of a design, and new designs with higher reconstruction error values are considered to be more novel.

In stage 8, new designs obtained from iterative design exploration are evaluated based on various design attributes that are critical to the design. Finally, trade-offs between attributes are represented by plotting the design on each axis of the design attributes and the appropriate design can be selected considering the relative importance of each attribute.

3.4 A Case for Intelligent Design of Fixtures Based on Knowledge Reuse

Fixture design is a design process that relies heavily on knowledge and experience, and high-quality fixture design requires the support of a variety of related knowledge to be realized. With the application of related technologies in fixture design, aerospace design institutes have found that knowledge-driven fixture design reuse is important for intelligent design. The establishment of the knowledge reuse platform and model for fixture design process is highly significant for achieving lean fixture research and development. It can provide comprehensive knowledge support for fixture structure design, clamping planning, and the fixture planning process. Ultimately, it enables smooth knowledge flow in design activities and ensures effective reuse and management of fixture design knowledge. Enhance the intelligent level of fixture design, improve the efficiency and quality of fixture design, shorten the production lead time of enterprises, and accelerate the development and manufacturing of new products.

3.4.1 *Intelligent Design Patterns for Fixtures Based on Knowledge Reuse*

Aiming at the fixture design knowledge reuse demand, we analyze the design process, propose an intelligent design mode, and construct a reuse system architecture integrating planning and structure design. Within this framework, the design knowledge of fixtures is sorted out and acquired, the concepts of fixture design domains and their relationships are analyzed, and the fixture design domain ontology is constructed. Aiming at different stages of fixture design and different design tasks, relevant knowledge models are constructed. Based on the product integration information model, we use the process rules to reason the processing constraints, configure the processing methods for features based on manufacturing resource knowledge, and conduct intelligent fixture planning. For the fixture planning requirements of specific process models, they are expressed as the functional requirements of fixture instances. Similar fixture design instances are obtained through semantic similarity calculation, and the fixture planning scheme is formed by adaptive correction and optimization of the fixture scheme. At the same time, in the fixture structure design process, fixture components can be intelligently selected based on the proposed component knowledge component model. It can realize automatic parametric creation of solid models, automatic assembly of components and clamping features, and ultimately achieve the rapid design of digital fixtures. The structure of the architecture is shown in Fig. 3.13, which mainly contains the following levels:

1. Fixture design knowledge resource layer. Fixture design knowledge resources are the framework of the support platform, mainly including the fixture design knowledge base and related databases like the process knowledge base, examples library, components library, and design database. The process knowledge base mainly includes manufacturing resources knowledge and process rules knowledge, etc. The design example library stores successful fixture design cases, with the 3D model and semantic information. The fixture component model library has various standard (GB, HB, JB, and enterprise) fixture components' solid models like guiding, support, clamping, and positioning ones. The design resource library consists of the design process and the design database directly related to it.The design resource library is composed of materials directly related to the design process, such as fixture design manuals, mechanical design taboo manuals, mechanical design manuals and so on.
2. Domain knowledge layer. In the knowledge reuse based fixture intelligent design framework, ontology is an effective way to express fixture design domain knowledge. In the domain knowledge layer, ontology is used to express and store the knowledge to form a clear definition of the fixture design domain.The construction process of fixture design ontology includes resource analysis, concept extraction, semantic relationship addition and instance addition and other steps.
3. Knowledge modeling layer. For different tasks of fixture design, the key link of fixture design knowledge reuse is how to associate domain knowledge with specific knowledge resources. By analyzing the fixture design process, the fixture

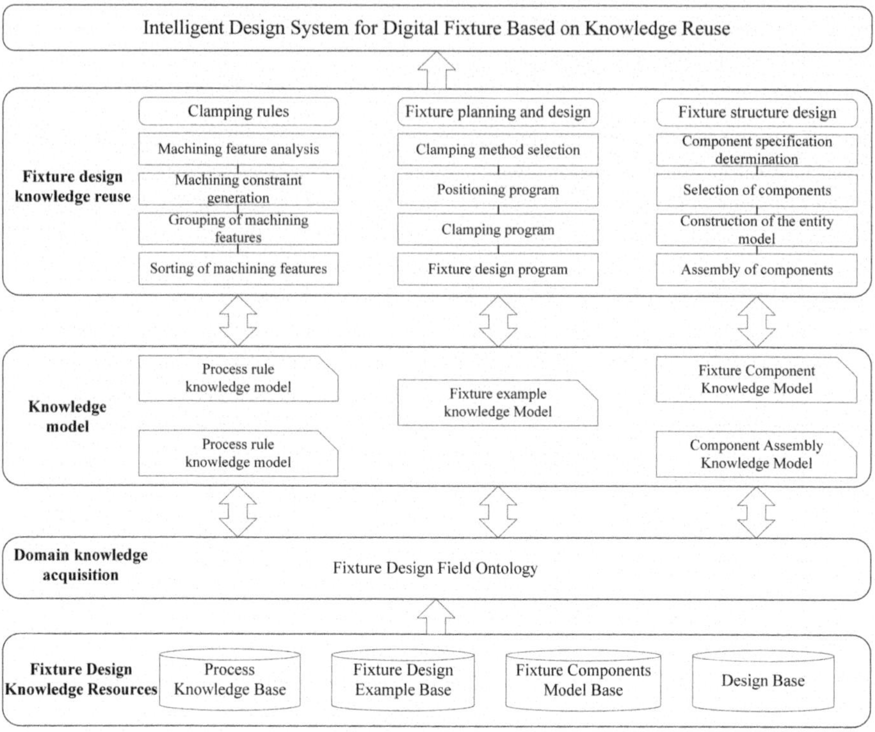

Fig. 3.13 Knowledge reuse based fixture intelligent design architecture framework

design process is modeled and decomposed into local knowledge-oriented design tasks. The knowledge related to the design task is coalesced in the domain ontology layer to construct a knowledge model for the fixture design task. Based on the knowledge model-driven fixture design process, there are various forms of knowledge resources, and the knowledge is effectively pushed to specific design tasks. In digital fixture design, the knowledge model includes: fixture component knowledge model, fixture example knowledge model, process rules knowledge model and manufacturing resources knowledge model. The use of these knowledge models can be effectively embedded in the corresponding knowledge resources to the fixture design process.

4. Fixture design knowledge reuse layer. The knowledge model offers an interface for knowledge reuse. The fixture design knowledge reuse layer is for the fixture design stages of planning, planning and design, and structure design, providing specific reuse methods. For example: Reason constraints of workpiece machining features through process rule knowledge. Configure machining methods based on manufacturing capability from the manufacturing resource knowledge model. Semantic retrieval and reuse of fixture instances based on the instance knowledge model. The fixture component knowledge model can support tasks like reasoning

and selecting component specifications, automatic creation of semantic model-driven solid modeling, and automatic assembly of fixture components for structural component design. In the fixture design process to design task-oriented knowledge model as the core, making full use of the existing fixture design experience and various knowledge, to achieve intelligent fixture design, improve the agility of the design.

5. Digital fixture design interaction layer. For designers to provide information input, interactive operation, system display and results output and other auxiliary design functions. Fixture designers through the interactive layer of knowledge selection and call, drive the flow of design knowledge in the design process, to complete the design task reasoning and decision-making. Based on the new design task's existing info, the designer acquires a reasonable fixture design scheme via the knowledge model and returns the design result, its 3D model and other info to support rapid generation of the scheme.

3.4.2 Reuse Process for Fixture Intelligent Design Knowledge

In the 3D design and manufacturing environment, with the help of the workpiece's integrated information model (including geometric and non-geometric info), the fixture design is oriented towards the clamping planning of the workpiece machining process to assist the 3D process planning. Prior to the clamping planning, the machining features of the workpiece are first analyzed and converted into the minimum machining elements with the machining features and clamping features. Secondly, the design task of fixture planning is decomposed to form the sequential constraint generation task and the machining method configuration task of the machining element. Based on the process planning knowledge constructed from the tolerance relationship, the machining sequence constraints of machining elements are reasoned. Based on the machining capability representation of the manufacturing resource knowledge model, the corresponding machining method is configured for different machining stages of each machining element. Based on the sequence constraints of candidate machining methods and machining features, a feasible clamping scheme is constructed. Finally, through the intelligent clamping planning method, a reasonable clamping scheme is obtained to guide the 3D process planning of the part. After that, various intermediate process models and related parameters are generated. Specific knowledge reuse process shown in Fig. 3.14.

Intermediate process model of the clamping demand information as the basis for the construction of the clamping demand information model of the fixture. And the model and the semantic information of fixture instances are used to calculate the semantic similarity between instances, realize the full matching from concept to attribute, and obtain a reasonable fixture planning scheme. In the adaptive adjustment process, based on the deformation constraints and the knowledge of fixture layout, the layout of the fixture features is optimized, and the fixture planning and design

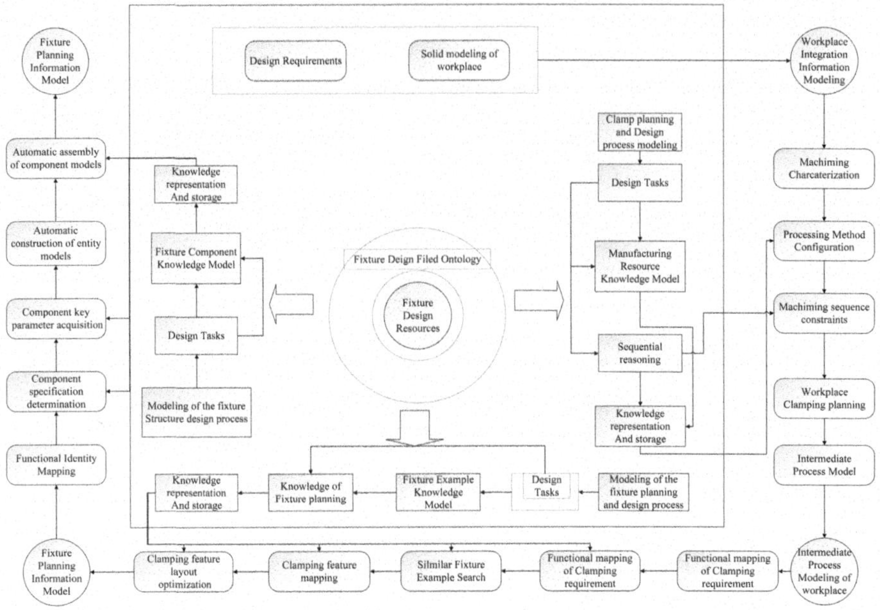

Fig. 3.14 Intelligent design flow of fixture based on knowledge reuse

scheme that meets the fixture requirements is constructed. In the fixture structure design, the clamping requirements of fixture planning turn into component identification. Based on the component knowledge model, the specific model is reasoned, parameters are obtained, and the solid model is built. At the same time, the assembly relationship is constructed with the clamping characteristics of the workpiece to realize the automatic assembly of the fixture components. The knowledge of fixture design can be effectively reused to realize the intelligent design of digital fixtures.

3.4.3 Fixture Design Example Verification

The above knowledge component model and algorithm are validated by taking the fixture structure design process of frame parts as an example. The rapid design process of fixture components driven by the knowledge model is: Through fixture planning and design, get specific clamping features of the workpiece like positioning points, clamping points and support points. Then, in the 3D model of the workpiece, the functional identity turns into a fixture component one (a specific set of points and lines) to represent the detailed clamping characteristics, providing a fixture planning program for component instantiation and assembly.

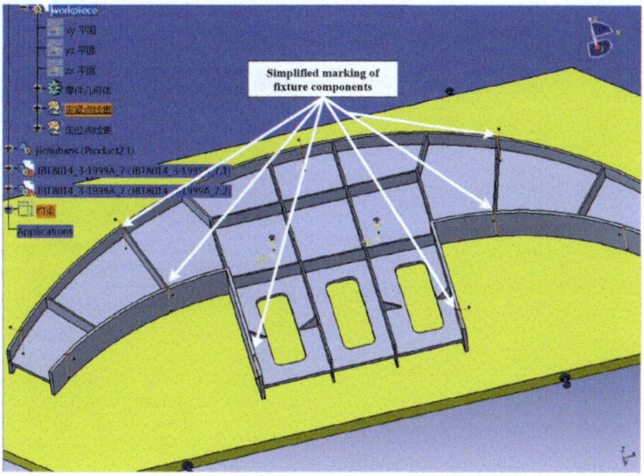

Fig. 3.15 Clamp structure design scheme

According to the fixture component identification (clamping points and vector line objects), the designer obtains these point and line sets, as shown in Fig. 3.15. The designer selects the fixture component objects to be instantiated and calculates the action distance according to the action distance algorithm. Based on the parameter selection rules, through this value to determine the grade of the fixture components.

During instantiation, according to the component type in the knowledge model, the entity model definition of each assembly is obtained, and the drive component model is parameterized by the parameter info.

In the automatic assembly process of fixture components, the assembly knowledge in the fixture component knowledge component model is invoked to introduce the fixture components into the assembly environment. At the same time, it analyzes how many kinds of assembly constraints there are in this kind of assembly relationship, and which assemblies and assembly features are composed of each kind of assembly constraint. The fixture component knowledge component model can assist in the selection of fixture components, and solid models can be created automatically. At the same time, the assembly relationship of fixture components is automatically created, ultimately achieving the rapid design of fixture components based on the knowledge component model driven. As shown in Fig. 3.16 for the design result of the fixture structure.

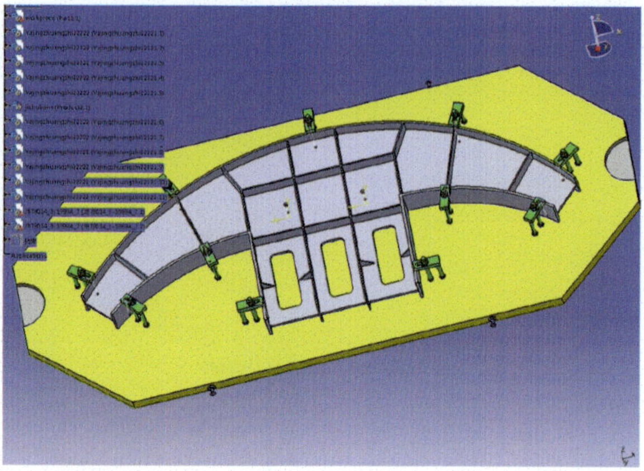

Fig. 3.16 The results of the design of the fixture structure

References

1. Zhou M, Jiang G (2016) Application of object-oriented hybrid knowledge representation method in integrated steel production. Mod Manufact Eng 07:30–34
2. Hua R, Wang J, Zheng Q et al (2000) Research and implementation of intelligent grouping method based on constraint satisfaction. Appl Comput Res 11:20–22
3. Yu C, Ji T (2019) A product prototyping method based on semantic hierarchical link structure. Mech Des 36(11):139–144
4. Jie J, Zhang J, Chen D (2009) Design and implementation of context-aware RBAC model based on inference. J Zhejiang Univ (Eng Ed) 43(09):1609–1614
5. Hu D, Lei R (2018) An engineering product design method based on instance reasoning and artificial immune system. Comput Integr Manufact Syst 24(01):146–153
6. Zhang T, Zhang F, Yan Y et al (2016) Intelligent design of fixture structure based on ontology and knowledge components. Comput Integr Manufact Syst 22(05):1165–1178
7. Chen W, Fan W, Li Z (2015) A large machine tool part cutting database system based on hybrid reasoning. China Mech Eng 26(07):923–928
8. Zhang Y (2016) Research on the application of case-based reasoning technique in emergency decision-making of environmental mass events. Saf Environ Eng 23(01):94–99
9. Xu Y, Yin G, Xu D et al (2014) Research on intelligent design of machine tool guideway based on CBR. J Southwest Univ (Nat Sci Ed) 36(12):177–186
10. Wang X (2005) Research on intelligent design system of E-commerce website based on CBR. Qingdao University
11. Yang BY (2004) Application of case-based reasoning in intelligent decision support systems. Jilin University
12. Qiu H, Liu H (2005) Part recognition technology in CBR system for intelligent design of blanking die. Forging Technol 03:76–79
13. Liu S (2014) Neural network surface reconstruction optimization technology based on CAA platform. Changchun University of Science and Technology
14. Zhang Z, Jaiswal P, Rai R (2018) FeatureNet: machining feature recognition based on 3D convolution neural network. Comput Aided Des 101:12–22
15. Li C, Pan H, Bousseau A et al (2020) Sketch2CAD: sequential CAD modeling by sketching in context. ACM Trans Graph 39(6):164:1–164:14

Chapter 4
Structural Analysis of Design Documents

Enterprises typically possess a vast repository of design documents. Establishing a knowledge management system based on these documents can significantly enhance organizational efficiency [1]. Such a system effectively leverages internal knowledge resources. Standard documents represent one of the most common forms of documents, usually stored as electronic documents within the internal knowledge management system database. This chapter introduces a structured analysis approach for aviation standard documents. It provides technical professionals with extensive referencing tools for various processes. These include product design, process design, technical reviews, and airworthiness reviews. When drafting the assembly outline for aviation products, technical professionals must consult relevant process specification documents. They reference specific content from standard documents and specifications, including part lists and process parameters. However, existing knowledge management systems lack intelligent applications for managing aviation standards. When technical professionals need specific standards, they must manually sift through the database. They do this by using document names or categories. Once the correct document is retrieved, professionals face the daunting task of browsing through lengthy standard documents, often spanning dozens of pages. They must pinpoint the specific location of the desired information, which is both inefficient and labor-intensive.

Aviation standard documents are primarily stored in electronic formats due to the diverse sources and issuing organizations involved. These documents vary widely in format, including WORD, PDF, and image files. The image files represent text in pictorial form, adding another layer of complexity. Some non-internal or early standard documents may be stored in image formats. Additionally, due to the lack of unified standards in the early stages of standard document creation, the formatting of electronic standard documents varies depending on the author and issuing organization. For example, documents prepared by different departments often have discernible differences in format. Text data in aviation standard electronic documents

T. Liu et al., *Industrial Intelligence: Methods and Applications*, Springer Series in Advanced Manufacturing, https://doi.org/10.1007/978-3-031-81477-8_4

often includes significant non-content-related elements like corporate logos, confidentiality statements, headers, footers, and garbled text. These documents differ significantly from general documents by featuring a distinct structure and extensive technical terminology. Additionally, they tend to be much longer than typical documents.

Extracting and structuring text data from knowledge documents and storing it in an appropriate data format is foundational for intelligent applications in manufacturing enterprises, such as semantic retrieval and personalized recommendations. In addition, standard documents differ significantly from commonly used texts in natural language processing tasks—such as online news, comments, medical records, etc. Standard documents are structured documents with clear article and section outlines, containing numerous specialized technical terms, lengthy content, and extensive charts. These characteristics make research on intelligent applications for aviation standard documents unique and valuable.

4.1 Research on the Current State of Domestic and Foreign Technical Developments

This subsection will review and analyze the related technologies from three aspects: text extraction and structuring of standard electronic documents, automatic keyword extraction technology for aviation standard electronic documents, and information retrieval technology for aviation standard electronic documents.

4.1.1 Research on Text Extraction and Structuring of Standard Electronic Documents

With the development of research in natural language processing, natural language processing algorithms are gradually being applied to document tasks across various industries. Research on the effective extraction of text from specific electronic documents has emerged. Zhugeng Zhu focused on electronic documents stored in Microsoft Office 2007 document format, designing a state machine-based method for text extraction. The method demonstrated a high accuracy in text extraction experiments on various Office electronic documents such as docx, xml, and ppt [2]. Xianying Liu conducted research on text extraction and structuring of medical-related PDF documents. Through PDF document parsing and generating XML text data from txt text, the research ultimately extracted text suitable for semantic analysis [3]. Xueju Li proposed a method combining rule matching and support vector machines for PDF paper text extraction. The method utilized support vector machine algorithms to discover text extraction patterns in PDF documents, combining established rules for text extraction. Experimental verification showed promising results in text extraction [4]. Botao Yu focused on the extraction of formulas in electronic documents, applying rule matching to locate mathematical expressions in the document's area [5]. The method leveraged the formatting characteristics of mathematical expressions to successfully reconstruct them. It demonstrated excellent extraction results in experiments. Jun Pan studied image documents containing complex tables. The detection of tables in image documents was optimized using region localization algorithms and adjacency graph algorithms. This approach achieved good results, particularly in medical record and express order image document [6].

Existing methods for extracting and structuring text from electronic documents are mostly tailored to specific application scenarios and document types. The abundance of specialized technical terms, varied document formatting, and a high prevalence of noise text in aviation standard electronic documents pose challenges for generic text extraction methods. There is a need to develop specialized research on text extraction and denoising methods for aviation standard electronic documents. These methods should build upon common text extraction tools and involve designing a rational data structure for storing text data.

4.1.2 Relevant Research on Automatic Extraction of Keywords from Aviation Standard Documents

After extracting and structuring text data from aviation standard electronic documents, it is necessary to establish appropriate models to automatically extract keywords from these documents. Keywords are words that can summarize the main content and characteristics of a document. Automatic keyword extraction algorithms can be divided into two categories. Unsupervised learning algorithms extract keywords by constructing models based on established patterns. In contrast, supervised algorithms build machine learning models and train them using annotated corpora.

Unsupervised algorithms for automatic keyword extraction typically involve constructing mathematical models for documents and vocabulary based on established rules. These models incorporate manually set quantifiable indicators to measure the importance of words in the document. Formulas are then devised based on these quantifiable indicators to calculate the weight of each word in the document. Using these weights, the algorithm ranks candidate words and selects the top k candidates, representing the final extraction result. Unsupervised methods, which do not require extensive annotation of well-labeled corpora, are a focal point in related research. Keyword extraction algorithms based on unsupervised methods can be broadly categorized into three models: topic generation models, graph-based models, and statistical models. Approaches utilizing topic generation models construct a model by building a thematic structure from the document corpus. In these models, themes represent probability distributions over sets of words. Topic generation models observe statistical quantities in the document corpus to estimate latent random variables [7]. They assume that the relationships between documents and themes, as well as themes and words, adhere to specific probability models. During the training process, different hyperparameters, such as those in Latent Dirichlet Allocation (LDA) and probabilistic latent semantic analysis (PLSA) models, are estimated [8]. Weng Wei employed the construction of an LDA topic model to compute the document's topic probabilities and the frequency of word assignments to different topics, calculating the importance of words. Ultimately, keywords were selected based on the sorted scores, and experimental results verified the feasibility and effectiveness of this method [9]. Graph-based models represent documents as graph structures, where vertices correspond to words appearing in the document, and distances between nodes measure their co-occurrence relationships. Subsequently, word scores are computed by evaluating indicators such as node centrality. TextRank, a classic graph-based model, follows a similar approach to PageRank in its construction [10]. TextRank algorithm quantifies the importance of words by measuring the centrality of nodes in the constructed graph [11]. Florian quantified the importance of words by measuring the centrality of nodes in the word graph and experimented with various methods for calculating node centrality. The quantification method based on graph node centrality achieved better results in the task of keyword extraction from short documents compared to TextRank [12]. Xia Tian constructed a word graph model and integrated statistical measures such as coverage influence, positional influence, and frequency influence into the calculation of the graph node probability transition matrix. Through iterative processing, the algorithm determined node importance scores. Keywords were then selected based on these scores [13]. Models based on statistical measures involve manually constructing formulas to quantify the importance of words in the document corpus. TFIDF (Term Frequency-Inverse Document Frequency) is an algorithm used to assess the importance of words. It calculates this importance by multiplying a word's term frequency by its inverse document frequency. TFIDF calculation is simple, convenient, and exhibits strong interpretability. Xia Tian et al. utilized the Pearson chi-square test to optimize the statistical measures in TFIDF calculation and demonstrated the effectiveness of the method through experiments [14]. Qin P. attempted to incorporate negative instance information into the TFIDF formula, proposing the NEG-TFIDF (negative sampling—TFIDF) method, which achieved better accuracy in keyword extraction in experiments [15]. Recent studies have explored combining different models. For example, Gu Yijun and colleagues integrated the advantages of TextRank and LDA to create a new model. Their experiments demonstrated a significant performance improvement, especially when the topic distribution in the text collection was clear [16]. Similarly, Liu Xiaojian initially used the LDA topic model to quantify the similarity of different words in the document. The model outperformed some existing keyword extraction models by quantifying node distances in the word graph model based on the calculated word similarity [17]. Unsupervised learning methods quantify the importance of words by building models and relevant statistical measures of words in the document, providing good interpretability. However, unsupervised learning algorithms designed based on certain rules may exhibit different application effects on different datasets due to variations in document features.

Supervised learning-based keyword extraction algorithms typically require constructing machine learning models and using a substantial amount of annotated text for training. During this process, the model learns to recognize patterns in keyword extraction [18]. Supervised learning algorithms for keyword extraction transform the task into a classification problem, specifically identifying the binary classification of "keyword" or "non-keyword." The input features of machine learning models typically include common statistical features or semantic features of words, such as part-of-speech. Suitable machine learning models encompass Support Vector Machine (SVM), Maximum Entropy (ME), Hidden Markov Model (HMM), Conditional Random Fields (CRF), among others [19]. Witte [20] proposed the KEA algorithm, which selects TFIDF values and the first appearance position of words in the document as input features, applying the Naive Bayes method as the classifier. Turney [7] introduced the GenEx algorithm, which uses word frequency and part-of-speech information as input features, employing a decision tree as the classifier. Hulth [21], building upon Frank [22], incorporated linguistic knowledge as features to train a keyword extraction model, and experimental results demonstrated a significant improvement in keyword extraction effectiveness in paper abstracts. In the early stages, supervised learning-based keyword extraction algorithms mostly utilized traditional statistical features such as word frequency, inverse document frequency (IDF), and part-of-speech as model inputs. These approaches struggled to effectively leverage the semantic features of words and contextual relationships. The emergence of word embedding models and advancements in deep learning have enhanced traditional statistical methods. Integrating these models allows for more effective use of semantic information and contextual relationships in automatic keyword extraction. Common word embedding models include Word2Vec [23], Glove [24], ELMo [25], BERT [26], and prevalent deep learning models for natural language processing comprise LSTM [27], Transformer [28], among others. Wen employed the Word2Vec model to compute word vectors for semantic information and improved the TextRank algorithm by using the similarity between word vectors as transition probabilities for node weights [29]. Chen Wei applied the BiLSTM-CRF model for automatic keyword extraction in shorter texts, achieving favorable results in the context of extracting keywords from short video titles [30]. Supervised learning algorithms for automatic keyword extraction typically leverage more comprehensive document information to construct more complex models, often resulting in higher extraction accuracy. However, these supervised learning-based algorithms generally require a large amount of high-quality annotated data, and the increased model complexity also raises the hardware and software requirements for algorithm application.

4.1.3 Research on Information Retrieval Techniques for Aerospace Knowledge Documents

The numerous, extensive, and highly technical nature of aerospace standard documents makes manual retrieval of information challenging for users through file names and document categorization. Therefore, it is crucial to explore and develop an appropriate information retrieval system tailored to aerospace standard documents. Information retrieval (IR) involves the mathematical representation, storage, and organization of information, aiming to optimize the speed and accuracy of user access to the required information [31]. In information retrieval, mathematical representations are applied to query and document information, and matching scores for each document are calculated based on these representations. The documents are then ranked according to the matching scores. Document information can encompass various types, such as commodity information, video content information, etc. Document retrieval, product recommendation, video content recommendation, and specifically, aerospace knowledge document information retrieval fall under the umbrella of information retrieval.

Information retrieval typically consists of three key processes. First, it involves the mathematical representation of user query information. Second, it includes the mathematical representation of retrieval information, such as document information, followed by the matching between these two sets of information. Traditional information retrieval models match words or characters in the user query with those in the document for information retrieval. These models include the Boolean model, vector space model, probability model, etc. The Boolean model constructs a Boolean expression representing query information based on the matching relationship between query terms and terms that have appeared in a document (1 for term presence, 0 for absence) [32]. The vector space model represents documents using a bag-of-words model, where a document is represented as a vector recording the occurrence frequency of different vocabulary words. Matching between documents and query text is performed by calculating the similarity between the query representation vector and the document representation vector. For example, the SMART retrieval system, developed by Salton et al., is based on this model and has achieved excellent results in practical applications [33]. The probability model represents document words and query words as probability models and calculates the probability of relevance between documents and query text based on these models [34]. Through continuous technological research and engineering applications, the rotated normalized vector space model [35], BM25 model [36], and Dirichlet prior language model [37] have shown good retrieval performance. After optimization, these three algorithms typically exhibit similar performance [38]. The BM25 algorithm is a probabilistic retrieval model-based algorithm. After splitting the user input text and documents, the system calculates the relevance of each query term to the document. It also assesses the inverse document frequency and the relevance of the query term to the query text. These calculations are then used to determine how relevant the user input is to a specific document. The BM25 algorithm is relatively simple yet has good application effectiveness, leading to numerous studies based on the BM25 algorithm. Géry et al. introduced the BM25t algorithm by incorporating XML tags into the original BM25 algorithm's term frequency calculation. This modification assigns different weights to structurally meaningful tags such as titles or tags with special semantics, such as bolded terms, to improve the BM25 algorithm [39]. Lv et al. proposed modifications to the BM25 algorithm by adjusting the calculation related to document length in the original BM25, aiming to avoid excessive penalties based on text length [40]. Relevance retrieval models based on word matching cannot address the issue of users inputting query text that may not precisely match documents during the information retrieval process. Semantic analysis-based models effectively tackle this problem. Latent Semantic Analysis (LSA) typically represents (high-dimensional) vector space of documents based on word frequency [41]. By reducing dimensions, high-dimensional data is projected onto a low-dimensional space, allowing documents and query text corresponding vectors in the low-dimensional space to have high similarity even without common vocabulary. Dimensionality reduction methods are based on Singular Value Decomposition (SVD) of the corresponding term/document matrix [42]. The Probabilistic Latent Semantic Analysis (pLSA) algorithm [8] defines an appropriate document generation model based on the likelihood principle. The factor representation obtained through pLSA can handle polysemous words, distinguishing different meanings and types of words explicitly.

The development of the Internet and information technology has generated vast amounts of information retrieval data, comprising user query texts, document sets, and user-clicked documents, with user clicks serving as labels. In scenarios with abundant annotated data, deep learning retrieval algorithms have emerged and achieved rapid development [43]. The DSSM algorithm [44], leveraging annotated user queries and click data, trains semantic models for information retrieval. Experimental results demonstrate that the model's performance far surpasses previous unsupervised algorithms. The DSSM algorithm and similar algorithms use neural networks to project query texts and documents into a common low-dimensional space. In this space, the texts are represented as vectors. The relevance between query texts and documents is then determined by calculating the distance between these vectors [45]. The Deep Crossing algorithm [46] introduces an embedding layer in the deep model, enabling the mapping of high-dimensional data to low-dimensional vectors. The low-dimensional feature vectors are concatenated and serve as the input to a deep network with residual fully connected units as the basic components. The DIN algorithm [47] incorporates an attention mechanism into the deep neural network, achieving better results in recommendation tasks by introducing attention between the model's embedding layer and feature merging layer. Information retrieval models based on machine learning and deep learning use large annotated datasets to achieve a deeper semantic understanding of user-input query and document representations. These models often offer more precise matching by employing more complex algorithms. However, methods that rely on networks to understand documents are constrained by NLP-related models, making it challenging to directly handle long texts and turning neural network-based full-text retrieval into a challenge [48].

4.2 Text Denoising and Structuring Methods Based on Rule Matching

Aviation standard documents are mostly classified and stored in electronic formats such as PDF and WORD. In tasks such as product design, process design, and airworthiness review, internal users within enterprises often need to extensively consult and reference documents in the aviation standard document repository. Electronic documents stored in formats like PDF, WORD, and images are challenging for direct application processing by software programs. Before applying intelligent solutions to aviation standard documents, it is necessary to use electronic document parsing tools to extract text data from these documents and store it in a suitable data structure. The diverse storage and writing formats in existing aviation standard electronic documents, combined with numerous noise elements like headers, footers, and garbled characters, pose challenges. Directly applying existing text extraction tools often leads to issues, such as garbled text and other noise-related problems. Additionally, these tools may not extract the original document structure information. Therefore, research on text denoising methods is needed on the preliminary results of text extraction. This paper focuses on PDF format documents, WORD documents, and documents stored in image formats. Suitable open-source tools are selected to perform initial document data extraction. Based on these preliminary results, noise text patterns in the extracted text data are identified and summarized. Rule-based matching is then applied to denoise the text data, resulting in more accurate and intact text information.

Simultaneously, aviation standard documents are mostly structured documents with clear and complete chapter structures. The document structure features are distinct. However, the extracted text data tends to lose the original structural information of the articles. Therefore, this paper addresses the structural characteristics of aviation standard documents and, based on a tree structure, designs a text tree to store the extracted data, thus preserving the original structural information of the document.

4.2.1 Description of Aviation Standard Electronic Document Data

Aviation standard electronic documents are primarily stored in PDF and WORD file formats. Some PDF documents, in which text cannot be extracted and is stored in the form of images, are referred to as image documents. Typically, electronic documents in PDF and WORD formats can be categorized into Chinese and English documents. Examples of some aviation standard documents are illustrated in Figs. 4.1, 4.2, and 4.3.

The dataset used in this paper consists of a total of 1000 aviation standard documents. The majority of electronic documents are stored in PDF format, while the

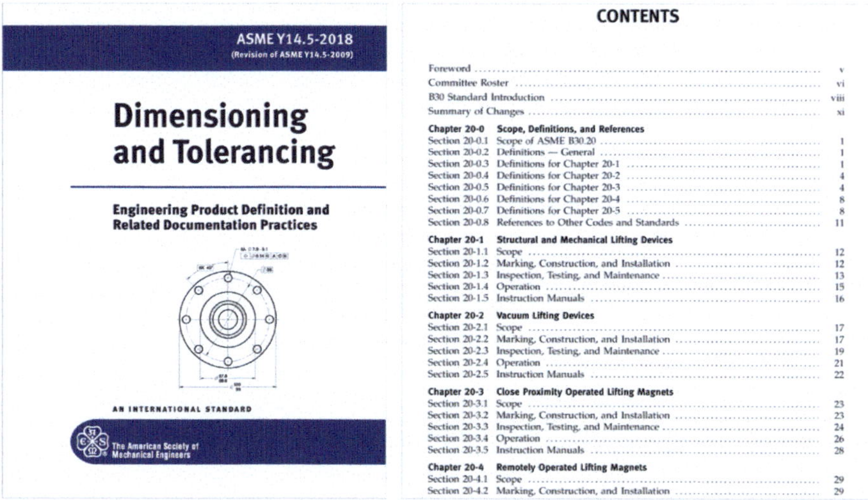
ASME Y14.5-2018
(Revision of ASME Y14.5-2009)

Dimensioning and Tolerancing

Engineering Product Definition and Related Documentation Practices

AN INTERNATIONAL STANDARD

The American Society of Mechanical Engineers

CONTENTS

Foreword		v
Committee Roster		vi
B30 Standard Introduction		viii
Summary of Changes		xi
Chapter 20-0	**Scope, Definitions, and References**	
Section 20-0.1	Scope of ASME B30.20	1
Section 20-0.2	Definitions — General	1
Section 20-0.3	Definitions for Chapter 20-1	1
Section 20-0.4	Definitions for Chapter 20-2	4
Section 20-0.5	Definitions for Chapter 20-3	4
Section 20-0.6	Definitions for Chapter 20-4	8
Section 20-0.7	Definitions for Chapter 20-5	8
Section 20-0.8	References to Other Codes and Standards	11
Chapter 20-1	**Structural and Mechanical Lifting Devices**	
Section 20-1.1	Scope	12
Section 20-1.2	Marking, Construction, and Installation	12
Section 20-1.3	Inspection, Testing, and Maintenance	13
Section 20-1.4	Operation	15
Section 20-1.5	Instruction Manuals	16
Chapter 20-2	**Vacuum Lifting Devices**	
Section 20-2.1	Scope	17
Section 20-2.2	Marking, Construction, and Installation	17
Section 20-2.3	Inspection, Testing, and Maintenance	19
Section 20-2.4	Operation	21
Section 20-2.5	Instruction Manuals	22
Chapter 20-3	**Close Proximity Operated Lifting Magnets**	
Section 20-3.1	Scope	23
Section 20-3.2	Marking, Construction, and Installation	23
Section 20-3.3	Inspection, Testing, and Maintenance	24
Section 20-3.4	Operation	26
Section 20-3.5	Instruction Manuals	28
Chapter 20-4	**Remotely Operated Lifting Magnets**	
Section 20-4.1	Scope	29
Section 20-4.2	Marking, Construction, and Installation	29

Fig. 4.1 Image documents

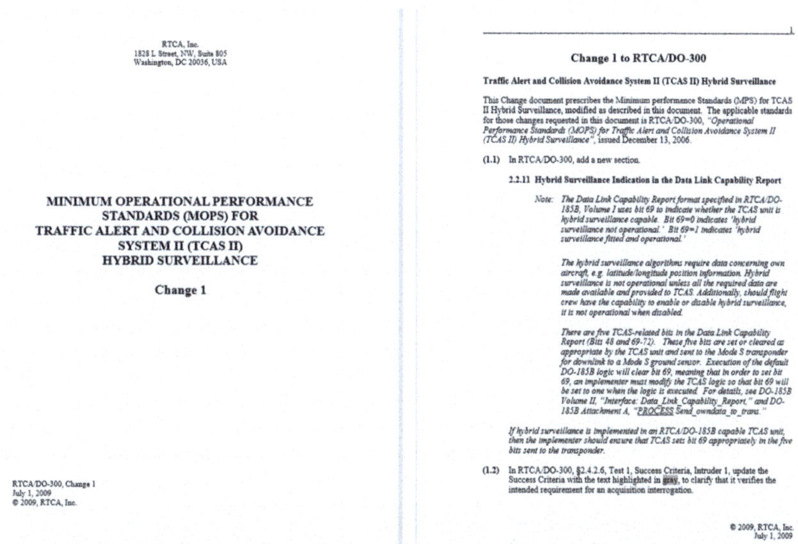
RTCA, Inc.
1828 L Street, NW, Suite 805
Washington, DC 20036, USA

MINIMUM OPERATIONAL PERFORMANCE STANDARDS (MOPS) FOR TRAFFIC ALERT AND COLLISION AVOIDANCE SYSTEM II (TCAS II) HYBRID SURVEILLANCE

Change 1

RTCA/DO-300, Change 1
July 1, 2009
© 2009, RTCA, Inc.

1

Change 1 to RTCA/DO-300

Traffic Alert and Collision Avoidance System II (TCAS II) Hybrid Surveillance

This Change document prescribes the Minimum performance Standards (MPS) for TCAS II Hybrid Surveillance, modified as described in this document. The applicable standards for those changes requested in this document is RTCA/DO-300, *"Operational Performance Standards (MOPS) for Traffic Alert and Collision Avoidance System II (TCAS II) Hybrid Surveillance"*, issued December 13, 2006.

(1.1) In RTCA/DO-300, add a new section.

2.2.11 Hybrid Surveillance Indication in the Data Link Capability Report

Note: *The Data Link Capability Report format specified in RTCA/DO-185B, Volume I uses bit 69 to indicate whether the TCAS unit is hybrid surveillance capable. Bit 69=0 indicates 'hybrid surveillance not operational.' Bit 69=1 indicates 'hybrid surveillance fitted and operational.'*

The hybrid surveillance algorithms require data concerning own aircraft, e.g. latitude/longitude position information. Hybrid surveillance is not operational unless all the required data are made available and provided to TCAS. Additionally, should flight crew have the capability to enable or disable hybrid surveillance, it is not operational when disabled.

There are five TCAS-related bits in the Data Link Capability Report (Bits 48 and 69-72). These five bits are set or cleared as appropriate by the TCAS unit and sent to the Mode S transponder for downlink to a Mode S ground sensor. Execution of the default DO-185B logic will clear bit 69, meaning that in order to set bit 69, an implementer must modify the TCAS logic so that bit 69 will be set to one when the logic is executed. For details, see DO-185B Volume II, "Interface: Data_Link_Capability_Report," and DO-185B Attachment A, "PROCESS Send_owndata_to_trans."

If hybrid surveillance is implemented in an RTCA/DO-185B capable TCAS unit, then the implementer should ensure that TCAS sets bit 69 appropriately in the five bits sent to the transponder.

(1.2) In RTCA/DO-300, §2.4.2.6, Test 1, Success Criteria, Intruder 1, update the Success Criteria with the text highlighted in gray, to clarify that it verifies the intended requirement for an acquisition interrogation.

© 2009, RTCA, Inc.
July 1, 2009

Fig. 4.2 English documents

ICS 81.040.20
CCS Q 33
GB
中华人民共和国国家标准
GB 11614—2022
代替 GB 11614—2009
平板玻璃
Flat glass
2022-07-13 发布 2023-08-01 实施
国家市场监督管理总局
国家标准化管理委员会 发布

GB 11614—2022
平板玻璃
1 范围
2 规范性引用文件
3 术语和定义
3.1
光学变形 optical distortion
3.2
点状缺陷 spot faults
3.3
断面缺陷 edge defects
3.4
厚薄差 thickness wedge

Fig. 4.3 Chinese documents

remaining are in WORD format, with PDF documents accounting for 85.6%. The electronic documents include Chinese, English, and image documents, with Chinese documents comprising 86.1%. Therefore, subsequent research on automatic keyword extraction and information retrieval methods primarily focuses on Chinese documents. The aviation standard electronic document library consists of 861 Chinese documents, 81 English documents, and 58 image documents.

In comparison to documents in other natural language processing tasks, such as web news, medical documents, legal files, etc., aviation standard documents exhibit distinct characteristics. The key features of aviation standard documents are summarized as follows.

(1) **Structured**

Aviation standard documents have a well-defined article structure, organized into a multi-level chapter hierarchy. The text within these documents includes document titles, immediate text, and various sub-chapter texts. Each sub-chapter text consists of primary subheading text, immediate text, and the subordinate sub-chapter texts belonging to a lower level. The document structure of aviation standard documents is illustrated in Fig. 4.4.

As shown in the figure, all text data can be divided into title text, consisting of document titles and subtitles, and main content text (body text).

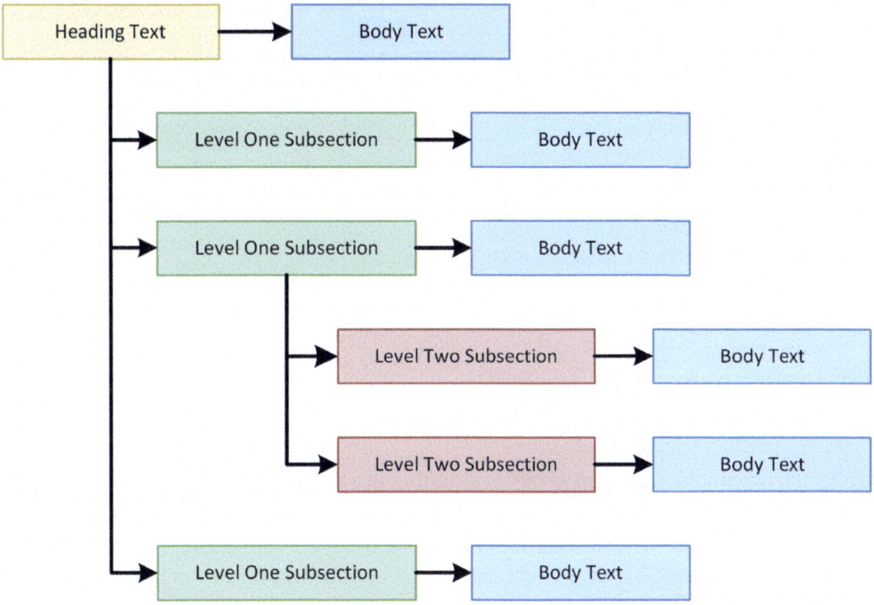

Fig. 4.4 Structure of aviation standard documents

(2) **Specialization**

Aviation standard documents are highly technical documents related to aviation and manufacturing. The structure and content of these documents differ significantly from general documents. One of the most notable features is the extensive use of professional technical terms in the text. Terms such as civil aircraft, models, composite materials, reliability, etc., are common in aviation standard documents. Conversely, everyday words like "life," "economy," and "expectations" are less common in professional technical documents. Aviation standard documents often contain numerous abbreviations in both Chinese and English. Examples include "SR" (Scheduled Reliability), "SIR" (Scheduled Interruption Rate), "MMEL" (Master Minimum Equipment List), and "MTBF" (Mean Time Between Failures), among others. As highly specialized and technical documents, aviation standard documents also contain numerous tables and images, each accompanied by corresponding captions.

(3) **Substantial Length**

Aviation standard documents have extensive lengths, often spanning tens or even hundreds of pages. The substantial length poses additional challenges for the design of algorithms and task requirements related to intelligent applications in aviation documents.

Fig. 4.5 Aviation standard documents in different formats

(4) **Inconsistency in Document Formatting**

Due to variations in standard types and authoring organizations, the formatting of electronic aviation standard documents differs significantly and can be quite disorderly. Taking Chinese aviation standard electronic documents (stored in PDF format) as an example, excerpts of the covers and appendices of two documents with different formatting are shown in Fig. 4.5.

(5) **Existence of a Large Volume of Noisy Text**

In comparison to other electronic documents, aviation standard electronic documents contain a significant amount of additional noise text. Noise text refers to the text in the document that is irrelevant to the main content or has little to no connection to understanding the document content, such as headers, footers, and document declarations. These noise texts do not contribute to the actual understanding of the document content and, when mixed in with extracted text data, can disrupt the continuity of the text. Some examples of noise text in aviation standard electronic documents include headers, footers, and copyright statements as shown in Fig. 4.6.

4.2.2 Text Denoising and Structuring Techniques for Aviation Standard Electronic Documents

Under the technical route of text denoising and structuring for aviation standard electronic documents as shown in Fig. 4.7, it mainly consists of three parts: text extraction based on open-source tools, text denoising based on rule matching, and text structuring based on tree structure. Initially, all aviation standard electronic documents are subjected to preliminary text extraction using open-source tools. The documents are preliminarily classified based on file types, and for PDF format documents, they are further categorized into image documents or regular PDF documents based on the ability to extract text. Subsequently, different text extraction tools are applied

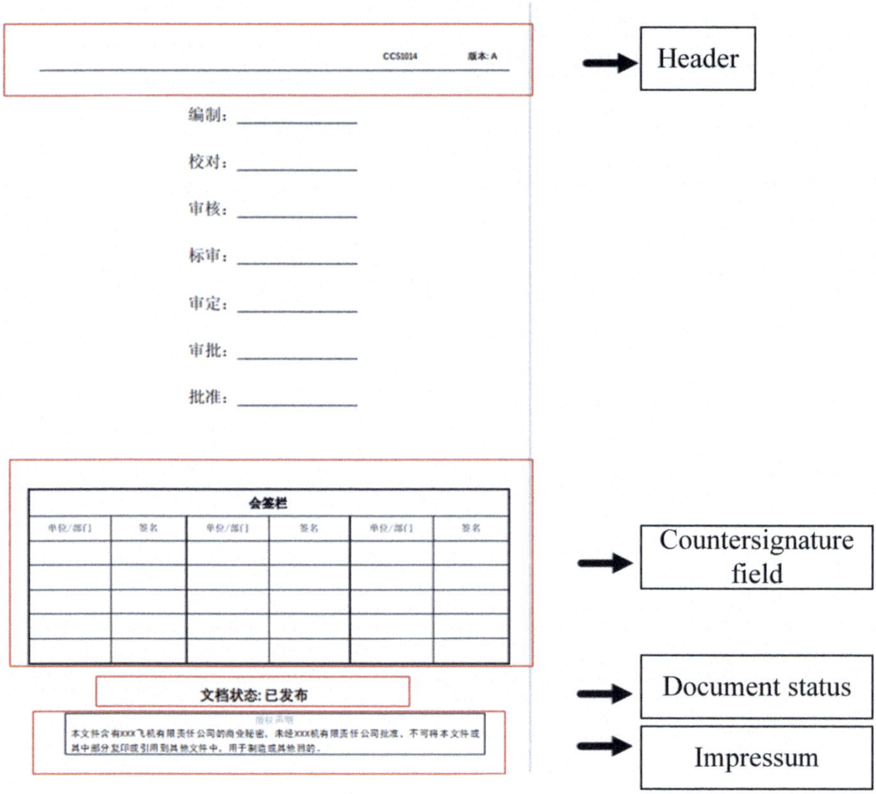

Fig. 4.6 Examples of noisy text

based on document types to obtain preliminary text extraction results. Considering the presence of a significant amount of noise text in the text extraction results, a rule-based text denoising method is applied to the extracted text data. The text is first categorized into Chinese and English based on the ratio of characters. Then, regular expressions are created to match and remove various types of noise text through rule matching, with a focus on Chinese documents in this study. This process leads to more accurate text data. To preserve the original document structure, a tree-based text structuring method is used. This method organizes the text data into a text tree, effectively representing the structure of aviation standard documents.

Aviation standard documents are stored in the enterprise's internal knowledge management platform in the form of PDF documents, Word documents, image documents, and other electronic formats. It is necessary to extract the text data from these electronic documents using appropriate text extraction tools tailored to different document types. Given that the preliminary extraction results contain a significant amount of noise text, a rule-based denoising method has been designed to remove the noise text through rule matching.

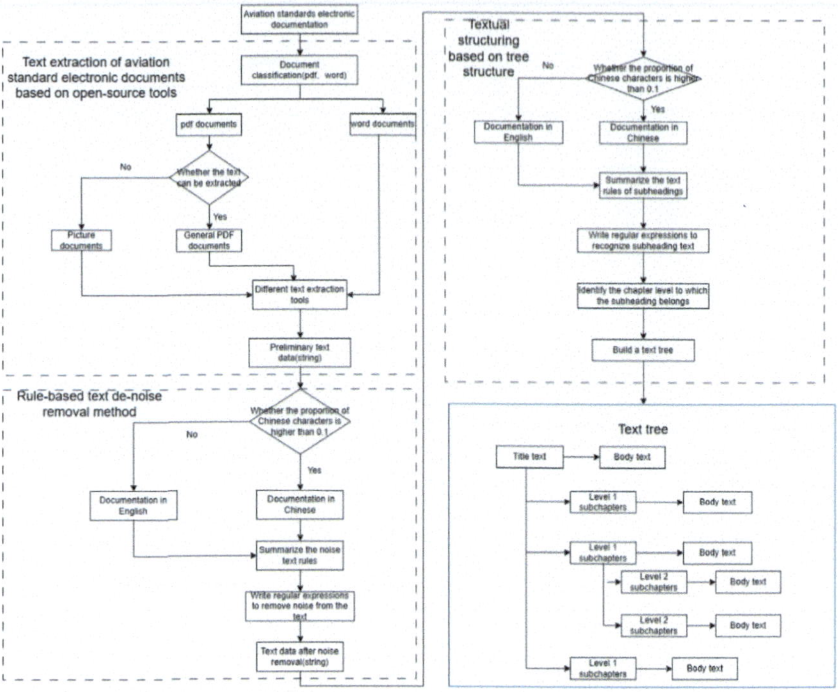

Fig. 4.7 Text denoising and structuring techniques for aviation standard electronic documents

4.2.3 Experimental Text Extraction of Aviation Standard Documents Based on Open-Source Tools

Text extraction from aviation standard documents refers to extracting valid text from electronic documents. Different programs and processing methods are employed for text extraction from PDF documents, Word documents, and image documents. This paper investigates existing Python and Java text extraction tools for specific issues regarding text extraction from PDF documents, Word documents, and image documents. Typical files of each document type are selected for experiments on text extraction from electronic documents, and the advantages and disadvantages of each tool are analyzed to select the most suitable text extraction tool.

The evaluation criteria for different text extraction tools are as follows:

1. Text extraction accuracy: Text extraction tools should be able to extract all valid text from documents accurately while minimizing issues such as garbled characters.
2. Text location: Considering subsequent research on related intelligent applications, text extraction should include recording the specific location of the text in the document, such as page numbers.

3. Extraction speed: Given the widespread application of text extraction tools in enterprise internal knowledge management platforms, they should be simple and efficient.
4. Open-source tools: To facilitate future optimization and modifications, text extraction tools should be open-source and support further development.

This paper conducts experiments using different text extraction tools for PDF document text extraction, Word document text extraction, and image text extraction. The selected text extraction tools and corresponding document extraction experiments are analyzed in Table 4.1.

Based on experimental analysis, the PDFminer tool can extract text excluding formulas and tables while unable to extract noise text such as headers, meeting the extraction requirements of this study. Therefore, the PDFminer tool based on the Python language was chosen to extract text from PDF documents. Both Tesseract and Python-docx are sufficient to meet the text extraction requirements of this study, hence selecting these two tools as the text extraction tools for image documents and Word documents respectively.

The experimental results of extracting text from PDF documents, image documents, and Word documents using PDFminer, Tesseract, and Python-docx are shown as follows:

(1) **Experiment of Extracting Text from PDF Documents Using PDFminer**

The text extraction using the PDFminer tool mainly involves the use of the PDFParser class and the PDFDocument class. The PDFParser is used to extract data from the file, while the PDFDocument class is used to store the data. In the experiment, the content of the PDF was extracted by reading page by page and the text data was obtained. An example of the experimental text is shown in Fig. 4.8.

The extracted text data is shown in Fig. 4.9.

(2) **Experiment of Extracting Text from Image Documents Using Tesseract**

The Tesseract algorithm analyzes connected components to identify text lines in images. It then locates baselines to segment text lines into words, and applies a trained word classifier to recognize each word in the image, thus extracting text from the image. Before applying the Tesseract algorithm, the PDF document needs to be converted into images using the PDF2image tool. An example of the experimental image document is shown in Fig. 4.10.

The extracted text data is shown in Fig. 4.11.

(3) **Experiment of Extracting Text from Word Documents Using Python-Docx**

The Python-docx tool, built on the Python language, allows for text reading and document editing in Word documents. Using Python-docx, each paragraph of the Word document is traversed, and the text is sequentially extracted from each paragraph to

Table 4.1 Analysis of experiments using different text extraction tools

Text extraction tool	Applicable document types	Programming languages	Experimental analysis
PDFminer	PDF document	Python	Text data excluding tables and formulas can be extracted. In the experiment, some documents cannot extract headers. In the extraction results, it is common to see the phenomenon of a single line of text being split into two lines before and after. The page number where each paragraph of text is located can be extracted
Itext	PDF document	Java	Text data excluding tables and formulas can be extracted. In the experiment, header text can be extracted. The page number where each paragraph of text is located can also be extracted
Free spire	PDF document	Java	All text data outside of tables and formulas can be extracted. In the experiment, header text can be extracted. Some documents may result in garbled text for table and image contents, and there may be a significant number of unnecessary blank lines in the extracted text
Python-docx	WORD document	Python	Essentially, text content excluding formulas and tables can be extracted. The page number where each paragraph of text is located can also be extracted
Tesseract	Picture documents	Python	For clear images, paragraphs can be extracted well, but there may be some typos. For blurry images, the recognition rate is low, resulting in more typos. The page number where each paragraph of text is located can still be extracted

complete the text extraction task. An example of the experimental Word document is shown in Fig. 4.12.

The extracted text data is shown in Fig. 4.13.

The experimental examples above demonstrate that the PDFminer, Tesseract, and Python-docx text extraction tools selected in this paper can effectively extract text data from aviation standard electronic documents.

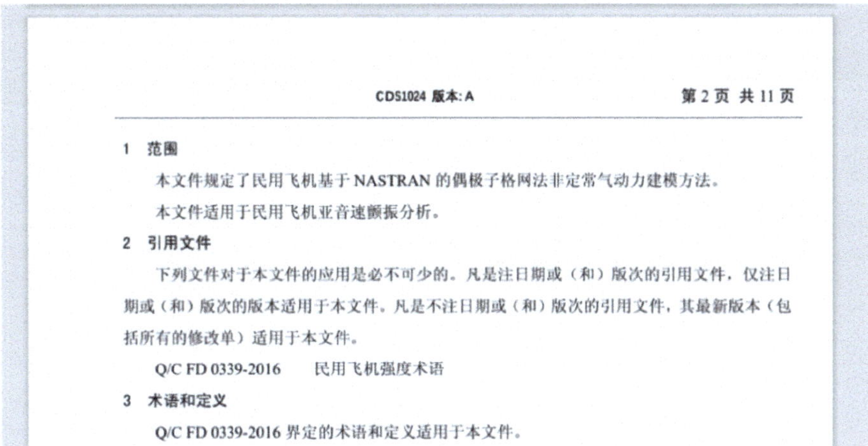

CDS1024 版本: A　　第 2 页 共 11 页

1 范围

本文件规定了民用飞机基于 NASTRAN 的偶极子格网法非定常气动力建模方法。

本文件适用于民用飞机亚音速颤振分析。

2 引用文件

下列文件对于本文件的应用是必不可少的。凡是注日期或（和）版次的引用文件，仅注日期或（和）版次的版本适用于本文件。凡是不注日期或（和）版次的引用文件，其最新版本（包括所有的修改单）适用于本文件。

Q/C FD 0339-2016　　民用飞机强度术语

3 术语和定义

Q/C FD 0339-2016 界定的术语和定义适用于本文件。

Fig. 4.8 Example of PDF document

CDS1027 版本: B
第 4 页 共 28 页
1 范围

民用飞机复合材料结构强度设计准则
本文件规定了民用飞机复合材料结构强度设计要求，给出了为满足复合材料结构完整性及
适航符合性所必需的验证项目。
本文件适用于民用飞机复合材料结构强度设计和适航符合性验证。
2 引用文件
下列文件对于本文件的应用是必不可少的。凡是注日期或（和）版次的引用文件，仅注日
期或（和）版次的版本适用于本文件。凡是不注日期或（和）版次的引用文件，其最新版本（包
括所有的修改单）适用于本文件。
CCAR-25-R4　　运输类飞机适航标准
AC 20-107B　　COMPOSITE AIRCRAFT STRUCTURE

Fig. 4.9 Example of extracted text from PDF document

4.2.4 Text Denoising Method Based on Rule Matching

Due to the formatting issues inherent in electronic aviation standard documents and limitations of text extraction tools, the preliminary text data extracted from these electronic documents often contain a significant amount of noise. This noise includes text garbling, formatting errors (such as incorrect line breaks and extraneous characters), as well as additional text content (such as headers and footers). The noise in the preliminary extraction results from aviation standard documents follows certain patterns. Therefore, it is feasible to consider using a rule-based matching method to develop regular expressions for identifying and removing this noise from the preliminary extracted text data. The process for rule-based text denoising is designed as shown in Fig. 4.14.

HB 8473—2014

民用飞机肺式氧气调节器

1 范围

本标准规定了民用飞机肺式氧气调节器(以下简称调节器)的技术要求、验证及交货准备等。
本标准适用于调节器的设计、制造和验收。

2 规范性引用文件

下列文件对于本文件的应用是必不可少的。凡是注日期的引用文件，仅所注日期的版本适用于本文件。凡是不注日期的引用文件，其最新版本(包括所有的修改单)适用于本文件。

Fig. 4.10 Example of image document

HB 8473-2014

民用飞机肺式氧气调节器

1 范围
本标准规定了民用飞机肺式氧气调节器（以下简称调节器）的技术要求、验证及交货准备等。
本标准适用于调节器的设计、制造和验收。
2 规范性引用文件
下列文件对于本文件的应用是必不可少的。凡是注日期的引用文件，仅所注日期的版本适用于本文件。凡是不注 日期的引用文件，其最新版本（包括所有的修改单）适用于本文件。

Fig. 4.11 Example of extracted text from image document

2 引用文件

下列文件对于本文件的应用是必不可少的。凡是注版本的引用文件，仅该版本文件适用于本文件。凡是不注版本的引用文件，其最新版本（包括所有的修改单）适用于本文件。

3 文件签署

3.1 签署流程

3.1.1 签署流程可分为三类：

Fig. 4.12 Example of word document

引用文件
下列文件对于本文件的应用是必不可少的。凡是注版本的引用文件，仅该版本文件适用于本文件。凡是不注版本的引 用文件，其最新版本（包括所有的修改单）适用于本文件。
中国商用飞机有限责任公司型号档案管理办法　中飞办（2011）252号
文件签署
签署流程

Fig. 4.13 Example of extracted text from word document

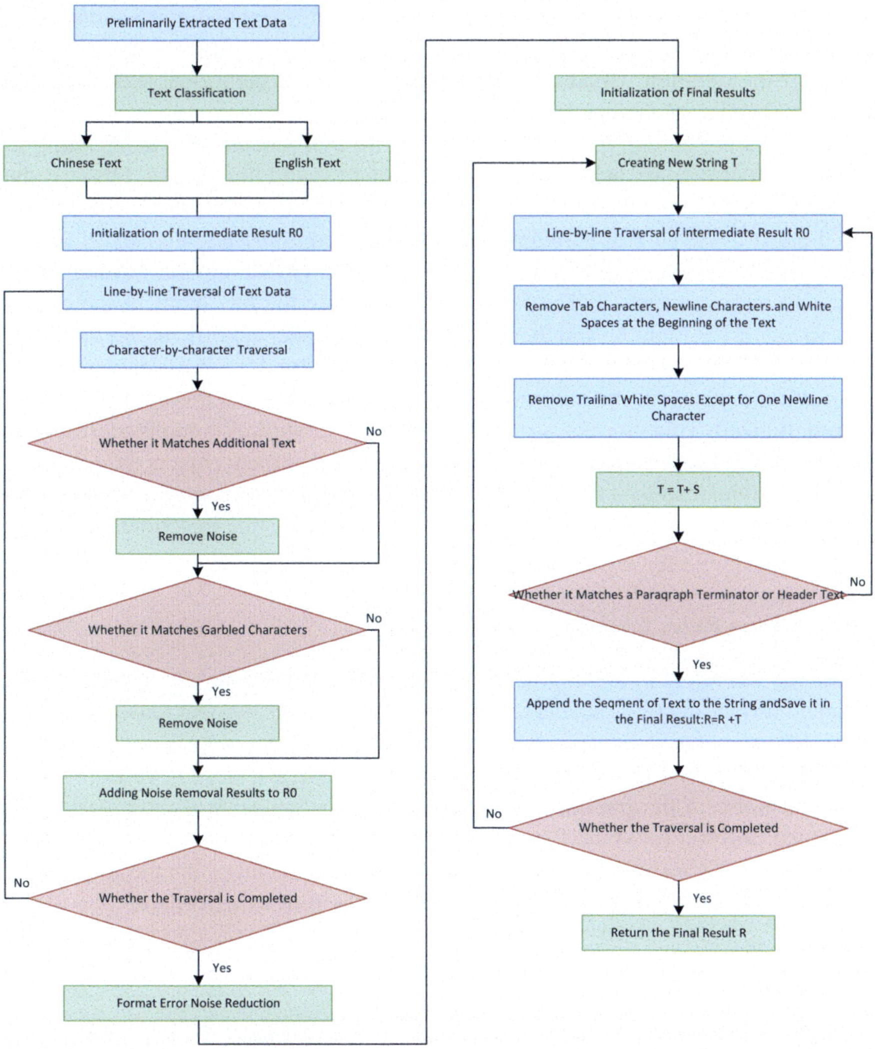

Fig. 4.14 Text denoising process flowchart

As depicted in the figure, due to significant differences in text formatting between Chinese and English texts, the noise present in the preliminary extracted text data of Chinese and English texts varies. Therefore, it is necessary to first categorize the text according to Chinese and English texts through a simple classification, which can be achieved by analyzing the proportion of Chinese characters in the text data. After categorization, noise patterns for different types of noise are summarized. Subsequently, an intermediate result R0 is initialized to record the noise of garbled characters and additional text removal. The text is then iterated line by line, where regular expressions are used to sequentially match additional text noise and garbled text noise. The preliminarily denoised text is added to the intermediate result R0 until the end of the iteration. Upon completing the removal of additional text noise and garbled text noise, the process transitions to the phase of removing format errors. Format errors are commonly found in text within PDF documents, manifesting as excessive tab characters, white spaces, tab characters, and line breaks. Initially, the final denoising result R is initialized, followed by the creation of a variable T to store the text in the current paragraph being processed. Subsequently, text S is read line by line from the intermediate result, where excess leading and trailing white spaces are removed from S. The first character before the line break at the end of S is evaluated for logic or whether S represents a header text. If these conditions are not met, indicating that the current line break is redundant, it is removed. If the conditions are satisfied, the current result T is saved, added to the final result, and T is reinitialized. This process is repeated until the iteration is completed, resulting in the removal of format error noise. Upon completion of the format error denoising process, the final denoised text data is obtained. The identification of different noise types primarily relies on designing regular expressions for rule matching. The noise text rules for additional text and garbled text are designed as follows:

(1) **Matching Rules for Additional Text Noise**

Additional text refers to text that is irrelevant to the content, such as text indicating document formatting (e.g., headers, footers, page numbers) and declarative text (e.g., copyright statements, document status, etc.). Additional text mainly includes headers, footers, signature blocks, change logs, copyright statements, etc. The rules for additional text noise in the preliminary extracted text data consist of three parts: text occurrence position (pos), text format (format), and repetition count (num). Here, pos represents the position of the text within the entire document and the appearance position of the text on its page. It is expressed by the following formula:

$$pos = \left(pos_d, pos_p\right) \tag{4.1}$$

where pos_d represents the page number where the text appears in the electronic document, and pos_p indicates the line number within its respective page. For example, pos = (3, 2) indicates that the text appears on the second line of the third page of the electronic document.

The text format (format) corresponds to the regular expression of the string. The repetition count (num) denotes the number of similar texts (where all characters are identical except for numbers) that meet the criteria of position and text format throughout the entire document, expressed as an integer (Tables 4.2 and 4.3).

The text matching rules for additional noise involve texts that simultaneously satisfy any one of the conditions corresponding to the position pos, format, and repetition count num for each type of additional noise.

Table 4.2 Classification of additional text noise and corresponding rules

Noise categories	Pos	Format	Num
Header	$pos_p = 1$ (excluding empty lines), where the header appears on the first line of text on each page	The total number of non-whitespace characters in the text is less than 15	Num ≥ 4 or num ≥ n_page – 3, where n_page represents the total number of pages in the electronic document, indicating that the text needs to be repeated on either more than 4.pages or on pages greater than the total number of pages minus 3
Footer	$pos_p = n\_row - 1$ (remove empty lines), where n_row represents the total number of lines in the page text	The total number of non-whitespace characters in the text is less than 15	Num ≥ 4 or num ≥ n_page – 3, n_page represents the total number of pages in the electronic document

Table 4.3 Classification of additional text noise and corresponding rules

Noise categories	Pos	Format	Num
Approval column, signature area, and revision log	$pos < pos_{ml}$, pos_{ml} represents the text before the table of contents	Extract only the titles of the articles above the table of contents, all other text can be discarded	Null
Copyright statement	$pos_p > n_{row} - 3$ (remove empty lines), where the header appears in the last two lines of text on each page	"Copyright statement" + newline at the beginning + "..." + newline, where "..." represents any text string	Null

(2) **Garbled Characters**

The main cause of garbled characters is the inability to recognize formulas, tables, or special characters in textual knowledge. In text data, this is mainly manifested as non-standard combinations of symbols, numbers, and letters. This type of text does not significantly contribute to the understanding and representation of documents. Therefore, garbled text is identified by checking if symbol combinations conform to conventional standards, and any garbled content is directly deleted. The garbled noise summarized in this paper mainly includes format garbled text and table garbled text. The format garbled text is caused by various reasons and is characterized by the unreasonable combination of punctuation marks, numbers, and letters. On the other hand, the presence of a large number of line breaks in tables in the document usually results in garbled text in the preliminary extracted text data, known as table noise. Furthermore, the content within table noise usually only has a significant impact on the understanding of the text for the table name. Therefore, after removing the table names, the table garbled text is selected for elimination. The garbled noise can be directly matched through text formatting (format), with specific matching rules and examples shown in Table 4.4.

Based on the text denoising methods described above, an example of text denoising is shown in Fig. 4.15. From the example, it can be observed that the text denoising method effectively removes noisy text.

Table 4.4 Classification and corresponding rules of garbled noise

Types of noise	Example	Format
Garbled formatting	'ta\uf065\nxya\uf067\n', '\n\uf06d\uf065\n', 'ca\uf065\n\n\uf06d\uf065\n', '4500\n\uf03d\n', '\uf03d\n', '6000\n', '\uf03d\n', '3300\n'	Irregular combinations of letters, numbers, and symbols, such as consecutive symbol-letter sequences or combinations of letters and numbers that do not end with a symbol (other than a period or comma)
Table garbled characters	'Table 4.1 composite material structure strength airworthiness related provisions \n applicable airworthiness regulations \n' ○ \n', 'Airworthiness certification requirements \n', 'Environmental design \n', 'Permissible and design values for materials \n', 'Static strength of the structure \n', 'CCAR25 \n Section 25.603 Material(c) \n'	"Table" + Number + " " + Non-numeric text + Text with no punctuation, less than 15 characters, and multiple line breaks

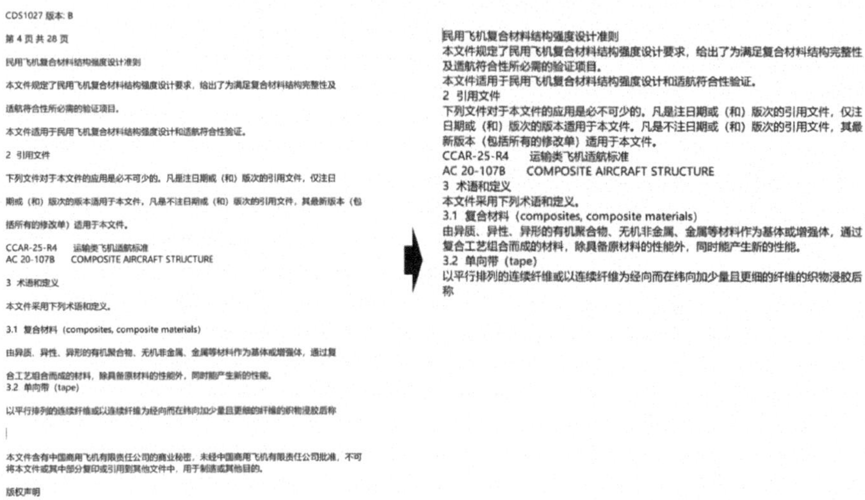

Fig. 4.15 Text denoising example

4.2.5 The Text Data Storage Structure of Aviation Standard Documents

Aviation standard documents are typically structured documents with a clear text structure (headings, multi-level subheadings) in the form of a hierarchical tree-like structure. The structural information of aviation standard documents is crucial for understanding their content. However, conventional storage of document text data as strings fails to capture the structural characteristics of aviation standard documents. Therefore, this paper proposes a text tree-based data structure for aviation standard document text data, referred to as a "text tree," to preserve the inherent structure of the document. Ultimately, text data is structured based on the text tree.

The structure of aerospace standard documents is typically hierarchical, resembling a tree. The document title and main text serve as the root node, with first-level sub-sections as child nodes, and second-level sub-sections as child nodes of the corresponding first-level sub-sections. Therefore, this paper stores text data as a tree structure data according to the structural characteristics of aerospace standard documents, referred to as a text tree. In the text tree, each node represents a level of text section, and the node value of each node represents the text sequence contained in that sub-section, serving as the title text of the section until the text sequence before the next title text. The text tree structure is illustrated in Fig. 4.16.

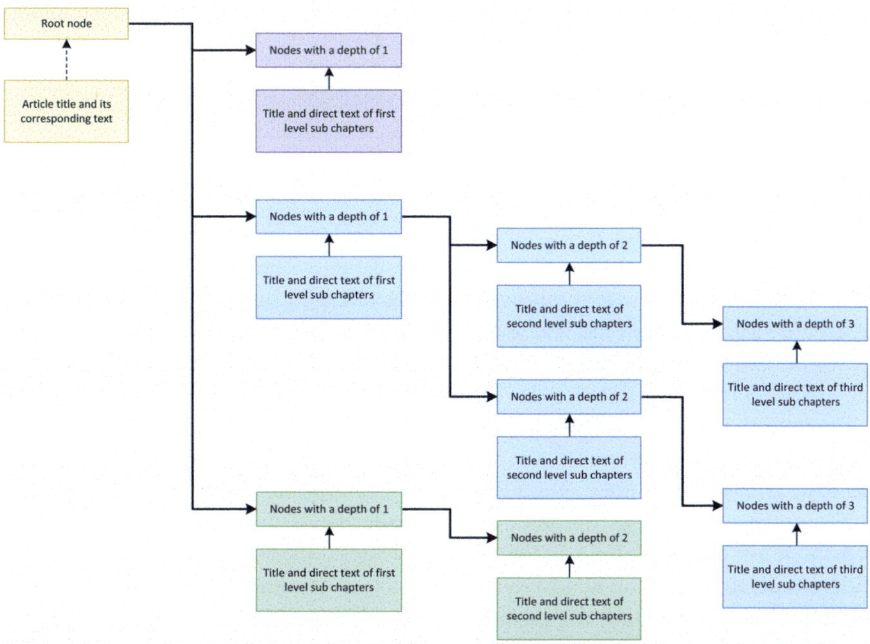

Fig. 4.16 Text tree

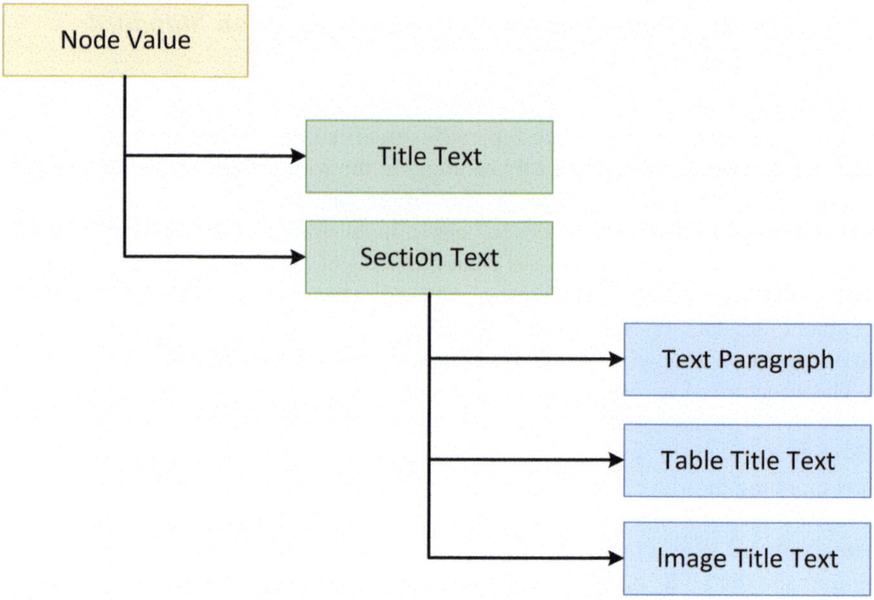

Fig. 4.17 Node values in the text tree

In the text tree, each node's value consists of the title text and the corresponding main content text (body text), as shown in Fig. 4.17.

4.2.6 *Method for Generating Text Trees*

After text extraction and noise removal, the obtained text data is a simple text sequence. The original structural information of the document is embedded in the semantic meaning of the text, allowing for the retrieval of title text to reconstruct the article structure from the text sequence. This process enables the extraction of a text tree containing the structure of the article from a simple text sequence. The construction process of the text tree is illustrated in Fig. 4.18.

Traverse the text sequence segment by segment. For each segment, apply rule-based matching to determine if the text is a title or sub-title. If it is a title or sub-title text, create a new node and add the text to the node's title text. At the same time, identify the level of the node through rule-based matching to determine its corresponding parent node. Add the node to the appropriate parent node. If it is not a title or sub-title, add the text to the current node's section text or main body text. Continue this process until the traversal is complete and the text tree is generated. The specific rules for matching titles and sub-titles in the text tree are as follows.

An example of the text used to build the text tree is shown in part in Fig. 4.19.

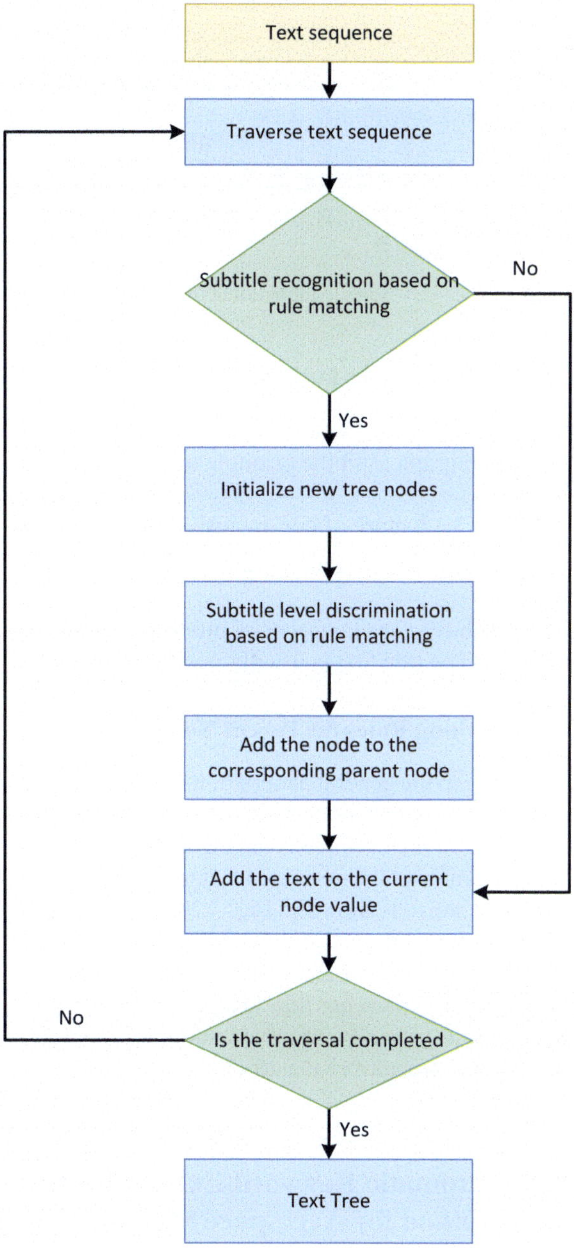

Fig. 4.18 The process of constructing a text tree

3 术语和定义
DESIGN CONSIDERATIONS FOR MINIMIZING HAZARDS CAUSED BY UNCONTAINED TURBINE ENGINE AND AUXILIARY POWER UNIT ROTOR FAILURE 产品结构创建与维护-定义视图 MBD 装配数据集组织要求下列术语和定义适用于本文件。
3.1 非包容性发动机转子失效（UERF）
当发动机发生转子失效时，发动机某些轮盘和叶片会断裂，产生不同尺寸的转子碎片，这些碎片具有不同的能量，相当一部分具有很高能量的碎片可击穿发动机舱，并沿不同的飞散角飞散出来，对飞机结构和系统形成威胁，这类事件被称为非包容性转子失效。根据其他机型的运营经验，尽管发动机供应商和制造商采用很多办法降低非包容性转子失效发生的概率，但是飞机实际运营期间非包容性转子失效事件仍然经常发生。

Fig. 4.19 The text example for building the text tree

(1) **Matching Rules for Title Text**

Based on induction, Title text = Sub-title level + Text title + line break symbol, such a paragraph as in the example text:

$$\begin{aligned}&\text{3.1 Failure of non-inclusive engine rotor UERF}\\&\quad = \text{3.1} + \text{Failure of non-inclusive engine rotor UERF} + \text{`}\backslash n\text{'}\end{aligned}$$

The sub-title level consists of numbers (Chinese or English) and "." between the numbers. The title text is usually non-numeric and no more than 20 characters.

(2) **Matching Rules for Parent Nodes**

According to the general pattern of the text, the parent node of the current subsection is the previous lower-level node. For example, the parent node of a second-level subsection in the example is the third section of the document.

An example of the generated text tree according to the text tree generation method described above is shown in Fig. 4.20.

On the left side of the figure is the document text (simplified), while on the right side is an example of a text tree generated based on the document on the left. In the figure, T represents title text, M represents body text, and None indicates an empty string. It can be observed that the proposed method effectively reconstructs the original structure of the article.

4.3 Automatic Keyword Extraction Data Preprocessing Method for Aerospace Standard Documents

The Research on Automatic Keyword Extraction Algorithm for Aerospace Standard Documents mainly focuses on Chinese documents. Before applying the keyword extraction algorithm to Chinese documents, the text must be transformed from a sequence of Chinese characters into a sequence of Chinese words. This is done through methods like word segmentation and stop word filtering.

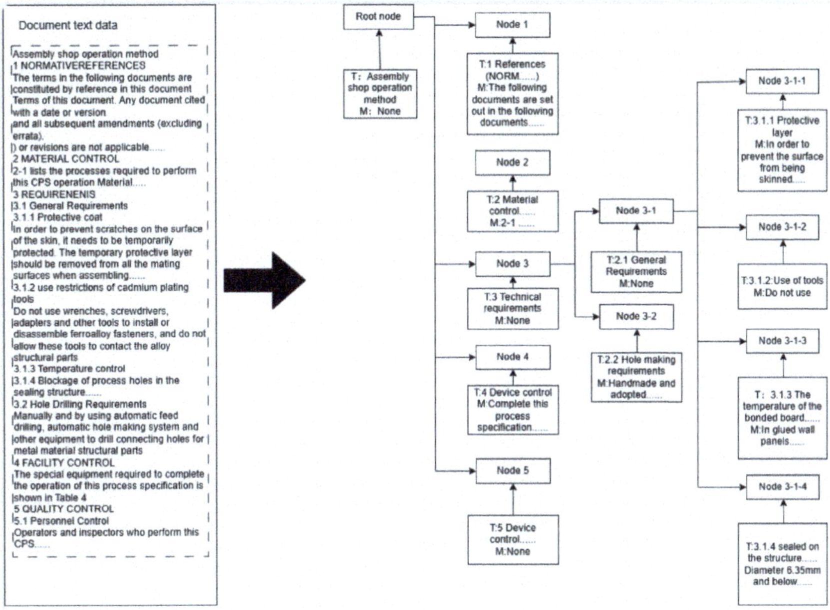

Fig. 4.20 Example of text tree generation

4.3.1 *Technical Route for Automatic Extraction of Keywords in Chinese Documents*

The technical roadmap for automatic keyword extraction from Chinese documents is illustrated in Fig. 4.21. Conventional Chinese text data is represented as a sequence of Chinese text. For Chinese text sequences, an appropriate Chinese word segmentation model is applied to segment the text into a sequence of Chinese words. Considering that the segmentation results often contain numerous irrelevant words such as particles and punctuation marks, it is necessary to utilize a suitable stop word library to filter out these meaningless words. The remaining words form the candidate keyword set. After preprocessing, a suitable keyword extraction algorithm is applied to assess the importance of each word in the set, determining its significance to the document. Finally, the words in the candidate keyword set are ranked based on their evaluation weights, and either the top-k words are selected as keywords or words with evaluation weights higher than a certain threshold are chosen as keywords.

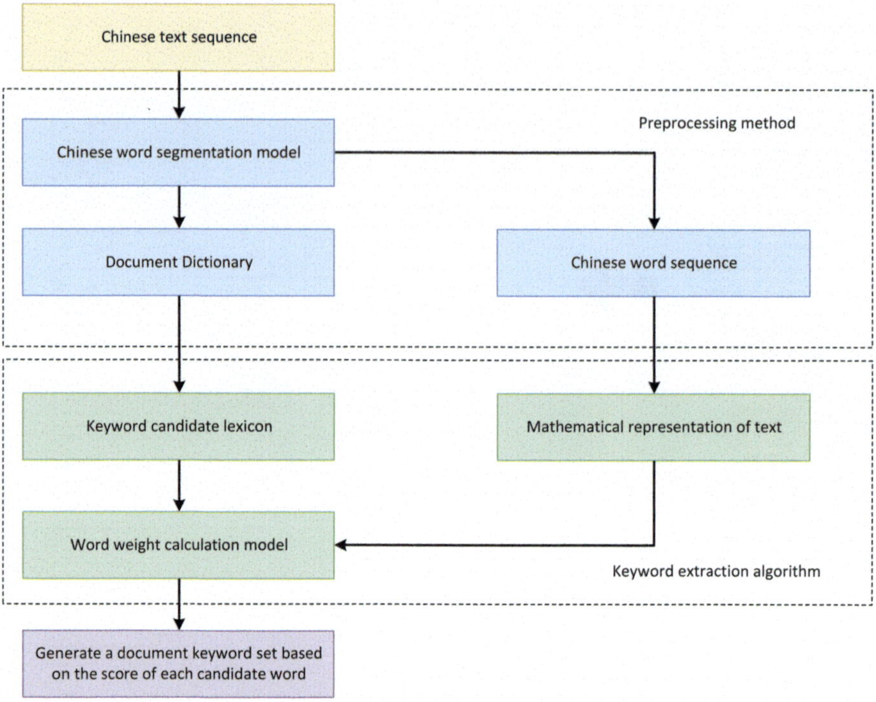

Fig. 4.21 Keyword extraction process

4.3.2 *Chinese Word Segmentation Model*

In Chinese text, characters are directly connected to form words, and a word is the smallest unit that represents semantics in Chinese (punctuation marks are also considered as individual words in this context). Words are directly connected to form Chinese text. Therefore, unlike English text where words are separated by spaces to form a text sequence, Chinese text needs to be segmented into words using a word segmentation model. A Chinese word segmentation model segments the original Chinese character sequence into the smallest semantic units—words, as understanding Chinese text requires segmenting the text into a sequence of words using such a model.

The Chinese word segmentation model is an important research focus in Chinese natural language processing tasks. It includes rule-based segmentation models and statistical-based segmentation models. Rule-based segmentation primarily relies on manually constructed dictionaries and follows predefined rules for matching and segmentation. Statistical-based segmentation models have shown better performance in scenarios such as discovering new words and have become a recent hot topic in related research. However, statistical-based segmentation heavily relies on the quality and quantity of corpora used for statistical methods. Considering that aviation standard documents may not meet the requirements in terms of quantity and quality for word segmentation algorithms, this study primarily selects existing mature algorithms. Existing Chinese word segmentation algorithms used in aviation knowledge texts often encounter the following issues:

(1) **Ambiguity in Chinese**

Chinese characters often have different meanings in different contexts, and even the same string may be segmented differently in different contextual backgrounds. Additionally, the choice of granularity for the words selected during segmentation can lead to different segmentation results using different segmentation tools. Therefore, it is necessary to conduct experiments to select the most suitable word segmentation algorithm.

(2) **Out-of-Vocabulary Words**

In aviation standard texts, there are a large number of technical terms, and with the advancement of science and technology, there will gradually be a large number of new words in aviation standard texts. However, the efficiency of including new words in dictionaries is low, including many technical terms and terminology. How to identify new words and technical terms will have a significant impact on the accuracy of word segmentation. Table 4.5 lists the common Chinese segmentation tools.

The installation and configuration of Harbin Institute of Technology's LTP (pyltp) and Hanlp tokenizers are more complex compared to the other three tokenizers

Table 4.5 Analysis of common Chinese word segmentation tools and their characteristics

Word segmentation tools	Characteristics analysis
Chinese Academy of Sciences NLPIR	As an open-source project, it supports languages such as Java, Python, and C++. Additionally, NLPIR supports functions such as extracting new words and importing local dictionaries
Jieba tokenizer	As an open-source project, it supports languages such as Java, Python, and C++. Jieba uses the prefix dictionary and word graph scanning. It also supports functions such as extracting new words and importing local dictionaries
Hanlp tokenizer	Based on the shortest path calculation, the original model used training data from the People's Daily. With abundant training data, it has certain advantages in recognizing entity boundaries
Tsinghua University THULAC	It is an open-source project that supports Java, Python, and C++ languages
Harbin Institute of Technology Word Segmenter	It can be accessed through the homepage interface, but it is not suitable for the research in this article due to limitations on interface access requests

Table 4.6 Testing the tokenization effectiveness of different tokenization tools

Tokenizer	Input sentence	Tokenized result
Chinese Academy of Sciences NLPIR	"Guidelines for civil aircraft pressure wind tunnel tests" "Operating procedure for assembly workshop"	"Civil aircraft", "pressure measurement", "wind tunnel", "test", "guidelines", "Assembly", "workshop", "operation method"
Tsinghua University THULAC		"Civil aircraft", "pressure measurement", "wind tunnel", "test", "guidelines", "Assembly", "workshop", "operating method"
Jieba tokenizer		"Civil aircraft", "pressure wind", "tunnel test guidelines", "assembly workshop ", "operating method"

mentioned. They have weaker portability, so this article does not consider including the Harbin Institute of Technology's LTP and Hanlp tokenizers as alternative tokenization tools. Instead, it will compare the application of the other three tokenization algorithms in experiments.

The test text is selected from excerpts of aviation standard texts provided by COMAC, and the comparison of tokenization results is shown in Table 4.6.

From Table 4.6, it can be seen that the NLPIR algorithm from the Chinese Academy of Sciences and the THULAC algorithm from Tsinghua University have a higher granularity in word segmentation. Most technical terms and compound nouns in aviation knowledge documents are segmented into simple words that are difficult to recover semantically, resulting in very unsatisfactory tokenization results. On the other hand, compared to these algorithms, the jieba tokenizer has a lower granularity and strong ability to discern compound words, making it more suitable for the application in this project.

Therefore, in the Chinese text tokenization stage of text preprocessing, this study adopts the jieba tokenizer, which demonstrates significantly better tokenization effectiveness.

4.3.3 Stopword Filtering

Before further exploratory analysis of the text, it is common to filter out certain words or symbols with low importance to understand the text in Chinese word sequences to improve efficiency. Words and symbols that are commonly filtered out in Chinese tasks are referred to as stop words. Generally, stop words can be classified into function words unrelated to the main content of the text and common words frequently used in all documents, such as modal particles and quantifiers. In the task of extracting

Table 4.7 Chinese stopword list

Type	Examples
Punctuation	"，", "。", "：", "！", "？"
Determiners	"的" (of), "地" (adverb marker), "着" (continuous action marker), "这" (this), "那" (that)
Modal particles	"哈哈" (haha), "啊" (ah), "呀" (ya)
Common words	"希望" (hope), "期望" (expect), "看见" (see), "得到" (obtain)

keywords from aviation standard documents, removing relevant stop words can effectively filter out words unrelated to the main content of the documents. Therefore, this study will use a Chinese stop-word list consisting of common punctuation marks, modal particles, quantifiers, and some common words for stop word filtering. The Chinese stop words used in this study are described in Table 4.7.

In the context of preprocessing methods for aviation knowledge documents, an experimental example is provided below. The original text, presented in italicized font, is as follows:

"针对一个已划分好的面元, 使用 MSC.Nastran 软件 CAERO1 卡片建立气动网格, 需要从结构数模测量获取关键点的坐标和关键边的弦长以定义该面元的位置和大小。"

The word sequence obtained after segmentation and removal of stopwords is as follows, presented in italicized font:

"划分", "面元", "使用", "MSC.Nastran", "软件", "CAERO1", "卡片", "建立", "气动网格", "结构数模", "测量", "获取", "关键点", "坐标", "关键边", "弦长", "定义", "面元", "位置", "大小".

4.4 Keyword Extraction Case

TFIDF is a common method for extracting keywords through statistical features. "TF" stands for Term Frequency, which quantifies a word's ability to summarize the main content of a text based on the number of times the word appears. On the other hand, "IDF" stands for Inverse Document Frequency, which represents the logarithm of the ratio of the total number of documents in the corpus to the number of documents containing the term, quantifying the term's ability to differentiate between texts. The TFIDF method quantifies the importance of a term to a document through both the TF and IDF values.

(1) **Algorithmic Enhancements Based on External Corpora and Positional Weighting**

In aviation standard documents, conventional methods of quantifying TF (Term Frequency) and IDF (Inverse Document Frequency) values often fail to account for the unique distribution of terms in specialized technical documents, which differs significantly from that in general texts. For instance, terms such as "aircraft," "assembly," and "bolt" are more prevalent in aviation standards, whereas terms like "log" and "expectation" are minimally distributed, diverging considerably from common texts. According to the traditional method of quantifying IDF, "log" and "expectation" would be assigned higher weights compared to "aircraft," "assembly," and "bolt," which clearly does not meet the requirements. Keyword extraction for aviation standard documents should favor terms relevant to the aviation knowledge domain. Therefore, this paper addresses this issue by refining the calculation of IDF values using an external corpus. Additionally, the structural information in aviation standard documents, such as the higher significance of title text over body text, is of paramount importance. Hence, this paper improves the calculation of TF values by designing positional weights that leverage the document's structural information, ultimately proposing a TFIDF algorithm enhanced by external corpora and positional weighting.

(2) **Mathematical Representation of Aviation Standard Documents**

Aviation standard documents stored for the text tree structures. Assuming that $\{D_i\}$ is a collection of documents in the aviation standard document library, T_i represents the text tree corresponding to the Article I document. The introduction of the text tree is shown in Chap. 2. The node value of each node in the text tree is the text of the section. Then the node value of each node is the word sequence. Suppose the set of node values of the text tree T_i is expressed as $\{N_{ij}\}$. Where a single node value N_{ij} is assumed to be the word sequence $N_{ij} = \left\{x_{ij1}, x_{ij2} \ldots x_{ijnum_{ij}}\right\}$. Where num_{ij} represents the length of the corresponding word sequence in the node values. In the improved TFIDF algorithm, the document is represented by the bag model, assuming the text vector $V_i \in R^n$, where V_i is the text vector corresponding to document D_i and the generation method of text vector V_i is as follows:

1. The set of words in the construction document set $\{D_i\}$ is called dictionary Dic: assuming that the set of words in each node value N_{ij} is dic_{ij}, then $\text{Dic} = \sum_i \sum_j dic_{ij}$. The sum here represents the merging operation of the set, and finally obtains the dictionary Dic;
2. Initialize the text vector $V_i = \{0, 0, \ldots\}$, $V_i \in R^n$;
3. Word x_{ijk} in the node values in the text tree T_i, $V_i\left[\text{index}\left(X_{ijk}\right)\right] = V_i\left[\text{index}\left(x_{ijk}\right)\right] + 1$, index($x_{ijk}$)represents the index of the word x_{ijk} in the dictionary Dic.

Suppose the dictionary is {"civil aircraft", "software", "measuring wind tunnel", "get" to"," the test "," guide, "}, statement" civil aircraft pressure wind tunnel test guide " after pretreatment corresponding word sequence for {" civil aircraft "," measuring wind tunnel "," test "", "guide"} ", the statement corresponding text vector V_i is [1, 0, 1, 0, 1, 1].

(3) **Calculation of TF Values Based on Positional Weighting**

The traditional calculation of TF (Term Frequency) values quantifies a candidate word's ability to summarize the content of a text solely based on the number of times it appears within the article, using simple word frequency as a metric. However, this approach overlooks the impact of a word's position on its importance. For example, text in titles and subtitles generally has a higher reuse value across documents, suggesting that words in these positions should be given higher TF values. Based on these considerations, this paper proposes a method for calculating TF values based on positional weighting:

1. For document D_i then the corresponding text vector V_i is built, and the same of the title text in the text tree is regarded as the corresponding title text vector Vt_i for a document;
2. Calculate the TF value of the t words in the document D_i:

$$TF_{t,i} = \frac{V_i[\text{index(t)}] + \varepsilon * Vt_i[\text{index(t)}]}{\text{SUM}(V_i)} \tag{4.2}$$

where SUM (V_i) represents the sum of the elements of the text vector V_i index (t) represents the index of the word t in D, and ε is the weight of the title position.

(4) **Calculation of IDF Values Based on an External Corpus**

The traditional calculation of IDF (Inverse Document Frequency) values involves taking the logarithm of the ratio of the number of texts in which a candidate word appears to the total number of texts. Since aviation standard documents focus on themes related to aviation and manufacturing, terms like "assembly," "workshop," and "composite materials" appear more frequently. In contrast, everyday words like "day" and "life" occur less often. According to the distribution of words in aviation standard documents, the conventional method of calculating IDF assigns lower weights to aviation and manufacturing-related vocabulary, while some common daily words are given higher weights. This approach does not align with the needs of the task, as the specificity among aviation knowledge documents is better measured through vocabulary pertaining to the aviation manufacturing domain. To address this issue, this paper proposes an IDF value calculation algorithm that incorporates an external corpus to reduce the weight of common daily vocabulary. During the calculation process, a general external corpus is added to the aviation standard document corpus, making the corpus for IDF calculation a combination of the external general corpus and the aviation standard document corpus. The formula for calculating IDF values with the addition of words from the external corpus is as follows:

$$IDF_t = \log \frac{|\{D_i\} \cup \{d_i\}|}{|D\colon t \in D\, and\, D \in \{D_i\} \cup \{d_i\}|} \tag{4.3}$$

where $\{D_i\}$ represents the collection of product development documents, and $\{d_i\}$ represents the collection of external corpus articles. Where $|\{D_i\} \cup \{d_i\}|$ represents the original corpus and external corpus and the total number of documents, $|D : t \in D \text{ and } D \in \{D_i\} \cup \{d_i\}|$ means the original corpus and external corpus documents in the number of t_i words.

(5) **Calculation of TFIDF Values**

The TFIDF value is the product of the TF (Term Frequency) value and the IDF (Inverse Document Frequency) value:

$$TFIDF_{t,i} = TF_{t,i} * IDF_t \tag{4.4}$$

(6) **Experiments and Results Analysis**

In this chapter, we implement the unsupervised keyword extraction algorithm designed in this section and conduct experiments on aviation knowledge documents.

1. **Corpus**

The corpus used in this study consists of 800 Chinese standard documents provided by Commercial Aircraft Corporation of China, Ltd. (COMAC) out of a total of 1000 aviation standard documents. Considering the original document data lacked labels, a subset of the documents (100 documents) was randomly selected for tagging by relevant experts to test the model's effectiveness. The corpus labels correspond to the number of keywords in the text, with an average of 5.87 keywords per document. Moreover, the methods proposed previously in this paper necessitate the use of an external corpus. This external corpus is utilized solely for the calculation of IDF values, to address the issue of aviation manufacturing-related vocabulary receiving lower IDF values due to the uneven distribution of words in aviation standard documents. The external corpus must encompass a wide range of topics and contain a substantial number of documents. Given these requirements, this study selects a portion of the THUCNews dataset as the external corpus, comprising over 100,000 news documents across multiple topics.

2. **Evaluation Methodology**

The evaluation of keyword extraction algorithm outcomes employs precision, recall, and F1 score to assess the effectiveness of the algorithm. Considering the limitations of the segmentation algorithm used, the keywords extracted are often simple words, whereas actual keywords are more complex. Therefore, this paper makes certain modifications as follows:

Precision (P) in this study refers to the proportion of extracted keywords that are accurate. In this context, if the extracted keyword is a subset of the actual keyword, it is considered to be accurately extracted. A higher P value indicates better performance of the keyword extraction improvement algorithm. The calculation formula is as follows:

$$P = \frac{\text{The number of correct keywords extracted}}{\text{The number of keywords extracted}} \tag{4.5}$$

Recall (R) in this study refers to the proportion of correct keywords from a document that are successfully extracted by the algorithm. In this context, if an extracted keyword partially matches an actual keyword, it is considered successfully extracted. A higher R value indicates a more effective keyword extraction algorithm. The formula for calculating recall is as follows:

$$R = \frac{\text{The number of correct keywords extracted}}{\text{The number of keywords in the text itself}} \tag{4.6}$$

The F1 Score: To mitigate potential discrepancies between precision and recall assessments, the F1 score is introduced as a metric that considers both evaluations. A higher F1 score indicates better algorithm performance. The formula for calculating the F1 score is as follows:

$$F1 = \frac{2 * P * R}{P + R} \tag{4.7}$$

3. **Analysis of Experimental Results for Keyword Extraction Algorithms**

The extraction algorithm introduced earlier was applied to train on all documents in the dataset, excluding the test corpus, and subsequently tested on a set of 100 documents in the test collection. Given that the majority of aviation knowledge documents contain between 4 and 8 keywords, the algorithm was configured to select the top 7 keywords for each document. Experiments were conducted to test both the traditional TFIDF and the improved TFIDF algorithms. In the improved TFIDF, the title position weight (α) was set to 4. The theme distribution for certain terms used in the algorithms is as follows:

electrical: (0.13885699, 0.58825904, 0.13668555, 0.13619839)

binding element: (0.5813283, 0.13910204, 0.1400031, 3, 0.13956657)

The extraction results for the above different algorithms are shown in Table 4.8.

In the experimental results, the TFIDF algorithm, enhanced with external corpora and position weighting, achieved better keyword extraction performance.

The keyword extraction results of some test documents are shown in Table 4.9.

Table 4.8 Comparison of the keyword extraction results

Metric	Modified TFIDF	TFIDF
Accuracy rate	0.428	0.366
Recall	0.546	0.453
F1 score	0.480	0.405

Table 4.9 Example of the keyword extraction results

Document name	Actual keywords	Modified TFIDF
A CAD or a CAM quality workflow	CAD, CAM, quality control, CMM, measuring machine, digital	Measurement, inspection, CAM, CAD, table, digital, measuring machine
120 °C cured composite material with structural adhesive	Structural adhesive, composite material, curing film, hot press, tank, bonding, material requirements, PCD	Test, PCD, materials, requirements, adhesives, suppliers, days
PMMEL preparation and revision management procedures	PMMEL, preparation, revision, management procedures, flow chart	PMMEL, Support, Flight, Department, Operations, Customer Service, Center

The examples also demonstrate that the improved TFIDF algorithm performs better in experiments. However, the discussed keyword extraction algorithms have limitations, such as the tendency to extract terms unrelated to aviation manufacturing, like "sky," "days," "table," and "center." While terms like "sky" can be effectively avoided by enhancing the algorithm with external corpora, filtering out terms like "days" and "table" remains more challenging. Furthermore, the improved TFIDF method, with its emphasis on position weighting, makes better use of words appearing in titles and subtitles, as seen in the example document "CAD or CAM Quality Workflow." However, it still does not fully exploit the structural information of the document itself.

References

1. Pan X, Wang J, Liu L (2006) A knowledge management system architecture based on the web knowledge service. Comput Integr Manufact Syst 12(8):1293–1299
2. Zhu Z, He H, Zhang H, Li Q (2012) A state-machine-based automatic document text extraction method. Comput Appl Softw 29(12):54–57
3. Liu X (2018) Design and implementation of the PDF text content extraction system for medical knowledge. Harbin Institute of Technology, Harbin
4. Li XJ, Wang Z, Lu Q (2017) A paper extraction method with rules combined with SVM. Comput Technol Dev 27(10):24–29
5. In the Rising Waves (2015) PDF mathematical expression information acquisition based on document properties. Hebei University, Shijiazhuang

6. Jun P (2017) Research on complex table document preprocessing and text extraction algorithm. Beijing Jiaotong University, Beijing
7. Turney PD (2000) Learning algorithms for keyphrase extraction. Inf Retrieval 2(4):303–336
8. Hofmann T (1999) Probabilistic latent semantic indexing. In: International ACM SIGIR conference on research and development in information retrieval, vol 51(2), pp 50–57
9. Weng W, Wang H (2010) LDA-based keyword ExMethod. In: Fifth national symposium on computational linguistics (YWCL 2010)
10. Page L, Brin S, Motwani R, Winograd T (1998) The Pagerank citation ranking: bringing order to the web. Stanford Digital Library Technologies Project
11. Mihalcea R, Tarau P (2004) TextRank: bringing order into texts. In: Proceedings of EMNLP-04.and the 2004.conference on empirical methods in natural language processing
12. Boudin F (2013) A comparison of centrality measures for graph-based keyphrase extraction. In: Proceedings of the 6th international joint conference on natural language processing
13. Tian X (2013) Keyword extraction study of word position-weighted TextRank. Mod Libr Inf Technol 29(9):30–34
14. Xia T, Chai Y (2011) An improvement to TF-IDF: term distribution based term weight algorithm. J Softw 6(3):252–255
15. Qin P, Xu W, Guo J (2016) A novel negative sampling based on TFIDF for learning word representation. Neurocomputing 177:257–265
16. Gu Y (2014) Keyword extraction study of fused LDA with TextRank. Mod Libr Inf Technol 30(Z1):41–47
17. Liu X, Xie F, Wu X (2016) Keyword extraction algorithm based on the graph and the LDA topic model. J Intell 035(006):664–672
18. Zhiyuan L (2011) Study on Keyword extraction method based on document topic structure. Tsinghua University, Beijing
19. Zhao J, Zhu Q, Zhou G et al (2017) Automatic keyword extraction research review. Softw J 9
20. Witten IH, Paynter GW, Frank E et al (1999) KEA: practical automatic keyphrase extraction. In: Fourth ACM conference on digital libraries. ACM
21. Hulth A (2003) Improved automatic keyword extraction given more linguistic knowledge. In: Empirical methods in natural language processing, pp 216–223
22. Frank E, Paynter GW, Witten IH et al (1999) Domain-specific keyphrase extraction. In: International joint conference on artificial intelligence, pp 668–673
23. Mikolov T, Corrado G, Chen K et al (2013) Efficient estimation of word representations in vector space. In: Proceedings of the international conference on learning representations
24. Pennington J, Socher R, Manning C (2014) Glove: global vectors for word representation. In: Conference on empirical methods in natural language processing
25. Peters M, Neumann M, Iyyer M et al (2018) Deep contextualized word representations. In: Proceedings of the 2018 conference of the North American chapter of the association for computational linguistics: human language technologies, vol 1 (long papers)
26. Devlin J, Chang MW, Lee K et al (2019) BERT: pre-training of deep bidirectional transformers for language understanding. In: Proceedings of the 2019 conference of the North American chapter of the association for computational linguistics: human language technologies, vol 1 (long and short papers)
27. Gers FA, Schmid Huber J, Cummins F et al (2000) Learning to forget: continual prediction with LSTM. Neural Comput 12(10):2451–2471
28. Vaswani A, Shazeer N, Parmar N et al (2017) Attention is all you need. arXiv
29. Wen Y, Yuan H, Zhang P et al (2016) Research on keyword extraction based on Word2vec weighted TextRank. IEEE international conference computer and communications, pp 2109–2113
30. Wei C, Wu Y, Chen CW et al (2018) Automatic keyword extraction based on BiLSTM-CRF. Comput Sci 45(S1):104–109+126
31. Baezayates R, Ribeironeto B (2004) Modern information retrieval. China Machine Press
32. Xueyan W (2020) Study on the retrieval method based on the location relationship within the document. Central China Normal University, Wuhan

33. Salton G (1971) The SMART retrieval system—experiments in automatic document processing. In: Prentice-Hall series in automatic computation, p 156
34. Maron ME, Kuhns JL (1960) On relevance, probabilistic indexing and information retrieval. J ACM (JACM) 7:216–244
35. Singhal A, Buckley C, Mitra M (1996) Pivoted document length normalization. In: Proceedings of the 1996 ACM SIGIR conference on research and development in information retrieval, pp 21–29
36. Robertson S, Walker S (1994) Some simple effective approximations to the 2-poisson model for probabilistic weighted retrieval. In: Proceedings of SIGIR'94, pp 232–241
37. Zhai C, Lafferty J (2001) A study of smoothing methods for language models applied to ad hoc information retrieval. In: Proceedings of SIGIR'01, pp 334–342
38. Zhai C (2020) A brief review of information retrieval models [EB/OL]. http://sifaka.cs.uiuc.edu/course/ds/irmod.PDF
39. Géry M, Largeron C (2012) BM25t: a BM25 extension for focused information retrieval. Knowl Inf Syst
40. Lv Y, Zhai CX (2011) When documents are very long, BM25 fails!. In: Proceeding of the 34th international ACM SIGIR conference on research and development in information retrieval, SIGIR 2011. ACM, Beijing, China, 25–29 July 2011
41. Deerwester S, Dumais ST, Furnas GW et al (1990) Indexing by latent semantic analysis. J Am Soc Inf Sci 41(6):391–407
42. Dillon M (1983) Introduction to modern information retrieval: G. Salton and M. McGill. McGraw-Hill, New York (1983). xv + 448 pp., $32.95 ISBN 0-07-054484.0. Inf Process Manag 19(6):402–403
43. Gao J, He X, Nie J-Y (2010) Clickthrough-based translation models for web search: from WORD models to phrase models. In: CIKM
44. Gao J, Toutanova K, Yih WT (2011) Clickthrough-based latent semantic models for web search. In: Proceeding of international ACM SIGIR conference on research & development in information retrieval. ACM
45. Huang P, He X, Gao J et al (2013) Learning deep structured semantic models for web search using clickthrough data. In: Conference on information and knowledge management, pp 2333–2338
46. Mitra B (2015) Exploring session context using distributed representations of queries and reformulations. In: Proceedings of the 38th international ACM SIGIR conference on research and development in information retrieval, pp 3–12
47. Mitra B, Craswell N (2015) Query auto-completion for rare prefixes. In: Proceedings of the 24th ACM international on conference on information and knowledge management, pp 1755–1758
48. Song Y, Elkahky AM, He X (2016) Multi-rate deep learning for temporal recommendation. In: Proceedings of the 39th international ACM SIGIR conference on research and development in information retrieval, SIGIR'16. New York, NY, USA, pp 909–912

Chapter 5
Intelligent Design of Assembly Process

Assembly is an important part of the manufacturing process, and the quality of assembly directly affects the performance of the product. In order to improve the quality of product assembly, it is necessary to carry out a reasonable design of the assembly process first. In view of this, this chapter first analyzes the progress and current situation of assembly process design. Secondly, it studies the intelligent design method of assembly process based on knowledge graph. Finally, it combines the assembly case of an aero-engine compressor rotor component to verify and illustrate the relevant technical methods.

5.1 Related Work of Assembly Process Design

Since the 1990s, with the vigorous development of automation and information technology, assembly execution efficiency and assembly inspection efficiency have been significantly improved. However, the time consumption in the assembly process mainly exists in the assembly process design stage, reaching 60–80% of the total assembly process time, and traditional automation and information technology have shown great limitations in the assembly process design stage. Therefore, the assembly process design needs to be improved with more intelligent technology. Currently, research on assembly process design mainly includes three aspects: assembly process information modeling, assembly process planning, and assembly process evaluation.

T. Liu et al., *Industrial Intelligence: Methods and Applications*, Springer Series in Advanced Manufacturing, https://doi.org/10.1007/978-3-031-81477-8_5

5.1.1 Related Work of Assembly Process Information Modeling

The product assembly process information encompasses both geometric and non-geometric data. The essence of the assembly process information model is how this data is organized. Creating a clear model is vital for reusing assembly knowledge and experience during the design stage. Currently, researchers focus on four main approaches: object-oriented thinking, XML language, Petri nets, and ontology.

Xu et al. [1] proposed an object-oriented assembly process information model. This model abstracts and encapsulates complex assembly relationships, simplifying descriptions for complex products and addressing combinatorial explosion issues. Cheng et al. [2] designed an assembly process meta-model using object-oriented thinking. They established an information object body for an aero-engine's assembly process based on this meta-model. Li et al. [3] proposed an object-oriented model with data units at its core. This model integrates assembly production preparation data and process data, improving data structuring.

Zhang et al. [4] proposed an XML-based assembly process information model to address information interaction between modules in virtual assembly systems, focusing on assembly structure expression. Tian [5] used XML to structurally describe assembly information in STEP models, making XML an intermediate format for assembly design and simulation, facilitating information exchange between these stages. Bao et al. [6] proposed an XML-based assembly process description method, which, due to its object-oriented foundation, combines the benefits of object-oriented models with solutions for information interaction.

Wang et al. [7] established a Petri net-based final assembly process model for optimizing the production line schedule in aircraft final assembly. Using Petri nets facilitates clear expression and optimization of the assembly process flow. Yang et al. [8] developed a virtual assembly model based on Petri nets to describe serial, parallel, and synchronous relationships between assembly processes, aiding process expression. Yang et al. [9] proposed a Petri net description method for the assembly/disassembly process, enabling assembly collision detection through node analysis.

The National Institute of Standards and Technology (NIST) in the United States proposed an ontology-based general model of assembly process (open assembly model) early on, using ontology language to describe assembly relationships and define assembly rules, enabling rule-based assembly process reasoning. Qiao et al. [10] established a data structure for product geometric information in the assembly process based on ontology, using the OWL-DL ontology language to describe product geometric information, assembly process information, and their relationships. Sayed et al. [11] integrated fault and error information into the ontology model to build an ontology-based expert knowledge system for diagnosing the state of manufacturing processes.

By analyzing the above literature, we can summarize the characteristics of various assembly process information models, as shown in Table 5.1. It is evident that the ontology-based assembly process information model is currently a promising

Table 5.1 Characteristics of the various assembly process information models

Assembly process information model type	Advantages	Disadvantages
Object-oriented assembly process information model	Abstraction, stability, reusability	Difficult to meet the information needs between heterogeneous systems
XML-based assembly process information model	Can meet the interaction requirements between heterogeneous systems	Only describes the structure of the assembly process, difficult to express the semantic information of the assembly process
Petri net-based assembly process information model	Stability, reusability, more intuitive model expression	Difficult to meet the interaction requirements between heterogeneous systems
Ontology-based assembly process information model	Has standardized descriptions, beneficial for describing assembly processes, beneficial for interaction between heterogeneous systems	Complex description language, expressing semantics through standardized representation

approach. However, the description form still requires improvement to move away from rule-based semantic expression and facilitate the automatic reuse of assembly process knowledge.

5.1.2 Related Work of Assembly Process Planning

Product assembly process planning includes assembly sequence planning and assembly path solving. For complex components, the design primarily involves assembly sequence planning, inferring the sequence of component assembly after the assembly structure design is complete. Current research on assembly sequence planning methods can be categorized into: assembly sequence planning based on assembly constraint priority, assembly sequence planning based on geometric constraint priority, assembly sequence planning based on knowledge, and assembly sequence planning based on disassembly.

Homen et al. [12] first proposed the assembly association AND/OR graph of assembly structure information, establishing assembly semantic constraints through human–computer interaction to obtain feasible assembly sequences, though the process was inefficient. Raju et al. [13] proposed an assembly sequence elimination algorithm based on the assembly association AND/OR graph, considering mating priority and assembly stability, and eliminating infeasible assembly sequences through human–computer interaction.

Raju et al. [14] established an assembly association graph based on geometric constraints and used a cut-set algorithm with dynamically changing priority queues for assembly sequence planning. Lin et al. [15] extended the geometric constraints to three-dimensional space and planned the assembly process according to assembly angles. However, this method divides three-dimensional space into thousands of directions, potentially causing computational explosion problems.

Zhong et al. [16] semantically abstracted the knowledge and experience required for assembly sequence planning into an ontology model, described using ontology language to form assembly rules. This method realizes the judgment of the rationality of the assembly sequence through ontology reasoning, solving the problem of expressing assembly knowledge and experience but needing improvement in generalization. Kashkoush et al. [17] proposed a knowledge-based mixed-integer programming (MIP) model, which helps to automatically generate assembly sequences by constructing an assembly sequence tree and searching for the most similar assembly sequence tree to the planned assembly sequence. Huang [18] adopted an assembly sequence planning method based on ontology reasoning, establishing a knowledge base of common assembly structures and assembly sequences, and using ontology language for assembly sequence reasoning.

Hu et al. [19] solved the disassembly sequence of products using a rapidly-exploring random tree (RRT) and obtained the assembly sequence by inverting the disassembly sequence. This method works for assemblies of any geometric shape but can lead to combinatorial explosion problems for assemblies with many parts. Pan et al. [20] combined a genetic algorithm with a disassembly method, solving feasible disassembly sequences first and then obtaining assembly sequences by inversion, which also faced combinatorial explosion problems. Sierla et al. [21] defined assembly priority relationships based on the idea of disassembly, reducing the generation of disassembly sequences and mitigating the combinatorial explosion problem.

By analyzing the above literature, the characteristics of various assembly sequence planning methods can be summarized, as shown in Table 5.2.

In summary, current assembly sequence planning faces challenges such as computational and combinatorial explosions. Knowledge-based assembly sequence planning is a promising approach, but it often relies on predefined assembly rules, limiting generalization. The issue lies in the inadequate description of assembly semantics. Therefore, improving the assembly semantic description method in knowledge-based assembly sequence planning is necessary.

5.1.3 Related Work of Assembly Process Evaluation

For the assembly process evaluation of complex components, the focus is on evaluating multiple assembly sequences. Current research on assembly sequence evaluation can be categorized into: qualitative evaluation, quantitative evaluation, and comprehensive evaluation considering both.

Zhang et al. [22] proposed an assembly performance evaluation function that combines overall and unit-level requirements for fixtures and jigs. This method grades

Table 5.2 Characteristics of the various assembly sequence planning methods

Assembly sequence planning method type	Advantages	Disadvantages
Assembly constraint priority-based sequence planning	Usually based on few constraints, simple computation, relatively reliable	Requires human–computer interaction, low efficiency, not suitable for complex assemblies
Geometric constraint priority-based sequence planning	Can avoid human–computer interaction, reduces human errors	When there are many constraints, or the assembly is complex, it easily leads to computational explosion issues
Knowledge-based assembly sequence planning	Utilizes process knowledge and experience, suitable for object-oriented models	Low efficiency in building the ontology library, limited inference methods based on rules
Disassembly-based assembly sequence planning	Applicable to any shape of assembly for sequence planning	When the assembly is complex, it easily leads to combination explosion issues

the requirements based on experience but is influenced by subjective factors. Ma et al. [23] established a qualitative evaluation system for assembly sequences based on a human factor simulation analysis model. Ma et al. [24] considered qualitative evaluation indicators like assembly operation convenience, process simplicity, and assembly feasibility to determine the optimal assembly sequence.

Muslim et al. [25] evaluated the assembly sequence by seeking the assembly time required for the shortest assembly path. Yuan et al. [26] realized the interference analysis of the assembly process by measuring and analyzing the point cloud data of the assembly, so as to evaluate the assembly sequence. Lazzerini et al. [27] obtained assembly accuracy simulation results through simulation software and evaluated the assembly sequence using genetic algorithms.

Wang et al. [28] obtained qualitative evaluation results for assembly performance stability, assembly process reliability, and assembly process simplicity, along with quantitative evaluation results for assembly accuracy and assembly cost through assembly simulation. They established an assembly quality evaluation system to evaluate the assembly sequence. Ni et al. [29] considered assembly path length and assembly accuracy to establish an evaluation matrix, solving it based on the ant colony algorithm to evaluate the assembly sequence. Zhao et al. [30] evaluated the assembly sequence based on assembly time and assembly accuracy, using an improved genetic algorithm.

By analyzing the above literature, the characteristics of various assembly sequence evaluations can be summarized, as shown in Table 5.3. In summary, qualitative evaluation criteria for assembly sequences typically include fixture and jig requirements, human factor engineering analysis, operation convenience, and process simplicity. These criteria come from design experience and assembly operations. Quantitative evaluation indicators generally include assembly path, assembly interference, assembly accuracy, and assembly time, and are usually obtained based on assembly simulation results. Each evaluation method has its strengths and weaknesses, and

Table 5.3 Characteristics of the evaluation of the various assembly sequences

Assembly sequence evaluation type	Advantages	Disadvantages
Qualitative evaluation of assembly sequence	Higher efficiency in evaluating assembly sequences for simple assemblies	Relies on human experience, subject to subjective influences, and the evaluation system is not complete
Quantitative evaluation of assembly sequence	Quantitative evaluation indicators and results are more objective	The solving process is generally complex and less efficient
Comprehensive evaluation of assembly sequence	Relatively complete evaluation system, more referential evaluation results	Complex evaluation system and less efficient solving process

balancing efficiency and effectiveness is crucial when addressing practical problems. For the evaluation of multiple assembly sequences of complex components, screening out unreasonable assembly process schemes can start by evaluating assembly interference.

5.2 Intelligent Design Method of Assembly Process Based on Knowledge Graph

Knowledge graph is essential for the representation and storage of assembly knowledge. This subsection first models the assembly process information based on knowledge graph. Secondly, the assembly process information is modeled based on Bi-LSTM. Finally, the assembly process is generated based on the graph embedding method, which then realizes the intelligent design generation of the assembly process.

5.2.1 Assembly Process Information Modeling Based on the Knowledge Graph

(1) **Model Architecture and Basic Definitions**

Definition 5.1 Assembly process information Model KGAM (Knowledge-graph Assembly Model): According to the information organization mode of knowledge graph, "class-relational class-class" and "class-attribute-value" are used to define the representation of assembly process information described in assembly process documents and CAD models.

According to Definition 5.1, assembly process information contained in assembly process documents and CAD models was analyzed, and the assembly information model (KGAM) based on knowledge graph was expressed as:

$$\text{KGAM} = \{E \cup P \cup O \cup T \cup D \cup C \cup S\} \tag{5.1}$$

$$\begin{aligned} S = \\ &\sum_{i,j=1}^{n} \text{Has}((E_i, P_j), (O_i, D_j), (O_i, T_j), (O_i, E_j), (O_i, P_j), (O_i, C_j), (O_i, O_j)) \\ &\quad \cup \sum_{i,j=1}^{n} \text{HasPart}((E_i, E_j)) \cup \sum_{i,j=1}^{n} \text{PartOf}((E_j, E_i)) \cup \sum_{i,j=1}^{n} \text{Close}((E_\text{i}, E_j)) \\ &\quad \cup \sum_{i,j=1}^{n} \text{HasAssembly}((E_i, E_j)) \cup \sum_{i,j=1}^{n} \text{SubAssemblyOf}((E_j, E_i)) \\ &\quad \cup \sum_{i,j=1}^{n} \text{Sequence}((O_i, O_j)) \cup \sum_{i,j=1}^{n} \text{Parallel}((O_i, O_j)) \cup \sum_{i,j=1}^{n} \text{Recycle}(O_i) \end{aligned} \tag{5.2}$$

where, E represents assembly element information, P represents assembly feature information, and E and P belong to CAD model information; O represents the operation information, T represents the tool information, D represents the equipment information, C represents the requirement information, and S represents the semantic relationship, which constitute the assembly process information. The semantic relationship S includes *Has*, *HasPart*, *PartOf*, *HasAssembly*, *SubAssemblyOf*, *Close*, *Sequence*, *Parallel and Recycle*. Table 5.4 lists their definitions and descriptions.

Based on Definition 5.1, Eq. (5.1), and Eq. (5.2), the composition structure of KGAM is designed. To ensure the independence and determinacy of assembly element information, assembly feature information, operation information, tool information, equipment information, requirement information, and semantic information, they are abstracted into seven classes: workpiece, feature, operation, relationship, requirement, equipment, and tool. Then, subclasses are established through inheritance to obtain the composition structure of KGAM, as shown in Fig. 5.1. Parts and assembly subclasses inherit from the Workpiece class; Assembly Features and Combination Features inherit from the Feature class; Steps and Processes inherit from the Operation class; Has, HasPart, PartOf, HasAssembly, SubAssemblyOf, Close, Sequence, Parallel, and Recycle inherit from the Relationship class; Geometric and Non-Geometric Requirements inherit from the Requirement class; Parts and Subassemblies inherit from both Base and Non-Base classes; Geometric Requirements inherit from Tolerance and Position Tolerance classes; Assembly Features and Combination Features inherit from Base and Non-Base Feature subclasses.

The attributes of the classes in KGAM include public attributes and private attributes. Private attributes are owned by each instance, while public attributes are

Table 5.4 Definition and description of semantic relationships

Relationship type	Relationship name	Relationship definition	Relationship description
Structural relationship	*Has*	The inclusion relationship between assembly elements and assembly features, as well as the inclusion relationship between operation and equipment, tools, assembly elements, assembly features, requirements and operations, for example: Has(Ei,Fj) means that assembly elements Ei contain assembly features Fj; Has(Oi,Dj) Indicates that operation Oi includes a device Dj	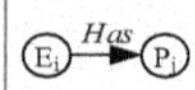  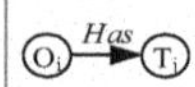 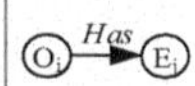 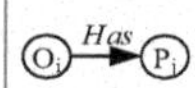 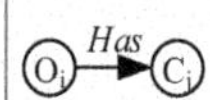 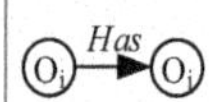
	HasPart	A type of inclusion relationship between assembly elements, for example: HasPart(Ei,Ej) indicates that the subassembly Ei contains the part Ej	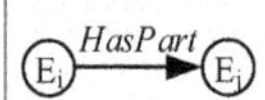
	PartOf	A type of inclusion relationship between assembly elements, for example: PartOf(Ej,Ei) means that part Ej is included in subassembly Ei	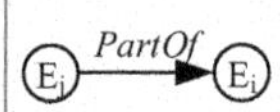

owned by the class and shared by all instances. The specially defined attribute values are shown in Table 5.5.

(2) **Model Composition**

Based on the architecture of the assembly process information model and the basic definitions, the KGAM pattern layer is designed, which is the core of the knowledge graph model. In order to clearly describe the complex semantic information of the assembly process, the pattern layer of KGAM is divided into the assembly structure pattern layer, the assembly step pattern layer, and the assembly process pattern layer.

1. **Assembly Structure Pattern Layer**

The assembly structure pattern layer describes the attribute information of the subassembly class, part class, base feature class, non-base feature class, combined feature class, and the assembly structure semantics between them, as shown in Fig. 5.2.

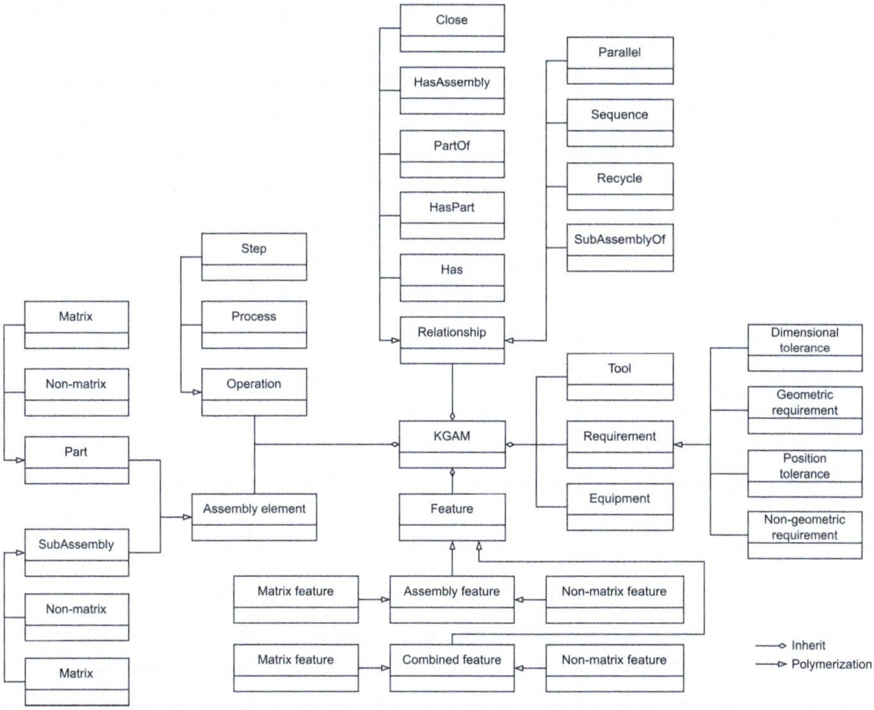

Fig. 5.1 Structural diagram of KGAM

Table 5.5 Value and definition of special attributes

Attribute value	Attribute definition
Base	Represents the part or subassembly with points, lines, or surfaces used to determine the assembly relationship
Non-base	Represents the part or subassembly without points, lines, or surfaces used to determine the assembly relationship
Virtual subassembly	Represents a subassembly composed of two assembly elements but not explicitly defined in the assembly process documents
Actual subassembly	Represents a subassembly composed of two or more assembly elements explicitly defined in the assembly process documents
Assembly feature	Represents a geometric feature directly related to establishing the assembly relationship
Combination feature	Represents a series of geometric features related to establishing the assembly relationship
Base Feature	Represents features with points, lines, or surfaces used to determine the assembly relationship
Non-base feature	Represents features without points, lines, or surfaces used to determine the assembly relationship

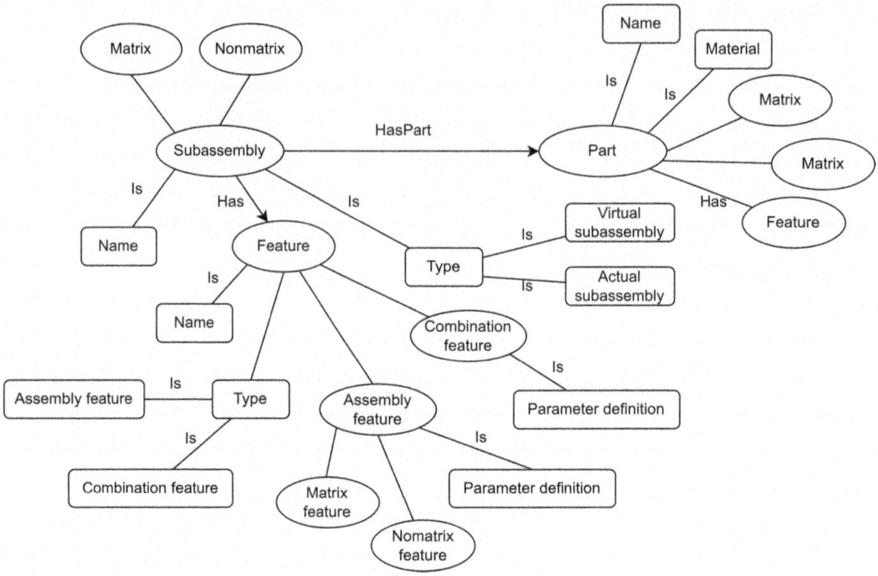

Fig. 5.2 Assembly structure model layer

2. **Assembly Step Pattern Layer**

The assembly step pattern layer mainly describes the attribute information of the step class and the assembly sequence semantics between step classes, as shown in Fig. 5.3.

3. **Assembly Process Pattern Layer**

The assembly process pattern layer mainly describes the attribute information of the process class and the assembly sequence semantics between process classes, as shown in Fig. 5.4.

5.2.2 Construction Method of Assembly Process Information Model Based on Bi-LSTM

Due to the fact that process information in the assembly process is mostly presented in natural language within assembly process documents, it is necessary to identify assembly semantic entities and relationships from unstructured text. These entities and relationships should then be linked to assembly elements and features in the CAD system to form an assembly process knowledge graph. As shown in Fig. 5.5, the assembly semantic recognition model based on Bi-LSTM includes

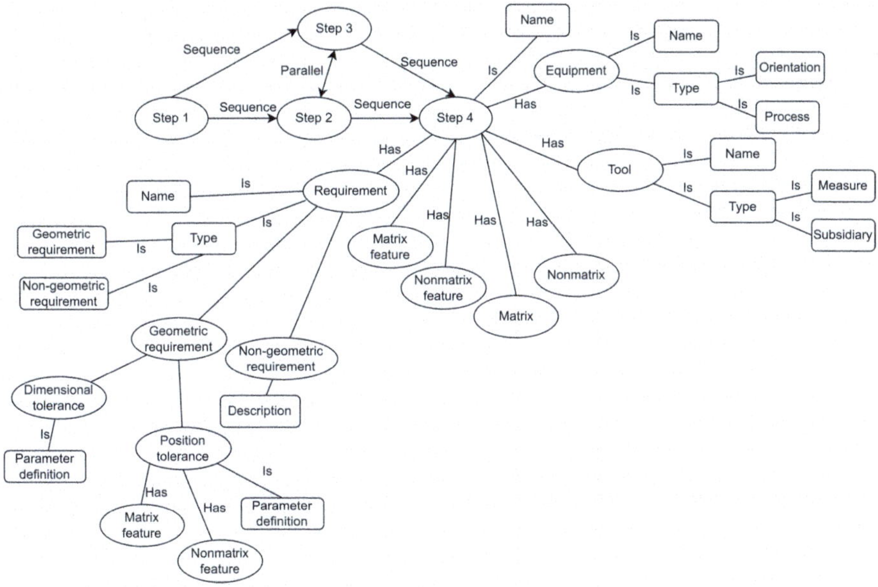

Fig. 5.3 Assembly-step mode layer

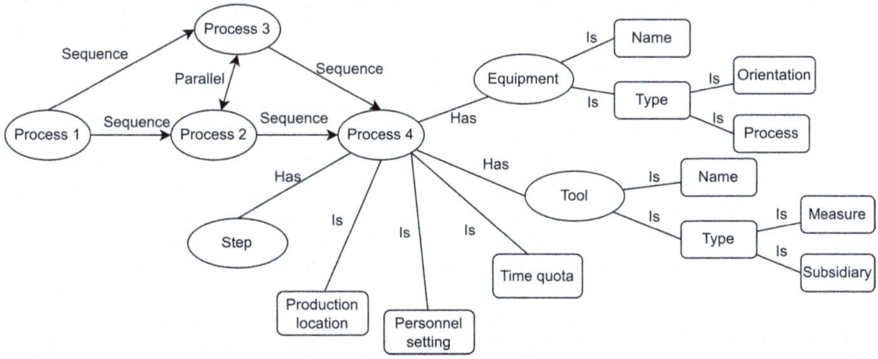

Fig. 5.4 The assembly process mode layer

assembly morpheme decomposition, assembly morpheme classification, and relationship construction. Assembly morpheme decomposition decomposes continuous assembly process statements into discontinuous word sequences; assembly morpheme classification identifies the categories of discontinuous word sequences, thus identifying "tool and equipment names", "assembly object names", "assembly action words", "assembly feature words", "assembly sequence semantic words", "assembly structure semantic words", and "measure words"; relationship construction establishes associations between the results of assembly morpheme classification and assembly element attribute information and assembly feature information.

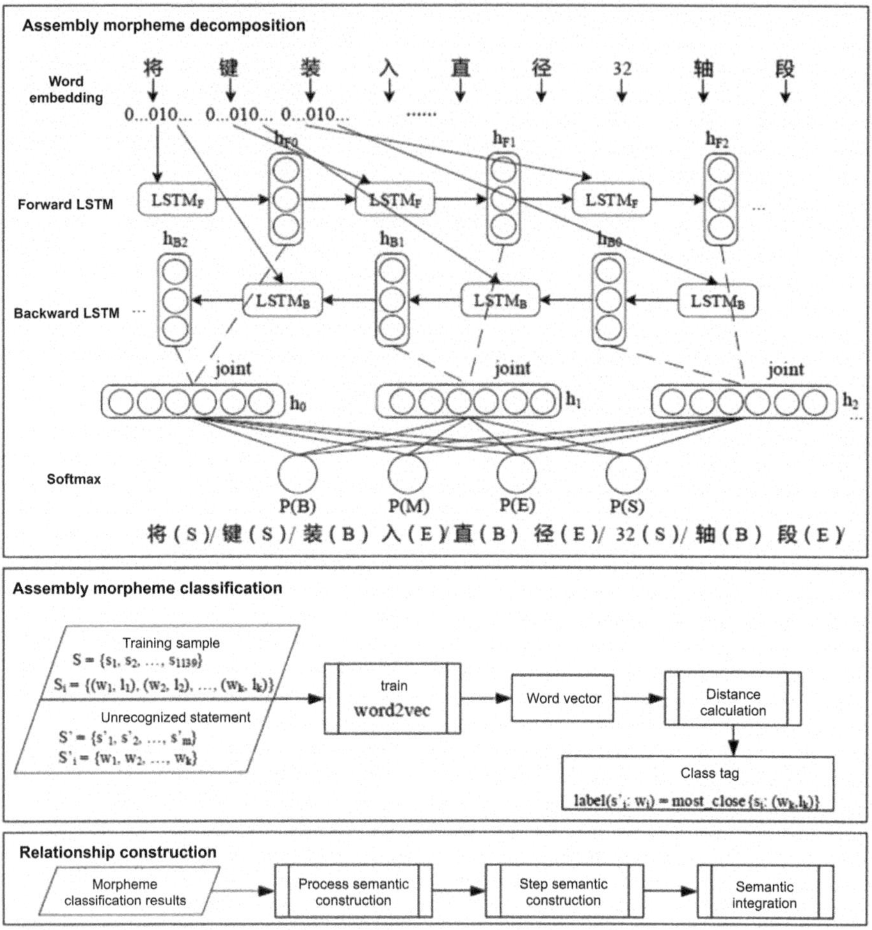

Fig. 5.5 Assembly semantic recognition model based on Bi-LSTM

(1) Assembly Morpheme Decomposition

The input of assembly morpheme decomposition is continuous assembly process statement, and the output is non-continuous word sequence, which is realized by word annotation. Mark "S, B, M, E" labels, "S" means the single Chinese word, "B" means the beginning of the word, "M" means the middle Chinese character, "E" means the Chinese character at the end of the word. For example, enter the assembly process statement "Install the key into the diameter 32 axis segment", and the output result is "Put (S)/key (S)/install (B) into (E)/straight (B) diameter (E)/32 (S) / axis (B) segment (E)". The detailed process is shown in Fig. 5.5. First, Chinese characters are converted into computable numerical encoding by word embedding. Break the sentences into sets of words, and then get the one-hot encoding of each

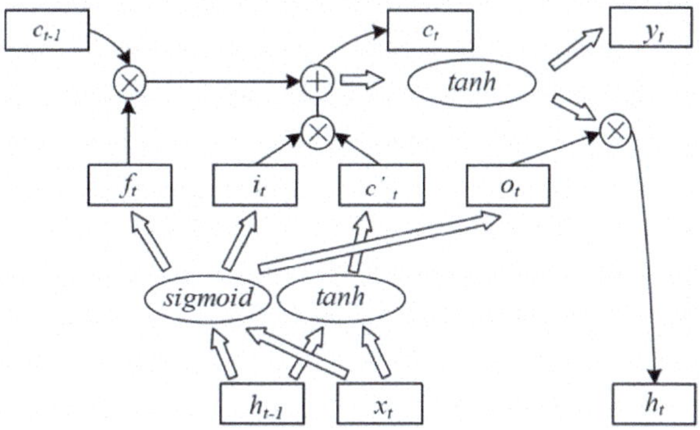

Fig. 5.6 LSTM, computing unit

word according to its frequency. This encoding is the word embedding result, serving as the input to the Bi-LSTM layer (Fig. 5.6).

Bi-LSTM is composed of forward LSTM and backward LSTM, with the basic unit being LSTM, as shown in Fig. 2.9. In the figure, f_t is the forget gate, i_t is the input gate, $c_t{\prime}$ is the temporary cell state, c_t is the current cell state, o_t is the output gate, and the hidden state h_t flows forward and backward. Their calculation process is as follows.

$$f_t = \sigma(w_f * [h_{t-1}, x_t] + b_f) \tag{5.3}$$

$$i_t = \sigma(w_i * [h_{t-1}, x_t] + b_i) \tag{5.4}$$

$$c_t' = \tanh(w_c * [h_{t-1}, x_t] + b_c) \tag{5.5}$$

$$c_t = f_t * c_{t-1} + i_t * c_t' \tag{5.6}$$

$$o_t = \sigma(w_o * [h_{t-1}, x_t] + b_o) \tag{5.7}$$

$$h_t = o_i * \tanh \sigma(c_t) \tag{5.8}$$

(2) **Classification of Assembly Terms**

The assembly morpheme classification is used to determine the category of a word sequence, as shown in Fig. 5.5. First, word vectors are obtained using word2vec, and then the spatial distance between word vectors is calculated to determine the similarity between words. The category of the training sample word with the smallest distance to the word to be recognized is taken as the category of the word to be recognized. When calculating the spatial distance between word vectors, both the direction and the absolute distance must be considered, defining the spatial distance between word vectors as shown in Eq. (5.9). Here, w_1 and w_2 represent the two word

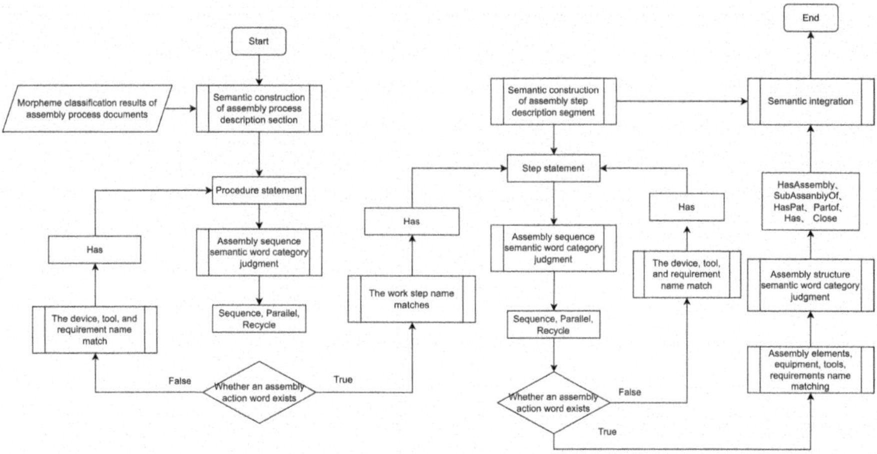

Fig. 5.7 Relationship building process

vectors, n is the dimension of the word vectors, and w_{1i} and w_{2i} represent the ith dimension of the first and second word vectors, respectively.

$$d = same(w_1, w_2) = \left[\sum_{i=1}^{n} (w_{1i} - w_{2i})^2\right]^{0.5} * \frac{\sum_{i=1}^{n} w_{1i}w_{2i}}{[(\sum_{i=1}^{n} w_{1i}^2)(\sum_{i=1}^{n} w_{2i}^2)]^{0.5}} \tag{5.9}$$

(3) **Assembly Relationship Construction**

Relationship construction associates the assembly process document morpheme classification results with assembly element attribute information and assembly feature information, as shown in Fig. 5.7.

5.2.3 Assembly Process Generation Method Based on Graph Embedding

Currently, artificial intelligence is progressing towards cognitive intelligence, and assembly process generation is a typical cognitive intelligence process. Traditional knowledge-based assembly process generation methods mostly rely on ontology to establish an assembly process knowledge base, and then use symbolic descriptions for attributes, rules, and other information. Although this symbolic representation can clearly denote nodes and edges, it fails to measure the semantic correlation between nodes or edges, making it difficult to utilize existing assembly knowledge and experience from the context of the assembly process knowledge graph. In light of this, this chapter first studies the graph embedding methods of the assembly process knowledge graph, representing different nodes and edges in the assembly process knowledge graph database as embedded vectors with corresponding semantic information.

Then, starting from the assembly structure tree of the design assembly, convert it into the "entity-relationship-entity" and "entity–attribute–value" triple formats. This results in an initial assembly process knowledge graph of the design assembly, which includes only part compositions, "Close" semantic relationships between parts, and "HasPart" semantic relationships between actual subassemblies and parts. Finally, compare the semantic similarity between the initial assembly process knowledge graph of the design assembly and the subgraphs of the assembly process knowledge graph database. Additionally, perform node and edge completion on the initial assembly process knowledge graph of the design assembly to achieve automatic generation of the assembly process, as illustrated in Fig. 5.8.

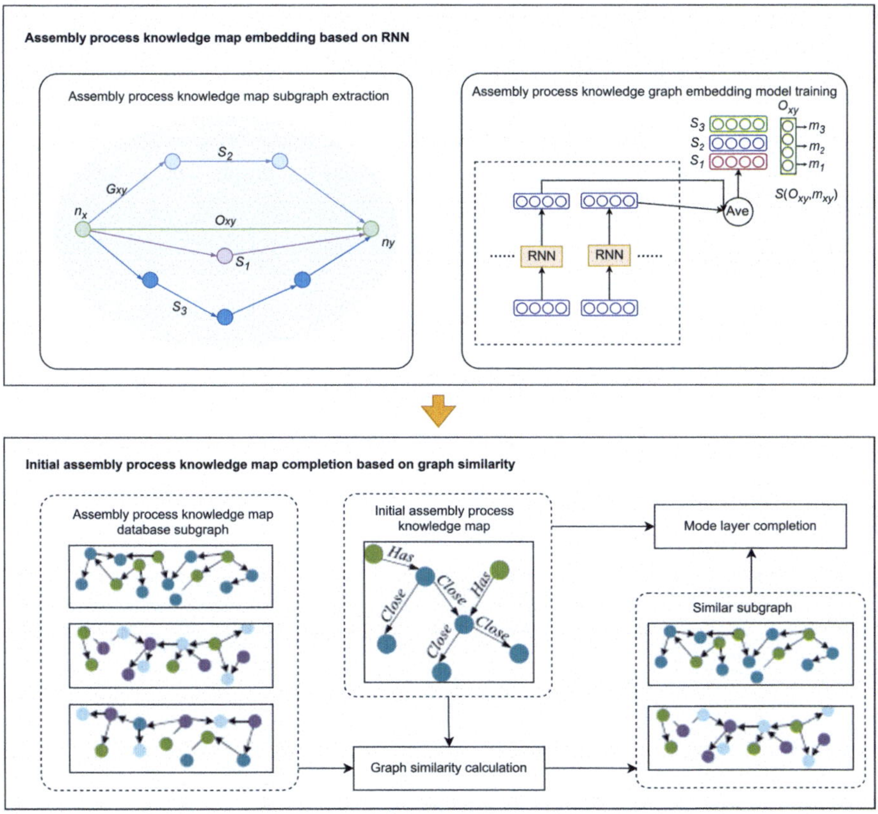

Fig. 5.8 Generation process of assembly process based on graph embedding

(1) Assembly Process Knowledge Graph Embedding Architecture Based on RNN

By analyzing the content and structure of KGAM, it is evident that the assembly process knowledge graph is characterized by its multi-relational nature, which is defined as follows. For graph embedding of the assembly process knowledge graph, since nodes and edges have different types, it is necessary to embed vectors for different types of nodes and edges simultaneously.

> **Definition 5.2** Multi-relational Nature of the Assembly Process Knowledge Graph:
>
> Let there be an assembly process knowledge graph G, $G = (N, L, T)$, where N is a set of n nodes, L is a set of l edges, and T is the set of t types of edges, with $1 < t \leq l$. This characteristic is referred to as the multi-relational nature of the assembly process knowledge graph.

The RNN-based graph embedding method for the assembly process knowledge graph simultaneously embeds the nodes and edges of the graph. The architecture is shown in Fig. 5.9.

(2) Subgraph Extraction

The key to extracting subgraphs from the assembly process knowledge graph lies in determining single-hop sequences and multi-hop sequences, defined as follows:

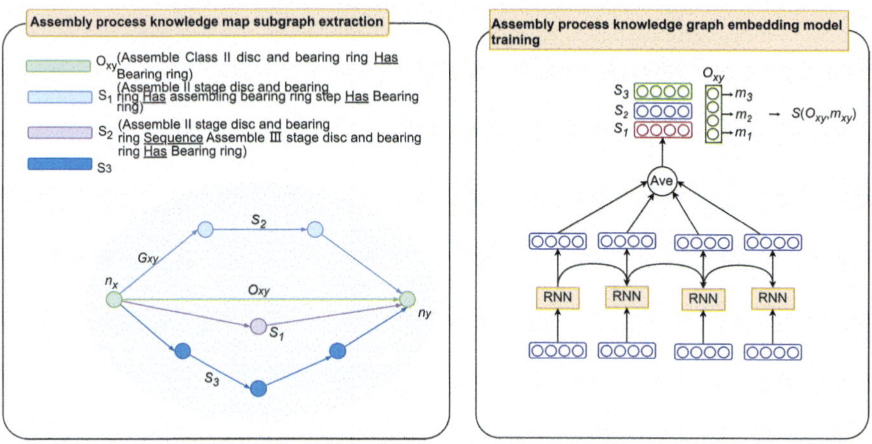

Fig. 5.9 Embedding architecture of assembly process knowledge graph based on RNN

Definition 5.3 Subgraph of the Assembly Process Knowledge Graph: Let G be the assembly process knowledge graph with nodes n_x and n_y. The subgraph $G_{xy} \in G$ is defined as $o_{xy} \cup m_{xy}$, where $o_{xy} = (n_x, r, n_y)$ is the single-hop sequence with direct edge type r, and $m_{xy} = \{s_1, s_2, \ldots, s_q\}$ is the set of multi-hop sequences between nodes n_x and n_y. In m_{xy}, a multi-hop sequence s_i is represented as $(n_x, r_1, n_1, \ldots, r_j, n_j, \ldots, n_y)$. Specifically, if there is no direct edge between nodes n_x and n_y, it is stipulated that $G_x y = \emptyset$, meaning that the subgraph between these two nodes does not exist.

For single-hop sequences, the relationship between nodes n_x and n_y can be obtained by direct search. For multi-hop sequences, the search begins from non-direct edges, obtaining multi-hop sequences of varying lengths by searching for the specified number of hops. Since obtaining all multi-hop sequences between nodes n_x and n_y is inefficient, and semantic relevance decreases with a higher maximum hop count, the variable $b_{\max}$ is set as the maximum hop count to keep the number of multi-hop sequences within a reasonable range. The process is detailed in Table 5.6.

(3) **Graph Embedding Model Training**

The training of the graph embedding model for the assembly process knowledge graph requires the sequential input of single-hop sequences and multi-hop sequences of the subgraph. By performing alternating operations on the nodes and edges of the single-hop and multi-hop sequences, different RNN hidden states are obtained through training. The mean value of these hidden states is then calculated, which serves as the graph embedding result for the subgraph. This graph embedding result considers the impact of each single-hop and multi-hop sequence in the subgraph on the embedding vectors, helping to preserve both the local structural characteristics and the global structural characteristics of the subgraph.

Single-hop sequence o_{xy} and multi-hop sequences $s_i \in m_{xy}$ are input into the RNN-based knowledge graph embedding model shown in Fig. 5.9. In this embedding model, the RNN computational unit is depicted in Fig. 5.10. Taking a multi-hop sequence $s_i = (n_x, r_1, n_1, \ldots, r_j, n_j, \ldots, n_y)$ as an example, the calculation of the tth order hidden state during forward propagation follows Eq. (5.10).

$$h_t = sigmoid(Ux_t + Wh_{t-1}) \tag{5.10}$$

where x_t is the embedding vector of the node or edge type, and U and W are weight matrices. Different elements of the input sequences o_{xy} and m_{xy} will result in different hidden states for each order, and the average value of each order's hidden state is calculated according to Eq. (5.11).

$$\overline{h} = (h_1 + \cdots + h_T)/T \tag{5.11}$$

where T represents the total number of elements in the sequences o_{xy} and m_{xy}.

Table 5.6 Algorithm for obtaining multi-hop sequences

Algorithm Function: Obtain multi-hop sequences between any two nodes in the assembly process knowledge graph
Algorithm Input: Assembly process knowledge graph, any two selected nodes
Algorithm Output: Multi-hop sequences between the selected two nodes in the assembly process knowledge graph

```
nx ← n1, ny ← n2, bmax ← 3, list ← List<MultiJump> // initialize
  for bi ← 1 to bmax do
if bi==1 do
      if GetNeighborNode(nx)!=ny do // Nearest neighbor node judgment
    n1 ← GetNeighborNode(nx)
    if GetNeighborNode(n1)==ny do
      r1 ← GetRelationship(nx,n1)
      r2 ←GetRelationship(n1,ny)
      list.Add(nx,r1,n1,r2,ny) // Multi-hop sequence storage
    end if
  end if
end if
if bi==2 do
      if GetNeighborNode(nx)!=ny do // Nearest neighbor node judgment
    n1 ← GetNeighborNode(nx)
    if GetNeighborNode(n1)!=ny do
      n2 ← GetNeighborNode(n1)
      if GetNeighborNode(n1)==ny do
        r1 ← GetRelationship(nx,n1)
        r2 ←GetRelationship(n1, n2)
        r3 ←GetRelationship(n2, ny)
      list.Add(nx,r1,n1,r2,n2,r3,ny) // Multi-hop sequence storage
      end if
    end if
  end if
end if
  end for
```

Fig. 5.10 RNN cell of the model

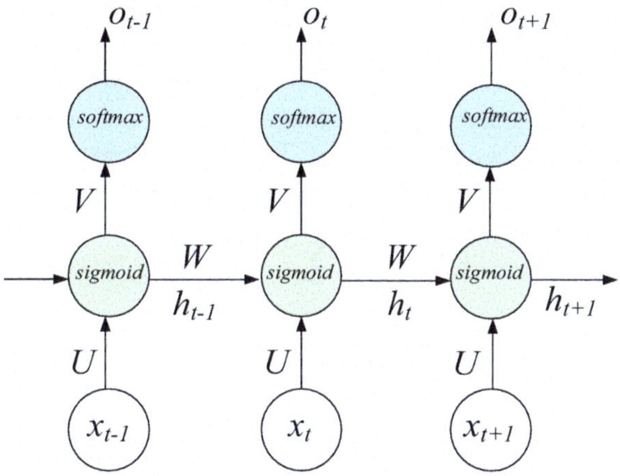

The above describes the forward computation process of the RNN-based graph embedding model. Next, the design of its loss function is discussed. The loss function can be used in the backpropagation process to optimize the hidden states, and the optimized hidden states are taken as the final embedding vectors. Due to the structural similarity between single-hop and multi-hop sequences in the subgraph, the loss function for the training process is designed based on maximizing the semantic similarity between single-hop and multi-hop sequences.

First, assume that there are single-hop sequences o_{xy} and multi-hop sequences $s_{xy} \in m_{xy}$. The similarity between single-hop and multi-hop sequences is calculated using Eq. (5.12), where o_i^T and s_i are the embedding vectors of the single-hop and multi-hop sequences, respectively.

$$s(o_{xy}, s_{xy}) = o_i^T s_i \tag{5.12}$$

For any subgraph, there are generally multiple multi-hop sequences. Suppose the subgraph G_{xy} has x multi-hop sequences $\{s_1, s_2, \ldots, s_x\}$, then the similarity between the single-hop sequence o_{xy} and the subgraph G_{xy} is calculated by Eq. (5.13).

$$g(o_{xy}, G_{xy}) = \log\left(\sum_{i=1}^{x} \exp(s_i)\right) \tag{5.13}$$

The similarity between the subgraph and the single-hop sequence (o_{xy}) is considered the probability of the single-hop sequence (o_{xy}) occurring under the condition of the multi-hop sequences$\{s_1, s_2, \ldots, s_x\} = m_{xy}$, which is calculated by Eq. (5.14). G_{ij} is the subgraph of graph$G$, o_{ij} is the single-hop sequence of subgraphG_{ij}, and μ represents the parameters to be optimized during the training process (embedding vectors of nodes and edges, and weight parameters of the RNN network).

$$p_u(o_{xy}/S_{xy}) = \frac{\exp(s(o_{xy}, S_{xy}))}{\sum_{G_{ij} \in G} \exp(g(o_{ij}, G_{ij}))} \tag{5.14}$$

The parameters (embedding vectors of nodes and edges, and weight parameters of the RNN network) in Eq. 3.5 are optimized, and the final designed loss function is shown in Eq. (5.15).

$$L = \sum_{S_{xy} \in G_{xy}} \sum_{o'_{xy} \in \Delta'_{o_{xy}}} \max[0, \alpha - p_u(o_{xy}/S_{xy}) + p_u(o'_{xy}/S_{xy})] \tag{5.15}$$

where: α is a constant representing the margin value between nodes, max[0,g] is the hinge loss, and $\Delta'_{o_{xy}}$ is the set of negative sample single-hop sequences, which can be obtained by replacing any element of o_{xy}.

In summary, the training algorithm for the RNN-based graph embedding model is obtained, as shown in Table 3.2. First, the nodes and edges of the assembly process knowledge graph are initialized. The training process is as follows:

1. Normalization of node embedding vectors and edge embedding vectors.
2. Randomly extract a batch of triplets and corresponding multi-hop sequences from the subgraph G_{xy}.
3. Solve for the negative samples of the triplets.
4. Update the loss function value to optimize the embedding vectors of the nodes and edges.

(4) **Assembly Process Knowledge Map Completion Based on Graph Similarity**

The graph embedding-based method for assembly process generation takes the designed assembly body to be processed and the assembly process knowledge graph database as input. The designed assembly body comes from the CAD system, with its parts and their interconnections expressed through an assembly structure tree. Therefore, the essence of inputting the designed assembly body is to input the assembly structure tree. The assembly structure tree is a graphical representation of the hierarchical relationships among the parts of an assembly. This tree is converted into a triplet form of "entity-relation-entity" and "entity–attribute–value," resulting in the initial assembly process knowledge graph for the designed assembly body, defined as follows.

Definition 5.4 Initial Assembly Process Knowledge Graph: Given a designed assembly body A, corresponding to an assembly structure tree A_T in the CAD system, the initial assembly process knowledge graph is formed by extracting the part name (ID) information, the HasPart structural semantics between the actual subassembly and the parts, and the Close structural semantics between the parts from the assembly structure tree A_T (definitions and attributes of the relevant semantics are shown in Tables 2.2 and 2.3). This graph is referred to as the initial assembly process knowledge graph.

For both the initial assembly process knowledge graph and the assembly process knowledge graph database, there exists a broad relationship of similarity between graphs, referred to as graph similarity. Graph similarity depends on the similarity of single-hop and multi-hop sequences within the graph. Before calculating sequence similarity, it is necessary to obtain the embedding vectors of nodes and edges in the initial assembly process knowledge graph using the RNN-based graph embedding method. Then, the similarity of single-hop sequences, multi-hop sequences, and single-hop versus multi-hop sequences are calculated. Graph similarity $same(G_1, G_2)$ reflects the similarity between the single-hop sequences $a1_{so}$ and multi-hop sequence sets $S1_{so}$ of graph G_1 and the single-hop sequences $a2_{so}$ and multi-hop sequence sets $S2_{so}$ of graph G_2. The calculation formula is as follows:

$$same(G_1, G_2) = \log\left(\sum_{i=1}^{q} \exp(f(a1_{so}, a2_{so})) * \sum_{i=1}^{q} \exp(f(a1_{so}, s2_i)) * \sum_{i=1}^{q} \exp(f(a1_{so}, s2_i)) * \sum_{i=1}^{q} \exp(f(s1_i, a2_{so}))\right) \tag{5.16}$$

where $f(a1_{so}, a2_{so})$ calculates the similarity between single-hop sequences $a1_{so}$ and $a2_{so}$, $f(a1_{so}, s2_i)$, $f(s1_i, s2_i)$, and $f(s1_i, a2_{so})$ calculate the similarity between the respective sequences, and $f(x)$ is the cosine similarity function, calculated as follows:

$$f(a1_{so}, a2_{so}) = a1_{so}^T a2_{so} \tag{5.17}$$

$$f(a1_{so}, s2_i) = a1_{so}^T s2_i \tag{5.18}$$

$$f(s1_i, s2_i) = s1_i^T s2_i \tag{5.19}$$

$$f(s1_i, a2_{so}) = s1_i^T a2_{so} \tag{5.20}$$

Although the initial assembly process knowledge graph contains "part-name (ID)-value" information, "part-Close-part," and "actual subassembly-HasPart-part" structural semantic information, it lacks process step information, process information, subassembly information, and other assembly process information. Therefore, starting from graph similarity, the structural semantic similarity between the initial assembly process knowledge graph and the subgraphs of the assembly process knowledge graph database is compared. By referencing the subgraphs of the assembly process knowledge graph database that are semantically similar to the initial assembly process knowledge graph, automatic completion of the initial assembly process knowledge graph is achieved, thereby completing the automatic generation of the assembly process.

In the process of knowledge graph completion, the basic logic is as follows: first, input the initial knowledge graph containing the basic nodes and relationships

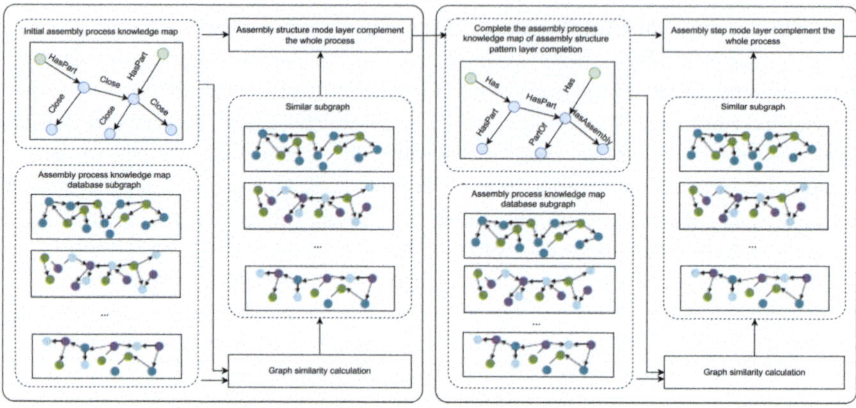

Fig. 5.11 Process for completing the knowledge graph of the initial assembly process

(the initial assembly process knowledge graph containing "part-name (ID)-value" information, "part-Close-part," and "actual subassembly-HasPart-part" structural semantic information). Next, complete a certain type of node, then complete the edges involving this type of node, and then determine whether all types of nodes are completed. When all types of nodes are completed, output the completed knowledge graph; if not, continue to complete the current knowledge graph. This cycle continues until all types of nodes are completed, as shown in Fig. 5.11.

5.2.4 Assembly Interference Detection Method Based on Point Cloud Deep Learning

The graph embedding-based assembly process planning method matches subgraphs in the assembly process knowledge graph database within a certain threshold range based on semantic similarity, thus obtaining multiple assembly process schemes. Not all of these schemes are feasible and need to be evaluated. Currently, the evaluation of multiple assembly process schemes mainly focuses on the feasibility analysis of the assembly sequence and is achieved through human–computer interaction, which is highly inefficient when there are numerous assembly sequences. In light of this, to reduce manual effort and improve the efficiency of assembly process evaluation, this chapter proposes a method for quickly filtering out infeasible assembly process plans. The method evaluates the feasibility of assembly sequences based on assembly interference.

(1) **Assembly Feature Recognition**

To achieve interference detection for a particular assembly process step, the assembly features of the base and the assembled body must first be determined. On the basis of 3D model point cloud data, assembly feature recognition can be converted into

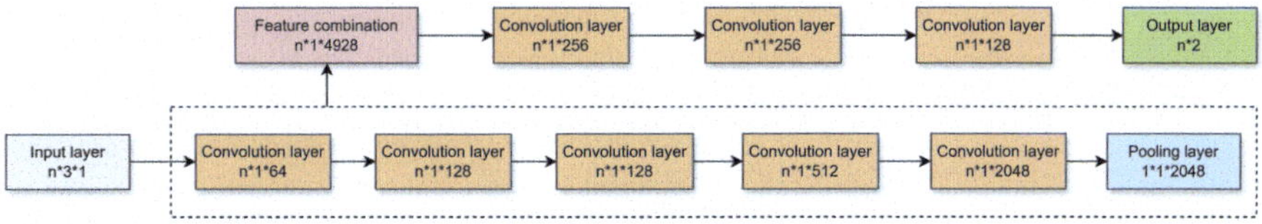

Fig. 5.12 Assembly feature recognition model structure

point cloud feature recognition of the part point cloud data, achieved by calculating the category of each point in the part point cloud data.

The core of the assembly feature recognition process is the point cloud feature recognition model based on Convolutional Neural Networks (CNN). The structure of this model is shown in Fig. 5.12. The input layer is a matrix of $n * 3 * 1$, where n represents the total number of points in the point cloud, 3 represents the coordinates (x, y, z), and 1 represents the number of channels in the original point cloud. The first convolutional layer uses 64 $1 * 3$ convolution kernels to perform convolution operations on the original point cloud $n * 3 * 1$, outputting a matrix of size $n * 3 * 64$. Other convolutional layers use multiple $1 * 1$ convolution kernels, and the pooling layer uses MaxPooling for feature sampling. The output layer uses Softmax to classify each point in the part point cloud. The loss function of the point cloud feature recognition model is the cross-entropy function, as shown in the following Eq. (5.21).

$$H_{y'} = -\sum_i y'_i \log(y_i) \tag{5.21}$$

(2) **Assembly Interference Detection**

Through assembly feature recognition of part point cloud data, the assembly feature point cloud can be obtained from the base and the assembled body point clouds. Solving the projection areas of the base assembly feature point cloud and the non-assembly feature point cloud in the xOy plane, yOz plane, and xOz plane completes the assembly interference detection, as shown in the flowchart in Fig. 5.13.

By analyzing the completed assembly process knowledge graph, it is evident that the assembly sequences in the assembly process schemes express the assembly relationship between two parts or subassemblies, with one viewed as the base and the other as the assembled body. During the assembly of the assembled body and the base, due to the unreasonable spatial position of the base mating area, there may be insufficient assembly space, leading to assembly interference, as shown in Fig. 5.14.

As the assembly sequence progresses, the assembly relationship between parts (subassemblies) gradually forms, with the base continually changing. The environmental point cloud of the base mating area also changes, treating the base before the current assembly relationship is established as the current base. Assembly interference detection is based on the point cloud data of the current base, as shown in Fig. 5.15. By recognizing its assembly features, the current base point cloud is

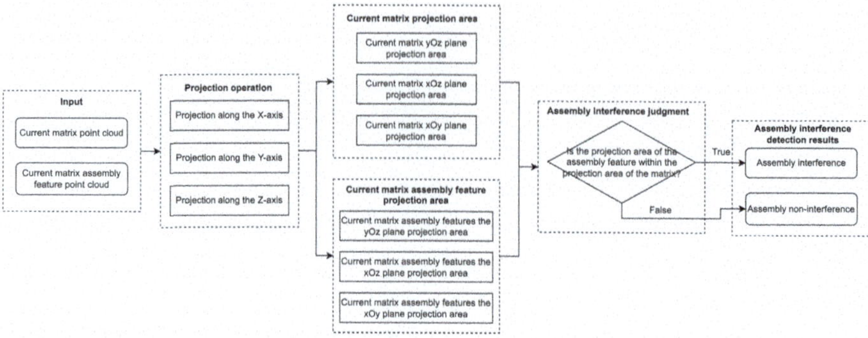

Fig. 5.13 Point cloud interference detection flowchart

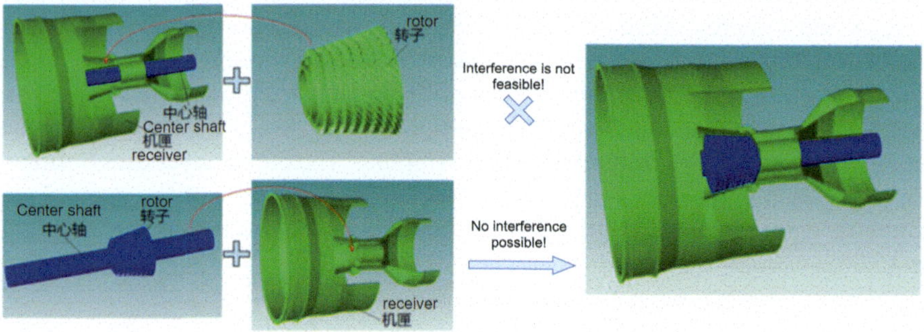

Fig. 5.14 Assembly interference generation

divided into assembly feature point clouds and non-assembly feature point clouds. The current base point cloud and the current base assembly feature point cloud are projected in the yOz plane, xOz plane, and xOy plane, obtaining the projection areas. The method for determining assembly interference is as follows: if the projection area of the current base assembly feature point cloud in the yOz plane, xOz plane, and xOy plane is always within the projection area of the current base point cloud in these planes, assembly interference will certainly occur; if the projection area of the current base assembly feature point cloud in the yOz plane, xOz plane, and xOy plane is never within the projection area of the current base point cloud in these planes, assembly interference will not occur. This method is suitable for quickly filtering unreasonable assembly sequences but cannot be used to determine the assembly interference if the projection area of the current base assembly feature point cloud is not always within the projection area of the current base point cloud in these planes.

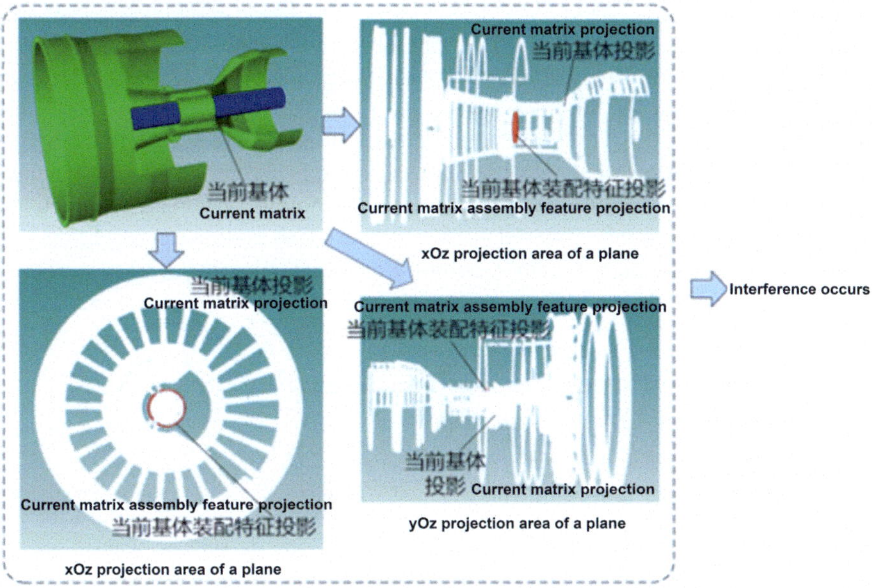

Fig. 5.15 Assembly interference determination method

5.3 Assembly Example and Verification of a Compressor Rotor Component in an Aircraft Engine

The compressor rotor component of an aircraft engine consists of parts such as the disc, shaft, and blades, and is one of the core components of the aircraft engine, as shown in Fig. 5.16. Currently, the assembly process is planned by assembly engineers based on existing 2D assembly documentation and 3D assembly models, resulting in a large workload and low efficiency. In evaluating the effectiveness of the assembly process, the assessment is primarily based on virtual assembly systems and trial assembly procedures. Virtual assembly relies on human–computer interaction, which can be labor-intensive and error-prone, especially with numerous assembly steps.

Fig. 5.16 Compressor rotor component of an aircraft engine

To address this, this chapter first uses a Bi-LSTM-based method for semantic recognition to automatically construct an assembly process knowledge graph database for an aircraft engine assembly workshop. Next, the assembly process knowledge graph and the initial graph of the example object undergo node and edge embeddings. Graph similarity calculations are then used to complete the initial graph, allowing for the automatic generation of multiple assembly process schemes for the compressor rotor component. Finally, an autonomous system is developed for evaluating assembly processes based on interference. This system quickly filters out unreasonable assembly schemes by inputting RDF/XML files and 3D part models.

5.3.1 Automatic Construction of the Assembly Process Knowledge Graph Database

(1) **Construction of the Assembly Process Corpus**

The assembly process corpus is built using documents from the turbine, transmission, and combustion chamber components of an aircraft engine assembly workshop. Each sentence in the assembly process is labeled with "S, B, M, E" character tags, creating a morpheme decomposition corpus. As shown in Fig. 5.1, "S" represents a single-character word, "B" marks the first character of a word, "M" is the middle character, and "E" is the last character (Fig. 5.17).

Each word's category is then annotated, with an explanation of the morpheme classification corpus provided in Table 5.7.

(2) **Training of the Morpheme Decomposition Model**

The morpheme decomposition module is trained with various character dimensions and sentence lengths, as shown in Fig. 5.18. The best results were achieved with a character dimension of 50, a sentence length of 20, and a batch size of 640. The training converged after around 100 iterations, taking about 81 min in total.

The morpheme decomposition model was compared with the classic CRF++ segmentation model * MERGEFORMAT and the LSTM segmentation model * MERGEFORMAT using the training corpus. Precision (P), recall (R), and F-values were evaluated. The model's training loss curve is shown in Fig. 5.19. Table 5.8

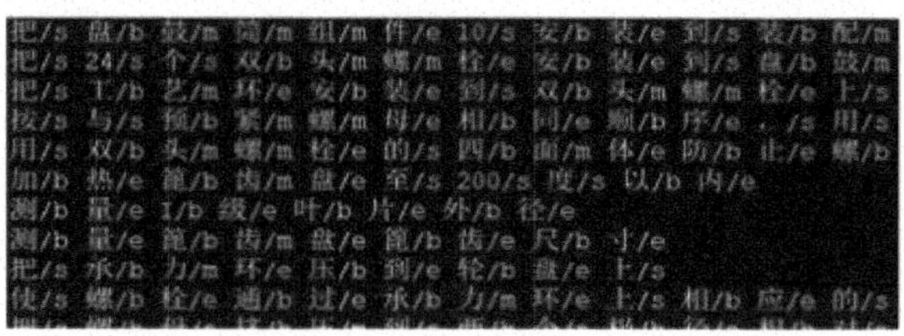

Fig. 5.17 Screenshot of the morpheme decomposition corpus

Table 5.7 Explanation of the morpheme classification corpus

Corpus context	Assembly process documents of an aviation company
Total number of sentences	1139
Total number of words	12,256
Number of tool/equipment names	162
Number of assembly object names	977
Number of assembly action words	1096
Number of assembly feature words	1557
Number of assembly sequence semantic words	1456
Number of assembly structure semantic words	593
Number of quantifiers	615

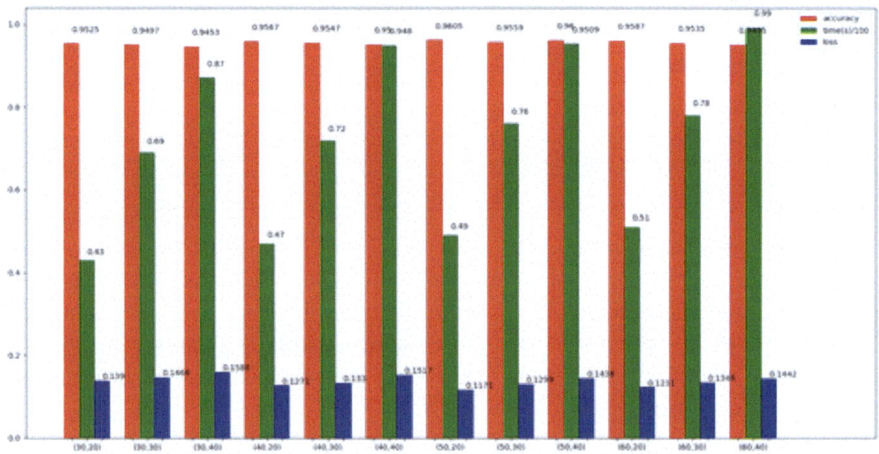

Fig. 5.18 Training results of the morpheme model with different parameters

presents the evaluation metrics, showing that the Bi-LSTM-based morpheme decomposition model has lower loss and better P, R, and F-values than the other methods.

(3) **Training of the Morpheme Classification Model**

Using the constructed morpheme classification corpus, word vectors are trained with Google's open-source project Word2Vec, using the Skip-gram model. The training is done on the same computer setup used for the morpheme decomposition model. The output word vector dimension is set to 128, with a training window size of 5, a learning rate of 0.001, and a minimum frequency of 3. Other parameters are left at default.

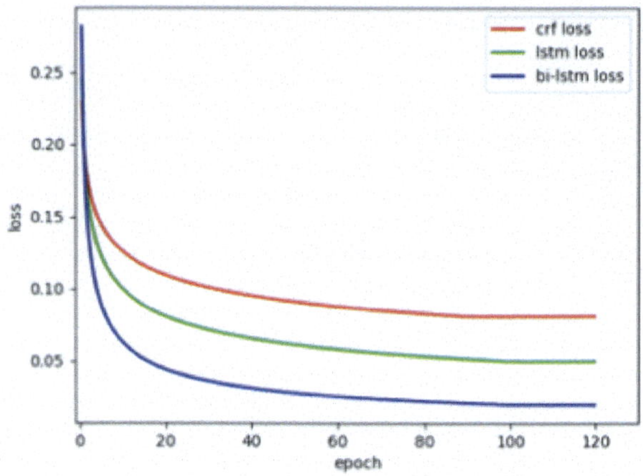

Fig. 5.19 Loss curve of the training process

Table 5.8 Comparison of segmentation performance metrics

Model	P	R	F
CRF++ [71]	0.933	0.929	0.931
LSTM [72]	0.958	0.953	0.955
Bi-LSTM	0.964	0.960	0.962

Figure 5.20 shows the distribution of word vectors in 2D space after dimensionality reduction. Black represents tool names, yellow for assembly object names, red for assembly actions, dark green for assembly features, blue for assembly sequence semantics, purple for assembly structure semantics, and light green for quantifiers. After training, word vectors for the test dataset are obtained through incremental training, with test results shown in Table 5.9.

(4) **Establishment of Assembly Entities and Assembly Relationships**

The trained morpheme decomposition and classification models are used to perform semantic recognition on assembly process documents from the aircraft engine assembly workshop. The recognition results are shown in Fig. 5.21, with some knowledge graphs visualized in Fig. 5.22. Table 5.10 presents the distribution of nodes and edges in the assembly process knowledge graph database.

5.3.2 *Automatic Generation of Assembly Processes Based on Graph Embedding*

(1) **Graph Embedding of the Assembly Process Knowledge Graph**

First, the multi-sequence acquisition algorithm from Table 5.6 is used to obtain 139 subgraphs from the assembly process knowledge graph database. The nodes and

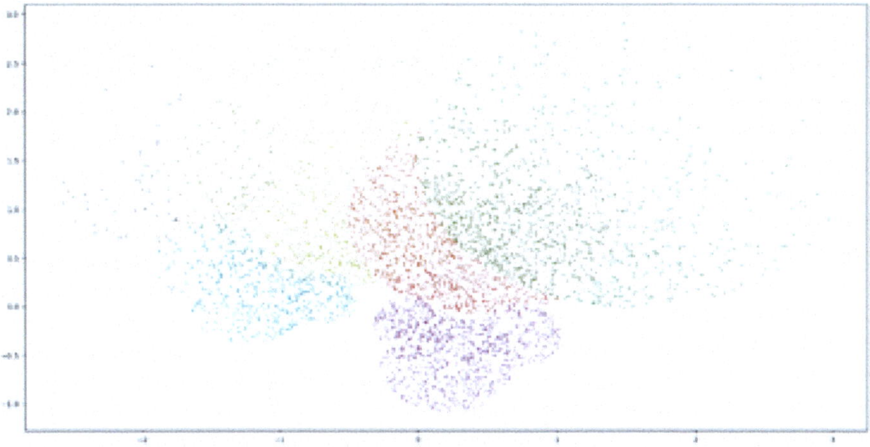

Fig. 5.20 Distribution of word vectors

Table 5.9 Morpheme classification test results

Total number of words participating in classification	Chinese word segmentation F-value	Number of correctly classified words	Q
3679	0.962	3587	0.975

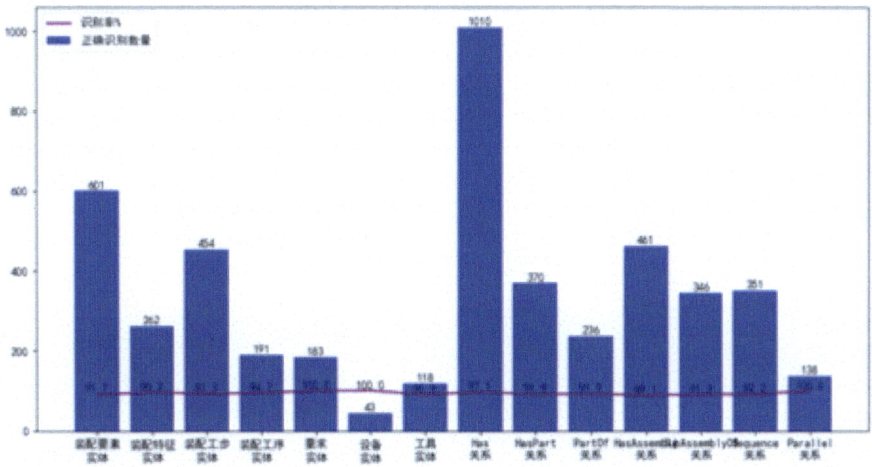

Fig. 5.21 Semantic recognition results of some assembly process documents in the aircraft engine assembly workshop

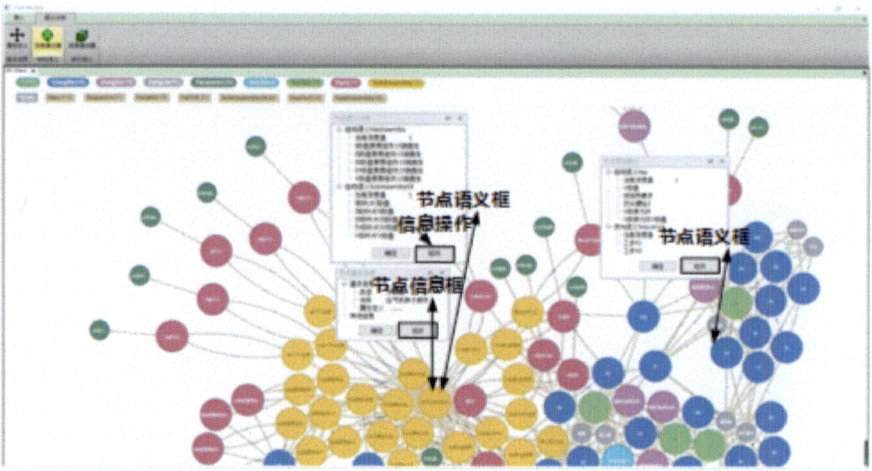

Fig. 5.22 Part of the assembly process knowledge graph database for the assembly workshop

Table 5.10 Distribution of nodes and edges in the assembly process knowledge graph database

Name	Nodes distribution		Relational distribution	
	Type	Scale	Type	Scale
Part assembly process knowledge map database of an aero engine assembly workshop	Part	1651	Has	1987
	sub-assembly	336	HasPart	551
			Part of	551
	Step	921	HasAssembly	214
	Procedure	344	SubAssemblyOf	214
	Requirement	53	Sequence	887
	Equipment	31	Parallel	34
	Tool	56		
	Total	3422	Total	4438

edges of each subgraph are then input into the graph embedding model for training. Hyperparameters include learning rate, margin values, and embedding vector dimensions, with settings detailed in Table 5.11. The training is performed on a computer with an Intel Xeon E5-2630 CPU and a GTX1080 8G GPU. The loss curve for the training process is shown in Fig. 5.23, and the embedding vectors for the nodes and edges are derived from this.

(2) **Completion of the Initial Assembly Process Knowledge Graph**

After obtaining node and edge vectors for both the initial and database graphs, the similarity between them is calculated to finalize the initial graph. This process

Table 5.11 Training parameters for the graph embedding model of the assembly process knowledge graph

Learning rate (**α**)	Margin value (**γ_1**)	Margin value (**γ_2**)	Embedding vector dimension (**d**)
0.001	0.3	0.3	50

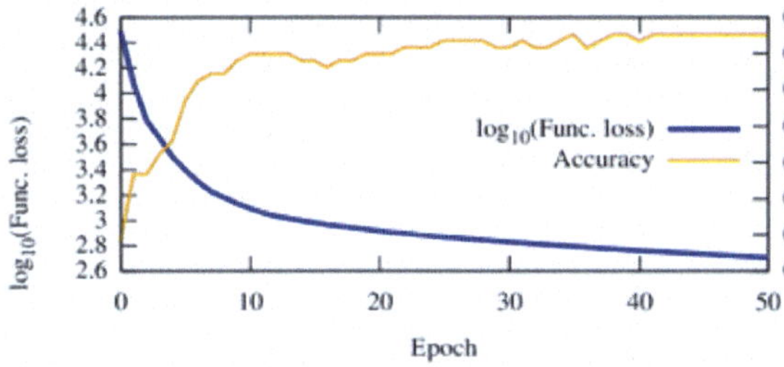

Fig. 5.23 Training process loss curve

involves sequentially completing nodes and edges for the assembly structure pattern layer, the assembly step pattern layer, and the assembly procedure pattern layer.

For the assembly structure pattern layer, virtual subassembly nodes and edges such as HasPart, HasAssembly, PartOf, and SubAssemblyOf are completed. For the assembly step pattern layer, step nodes and edges like Has, Sequence, and Parallel are completed. Similarly, for the assembly procedure pattern layer, procedure nodes and edges like Has, Sequence, and Parallel are completed.

Since the initial graph has only partial information, "Close" edges between parts, and "HasPart" edges between subassemblies and parts, similarity is calculated based on part nodes, subassembly nodes, and sequences with these edges. The two most similar subgraphs are selected from the 139 subgraphs, with their node and edge distributions shown in Table 5.12. Based on these distributions, two completion results for the assembly structure pattern layer of the initial graph are obtained, as shown in Fig. 5.24. In these results, the completed subassembly nodes are yellow, HasPart edges are brown, PartOf edges are green, HasAssembly edges are light blue, and SubAssemblyOf edges are dark blue.

These results form the current assembly process knowledge graph. Similarity is then computed between this graph and subgraphs from the database, focusing on part nodes, actual and virtual subassembly nodes, and sequences with edges like "Close," "HasPart," "HasAssembly," "PartOf," and "SubAssemblyOf." The two most similar subgraphs are selected from the 139 available. With two completion results for the assembly structure pattern layer, this yields four results for the assembly step pattern layer, as shown in Table 5.13.

Using a similar approach to the assembly step pattern layer, we focus on part nodes, subassembly nodes, step nodes, and sequences with "Close," "HasPart," "HasAssembly," "PartOf," "SubAssemblyOf," "Has," "Sequence," and "Parallel" edges. The initial assembly process knowledge graph is then completed through the pattern layers, resulting in 8 assembly process schemes, as shown in Fig. 5.25. The

Table 5.12 Node and edge distributions of the two most similar subgraphs

Name	Assembly structure mode layer node distribution		Assembly structure mode layer relational distribution	
	Type	Scale	Type	Scale
Subgraph 1	Part	43	Close	74
			HasPart	84
			PartOf	84
	Sub-assembly	50	HasPart	551
			HasAssembly	48
			SubAssemblyOf	48
Subgraph 2	Part	43	Close	74
			HasPart	84
			PartOf	84
	Sub-assembly	50	PartOf	84
			HasAssembly	49
			SubAssemblyOf	49

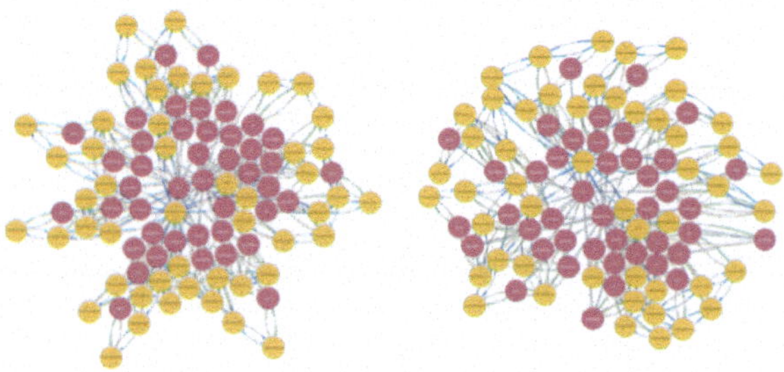

Fig. 5.24 Completion results for the assembly structure pattern layer of the initial assembly process knowledge graph (left: completion result 1, right: completion result 2)

completed procedure nodes are dark green, Has edges are purple, Sequence edges are light yellow, and Parallel edges are red.

5.3.2.1 Feasibility Evaluation of Multiple Assembly Process Schemes

(1) **Assembly Process Simulation System**

Figure 5.26 illustrates the architecture of the autonomous assembly process analysis system based on HOOPS.

The system includes several layers:

- Foundation Layer: This core layer comprises HOOPS/3dAF, RDF/XML file parsing, assembly feature recognition, point cloud interference detection, and assembly relationship construction. HOOPS/3dAF supports model reading and point cloud generation. RDF/XML parsing enables autonomous reading of

Table 5.13 Node and edge distributions of the four completion results for the assembly step pattern layer

Assembly step pattern layer completion result	Assembly structure mode layer node distribution		Assembly structure mode layer relational distribution	
	Type	Size	Type	Size
Based on assembly structure pattern layer completion result 1)	Step	42	Has	84
			Sequence	41
			Parallel	7
Based on assembly structure pattern layer completion result 1	Step	44	Has	90
			Sequence	43
			Parallel	9
Based on assembly structure pattern layer completion result 2	Step	43	Has	85
			Sequence	42
			Parallel	9
Based on assembly structure pattern layer completion result 2	Step	43	Has	87
			Sequence	45
			Parallel	8

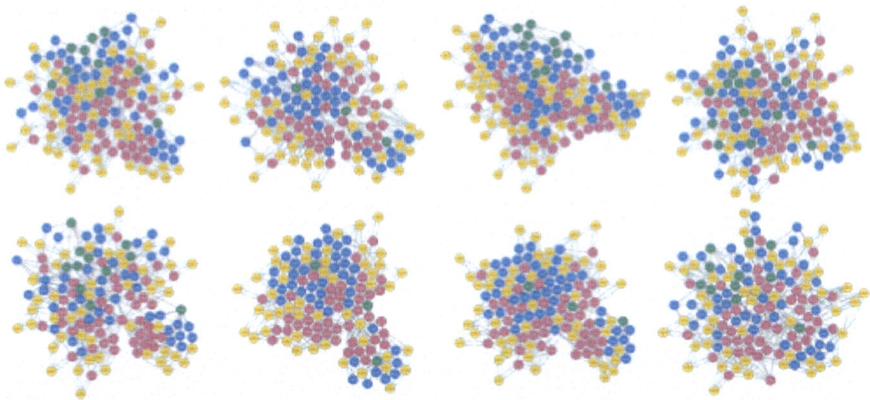

Fig. 5.25 completion results for the initial assembly process knowledge graph

assembly sequences. Assembly feature recognition identifies assembly feature point clouds in part data. Point cloud interference detection analyzes assembly interference using base point cloud data. The assembly relationship construction method stitches point clouds to build assembly relationships automatically.

- Interface Layer: The data interaction channel among modules in the business layer. RDF/XML is the interaction channel for assembly sequences and system analysis processes, and the HOOPS Parasolid Bridge converts data between part CAD models and HOOPS triangular mesh rendering models.

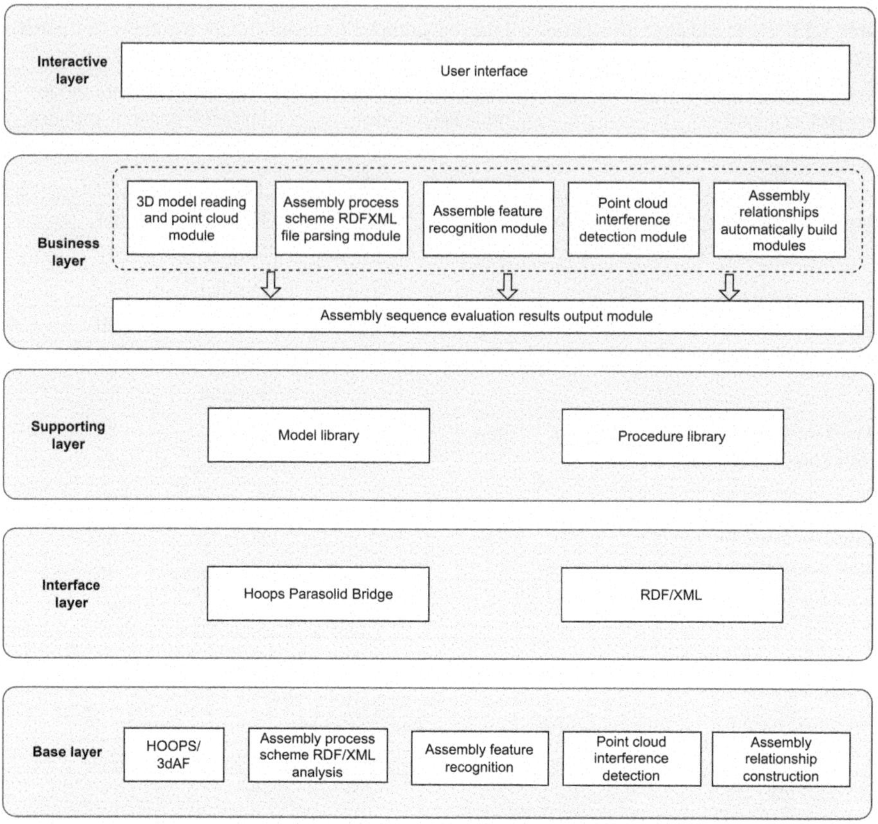

Fig. 5.26 Autonomous assembly simulation system framework based on HOOPS/3dAF

- Support Layer: Comprising the model library and process library. The model library stores 3D models needed for assembly simulation processes, including parts and standard components. The process library stores the assembly results of each assembly sequence.
- Business Layer: Includes all functional modules of the autonomous assembly process analysis system, such as the 3D model reading and point cloud generation module, RDF/XML file parsing module, assembly feature recognition module, point cloud interference detection module, automatic assembly relationship construction module, and assembly sequence evaluation results output module.
- Interaction Layer: The user interface.

(1) **Evaluation Process of Multiple Assembly Process Schemes**

The autonomous assembly process analysis system retrieves assembly sequence information from the RDF/XML file and then calculates interference for each assembly sequence step-by-step. This process yields the final analysis results.

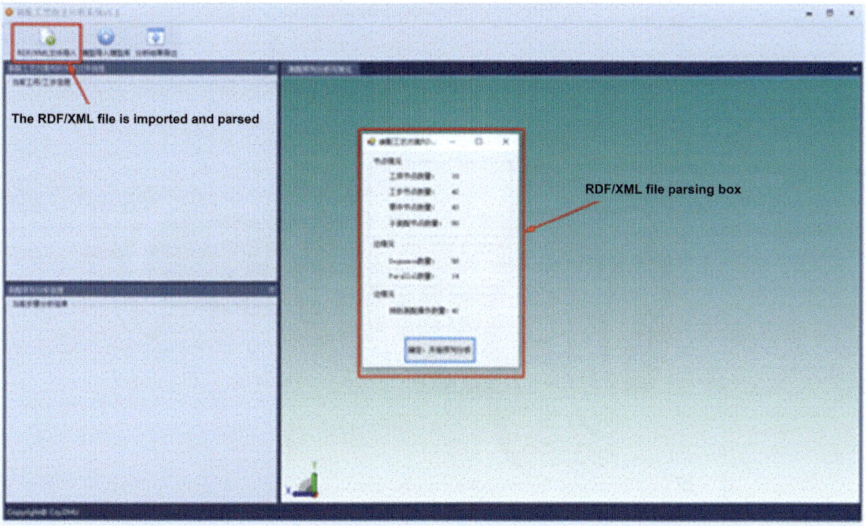

Fig. 5.27 RDF/XML file parsing interface for the assembly process scheme

For example, using one of the eight assembly process schemes, the evaluation process is as follows: First, store the assembly procedure nodes and assembly step nodes in a linked list structure to parse the RDF/XML file. Figure 5.27 displays the interface for parsing the RDF/XML file.

Figure 5.28 shows the interface for analyzing interference in a specific assembly operation step. Once the interference analysis is complete, the current assembly point cloud is automatically updated. The 3D model of the part is then read to analyze the next step in the sequence. By sequentially analyzing each step, the interference vector of the entire assembly sequence is obtained, as detailed in the "Autonomous Evaluation Process of Assembly Sequences." Inputting the RDF/XML files of the eight assembly process schemes into the system yields the interference vectors for all eight sequences.

$$\text{matrix}=\begin{bmatrix} 1111111111111111111111111111111111111111 \\ 1101101110111011111001111101110110111011 \\ 1111111111111111111111111111111111111111 \\ 1110101110111111111101110111010111111101 \\ 1101011110111110111101111111110111111011 \\ 1111101111011111011111010110111110111011 \\ 1101010110111011110110110111011101110111 \\ 1110101110111011111011110101011111101101 \end{bmatrix}$$

From the matrix, it is clear that process schemes 1 and 3 are feasible. To verify their feasibility, both schemes were tested in the workshop. The results showed that both schemes successfully completed the assembly of the aircraft engine compressor rotor component.

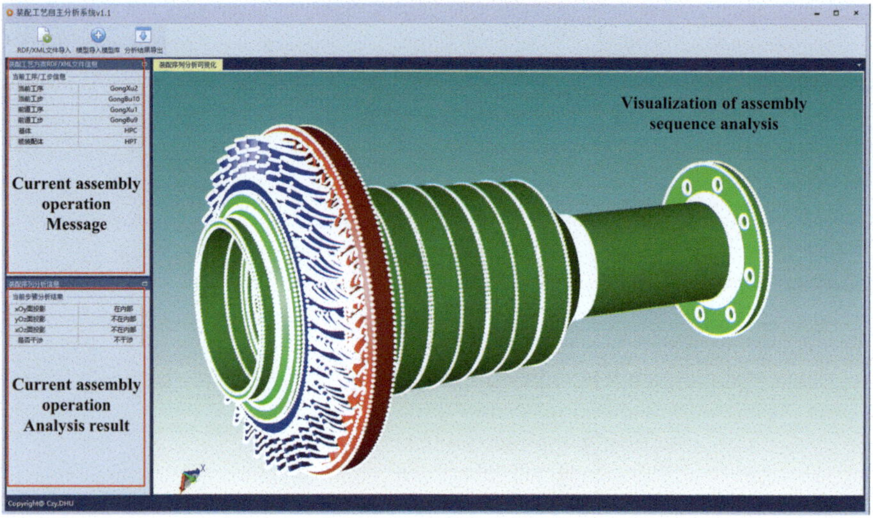

Fig. 5.28 Interference analysis interface for a specific assembly operation

References

1. Da Xu L, Wang C, Bi Z et al (2013) Object-oriented templates for automated assembly planning of complex products. IEEE Trans Autom Sci Eng 11(2):492–503
2. Wang C, Yu H, Zhang W et al. Object-oriented aero-engine assembly models[J]. Computer Integrated Manufacturing System, 2010, 16(05): 0.
3. Li K, Mo R (2014) Structured modeling and application of aero engine assembly data. Aerosp Manuf Technol 57(8):58–61
4. Zhang Y, Yang Z, Ding H et al (2014) Virtual assembly information expression based on XML technology and its application. Mech Design Manuf 2014(9):205–207
5. Tian L (2015) Research on information modeling and integrated simulation technology for assembly design. Shanghai Jiao Tong University, pp 10–26
6. Bao JS, Wu DL, Cheng QH et al (2014) Information modeling and visualization of assembly fat model for large-scale product. Key Eng Mater 579:711–718
7. Wang Q, Wen L, Li J et al (2015) Aircraft general assembly production line modeling and optimization method based on Petri net. J Zhejiang Univ (Eng Ed) 49(7):1224–1231
8. Yang L, Jiao ZG, Lin HB (2012) Modeling and applied research in petri net of virtual assembly program control. Adv Mater Res 482:264–269
9. Yang X, Han J, Pan Y (2013) Virtual training system of assembly and disassembly based on petri ne. In: Proceedings of international conference on soft computing techniques and engineering application: ICSCTEA 2013, 25–27 Sept 2013, Kunming, China. Springer India, pp 205–212
10. Qia L, Zhu Y, Anwer N (2015) Geometrically enhanced ontology modeling for assembly processes. J Mech Eng 51(22):11
11. Sayed MS, Lohse N (2014) Ontology-driven generation of Bayesian diagnostic patterns for assembly systems. Int J Adv Manuf Technol 74(5–8):1033–1052
12. De Mello LSH, Sanderson AC (1990) AND/OR graph representation of assembly plans. IEEE Trans Robot Autom 6(2):188–199
13. Raju Bahubalendruni MVA, Biswal BB (2016) Liaison concatenation—a method to obtain feasible assembly sequences from 3D-CAD product. Sadhana 41:67–74

14. Kardos C, Váncza J (2018) Mixed-initiative assembly planning combining geometric reasoning and constrained optimization. CIRP Ann 67(1):463–466
15. Xin L, Jianzhong S, Yujun C (2017) An efficient method of automatic assembly sequence planning for aerospace industry based on genetic algorithm. Int J Adv Manuf Technol 90:1307–1315
16. Yanru Z, Chaohao J, Yuchu Q et al (2018) Automatic generation of assembly sequences based on ontology. Comput Integr Manuf Syst 24(6):1345
17. Kashkoush M, Elmaraghy H (2015) Knowledge-based pattern for constructing master assembly sequence. J Manuf Syst 34:43–52
18. Huang Xi (2017) Knowledge-driven research on modeling and simulation of aircraft pulsating assembly line. Nanjing University of Aeronautics and Astronautics, pp 10–38
19. Hu L (2015) Research on computer-aided assembly process planning methods. Hefei University of Technology, pp 21–53
20. Pan F (2017) Research on virtual assembly technology of medical display stent based on genetic algorithm. Hefei University of Technology, pp 24–43
21. Sierla S, Kyrki V, Aarnio P et al (2018) Automatic assembly planning based on digital product descriptions. Comput Ind 97:34–46
22. Zhang JY, Wang CE, Ma MX et al (2009) Modeling of product assembly sequence evaluation methods. J Mech Eng 45(11):218–224
23. Ma HZ, Chu XN, Liu ZH et al (2015) Evaluation model and application of assembly sequence based on human factors simulation analysis. China Mech Eng 26(05):652–657
24. MA Shaoxing, HE Zhaohui, ZHANG Risheng, et al. Comprehensive evaluation of assembly process of molecular pump. Comb Mach Tools Autom Mach Technol 2016(9):145–148
25. Muslim MT, Selanmat H, Mukped JAA (2013) Optimizing assembly sequence time using particle swarm optimization (PSO). Appl Mech Mater 315:88–92
26. Baoxun YUAN, Xuening CHU, Yupeng LI et al (2014) A quantitative evaluation method for assembly sequence based on product design data. Comput Integr Manuf Syst 20(04):807–816
27. Lazzerini B, Marcelloni F (2000) A genetic algorithm for generating optimal assembly plans. Artif Intell Eng 14(3):319–329
28. Wang Y (2016) Research on comprehensive evaluation technology of aircraft assembly process. Nanjing University of Aeronautics and Astronautics, pp 8–36
29. Ni J, Tang WC, Pan M et al (2018) Assembly sequence optimization for minimizing the riveting path and overall dimensional error. Proc Inst Mech Eng 232(14):1–11
30. Zhao LP, Li BH, Chen HG et al (2018) An assembly sequence optimization oriented small world networks genetic algorithm and case study. Assem Autom 38(4):387–397

Chapter 6
Intelligent Detection of Industrial Defect Images

Users and manufacturers now demand higher product quality, seeking not only functional performance but also an appealing appearance, particularly in surface quality. However, surface defects are often unavoidable during manufacturing. These defects vary by product type but generally refer to areas where the surface's physical or chemical properties are inconsistent. Examples include scratches, spots, and holes on metal surfaces, color differences and indentations on paper, and inclusions, breakages, and stains on glass and other non-metal surfaces. Surface defects not only detract from the product's appearance and comfort but also negatively impact its performance. Therefore, manufacturers prioritize surface defect detection to ensure product quality. Detection results can also help identify and address production issues, reducing defective products and avoiding potential trade disputes, thereby safeguarding the company's reputation. Today, machine vision-based surface defect detection equipment has largely replaced manual inspection across various industries, including automotive, home appliances, machinery, semiconductors, electronics, chemicals, pharmaceuticals, aerospace, and light industry.

6.1 Progress and Current Status of Industrial Defect Image Detection

Industrial defect detection is an important application scenario for machine vision technology. This subsection firstly provides a strict definition of the industrial defect detection problem, secondly analyzes the three development stages of industrial defect image detection, and finally provides a comprehensive overview of deep learning-based industrial defect image detection techniques in terms of enriching the sample space and limiting the hypothesis space.

T. Liu et al., *Industrial Intelligence: Methods and Applications*, Springer Series in Advanced Manufacturing, https://doi.org/10.1007/978-3-031-81477-8_6

6.1.1 Definition of Industrial Defect Image Detection

In machine vision tasks, defects tend to be concepts based on human experience rather than purely mathematical definitions. Different perceptions of defect patterns lead to two distinctly different detection methods. Taking fabric surface defect detection as an example, as shown in Fig. 6.1, the first supervised method involves using labeled defect images (including categories, rectangular boxes, or pixel-level labels) input into the network for training. In this context, "defect" means the labeled areas or images. Therefore, this method focuses more on defect features, such as marking regions or images with large black areas as "color difference" defects during the training phase for network training. In the testing phase, when large black features are detected in the fabric image, it is considered a "color difference" defect.

The second method is unsupervised defect detection, which typically requires only normal, defect-free samples for network training, also known as one-class learning. This method focuses more on the features of defect-free (i.e., normal) samples. When previously unseen features (anomalous features) are found during the defect detection process, they are considered defects. In this context, "defect" means an anomaly, so this method is also known as anomaly detection.

Compared to the clearly defined tasks of classification, detection, and segmentation in computer vision, the requirements for defect detection are more general. These can be divided into three levels: identifying what the defect is, determining where the defect is, and assessing how much defect there is.

The first level, "what the defect is," corresponds to the classification task in computer vision. As shown in Fig. 6.1, this task involves categorizing defects,

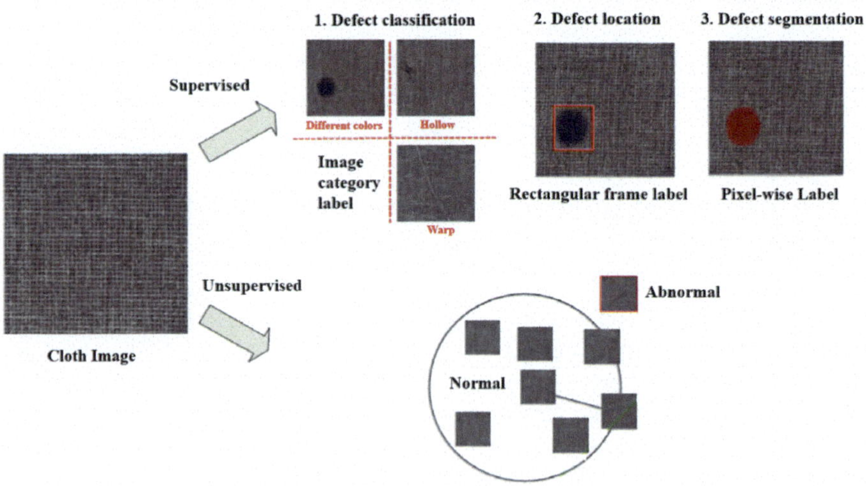

Fig. 6.1 Definition of defect detection problems

such as color differences, hollows, and warps. This process is known as "defect classification," which only provides the category information of the image.

The second level, "where the defect is," corresponds to the localization task in computer vision. Here, defect localization involves not only identifying the types of defects in the image but also pinpointing their specific locations, such as marking a color difference defect with a rectangular box in Fig. 6.1.

The third level, "how much defect there is," corresponds to the segmentation task in computer vision. This involves segmenting the defect pixel by pixel from the background, as shown by the segmented defect areas in Fig. 6.1. It provides additional information, such as the length, area, and position of the defect, which can assist in higher-level quality assessments like determining the quality grade.

Although the functional requirements and objectives of these three levels differ, they overlap and can be interconnected. For example, defect localization (second level) includes defect classification (first level), and defect segmentation (third level) can achieve both localization and classification. Conversely, defect classification can also fulfill the objectives of localization and segmentation through specific methods. Therefore, in the following text, we will collectively refer to these tasks as defect detection, distinguishing only when discussing different network structures and target functions.

6.1.2 Development History of Industrial Defect Image Detection

The development of vision-based industrial defect detection systems has mainly gone through the three stages shown in Fig. 6.2.

The first stage involves the human eye capturing visual signals and the human brain processing them. Although this stage offers high flexibility due to human involvement, it has low efficiency and consistency in quality.

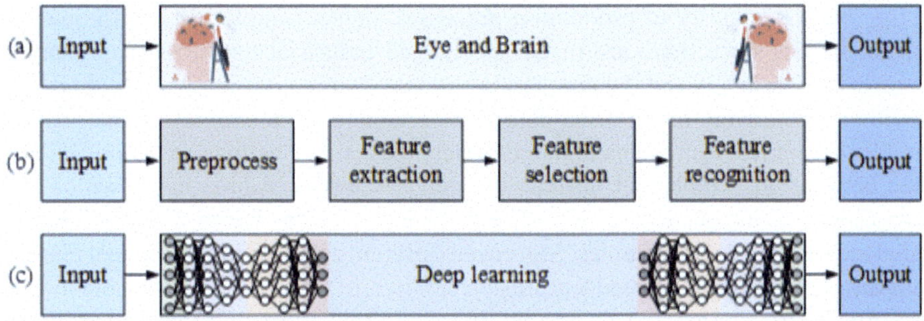

Fig. 6.2 Development stages of vision detection systems

The second stage is based on traditional machine learning methods for industrial image pattern recognition. This stage primarily follows a process from image preprocessing to feature extraction, feature selection, and feature classification. Since the design, extraction, and selection of features are human-led processes, the model has high interpretability and reliability in fixed scenarios. However, this stage has several drawbacks due to the numerous intermediate steps. First, the outcome of each intermediate task affects the final result, and the optimal solution for sub-tasks does not necessarily mean the optimal solution for the overall problem. Second, each step is an independent sub-task, severely affecting the overall efficiency of industrial image pattern recognition and making it difficult to adapt to complex and changing production environments. Third, due to limited human prior knowledge, the value of the data cannot be fully realized.

The third stage involves industrial image pattern recognition based on deep learning. The main idea of this stage is to select a hypothesis function from the hypothesis space as the final model based on the learning algorithm after obtaining the image dataset. Deep learning models can capture deep abstract information of industrial images, leading to better representation of image patterns. This end-to-end modeling approach also greatly improves the efficiency of industrial image pattern recognition.

6.1.3 *Advancements in Industrial Defect Image Detection Based on Deep Learning*

(1) **Challenge Analysis**

Although deep learning methods have been widely applied in industrial image detection, certain characteristics of industrial images limit their deeper application. Through analysis, it can be observed that industrial images mainly have the following characteristics. (1) Small sample size. Firstly, due to Industry 4.0 just starting, the industry is only beginning to pay attention to and accumulate data. Secondly, because of the relative maturity of production processes, defect samples do not frequently appear. Lastly, since there are many specialized industrial fields, data annotation requires professionals and the annotation process is time-consuming and laborious, resulting in relatively few labeled data. (2) Imbalance. As mentioned above, defect samples do not frequently appear in industrial production, leading to a serious imbalance between the number of normal and defect samples. Additionally, within the defect samples, the occurrence of different defect types also varies, causing internal imbalance among defect samples. Moreover, different categories have distinct feature patterns, leading to imbalanced learning capabilities of the same deep learning model for different categories. (3) Strong interference. Industrial environments are relatively complex, with the production process generally involving complex physical and chemical reactions, accompanied by issues such as lighting changes, vibration changes, and motion blur. These interferences significantly hinder the deep learning model's ability to extract effective patterns. (4) Small targets. Industrial products are generally composed of many small components, and the defects of industrial products are usually small relative to the product itself. Since the backbone networks of deep learning (VGG series and ResNet series) have multiple downsampling processes, the size of small targets in the feature map is usually reduced to only a few pixels, severely limiting the model's ability to describe small targets. (5) Fine granularity. Since image analysis in industrial production is generally applied within a specific field, this leads to minimal macro differences between different images. However, due to occasional circumstances, the differences within the same category can be significant. Previously, this required skilled workers to carefully distinguish subtle features. For deep learning models, this is a significant challenge, as they must find the most discriminative abstract features among fine-grained images. Unfortunately, in practical applications, these issues often appear together.

A comparative analysis of industrial image characteristics shows that issues like small sample size and imbalance are related to the sample space. In contrast, problems such as strong interference, small targets, and fine granularity pertain to the samples themselves. According to computational learning theory, small sample sizes result in insufficient information within the sample space, causing deep learning models to lack inductive bias. Imbalance in the dataset leads to inductive bias that does not accurately reflect the real problem. Data characteristics themselves create complex image patterns, requiring a broader hypothesis space to find optimal solutions.

Characteristics related to the sample space lead to deep learning models having near-zero empirical error but a significant gap between generalization error and empirical error. On the other hand, characteristics inherent to the samples themselves make it difficult for the models to minimize empirical error, resulting in a large difference between test error and training error. The core issue here is that these factors reduce the effectiveness of deep learning, ultimately degrading generalization performance in practical applications.

(2) **Countermeasures**

To enhance the feasibility and generalization ability of deep learning methods in industrial image detection, enriching the sample space and constraining the hypothesis space are two basic approaches.

1. **Enriching the Sample Space**

Samples are the cornerstone of deep learning. Broadly speaking, enriching the sample space means expanding the amount of information it contains. This is a fundamental method to improve the feasibility and generalization ability of deep learning-based industrial image detection. Current approaches mainly enrich the sample space from the perspectives of sample quantity and sample characteristics, thereby enabling deep learning models to have more explicit and reasonable inductive biases.

Cross-Validation. In deep learning, the dataset is generally divided into three parts: training set, validation set, and test set. The validation set is primarily used for observing the training process and optimizing hyperparameters. Common cross-validation methods in deep learning include K-fold cross-validation, leave-one-out

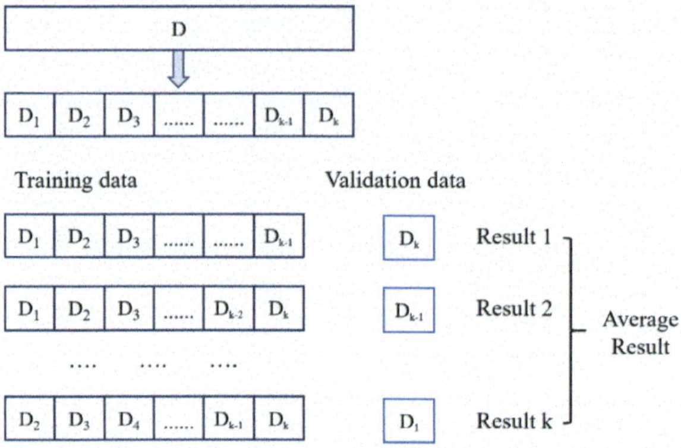

Fig. 6.3 k-fold cross validation

cross-validation, and stratified K-fold cross-validation. K-fold cross-validation is most commonly used in the industrial field. As shown in Fig. 6.3, K-fold cross-validation divides the original dataset into k subsets, with one subset retained for validating the model and the remaining subsets used for training. Cross-validation is repeated k times, with each subset used for validation once. The final result is an average of the k iterations or another combination method, achieving maximum utilization of the limited sample space and eliminating data interference to the greatest extent.

Semi-Supervised Learning. Due to the need for specialized knowledge to annotate industrial images, obtaining large-scale labeled data is difficult and costly. Although this issue severely limits the application of deep learning methods in industrial image pattern recognition, it also implies the existence of many unlabeled samples in practice. The semi-supervised learning method, which integrates labeled and unlabeled samples, provides a feasible solution. Labeled data is used for model training, while unlabeled samples help improve model performance. In deep learning, semi-supervised learning methods are categorized into several types as shown in Fig. 6.4. However, their application in industrial image pattern recognition is still limited.

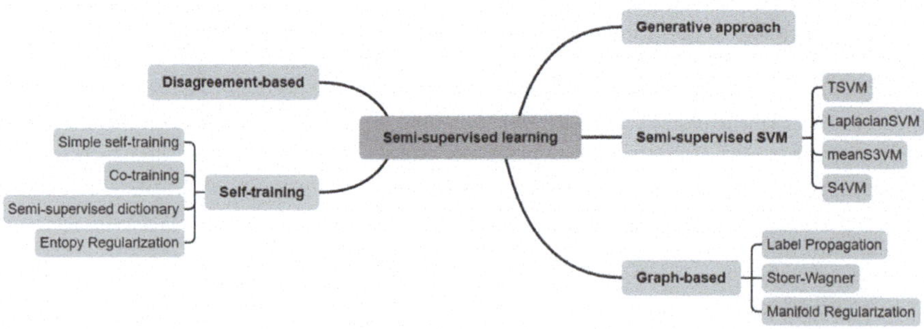

Fig. 6.4 Unsupervised learning classification

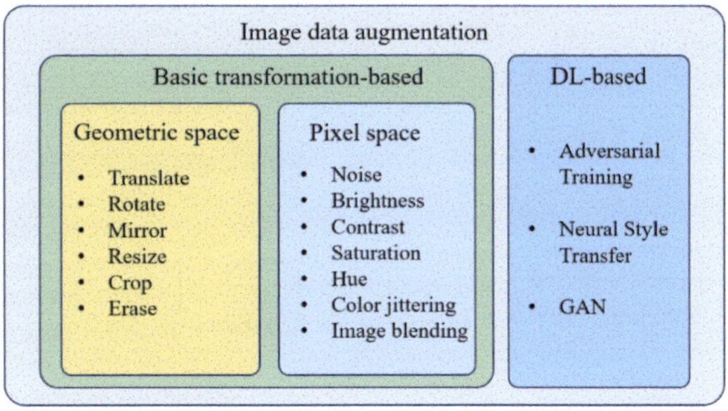

Fig. 6.5 Classification of data augmentation methods

Data Augmentation. Data augmentation is a fundamental and direct method to address the insufficient information problem in the sample space. As shown in Fig. 6.5, common image data augmentation methods in deep learning include basic image processing methods and deep learning methods. Basic image processing methods involve geometric and pixel space operations.

Multi-scale Learning. By representing image patterns at multiple scales, multi-scale learning introduces multi-scale information into the sample space, granting the trained deep learning model scale invariance, which is crucial for improving image pattern recognition. As shown in Fig. 6.6, commonly used multi-scale representations in industrial images include geometric and Gaussian spaces. Different geometric scales can be represented by inputting images of different sizes (Fig. 6.6a), pooling operations (or convolution operations) to obtain features at different resolutions (Fig. 6.6b), or using convolution kernels of different sizes to obtain various receptive fields (Fig. 6.6c). Different Gaussian scales are achieved by applying varying degrees of Gaussian blur to the original image (Fig. 6.6d).

Attention Mechanism. The attention mechanism, inspired by human cognitive processes, addresses the issue of redundant expressions in image spatial pixels and feature channels obtained through multiple convolution kernels. As shown in Fig. 6.7, the flexible attention mechanisms commonly used in industrial image fields include spatial and channel attention mechanisms. The attention mechanism can be considered as adding prior information about the importance of different spatial positions in the image and different channels in the feature tensor to the sample space, thus enhancing the inductive bias of the deep learning model and guiding it towards the direction of prior knowledge during training.

Transfer Learning. As shown in Fig. 6.8, transfer learning involves a source domain and source task, a target domain and target task, and uses knowledge obtained in the source domain to improve the target task. Transfer learning relaxes the assumption that training data must be independently and identically distributed (i.i.d.) with test data. Knowledge acquired from the source domain enriches the target domain. Transfer learning methods in deep learning include sample-based, model-based, feature-based, and relation-based transfer. In the industrial field, feature-based

Fig. 6.6 Multi-scale learning method

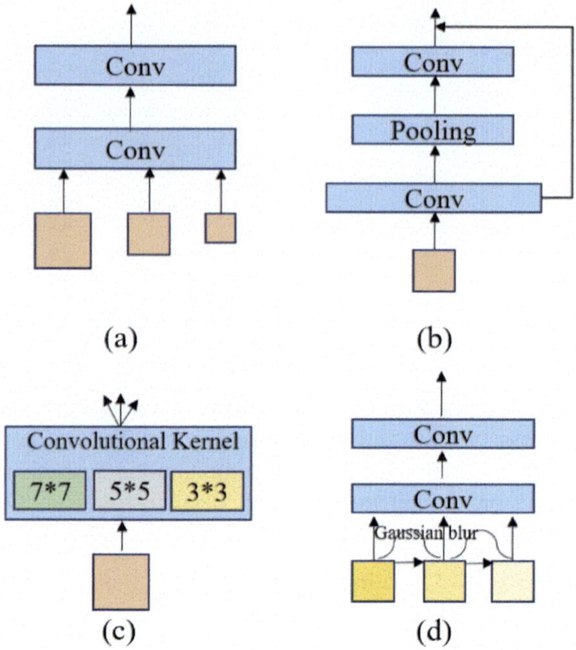

Fig. 6.7 Spatial attention and channel attention mechanism

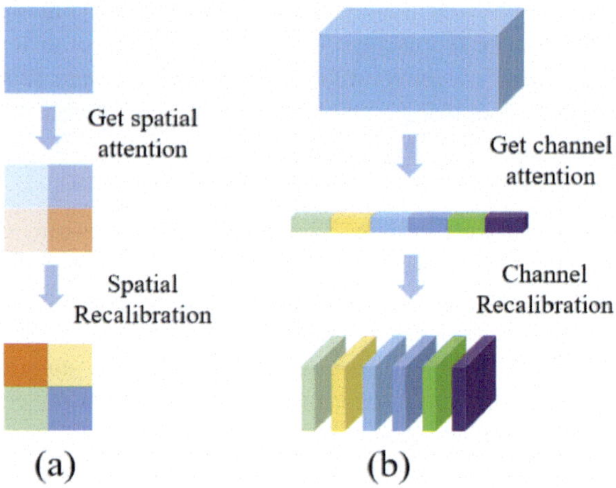

transfer methods are widely applied, further divided into homogeneous and heterogeneous transfer learning based on the similarity of data characteristics between the source and target domains.

Feature Priors. Deep learning involves extracting feature patterns from images layer by layer, with lower layers capturing basic features like edges and colors, and

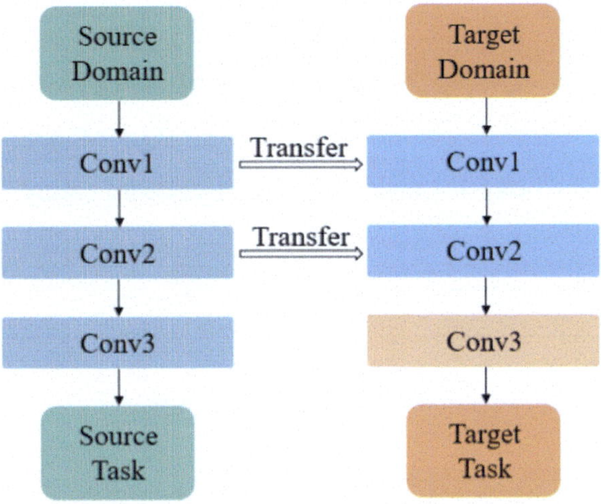

Fig. 6.8 Connotation of transfer learning

higher layers identifying deeper, abstract features. When the sample space lacks sufficient information, extracting reliable features becomes challenging. To address this, feature priors can be introduced, where additional feature information is manually added to the sample space to simplify the network's learning process.

Feature priors can come from either handcrafted features or those automatically extracted through deep learning. These priors mainly contribute during the feature extraction and classification stages. In industrial image analysis, three main approaches for introducing prior features are illustrated in Fig. 6.9: deep annotation of images (Fig. 6.9a), using handcrafted features as CNN inputs, and directly incorporating handcrafted features into CNN-based feature classification.

2. **Constraining the Hypothesis Space**

Parameter Regularization. Parameter regularization is a common method to constrain the hypothesis space. Essentially, it restricts the distribution of weight parameters, thereby reducing the hypothesis space. Formally described as: Loss = L0 + α*f(w), where α is the regularization coefficient, and f(w) is the parameter regularization term, an expression about the parameter w. Common forms include L1 norm and L2 norm regularization. L1 regularization introduces prior knowledge that the weight distribution follows a Laplace distribution, while L2 regularization introduces prior knowledge of a Gaussian distribution. As shown in Fig. 6.6.10, L1 regularization makes the weight parameters sparse, and L2 regularization makes the weight parameters smaller and smoother.

Random Dropout. Dropout is a stochastic technology used in neural networks. Common dropout methods in deep learning include standard dropout, dropout for CNNs, and dropout for RNNs. As shown in Fig. 6.11, standard dropout involves randomly stopping the activation value of certain neurons during forward propagation with a probability of p, and then updating the network based on the resulting loss

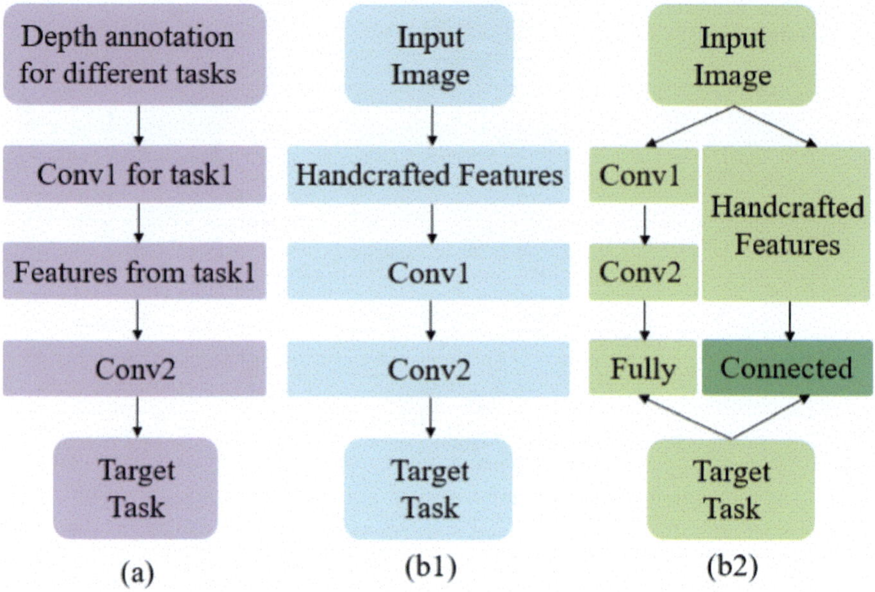

Fig. 6.9 Feature prior

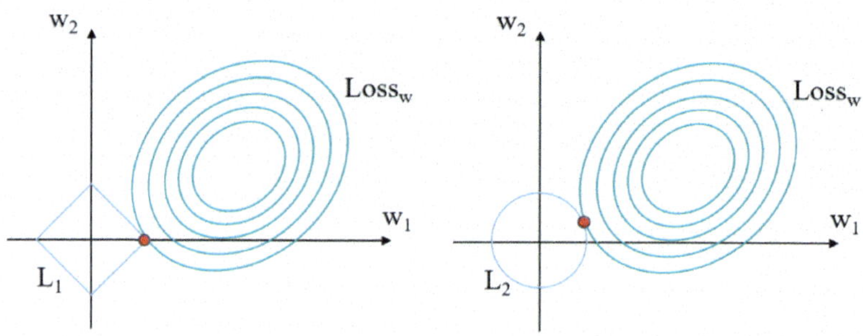

Fig. 6.10 Parameter regularization method

through backpropagation. This method significantly reduces the parameter search space. The weight updates no longer depend on the collective effect of fixed related hidden nodes, preventing certain features from only being effective under specific conditions. Each time different neurons are dropped, the resulting model is equivalent to an average of multiple simple models, enabling the network to learn more robust features.

Efficient Modules. As shown in Fig. 6.12, depthwise separable convolutions decompose the standard convolution operation into a depthwise convolution and a pointwise convolution. This method implicitly assumes that the spatial positions of data are highly correlated while different channels are independent. Depthwise separable

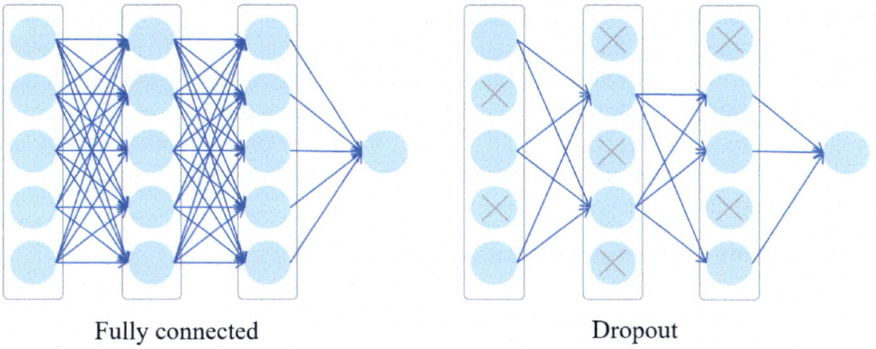

Fig. 6.11 Classic random drop method

convolutions significantly reduce parameters compared to ordinary convolutions, thus decreasing the size of the hypothesis space while maintaining network accuracy.

Fig. 6.12 Depthwise separable convolution

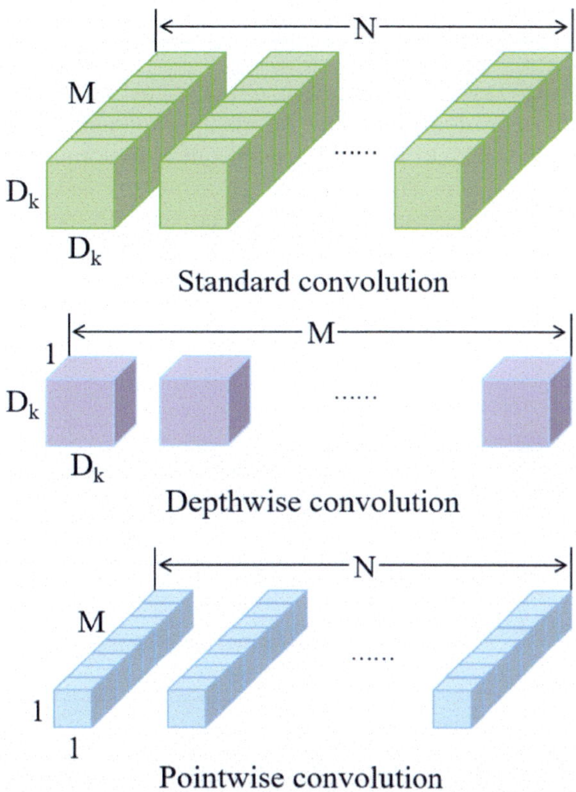

Image Normalization. Each pixel in an image can be considered a feature point, with varying ranges of fluctuation across different pixels. Normalizing the input image ensures that the feature scales of each pixel fall within a specific range, preventing significant differences in backpropagation errors among features, thereby compressing the hypothesis space and allowing the deep learning model to converge more easily to the optimal solution, implicitly serving a similar function to L2 parameter regularization.

Early Stopping. For many optimization algorithms used in training deep learning networks (e.g., gradient descent), the loss function is a decreasing function of the number of iterations. However, the error on an independent validation set usually decreases initially and then gradually increases due to model overfitting. As shown in Fig. 6.13, the basic idea of Early Stopping is to stop training when the model's performance on the validation set starts to decline (user-defined). This yields a model with good generalization performance. Common stopping rules include stopping when the generalization loss exceeds a threshold, stopping when the generalization loss increases for s consecutive cycles, and stopping based on both the generalization loss and its rate of change. From the perspective of restricting the solution space, this method confines the optimization process to the initial parameter's vicinity, thereby achieving an effect similar to L2 parameter regularization.

Multi-Task Learning. The primary goal of Multi-Task Learning (MTL) is to utilize the domain-specific information contained in the training signals of multiple related tasks to improve generalization performance. The learned model emphasizes common representations across multiple tasks. As shown in Fig. 6.14, common multi-task learning patterns in deep learning include hard parameter sharing and soft parameter sharing mechanisms. Shared representations in related tasks provide stronger inductive biases for the deep learning model regarding certain assumptions, thereby constraining the hypothesis space and enhancing the generalization ability of the model.

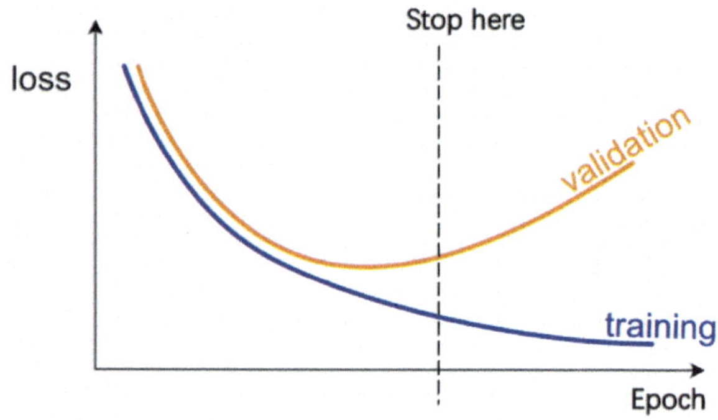

Fig. 6.13 Early stopping method

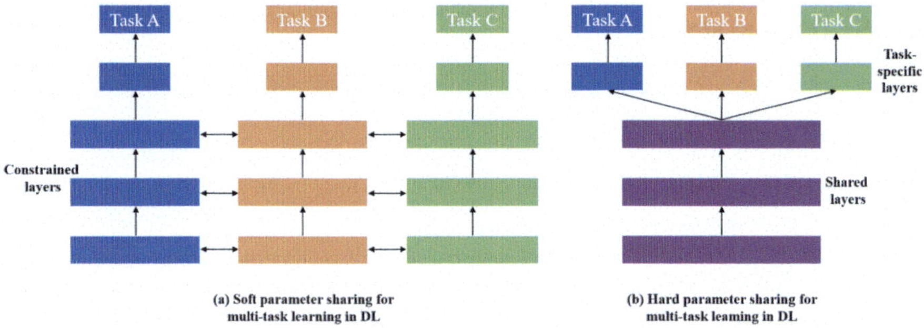

Fig. 6.14 Multi-task learning connotation

6.2 Casting Defect Detection Model Based on Convolutional Neural Networks

In traditional welding processes, flaw detection is typically performed on the processed area after welding is completed. However, this detection is often selective. Some areas are deemed less critical, and flaw detection results are usually judged manually to identify welding defects and their types. This process requires a high level of expertise from the inspector, consumes significant manpower and time, and is prone to errors, including omissions and misjudgments.

In industrial fields, especially in safety–critical areas like aerospace and autonomous driving, there are extremely high-quality requirements for products, making comprehensive flaw detection essential. To address these challenges, the author proposes a casting defect detection model based on convolutional neural networks (CNNs) to automate the interpretation of casting radiographic images. The overall framework of the proposed model is illustrated in Fig. 6.15.

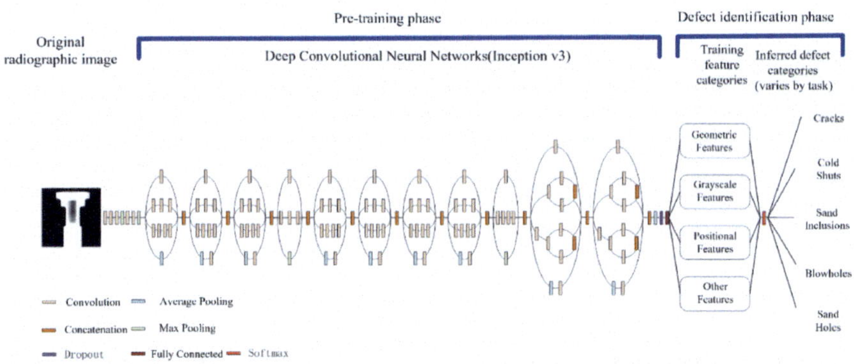

Fig. 6.15 Welding defect recognition algorithm framework

(1) **Pre-training Stage**

In traditional radiographic image defect identification methods based on machine learning, there are two main approaches to defect feature extraction. The first approach involves manually extracting features, pre-processing them through techniques like dimensionality reduction, and then feeding them into a neural network classifier. The major drawback of this approach is that it relies on manually defined features and prior knowledge, which limits the description of high-level image features and reduces identification accuracy and efficiency.

The second approach utilizes the strong feature extraction capabilities of CNNs to automatically obtain image features, which are then fed into a classifier for identification. While this method significantly improves identification efficiency compared to manual feature extraction, its accuracy is constrained by the network's width and depth.

To address these limitations, this paper employs a transfer learning method during the pre-training stage, using the GoogleNet InceptionV3 network to map features from the original radiographic images. By using different sizes of convolutional kernels, features of various scales are extracted and fused, then mapped to a high-dimensional space to create a more abstract feature set.

(2) **Welding Defect Identification Stage**

Historical datasets with manually identified data are divided, with some used for training and others for cross-testing to optimize the parameters of this model. The predicted classification results are compared with historical data to calculate the error value. The model with the smallest error value is considered the optimal model. Through this deep convolutional neural network model, the flaw detection results of laser welding products can be predicted, providing reference and support for manual decision-making. The key points of using deep convolutional neural networks for intelligent classification of welding defects include:

Initially classifying the feature information extracted during the network pre-training stage, which can be roughly divided into geometric features, grayscale features, positional features, and other high-dimensional abstract features. Using a SOFTMAX classifier to infer the feature information extracted by the convolutional network and expressing the inference results, i.e., defect types (cracks, cold shuts, sand inclusions, blowholes, sand holes), in probabilistic form. Assume there is an array V, and V_i represents the ith element in V, the SOFTMAX value of this element is:

$$S_i = \frac{e^{V_i}}{\sum_j e^{V_j}}$$

Compared with other methods, this deep convolutional neural network model requires less prior intervention, effectively avoiding the "cold start" problem and minimizing human factors' impact on the prediction results during the learning process. Additionally, for welding data with certain complexity and redundancy, deep convolutional neural networks help capture key samples and eliminate a large number of redundant samples. This method has a clear framework and good robustness.

The welding defect identification results can serve as a reference for manual judgment and can also be used for pre-evaluation of welding technology, providing effective data support for technological improvements. Furthermore, the results achieved by experts can be defined as business rules, stored in historical data and models, serving as important references for future analysis and application.

6.3 Fine-Grained Casting Defect Recognition Method Based on Attention Mechanism

Compared to standard object detection tasks, detecting defects in casting X-ray images demands greater precision. This task involves smaller detection targets, finer features, stronger fine-grained characteristics, and stricter credibility requirements. In computer vision, tasks involving small targets and fine-grained details often rely on techniques like multi-scale training, attention mechanisms, and data augmentation to enhance performance. Among these, attention mechanisms provide the most significant improvement, making them the primary method for achieving high credibility in fine-grained image tasks.

The attention mechanism mimics the human brain's signal processing in vision. Human vision scans an entire image quickly to identify target areas, suppressing irrelevant information and improving the efficiency and accuracy of visual processing. Attention models in image processing emulate this by continuously focusing on regions of interest, enhancing the extraction of effective information. In the context of image feature extraction, attention mechanisms help by assigning importance to different local areas. Mathematically, the attention mechanism operates as a weighted sum, often using activation functions like Softmax and Sigmoid. By introducing new parameters, it enhances the model's fitting ability. Commonly used attention mechanisms are categorized into spatial attention and channel attention.

(1) **Spatial Attention Mechanism**

The spatial attention module focuses more on the spatial positional information of discriminative parts, such as the image features of the area where the defect is located. The fusion process is illustrated in Fig. 6.16.

Take the convolutional feature map as the input, G, $\mathbf{G} \in \mathbf{R}^{w \times h \times t}$, where $w \times h$ is the size of spatial dimensions of G, t is the number of channels. Compress G along the channel axis to extract spatial attention information. A column of channel values is compressed into a single channel. This step is achieved through pooling along the channel dimension.

Using the same multi-scale pooling method: apply the max pooling function p_m and the average pooling function p_a to reduce the dimensions of G, obtaining two feature maps of size $w \times h \times 1$. Concatenate these two feature maps along the

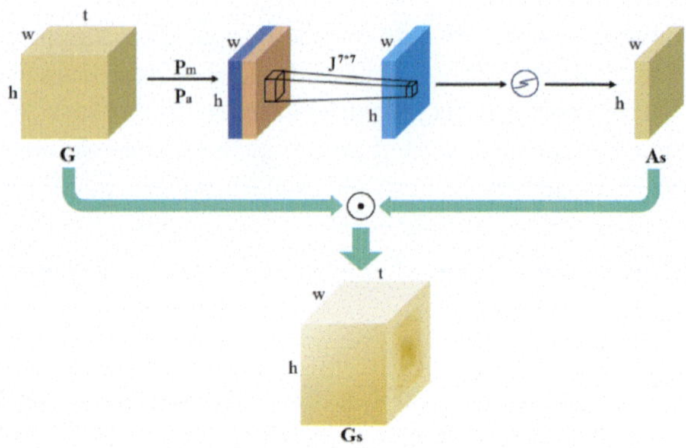

Fig. 6.16 The spatial attention mechanism

channel axis by summing corresponding elements, resulting in a new feature map of size $w \times h \times 2$.

Using a 7×7 convolutional kernel on the concatenated feature map to perform convolution, reducing its size to $w \times h \times 1$ once again. Apply the Sigmoid activation function to the convolved feature map to generate the spatial attention map A_s, $A_s \in \mathbf{R}^{n \times h \times 1}$.

Finally, perform feature fusion by element-wise multiplication of the spatial attention map A_s and the original feature map G to obtain the fused spatial attention feature map $\mathbf{G}_s$, $\mathbf{G}_s \in \mathbf{R}^{w \times h \times t}$, Replace the original input feature G with $\mathbf{G}_s$ to achieve spatial dimension attention extraction.

(2) **Channel Attention Mechanism**

In fine-grained image pattern recognition problems, each feature channel may represent different information in the image, with some channels containing irrelevant background information and redundancies. Therefore, concentrating attention on feature channels that contain discriminative spatial information and assigning them a higher weight distribution can effectively enhance fine-grained classification performance. The fusion process is illustrated in Fig. 6.17.

Taking the convolutional feature map as the original input F, with $\mathbf{F} \in \mathbf{R}^{w \times h \times t}$ where $w \times h$ represents the spatial dimension size of F and t represents the number of channels. In order to effectively extract channel attention, F is compressed along the spatial dimensions, with the features of the same channel compressed into a real number. This step can be achieved through pooling operations.

Adopt a multi-scale pooling approach: apply the max pooling function p_m and the average pooling function p_a to F to reduce its dimensions, resulting in two feature vectors of size $1 \times 1 \times t$. Feed both vectors into a shared network to obtain the attention weight distribution along the channel dimension. The shared network consists of a multi-layer perceptron with one hidden layer.

Perform element-wise summation on the two output vectors after reassigning the attention weights, and then map the merged feature vector using the Sigmoid activation function to generate the channel attention weights A_c, $A_c \in \mathbf{R}^{1 \times 1 \times t}$.

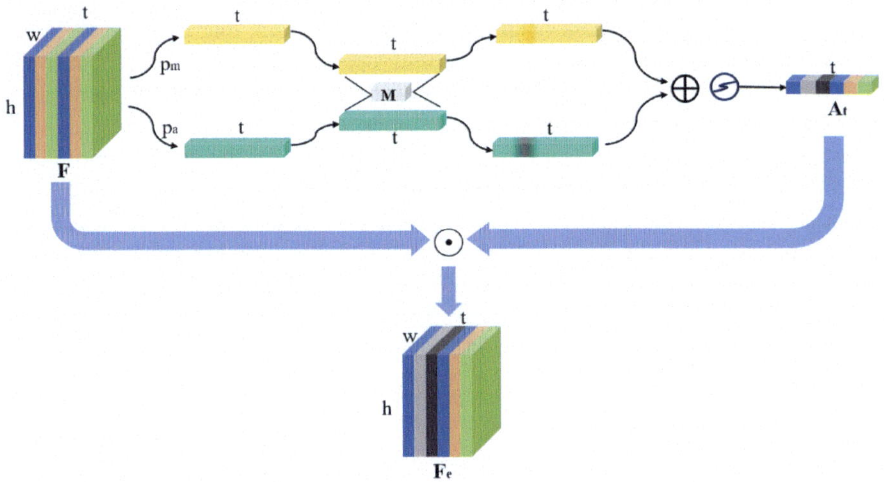

Fig. 6.17 The channel attention mechanism

Finally, integrate the attention weights A_c with the original feature map F using an element-wise multiplication method. This results in the fused attention feature map F_c, $F_c \in \mathbf{R}^{w \times h \times 1}$, achieving channel-wise attention extraction.

6.4 Case Study of Casting Radiographic Image Detection Using Deep Learning

This subsection demonstrates the practical application of machine vision technology by taking casting ray image inspection as an example. Firstly, it introduces the input data of the visual inspection model. Secondly, it introduces the training process of the casting image inspection model based on deep learning. Finally, it introduces the output data of the visual inspection model and the evaluation method.

6.4.1 Input Analysis

This project adopts a deep learning approach as the foundation model for casting defect detection. Since digital image reading is commonly used in deep learning for image processing, such as with.jpg,.png, and.bmp formats, the original X-ray image format (.dcm) is standardized to.jpg format to ensure that all collected casting X-ray images can be batch-read and processed.

During the training process of the deep learning model, due to the constraints of convolutional kernel operations and pooling operations, it is required that input images have uniform specifications. Images of different sizes cannot be simultaneously input into the deep learning model. In order to achieve batch loading of input endoscope images, it is necessary to ensure that the size of the endoscope images fed into the deep learning model is consistent.

This format includes the information necessary for any image: the casting X-ray images obtained from the X-ray detection system are converted to.jpg digital image format, with the pixel size standardized to 1024 * 1024 and a channel number of 1, resulting in a file size of approximately 130 KB. In addition, there should be an XML file corresponding to the image file name, containing the category and defect position attributes of the corresponding image, as shown in Fig. 6.18.

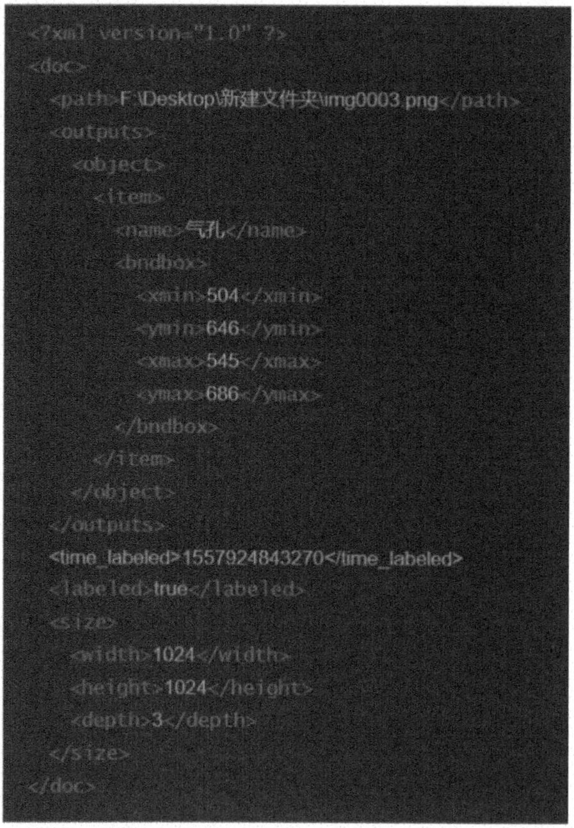


```
<?xml version="1.0" ?>
<doc>
  <path>F:\Desktop\新建文件夹\img0003.png</path>
  <outputs>
    <object>
      <item>
        <name>气孔</name>
        <bndbox>
          <xmin>504</xmin>
          <ymin>646</ymin>
          <xmax>545</xmax>
          <ymax>686</ymax>
        </bndbox>
      </item>
    </object>
  </outputs>
  <time_labeled>1557924843270</time_labeled>
  <labeled>true</labeled>
  <size>
    <width>1024</width>
    <height>1024</height>
    <depth>3</depth>
  </size>
</doc>
```


Fig. 6.18 Content of the XML file

6.4.2 Training Process Analysis

(1) **Training Methods**

To enable the model to have a good ability to recognize features and classify sample images, a sufficient number of sample images is essential. Additionally, it is necessary to use sample images with known labels for detection to better verify the model's performance and prediction accuracy. Given the limited annotated data for casting radiographic images, to prevent the deep learning model from overfitting during training, this project adopts a ten-fold cross-validation method: that is, using ten different training/validation set partitions to perform multiple sets of different training/validation on the model. This approach addresses the issues of overly one-sided individual test results and insufficient training data.

(2) **Training Content**

For the location recognition task of casting defects, the training content should include the following parts:

Unsupervised pre-training: Due to the lack of labeled data for casting defects, this paper uses transfer learning method to obtain an initial set of weights during the pre-training phase.

Supervised fine-tuning training: In this phase, the network architecture and initialization parameters from the previous step are used, and labeled data is added for network fine-tuning training.

Feature classifier training: Each proposed region in the to-be-detected image will receive a high-dimensional feature vector after the feature extraction step. This vector is placed into each classifier for binary classification, and the score for each proposed region is output.

Position regressor training: For each type of target, a linear regressor should be trained to refine the position. For windows, a four-dimensional vector (x, y, w, h) is usually used to represent, which respectively indicates the center coordinates and width and height of the window. As shown in Fig. 6.19, the red box P represents the original proposed box, and the green box G represents the original position box of the target. Our goal should be to find a relationship that maps the input original window P to a regression window G' that is closer to the real window G'. That is: given (Px, Py, Pw, Ph), seek a mapping f such that:

$$f\left(P_x, P_y, P_w, P_h\right) = \left(G'_x, G'_y, G'_w, G'_h\right) \approx \left(G_x, G_y, G_w, G_h\right)$$

(3) **Training Stopping Condition**

To ensure the performance of the deep learning model, it is necessary to ensure that the model has a certain number of training iterations. However, too many training iterations may lead to overfitting of the model. Choosing an appropriate number of training iterations not only helps control the training time but also better ensures

Fig. 6.19 Detection Box Position Regression

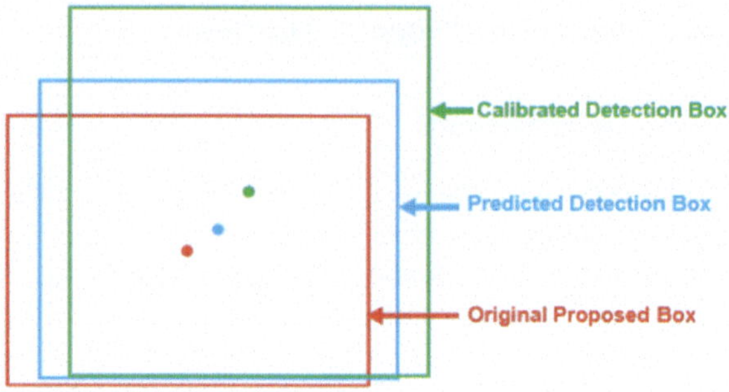

the performance of the model. Considering relevant factors, this paper adopts 1000 iterations for the training of the deep learning model.

During the training of the deep learning model, in order to continuously improve the performance of the model, it is necessary to evaluate the performance of the model. We refer to the method of evaluation as the loss value (loss), that is, the smaller the loss value of the model, the higher its performance. The commonly used loss function is cross-entropy loss, whose calculation formula is as follows:

$$\mathrm{L}(y, f(x)) = \log\left(1 + e^{-yf(x)}\right)$$

The loss value of the trained model continues to decrease with the increase in training iterations. Typically, the model is trained for 1000 iterations. However, if the sample data is relatively simple and does not require training for 1000 iterations, the model can be judged for further training based on the loss value. When the loss value of the model is below 0.05, the model performs well, and there is no need for further training. At this point, you can stop further training.

(4) **Training Time**

To ensure deep learning models are effectively applied in real-world detection environments, it is essential to manage their training time. Frequent retraining is often necessary due to changes in these environments. To maintain the practical value of these models, it is important to limit the training duration. Excessively long training times can reduce the models' practical value and their ability to meet real-world demands.

Given that the casting radiographic image detection system involves both image classification and specific location framing, the training time for the casting defect recognition model in this study is kept within 120 min.

(5) **Training Environment**

In real-world applications, diverse computational capabilities necessitate a heterogeneous training environment for casting defect recognition models. Currently, the primary factor affecting deep learning model training on standard computers is memory utilization. The high computational demands and large feature dimensions of deep learning models make the computer's memory performance crucial for training speed.

Given that modern computers are generally well-equipped, it is recommended that memory usage for training deep learning models should not exceed 45% of the total memory. This ensures both efficient training and the smooth operation of other programs. Running deep learning training alongside other applications can also enhance overall computer efficiency.

6.4.3 Output Analysis

(1) **Output Category**

According to the actual requirements of casting defect detection, the output of the deep learning model during the detection process is composed of three parts: defect location, defect category, and defect area.

Defect location includes four numerical values: the horizontal coordinate of the predicted box's center point, the vertical coordinate of the predicted box's center point, the length of the predicted box, and the width of the predicted box. Defect category includes two types: porosity and crack.

(2) **Output Method**

In the detection of casting defect types, the SOFTMAX function is applied to express the detection results in probabilities. The category with the highest probability is the final prediction. As shown in Fig. 6.20, the left side of the equal sign represents what the fully connected layer does. W is the parameter of the fully connected layer, and X is the input of the fully connected layer (also known as the feature). This feature is obtained by processing the input of the fully connected layer with multiple convolutional and pooling layers. Assuming the convolutional layer before the fully connected layer outputs 100 feature maps, each with a size of 4×4, before feeding these features into the fully connected layer, they are stretched into an $N \times 1$ vector (at this point, N is $100 \times 4 \times 4 = 1600$). W is a TxN matrix, where N corresponds to the N in X, and T represents the number of categories, which in this paper corresponds to two classifications: porosity and crack. Therefore, T is 2. The input to SOFTMAX is a Tx1 vector, and the output is also a Tx1 vector (i.e., the probability vector prob[Tx1] in the figure, where each value in this vector represents the probability of the sample belonging to each class). The calculation formula for SOFTMAX is shown in Eq. 1. It can be observed that each value in the output vector of SOFTMAX ranges from 0 to 1, and the sum of probabilities for all categories equals 1.

$$S_j = \frac{e^{a_j}}{\sum_{k=1}^{T} e^{a_k}}$$

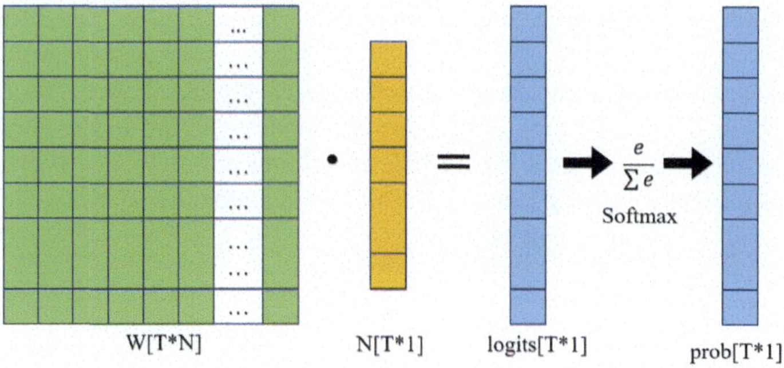

Fig. 6.20 SOFTMAX output format

6.4.4 *Model Evaluation*

(1) **Precision evaluation**

The quality of the casting defect detection model requires a series of quantitative indicators for evaluation. During model evaluation, metrics such as accuracy, sensitivity, and specificity will be used as evaluation indicators.

The casting defect detection classification module is a binary classification model, where the detection result can be one of two types: porosity or crack. When performing classification, it is essential to ensure both overall accuracy and the accuracy of each class.

Overall Accuracy: $P = \frac{\text{Number of Correct Judgments}}{\text{Total Number of Samples}}$, This represents the proportion of correctly predicted categories in the entire sample within the casting defect detection classification module. According to actual diagnostic needs and historical diagnostic statistics, this paper standardizes it to be above 95%.

Accuracy of Each Class: $P = 0\frac{\text{Total Number of Samples in This Class}}{\text{Number of Correctly Judged Samples in This Class}}$, This represents the proportion of correctly predicted samples for a specific category within the casting defect detection classification module. According to actual diagnostic needs and historical diagnostic statistics, this paper standardizes it as shown in Table 6.1.

At the same time, it is necessary to quantitatively evaluate the predicted position, as shown in Fig. 6.21: The IOU value is used as the evaluation indicator for the position, which represents the overlap rate between the generated proposal box and the original marked box. Its calculation formula is as follows:

$$\text{IOU} = \frac{G \cap G'}{G \cup G'}$$

Table 6.1 Requirements for casting defect identification

Defect Category	Recognition accuracy
Porosity	$\geq 95\%$
Cracks	$\geq 95\%$

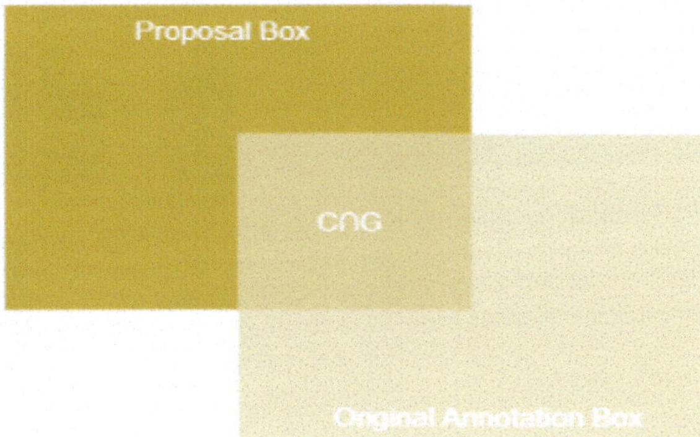

Fig. 6.21 IOU diagram

In the above formula, G represents the true position, and G' represents the predicted position.

According to the actual detection requirements, the IOU value is standardized to be above 90%.

(2) **Stability Assessment**

The process of establishing and applying a casting defect detection model is a continuous process of training and updating, during which the stability of the model is particularly important. Therefore, during the training process, the model needs to be trained multiple times. This article provides the variance of the results of 30 model training sessions as an indicator of the model's stability. The calculation process is as follows:

$$S = \frac{\sqrt{\sum X - \overline{X}}}{N}$$

According to the actual detection requirements, this article standardizes it within 0.05.

6.4.5 *Results Presentation*

The main functions of the system developed by the author include the following:

1. Reading the radiographic image of the drug pellet to be interpreted through the open button.
2. The confidence level is used to adjust the probability of an object being considered a defect.
3. The Intersection over Union (IoU) is used to adjust the overlap ratio between two defect detection boxes.

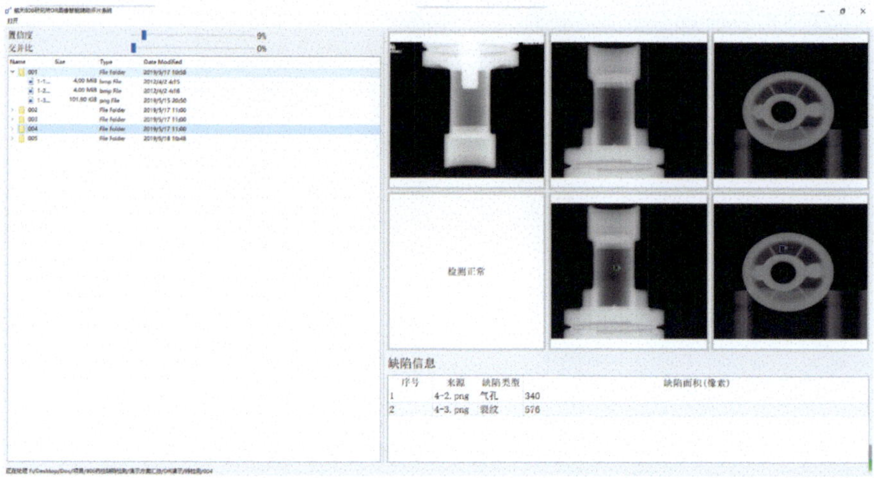

Fig. 6.22 Operational interface of the casting defect image detection system

4. The detection results area is used to display the defect categories and their locations.
5. The defect information area is used to count defects and display the defect area.

As shown in Fig. 6.22, the main function of the system is to select the radiographic image of the drug pellet to be interpreted on the left side of the interface. The detailed detection results of the drug pellet's radiographic image will then be displayed on the right side of the interface, and the statistical information such as the area and number of defects will be displayed at the bottom of the interface.

References

1. Simonyan K (2014) Very deep convolutional networks for large-scale image recognition. arXiv preprint arXiv:1409.1556
2. Szegedy C, Liu W, Jia Y et al (2015) Going deeper with convolutions. In: Proceedings of the IEEE conference on computer vision and pattern recognition, pp 1–9
3. He K, Zhang X, Ren S et al (2016) Deep residual learning for image recognition. In: Proceedings of the IEEE conference on computer vision and pattern recognition, pp 770–778
4. Huang G, Liu Z, Van Der Maaten L et al (2017) Densely connected convolutional networks. In: Proceedings of the IEEE conference on computer vision and pattern recognition, pp 4700–4708
5. Hu J, Shen L, Sun G (2018) Squeeze-and-excitation networks. In: Proceedings of the IEEE conference on computer vision and pattern recognition, pp 7132–7141
6. Zhang X, Zhou X, Lin M et al (2018) Shufflenet: An extremely efficient convolutional neural network for mobile devices. In: Proceedings of the IEEE conference on computer vision and pattern recognition, pp 6848–6856
7. Howard AG (2017) Mobilenets: efficient convolutional neural networks for mobile vision applications. arXiv preprint arXiv:1704.04861

8. Chen FC, Jahanshahi MR (2017) NB-CNN: Deep learning-based crack detection using convolutional neural network and Naïve Bayes data fusion. IEEE Trans Industr Electron 65(5):4392–4400
9. Park S, Bang S, Kim H et al (2019) Patch-based crack detection in black box images using convolutional neural networks. J Comput Civ Eng 33(3):04019017
10. Feng C, Liu M Y, Kao CC et al (2017) Deep active learning for civil infrastructure defect detection and classification. In: Computing in civil engineering, pp 298–306
11. Long J, Shelhamer E, Darrell T (2015) Fully convolutional networks for semantic segmentation. In: Proceedings of the IEEE conference on computer vision and pattern recognition, pp 3431–3440
12. He K, Gkioxari G, Dollár P et al (2017) Mask R-CNN[C]. In: Proceedings of the IEEE international conference on computer vision, pp 2961–2969
13. Singh J, Shekhar S (2018) Road damage detection and classification in smartphone captured images using mask R-CNN. arXiv preprint arXiv:1811.04535
14. Ferguson MK, Ronay AK, Lee YTT et al (2018) Detection and segmentation of manufacturing defects with convolutional neural networks and transfer learning. Smart Sustain Manuf Syst 2
15. Taheritanjani S, Schoenfeld R, Bruegge B (2019) Automatic damage detection of fasteners in overhaul processes. In: 2019 IEEE 15th international conference on automation science and engineering (CASE). IEEE, pp 1289–1295
16. Liong ST, Gan YS, Huang YC et al (2019) Automatic defect segmentation on leather with deep learning. arXiv preprint. arXiv:1903.12139
17. Racki D, Tomazevic D, Skocaj D (2018) A compact convolutional neural network for textured surface anomaly detection. In: 2018 IEEE winter conference on applications of computer vision (WACV). IEEE, pp 1331–1339
18. Mayr M, Hoffmann M, Maier A et al (2019) Weakly supervised segmentation of cracks on solar cells using normalized L p norm. In: 2019 IEEE International conference on image processing (ICIP). IEEE, pp 1885–1889
19. Tabernik D, Šela S, Skvarč J et al (2020) Segmentation-based deep-learning approach for surface-defect detection. J Intell Manuf 31(3):759–776
20. Silvén O, Niskanen M, Kauppinen H (2003) Wood inspection with non-supervised clustering. Mach Vis Appl 13:275–285
21. He Y, Song K, Meng Q et al (2019) An end-to-end steel surface defect detection approach via fusing multiple hierarchical features. IEEE Trans Instrum Meas 69(4):1493–1504
22. Ferguson M, Ak R, Lee YTT et al (2017) Automatic localization of casting defects with convolutional neural networks. In: 2017 IEEE international conference on big data (big data). IEEE, pp 1726–1735
23. Huang Y, Qiu C, Yuan K (2020) Surface defect saliency of magnetic tile. Vis Comput 36(1):85–96
24. Gan J, Li Q, Wang J et al (2017) A hierarchical extractor-based visual rail surface inspection system. IEEE Sens J 17(23):7935–7944
25. Yang X, Li H, Yu Y et al (2018) Automatic pixel-level crack detection and measurement using fully convolutional network. Comput Aided Civil Infrastruct Eng 33(12):1090–1109
26. Li LF, Ma WF, Li L et al (2019) Research on detection algorithm for bridge cracks based on deep learning. Acta Automatica Sinica 45(9):1727–1742
27. Bergmann P, Fauser M, Sattlegger D et al (2019) MVTec AD—a comprehensive real-world dataset for unsupervised anomaly detection. In: Proceedings of the IEEE/CVF conference on computer vision and pattern recognition, pp 9592–9600
28. Carvalho P, Durupt A, Grandvalet Y (2022) A survey of machine learning approaches for visual inspection on the DAGM dataset. In: Advances in manufacturing technology XXXV. IOS Press, pp 255–260

Chapter 7
Personnel Operation Behavior Detection and Monitoring

As the automation and intelligent transformation of the manufacturing industry advances, more and more automated equipment and industrial intelligent robots are participating in workshop production operations, gradually replacing some of the work done by workshop workers. They are playing an increasingly important role in modern production. Although the number of workshop workers has actually decreased to some extent, this does not mean that robots can significantly or completely replace the role of humans in workshop operations in the short term.

7.1 Progress and Current Status of Behavior Detection and Monitoring

Personnel operation behavior involves the application of computer vision, and this chapter introduces a method of personnel operation behavior based on personnel posture estimation. At present, most types of automated robots can only perform work according to pre-arranged instructions, such as palletizing robots and welding robots. As shown in Fig. 7.1, the work these robots engage in generally has the characteristics of large-scale, monotony, mechanization, and repetitiveness, which can serve as a substitute for personnel in high-intensity or hazardous and harsh work environments. There are also some intelligent robots that have acquired a certain degree of "perception" and "judgment" ability through the integration of multi-source sensors and embedded intelligent algorithms, such as intelligent sorting robots and cooperative assembly robots used in automated flexible production lines. However, the degree of intelligence and flexibility in their operation is still significantly different from that of humans. In some manufacturing scenarios with relatively complex processes and high precision requirements, they are still unable to complete high-quality work.

The assembly workshops for automotive products are currently one of the most highly automated manufacturing areas in the industrial sector. The classic automated

T. Liu et al., *Industrial Intelligence: Methods and Applications*, Springer Series in Advanced Manufacturing, https://doi.org/10.1007/978-3-031-81477-8_7

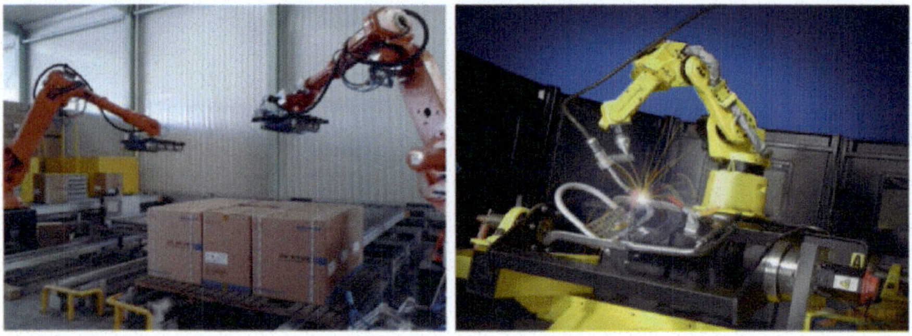

Fig. 7.1 Palletizing robot and welding robot operating in the workshop

production line model originated from the automotive manufacturing industry. The standardization and modularization of components, as well as the advanced assembly processes, have greatly promoted the level of automation in this field. However, even so, over 20% of the assembly work still needs to be completed manually [1]. In the short term, robots are still unable to match the quality level of human operations, and these tasks that can only be performed manually (such as the debugging of assembly modules, as shown in Fig. 7.2) are often the core process steps in the production process. For example, in the production workshops of pyrotechnic products, since most of the process operations require direct contact with gunpowder and other materials, the complex electrical equipment required for robotic operations poses a major safety hazard, resulting in the pyrotechnics industry still relying mainly on manual labor.

So even though the level of industrial intelligence has developed to a fairly high degree, humans are still the most advanced intelligent entities and still dominate the production operations in most workshops at the current stage. Especially in production lines or assembly lines where the process flow is discrete, the process

Fig. 7.2 Manual adjustment and calibration work in the automotive assembly workshop

flow is complex, the degree of refinement is high, the automation cost is high, or the process requirements are high, human labor is irreplaceable [2].

When robots perform work within their capability range, even in continuous high-intensity work situations, they can produce output of consistent quality. In comparison, human workers have unavoidable physiological defects, which can lead to problems such as distraction, physical and mental fatigue, and emotional fluctuations after prolonged work, making worker behavior more diverse and uncontrollable [3]. Although workers are required to strictly follow the established production behavioral norms, they may still fail to operate according to the standard due to the aforementioned subjective or objective factors, leading to problems with non-standard operation.

The importance of whether workers' behavior in production operations is standardized or not is evident: on the one hand, non-standard operation behavior can largely lead to a decline in product quality, or even non-conformity; on the other hand, in some high-risk operation workshops, such as the pyrotechnics production workshop mentioned above, non-standard worker movements are also potential safety hazards that may lead to the occurrence of high-risk accidents such as fire and explosion.

In summary, on the one hand, the role of manual operation is irreplaceable in the core production processes or operational steps in most industries; on the other hand, the inherent defects of humans may lead to irregular physical behaviors or procedural issues during the operation, thereby causing quality or safety problems. Given this, it is necessary to study the regularity of workers' operational behaviors in the production workshop, which is also an important research direction of production management science [4].

Against the background of the process/operational requirements in the workshop production, with the workers' operational behaviors as the research object, the analysis and judgment of their regularity can be divided into two types:

The first type is the regularity analysis of individual behaviors. That is, for a known operational behavior or an identified unknown behavior, its regularity is analyzed and judged based on the standard degree of the action or the degree of execution. This type of research can be used for teaching demonstrations of operational behaviors, Key Performance Indicator (KPI) analysis, etc.

The second type is the regularity analysis of continuous behaviors. In many production scenarios, due to the existence of production process specifications or operational procedures, there are also procedural requirements for the continuous behaviors of workers. Irregular operational procedures are also likely to lead to product quality problems, and in the production of explosives and chemicals, irregular procedures are also potential triggers for accidents. Therefore, the judgment of the regularity of this sequence-type task is equally important, and the requirements and practical characteristics of procedural regularity often make it strict, irregular, and redundant, posing certain difficulties in judgment, as follows:

Strictness: Workers need to strictly follow the order of the established process or operational procedures to perform various operational behaviors, and for the core process behaviors, there can be no omission or disorder.

Irregularity: The common rework phenomenon in production processes can lead to the complexity of the operational behavior process, resulting in the irregularity of partial behaviors. The existence of rework behaviors makes the operational process more flexible and changeable, making the definition of its regularity more complex.

Redundancy: Workers will also intersperse some transitional behaviors, such as walking, sitting down, stretching, etc., during the production operation, and these transitional behaviors cause the redundancy of the operational behavior process. In most cases, transitional behaviors are of little significance in the analysis of process regularity, but they cannot be completely ignored at times.

7.2 Overall Framework of Operation Behavior Detection and Monitoring Algorithms

To achieve personnel operation behavior detection and monitoring, the three main steps are human pose estimation, behavior detection, and operation safety monitoring.

7.2.1 Pose Estimation

Pose serves as high-level semantic information for human behavior, and plays a critical role in human behavior recognition. Behavior recognition methods based on human pose information often have higher accuracy compared to image-based behavior recognition methods. Currently, human skeleton information is generally obtained through three main approaches: wearable device capture, depth camera capture, and human pose estimation technology.

(1) **Capture Techniques Based on Wearable Devices**

The technology of capturing human skeleton information using wearable devices is also commonly referred to as motion capture technology. The basic idea is to install sensors or special markers on the key parts of the human body (usually at the joint locations) to track the movements, which can provide data such as joint coordinates, motion speed, angular velocity, and motion direction. This data can then be used to reconstruct the behavior, for applications like human–computer interaction, virtual reality, and movie special effects.

The motion capture systems commonly used can be categorized based on the sensor type, such as optical, mechanical, acoustic, and inertial navigation systems.

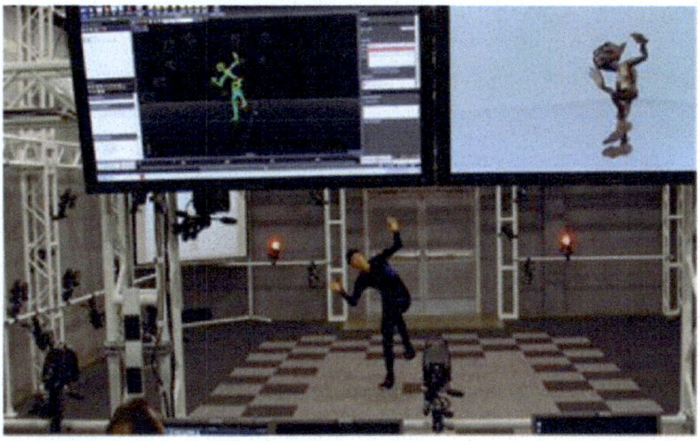

Fig. 7.3 Optical motion capture system capturing human poses and applying them to visual effects animation

Among them, the optical motion capture system uses stereo vision and marker tracking to identify the spatial motion trajectories of each key body node in a continuous image sequence (as shown in Fig. 7.3). This approach has high accuracy, good real-time feedback, and is currently one of the most mature and widely-applied solutions.

However, these human skeleton capture systems have two main drawbacks:

Firstly, they are only suitable for use in relatively ideal research settings or photography studios, and the wearing of the sensor devices imposes certain restrictions on the personnel's behavior and attire, making it difficult to apply in industrial workshop environments.

Secondly, the cost of these systems is relatively high, generally used by large tech companies or research institutions, and ordinary enterprises would find it difficult to afford, let alone deploy them for large-scale worker monitoring in workshops.

(2) **Capture Techniques Based on Depth Cameras**

Depth cameras can obtain human skeleton joint information without the need for wearable devices. The principle is: using distance measurement technologies such as infrared pulses, depth information (Depth Map) is added to the traditional optical camera's RGB image to obtain an RGB-D depth image. Algorithms are then used to aggregate the joint points and obtain the human skeleton.

Kinect is a popular depth camera product, with applications in many fields such as entertainment and education. Shotton et al. have detailed the Kinect's working principle: First, the human body contour is segmented out based on RGB information using edge detection algorithms. Then, body part segmentation is performed using the depth information from the RGB-D data, by calculating the distances between pixels to determine the distribution of body parts.

Next, each pixel is classified, with the original paper using a random forest classifier, assigning labels corresponding to 32 body parts. This transforms the human pose capture problem into a pixel classification problem in the RGB-D image. Finally, by grouping the pixels with the same body part labels, the corresponding joint coordinates can be obtained. Figure 7.4 llustrates this process.

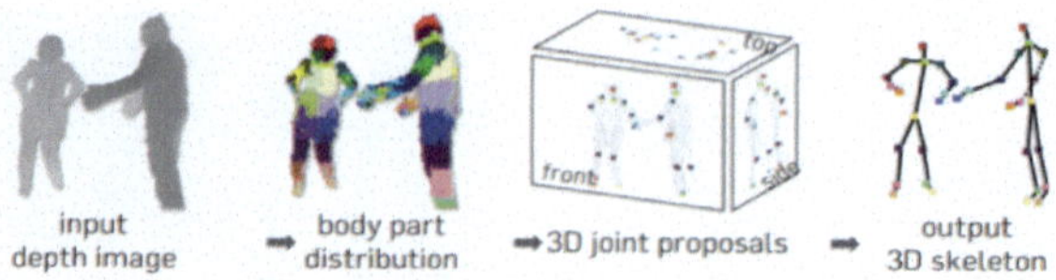

Fig. 7.4 Process of extracting human skeleton data using Kinect

Depth cameras also have their limitations. The capture range of Kinect is restricted, with a comfortable distance of 0.8–2.5 m, and a physical limit of no more than 3 m. If deploying depth cameras in an industrial workshop, they would need to be placed close to the workers, which may impact the workers' normal operation behavior, and even pose safety risks in high-hazard workshops (as depth cameras emit infrared light). Additionally, the workers' operational range may exceed the physical distance limits of the depth cameras, leading to missing information in the captured skeleton sequence.

(3) **Capture Techniques Based on Pose Estimation**

Pose estimation (PE) refers to the direct localization of human joint coordinates from regular images or videos, to form skeleton information. The effect is as shown in Fig. 7.5.

Traditional pose estimation algorithms are primarily based on graphical models, where the main idea is to first detect the appearance features of different body parts, then model the associative constraints between them, and finally construct an energy function for optimization to obtain the coordinates of the joint points.

The main problems with this approach are twofold: 1. It relies on handcrafted features, prior knowledge, and manual parameter tuning, without effectively utilizing the global information in the image, leading to limited generalization capability; 2. For relatively complex human poses, its solving capability and expressiveness are severely limited, but human poses have high degrees of freedom, so if the pose modeling cannot be done accurately, the practical value will be greatly discounted.

Fig. 7.5 Illustration of pose estimation technology

Due to the limitations of traditional pose estimation algorithms, which are difficult to overcome, scholars have started to try to transfer deep learning methods to this task, aiming to extract better explicit or implicit features from the image through the self-feature extraction capability of deep learning, avoiding the process of manual feature selection and parameter tuning.

With the emergence of convolutional neural networks (CNNs), there have been major breakthroughs in pose estimation technology: Toshev et al. proposed the DeepPose model, which first used CNNs for joint coordinate regression and performed multi-stage refinement, surpassing all traditional methods and achieving optimal performance at the time. Subsequently, Tompson et al. proposed using a regression heatmap localization method instead of direct joint regression, where the heatmap locates the joints indirectly through probabilities, avoiding the large number of parameters in the fully connected layers while also making the training process faster and more accurate. Most subsequent research has also been based on the heatmap regression approach. In 2016, Newell et al. proposed the Stacked Hourglass network, which interleaves convolutional and deconvolutional layers, allowing the model to better integrate global and local information, further improving the prediction accuracy. Overall, deep learning-based pose estimation techniques currently far outperform traditional algorithms. pose estimation technology can directly extract skeletal information from video images, without the need for wearable devices and avoiding the distance limitations of depth cameras. In the context of this research, production workshops generally already have video surveillance systems, so using pose estimation technology to obtain worker pose skeletons does not require additional hardware, saving costs and ensuring that there are no new uncertain factors in the workshop due to monitoring task upgrades. Therefore, in this research, a pose estimation-based solution will be used to obtain and model worker skeleton information. The team at Carnegie Mellon University has combined multiple pose estimation technology research to develop the OpenPose open-source framework, which supports online multi-person skeleton joint extraction and tracking. Therefore, in this research, OpenPose will be used to process and obtain worker pose skeleton information (see Sect. 7.3.1 for details).

7.2.2 Behavior Detection

Behavior detection is another important step, as it is necessary to detect the various work behaviors that workers execute in sequence during the production process/procedures in order to determine the standardization of the work process. However, due to the continuity of the work process, this problem is different from a simple classification problem of whether a behavior is abnormal or not. Instead, it requires high-precision recognition of the behavior at every moment, and then serializing and consolidating them to obtain accurate information about the work behavior flow.

(1) **Semantic Recognition**

Semantic recognition of human behaviors has a wide range of applications in the real world, such as human–computer interaction, education, healthcare, and the topic of this study—intelligent surveillance. It is for this reason that it has been a hot research area in computer vision for decades [5–7]. Currently, depending on the input features, behavior semantic recognition methods can be divided into RGB-based methods and skeleton-based methods.

1. **Methods Based on RGB Information**

Since the emergence of cameras and video recording devices, RGB-based behavior semantic recognition methods have been emerging one after another. These traditional methods mostly inherit two types of ideas.

Methods based on spatio-temporal features, which extract the spatial position information of human posture and the angular change information in the temporal dimension to form spatio-temporal feature descriptions, and then perform classification and recognition based on them.

Methods based on motion trajectory features, which directly extract motion trajectory features from the video stream or even the compressed video, form trajectory flows, and use them for behavior semantic recognition. However, human behavior recognition based on RGB information has many defects, such as great difficulties in complex background scenes, self-occlusion or mutual occlusion, changes in lighting conditions and scale, which limit the recognition performance of RGB-based methods.

Later, with the rapid development of sensor technology, the application of depth cameras gradually became popular, which can provide RGB-D depth images. The ordinary RGB information can be seen as a projection of the three-dimensional space onto the two-dimensional plane, while RGB-D contains depth information, providing the spatial geometric information of the object in a 2.5D form.

The traditional machine vision algorithm approach is to extract high-dimensional local features from the sequence of RGB-D images, thereby obtaining dense or sparse interest points in each frame, and then aligning the features in the time domain to form a visual bag-of-words (BoW), and finally training and predicting through a classifier.

In 2013, Shuiwang Ji et al. first proposed the use of 3D convolution operations for behavior recognition, which, although computationally expensive, introduced deep learning to this field. In 2014, Karpathy et al. proposed a method of 2D convolution pre-training and fusion of continuous temporal frames, which achieved certain effects. Based on this, Simoyan and Zisserman creatively proposed a spatial–temporal dual-stream convolutional network framework, which includes two independent networks for extracting spatial context information and temporal context information, and finally fuses optical flow for classification prediction. Subsequently, there have been many improved studies based on this, such as Feichtenhofer et al. proposing an early fusion convolutional architecture, Yao et al. proposing a 3D-CNN

+ LSTM basic architecture and first introducing an attention mechanism, and Zhu et al. experimenting with various network models targeting runtime speed, proposing to add an unsupervised MotionNet for optical flow preprocessing before the temporal module.

It can be said that in the field of behavior recognition based on video streams, the deep learning schemes that integrate the temporal and spatial features of RGB-D images have high accuracy and strong generalization, representing the current state-of-the-art. However, there are still defects that cannot be avoided in the short term:

High training cost: Although the RGB-D feature maps reduce the input image size, the number of parameters is still huge. Training a 3D-CNN model on the UCF101 dataset takes 3–4 days, making it difficult to optimize the model framework and easily leading to overfitting.

Lack of real-time performance: Similarly, due to the massive scale of the input image vectors and the huge number of network model weight parameters, even after the network training is completed, the online inference or prediction process on general computing hardware is still prone to stuttering and delay, making it difficult to be deployed in actual industrial production.

2. **Methods Based on Skeleton Information**

Some biological research has found that the description of human body actions can be based solely on the dynamic changes of a few key body nodes, and the combination and deformation of these key nodes forms the skeletal information, which facilitates the rapid and effective recognition of behaviors, as shown in Fig. 7.6, the skeletal behavior description of a golf action.

For general human body action movements, capturing just 10–20 key nodes can accurately represent the static posture at any given moment. By tracking the continuous static posture information, we can form the temporal dynamic behavioral characteristics and accurately describe the action, and it is not easily affected by changes in scale, lighting, etc., with stronger robustness in expression.

Compared to RGB, the skeleton information is merely a collection of joint coordinate sets, with a much smaller data dimension, greatly reducing the computational complexity of behavior recognition, which is a high-level representation of human behavior. Compared to RGB-D data, behavior recognition based on skeleton data

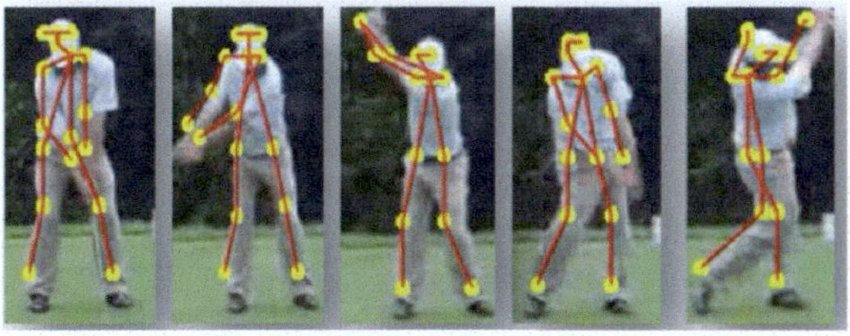

Fig. 7.6 Behavior expressed using skeleton information

may sacrifice some spatial context information in the depth map, but it still has unique advantages in many aspects:

1. Significant reduction in data scale, concise feature description, reduced computational complexity, higher processing speed, and closer to real-time requirements, which are its biggest advantages;
2. The skeletal sequence information can more accurately reflect the dynamic changes of the human limb geometry over time, with better recognition effects for specific actions, and is less affected by lighting and scene changes;
3. The refined features of skeletal data are suitable for further behavior evaluation research, and the evaluation system based on joint geometric feature information (such as position, velocity, trajectory) is more closely aligned with the evaluation methods of human visual perception.

The semantic recognition methods of behavior based on skeletal features have also received much attention from many scholars. These methods can be divided into two types according to the feature extraction approach:

1. Methods based on manually extracted features, where several feature values or feature vectors are manually designed to capture the dynamic behavior of the joints. For example, Hussein et al. used the covariance matrix of the time-varying joint positions as the sequence recognition descriptor, and deployed a hierarchical structure of multiple covariance sub-sequences; Wang proposed the concept of "Actionlet" small actions to address the intra-class differences between behaviors, and established a behavior classification model based on local all information.
2. Deep learning-based methods, which can be further divided into RNN-based and CNN-based approaches. Recurrent neural networks (RNNs) and their variants are widely used for feature mining of temporal information, and can effectively learn contextual relationships. However, they often over-emphasize temporal information and suffer from severe overfitting when training data is insufficient. Convolutional neural networks (CNNs) can also perform contextual information mining, and have even outperformed RNN models in recognition tasks. Researchers have explored various CNN-based input formats and network architectures to further optimize the performance of behavior semantic recognition.

(2) **Human Behavior Detection**

Behavior detection is also a major research hot spot in the field of video understanding, as this task is more closely related to real-life applications, and has immense potential value in surveillance and security. However, compared to behavior classification, behavior detection is more challenging. Not only does it require locating the video segments that may contain behavioral actions, but it also needs to classify them. Locating the video segments containing behavioral actions is an even more arduous task.

The behavior detection task is somewhat similar to the object detection task, as both require first localizing the target, then recognizing it. The existing behavior detection recognition methods can be mainly divided into two categories:

1. **Sliding Window Algorithm**

By defining an explicit fixed-size temporal window, the sequence frame information of the overall process can be traversed, and recognition can be performed based on the information (RGB, skeleton) within the window. However, the definition of the window size is extremely challenging, and the computational redundancy is very serious.

2. **Proposal of Behavior Candidate Regions**

Many current behavior detection methods are inspired by object detection techniques. The main idea is typically Temporal Proposal extraction (as shown in the image), followed by classification and regression operations. This category of methods includes those based on the Faster R-CNN framework, the SSD framework, and the TAG network.

Another type of approach is based on C3D for frame-level classification (Frame Label), then predicting the video segments where the behavior occurs and classifying them, such as the CDC network from ICCV 2017.

Beyond these two broad approaches, leveraging the advantages of RNN (LSTM) in sequential modeling, behavior classification at the frame level can also be used to effectively accomplish the combined detection and recognition task.

Given the characteristics of worker behavior, this study is based on RNN to extract behavior features, and uses Softmax for single-frame behavior recognition, with further sequence-level processing of the multi-frame recognition results. The detailed content of this part is introduced in Sects. 7.3.2 and 7.3.3.

7.2.3 Operation Safety Monitoring

For the safety monitoring of operations, it is known that in the given process/procedure context, the workflow of operational behaviors is generally serialized by multiple relatively fixed operational behaviors in chronological order, which simultaneously have the characteristics of rigorousness, irregularity, and redundancy. Therefore, judging their normativity is not simply a sequence matching problem. The analysis and judgment of such complex sequential information relies on human experience, and intelligent judgment methods need to learn the deep-seated mechanism knowledge contained in the experience, and extract strongly discriminative features oriented to process normativity.

To address this issue, we use Gated Recurrent Unit (GRU) for the basic encoding of the operational behavior process. Furthermore, the normativity judgment of the operational behavior process has similar characteristics: the importance of each type of production behavior executed by workers in a specific process may differ, and the importance of the same type of operational behavior that appears multiple times may also be inconsistent. This study proposes a Normativity-Aware Network (NAN) to learn and extract the normativity semantic weights of each operational behavior in the operational behavior process, and obtain high-level features based on this for efficient normativity judgment of the operational behavior process. Additionally, considering that the attention vector v is a high-level representation of the entire operational behavior process in the normativity discrimination task, it can be used as a strongly discriminative feature for normativity judgment and can be used for high-precision normativity judgment. In this study, we use the logistic regression algorithm, taking the attention vector as the input feature, to perform the classification of normativity, i.e. normative or non-normative. The detailed content of this part will be introduced in Sects. 7.3.4–7.3.6.

7.3 Operation Behavior Detection and Monitoring Methods Based on Deep Learning

Deep learning-based job behavior detection and monitoring contains the following elements: skeleton extraction and modeling based on OpenPose, job behavior feature extraction based on two-stream LSTM fusion network, job behavior detection and identification, job behavior process sequence coding based on GRU, job behavior normative weight mining based on normality-aware network, and normative judgment based on logistic regression. This section will introduce them one by one next.

7.3.1 Skeleton Extraction and Modeling Based on OpenPose

(1) Principles of Skeleton Extraction Using OpenPose

OpenPose [8] is a typical bottom-up type pose estimation algorithm, meaning it first detects the coordinates of all the key points for every person in the image. Then, it uses an assignment algorithm to correctly associate each key point with the appropriate person.

The OpenPose network first uses a backbone network (the original paper used VGG-19) to extract features from the input image. It then splits the extracted feature maps into two parallel branches. One branch is used to predict the locations, types, and confidence scores of all the body joints in the image, outputting a set of Part Confidence Maps (PCMs), with each heatmap containing the information for one type of joint. The second branch predicts the Part Affinity Fields (PAFs), which encode the degree of association between different body joints. The algorithm treats each body limb as a two-dimensional vector field, with the vector representing the x and y direction from one joint to its adjacent joint, and the vector values being zero for pixels not on the limb. This vector field is also learned by the deep neural network.

With the joint location set and the association degree between joints obtained from the two branches, the next step is the matching problem. The authors designed a minimum-spanning-tree-based approach constrained by bipartite graph and the Hungarian algorithm. During the joint association, the results from the PCMs and PAFs are used to perform pair-wise matching, which allows the joints to be assigned to the correct people and the final pose skeletons to be constructed by OpenPose [9].

(2) **Worker Skeleton Feature Modeling**

A dynamic operational behavior of a worker is composed of a sequence of static postures arranged in a temporal order. In this work, the static skeleton information of the worker in each frame is defined as the posture feature vector P.

The human posture features extracted using different methods and skeleton formats can vary. For example, the Kinect depth sensor can capture the coordinates of 25 body joints [10]. In this study, the MS-COCO data annotation format is adopted, extracting 18 body joints.

Using OpenPose, the pixel position information $(\tilde{x}, \tilde{y})$ of each body joint in every frame of the image is extracted. The origin is the point in the top-left corner of the image. Due to the fact that different video frames may have different canvas sizes, this can affect the judgment of joint positions. Additionally, the original position information has relatively large numerical values, which could lead to issues such as numerical overflow during training.

To address these two points, this study proposes using relative position instead of the original position. Specifically, for an image with size w × h, the following transformation is performed:

$$\begin{cases} x = \tilde{x}/w \\ y = (h - \tilde{y})/h \end{cases}$$

After the transformation, the origin is now at the bottom-left corner, and the resulting normalized coordinate values (x, y) are between 0 and 1. This normalization is independent of the image size, and depends only on the spatial position of the joint locations. Experiments have shown that this method is quite effective.

In this case, the worker's pose feature vector P can be represented as:

$$P = \{(x_1, y_1), (x_2, y_2), \ldots, (x_{18}, y_{18})\}$$

where (xi, yi) represents the coordinates of the i-th joint. We can define the worker's pose feature sequence S within the overall production behavior process as:

$$S = [P_1, P_2, P_3, \ldots, P_t]$$

where t represents the total number of time frames in the overall production behavior process.

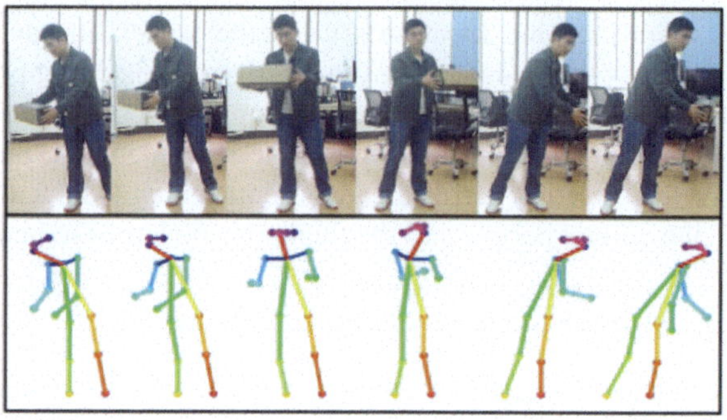

Fig. 7.7 Skeleton pose feature sequence for a material handling behavior [11]

The image in Fig. 7.7 illustrates an example of the pose feature sequence representing the worker's skeletal information during a material handling task. This demonstrates the capability to fully express the work behavior and the worker's motion actions within the workflow process.

7.3.2 Operation Behavior Feature Extraction Based on Dual-Stream LSTM Fusion Network

After obtaining the complete and accurate skeletal pose sequence of the worker during the production process in the previous step, the next step is to extract the behavioral feature characteristics from this data to enable the final detection and recognition. Depending on the feature extraction approach and method, the main categories are as follows.

(1) **Sliding Window Algorithm**

By defining an explicit fixed-size temporal window, the entire sequence of frames can be traversed, with the information within each window treated as a whole for the purpose of recognition. This method has two issues:

1. Difficulty in defining the window size: Different behavioral actions have varying durations, so using a fixed window size may fail to encompass the full duration of certain actions, or may include extraneous information beyond the action, leading to feature deficiency or redundancy.
2. Computational redundancy: Since the window moves by only one frame at a time, much of the computation is repeated, resulting in unnecessary overhead.

(2) **Region Proposal**

An approach is to directly extract semantically meaningful segments from the overall sequence, and then perform recognition on the behaviors within those segments. Heilbron used a sparse dictionary for behavior segmentation, while Zhou et al. proposed

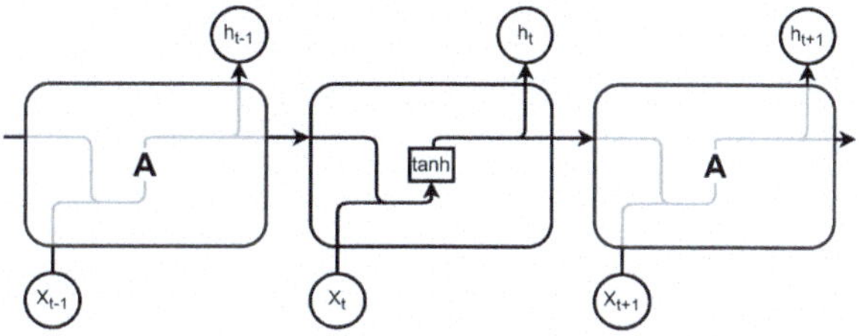

Fig. 7.8 Structure of the RNN model

a multi-stage convolutional method [12]. This approach fully leverages the global temporal context information for feature extraction. However, the quality of the candidate segment extraction cannot be reliably guaranteed, and the temporal boundaries of the behavioral actions are often not clearly defined.

(3) **Recurrent Neural Network (RNN)**

In addition to the two approaches mentioned above, the RNN utilizes a structure similar to traditional neural networks, with an input layer, hidden layers, and an output layer. The distinguishing feature is that the hidden layer state information is influenced not only by the current input, but also by the previous hidden state information. The structure of an RNN is shown in Fig. 7.8. When modeling sequential information, this structure can effectively leverage temporal context relationships, organically combining current and historical information. Through frame-wise behavior classification, it can also synchronously accomplish detection and recognition tasks with high quality [13].

However, RNN networks can suffer from issues such as gradient vanishing during the backpropagation process. The variant of the RNN unit structure, the Long Short-Term Memory (LSTM) network, has significantly improved upon this problem. LSTM introduces the concept of gates to the basic RNN structure, which helps avoid the gradient vanishing issue.

Building upon this LSTM structure, this paper proposes a novel Two-LSTM Fusion Network (TFN) model. This model not only effectively encodes the current input pose information, but also efficiently learns the long-term and short-term sequence relationships, including "semi-predictable sequences". The model performs online information fusion to generate frame-level feature outputs, which can then support the final detection and recognition module.

1. **Long Short-Term Memory Model**

Long Short-Term Memory (LSTM) is a type of recurrent neural network proposed by Sepp Hochreiter and Jurgen Schmidhuber. It is well-suited for handling sequence problems with time dependencies. Through its carefully designed three-gate structure, LSTM can selectively forget, input, and output feature tensors. It can filter and integrate blank information, similar information, and unnecessary information in the sequence, thereby storing effective feature information in the cell state for long-term storage [14]. The cell structure of LSTM is shown in Fig. 7.9.

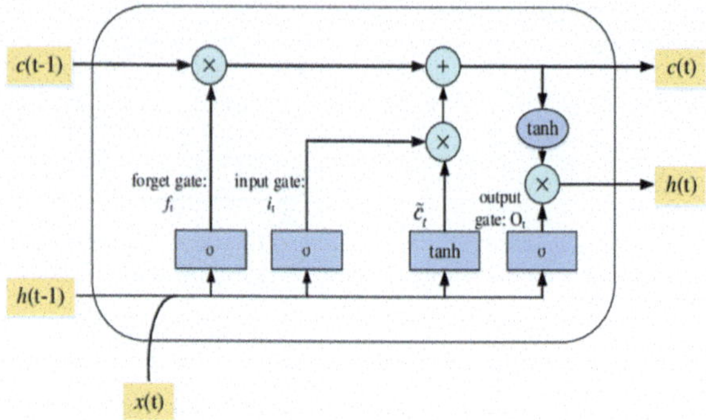

Fig. 7.9 Structure of the LSTM unit

The meaning of Long Short-Term Memory (LSTM) refers to: ct corresponds to long-term memory, and c̃_t corresponds to short-term memory. The forget gate ft determines how much of the historical information can be retained in the long-term state ct; c̃_t describes the short-term state of the current input; the input gate it determines how much of the current network input information can be added to the long-term state ct; the output gate controls how much of the summarized information can be used as the current output.

In summary: the input gate is responsible for introducing important information from the current input; the forget gate is responsible for retaining important information in the historical state; the output gate summarizes the two and determines the current output and the next time step and historical state. Their computational principles are as follows:

$$
\begin{aligned}
i_t &= \sigma\left(W_i \cdot \left[h_{t-1}, x_t\right] + b_i\right) \\
f_t &= \sigma\left(W_f \cdot \left[h_{i-1}, x_t\right] + b_f\right) \\
o_t &= \sigma\left(W_o \cdot \left[h_{t-1}, x_t\right] + b_o\right)
\end{aligned}
$$

The computational principles of the short-term memory unit and the long-term memory unit are as follows:

$$
\begin{aligned}
\tilde{c}_t &= \tanh\left(W_c \cdot \left[h_{t-1}, x_t\right] + b_c\right) \\
c_t &= f_t \circ c_{t-1} + i_t \circ \tilde{c}_t
\end{aligned}
$$

The prediction result from the hidden layer is:

$$
h_t = o_t \circ \tanh(c_t)
$$

The above formula represents the key stages in the process of information propagation. The "•" symbol denotes matrix multiplication, the "o" symbol represents element-wise multiplication, xt is the information input at time step t, ht is the hidden layer information from the previous time step, W and b are the weight matrix parameters that are continuously updated during training, and σ represents the sigmoid function.

At each time step, the LSTM model makes a prediction based on the current input and historical information, and the final prediction result is taken as the sequence of outputs. During the training phase, the network uses the backpropagation through time (BPTT) algorithm to iteratively update the weights and biases until the specified number of iterations is reached.

2. **Dual-Stream LSTM Fusion Network**

Theoretically, LSTM networks are well-suited for feature extraction of sequential information, but they have certain limitations when applied to the task of behavior detection and recognition in this study.

When processing sequential information, LSTM requires two key aspects of construction: one is the parsing of the input vector at the current time step (capturing the current information), and the other is the exploration of the long-term and short-term temporal associations within the sequence (selecting the historical memory information). Roeland's research has found that the performance of conventional LSTM is not always ideal: for tasks with relatively simple and low-dimensional input features (such as temperature prediction), LSTM can reasonably handle both aspects, as the input at each time step requires fewer neural computational resources for parsing, and the unit structure can capture both short-term and long-term temporal relationships. However, for more complex and high-dimensional input representations like human skeleton information (36-dimensional input vector with 18 skeleton joints in this case), the parsing of the input information becomes the "main focus" of the LSTM unit, while the exploration of the associations within the sequence is relatively "neglected" [15].

It is known that in a given production/process workflow, the worker's operational behavior follows a certain "semi-predictable order" to some extent, and the execution of various operational behaviors has a certain interdependence, which is a relatively unique sequential association. This special contextual relationship between operational behaviors is valuable information, so in addition to the conventional long-term and short-term dependencies in the sequence, the exploration of the "semi-predictable order" is also important during feature extraction.

In summary, when processing sequential information, it is generally necessary to perform feature modeling and extraction on both the input vector and the sequence associations. However, in this task, the online parsing of high-dimensional skeleton information and the exploration of the "semi-predictable order" within the sequence are difficult to be addressed simultaneously in a conventional LSTM network. If relying solely on LSTM for feature extraction of operational behaviors, the performance may be limited. To solve this problem, this research designed a Two-LSTM Fusion Network (TFN), where two LSTM layers are used to separately handle the input parsing and sequence relationship exploration, and online information fusion is performed to obtain frame-level output features that contain high-quality current and historical information [16, 17].

After obtaining the skeleton feature sequence through pose encoding, the sequence is first sent to the first (lower) LSTM layer in the temporal order, which is responsible for parsing the input skeleton pose information at each time step. The input to the second (upper) LSTM layer at each time step is the fusion vector obtained by fusing the hidden output of the first and second layers, which is used to learn the long-term and short-term temporal associations within the sequence. The hidden outputs of the two LSTM layers are online fused and output as the feature output of the current time step, which is then used for subsequent operational behavior recognition.

In Fig. 7.10, hi,t represents the t-th hidden state output of the i-th LSTM layer, h0 is an initial value, and C represents the information fusion unit. The operational behavior features are obtained by importing the corresponding hidden state outputs of the two LSTM layers into the information fusion unit C and calculating them. The fusion is performed through a weighted summation, as shown in the following formula:

$$C_t = W_1 h_{1,t} + W_2 h_{2,t}$$

The variables W_1 and W_2 represent the merged weight parameters of the output states from the respective LSTM layers, which are learned simultaneously during the training process. C_t denotes the merged result at the t-th time step.

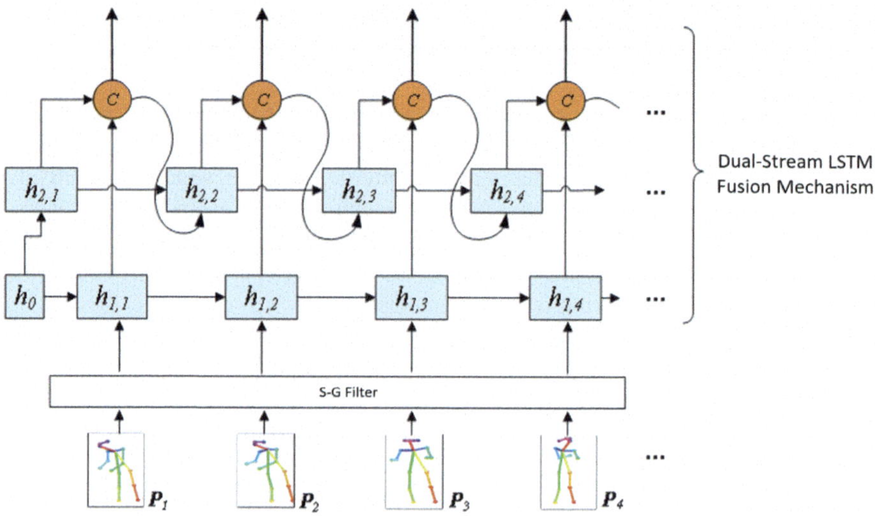

Fig. 7.10 Architecture of the two-stream LSTM fusion network

3. **Feature Extraction Pattern of Dual-Stream LSTM Fusion Mechanism**

The modified pose sequence S′ is fed into the first LSTM layer, where each pose vector is input sequentially according to the time steps. This LSTM layer maps the worker's pose information to a different feature dimension, performing an interpretable representation (determined by the number of neurons and the activation function within the LSTM unit). This process also enables a certain degree of short-term temporal correlation learning.

The second LSTM layer is responsible for exploring the long-term and short-term associations within the overall pose information sequence, including the "semi-predictability" aspect. The input to this layer at each time step is the high-level feature extracted from the previous time step's work behavior feature extraction. This input information has a lower dimensionality and higher interpretability, making it well-suited for modeling the associative information in the sequence context.

In this model, the fusion unit's weighted summation operation allows it to "autonomously" allocate the computation weight between the current input features (including a certain degree of short-term temporal relationships) and the long-term and short-term sequence relationships, including the "semi-predictability" aspect. If the input features are complex and the sequence-internal associations are simple, the data flow from the second LSTM layer will be relatively "ignored," and the network learning will focus on the input pose information analysis. Conversely, if the input features are simple and the sequence association features are relatively complex, the data flow from the first LSTM layer will be "ignored" to a certain extent, and the network will emphasize the exploration of the complex dependent relationships within the sequence [18].

Through this mechanism, the model can self-select during the training process to focus more on either the input information modeling or the feature extraction of the sequence-related relationships. This organic fusion of the state information from the two LSTM layers allows the model to fully explore the importance and interdependence of these two types of features during the training process, resulting in more robust and representative work behavior features.

7.3.3 Operation Behavior Detection and Recognition

(1) **Single-Frame Behavior Recognition Based on Softmax**

The Softmax function, also known as the normalized exponential function, can provide classification results in a probabilistic manner. It is a generalization of the commonly used sigmoid activation function for binary classification tasks to the multi-classification domain. The mathematical expression for the Softmax function is:

$$p(y \mid x) = \frac{\exp(W_y, x)}{\sum_{n=1}^{N} \exp(W_n \cdot x)}$$

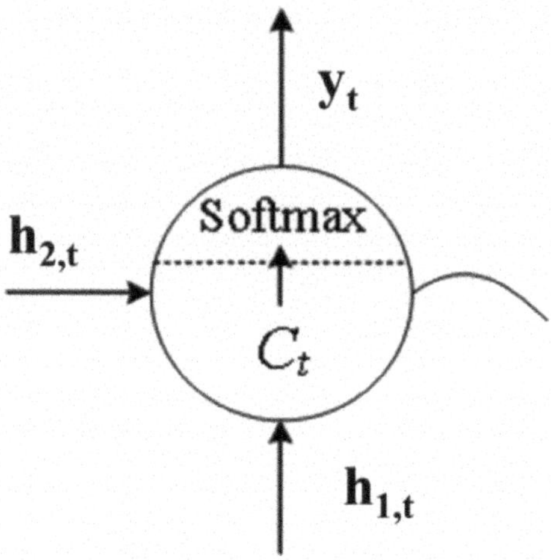

Fig. 7.11 Structure of the information fusion unit C

In the expression, x represents the multi-class classification results, W_y is the parameter weight for the specified class y, and W_n are the weights for the various classes. Both W_y and W_n are obtained through training.

The Softmax function's transformation of the multi-class classification results into probabilities can be seen in two steps. First, the exponential function is used to map each raw classification output (in real-number form) to the range $(0, +\infty)$. Then, all the mapped results are summed, and each result is normalized [19].

In the dual-stream LSTM network, the fusion unit C performs its internal calculation in two sub-steps. The first step is the weighted summation fusion operation mentioned earlier. The second step is the work behavior recognition for the current time step (frame) based on the fused information, using the Softmax function for prediction to obtain the frame-level classification results, as shown in Fig. 7.11.

The classification of the work behavior for each frame can be represented as:

$$y_t = \mathrm{softmax}(W_c C_t + b_c)$$

The W_c and b_c are also training parameters, and y_t is the result of task behavior recognition for the t-th frame.

(2) **Serialization Processing of Multi-frame Recognition Results**

Since the classification result is at the frame level, the consecutive frames with the same classification result need to be combined into one work behavior, in order to obtain the work behavior process information, as illustrated in Fig. 7.12.

The work behavior sequence can be represented as:

$$[WA_1, WA_2, WA_3, \ldots, WA_n] = f([y_1, y_2, y_3, \ldots, y_t])$$

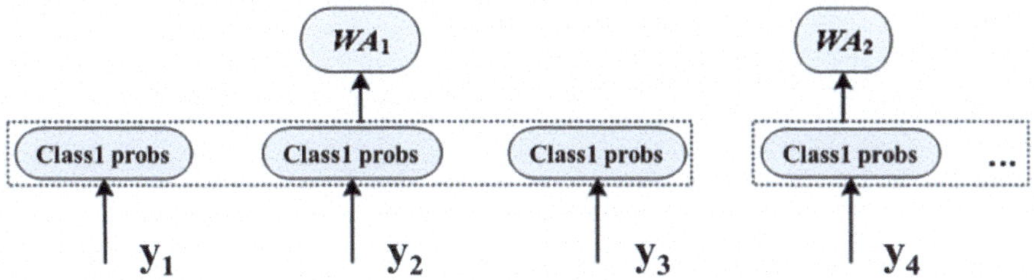

Fig. 7.12 Merging process

where f represents the merging operation, and WA_i represents the i-th work behavior in the work behavior process (there are a total of n, where n < t).

Additionally, considering the work behavior recognition process using the two-stream LSTM network, there may also be recognition errors, such as a frame being recognized as a different work behavior during the execution time frames of work behavior WA1. Undoubtedly, this is noise. To further eliminate noise, based on research and testing, this paper stipulates that only the recognized results that appear for 5 or more consecutive frames are considered as a work behavior, and work behaviors with less than 5 frames are directly filtered out. The resulting set of work behaviors is then arranged in sequence to obtain the sequence-form work behavior process information.

7.3.4 Operation Behavior Sequence Encoding Based on GRU

The basic work behavior process encoding is performed using Gated Recurrent Unit (GRU) [20]. GRU is another evolution of the RNN model, and the unit structure is shown in Fig. 7.13.

Unlike the Long Short-Term Memory (LSTM) network model, GRU has only two gating mechanisms—the update gate r_t and the reset gate z_t, and is commonly used for sequence encoding tasks in the field of natural language processing (NLP). Its working principle is as follows:

The current state feature h_t is obtained by combining the current input x_t and the hidden state information h_(t − 1) from the previous time step, which is the result of selective memory of current information and historical information [21].

The reset gate z_t can weigh the importance of the two, and selectively discard unimportant historical information based on the relationship between the current state information and the historical information. The update gate r_t controls the final selective output between the hidden state information at time t − 1 and the updated hidden state information at the current time t. The closer the output value of the update gate is to 1, the greater the weight of the historical information in the current state output. Conversely, the closer the output value of the update gate is to 0, the greater the weight of the current input information in the current state output.

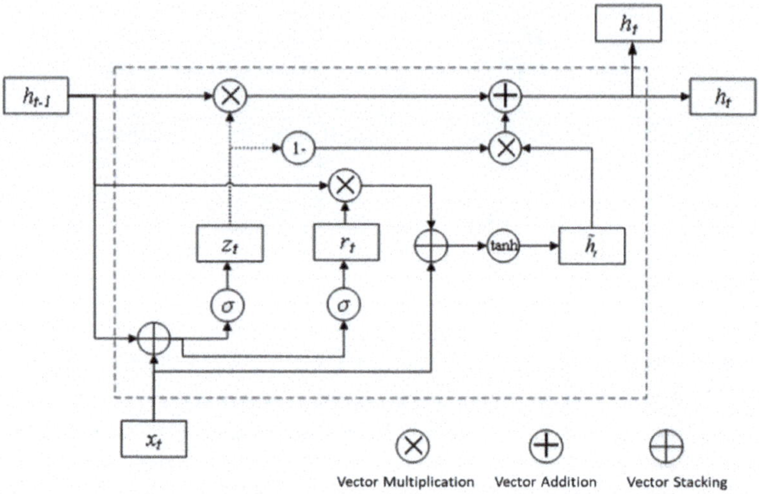

Fig. 7.13 Internal structure of the GRU unit

This mechanism not only avoids the problem of gradient disappearance, but also allows for faster convergence of the training process due to fewer parameters.

For the sequence information of the work behavior process, the specific encoding process is shown in Fig. 7.14.

$$h_t = (1 - z_t) \circ h_{t-1} + z_t \circ \tilde{\boldsymbol{h}}_t$$

"⊙" denotes element-wise matrix multiplication. h_(t − 1) is the state information from the previously executed work behavior set output. h̃_t is the new candidate

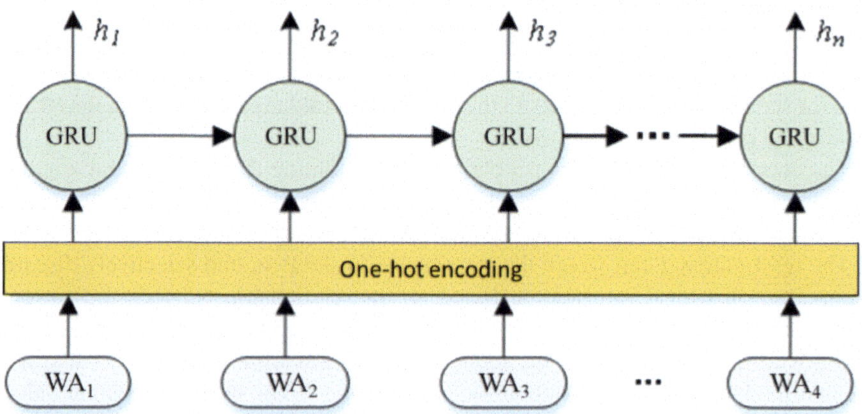

Fig. 7.14 Sequence encoding process for work behavior workflows

state information calculated after the current input work behavior information. The update gate z_t determines the trade-off between historical information and the new information from the current input. The update methods for *z_t* and h̃_t are:

$$z_t = \sigma(W_z \cdot WA_t + U_z \cdot h_{t-1} + b_z)$$
$$\tilde{h}_t = \tanh(W_h \cdot WA_i + r_t \circ (U_h \cdot h_{t-1}) + b_h)$$

The input vector at time step t is denoted as a_t. The weight matrices W_z and U_z, as well as the bias parameter b_z, are trained through backpropagation through time (BPTT). The symbol σ represents the sigmoid function.

The reset gate r_t determines the extent to which the history of past actions influences the candidate state information. If r_t is set to zero, it means ignoring the historical actions and directly resetting the candidate state based only on the current action information. This allows discarding information that is deemed unnecessary for future time steps.

The update formula for the reset gate r_t is:

$$r_t = \sigma(W_r \cdot WA_t + U_r h_{t-1} + b_r)$$

Similarly, the weight matrices W_r and U_r, as well as the bias parameter b_r, are trained during the process.

Through the above procedure, the information flow of the action sequence is encoded. Each individual action is encoded and parsed into a state information vector h_t, while also learning some temporal associations between the actions.

7.3.5 Normative Weight Mining of Operation Behavior Based on Normative Perception Network

(1) **Attention Mechanism**

The attention mechanism was inspired by the information processing mechanism of the human brain. When humans are presented with a large amount of information, they can quickly filter out the key information and ignore irrelevant information. This mechanism helps to reduce ineffective computation and improve accuracy when analyzing problems [22] (Fig. 7.15).

For example, when a human sees the image on the left, their visual attention focuses on the "two birds" and ignores the "sky" as background information. The brain identifies the birds as the focal point of the image, even though they occupy a smaller space than the sky.

However, convolutional neural networks and other intelligent algorithms treat each region equally, which limits their ability to understand the image fully. Similarly, when classifying movie reviews, humans focus on adjectives or adverbs based on

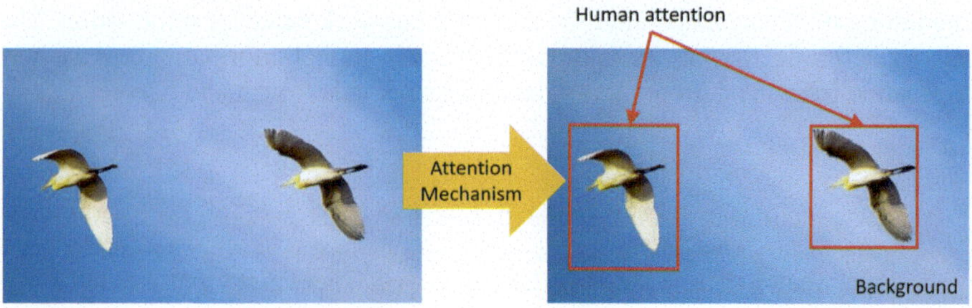

Fig. 7.15 Schematic of the human brain's attention mechanism

sentiment, while the subject and predicate are generally less important and less relevant.

Since the attention mechanism was proposed, its conceptual sophistication and practicality have attracted the attention of scholars in many fields. As deep learning research has continued to progress, the attention mechanism has been successfully integrated into deep learning methods, and has made advances in various tasks, such as text translation, image classification, and speech recognition (Fig. 7.16).

The mathematical idea behind the attention mechanism is illustrated by the diagram above. In this diagram, Query represents the query, Key represents the keywords, Value represents the weight values, Attention represents the attention value, and Source represents the input data source.

The essence of the attention mechanism is to map a query (Q) to a series of key (K)—value (V) pairs, and the computation process can generally be divided into three steps:

1. Calculate the similarity between each Q and the various Ks, to obtain the weight coefficients of the Ks and Vs. Common similarity measurement functions include dot product and cosine similarity [23].

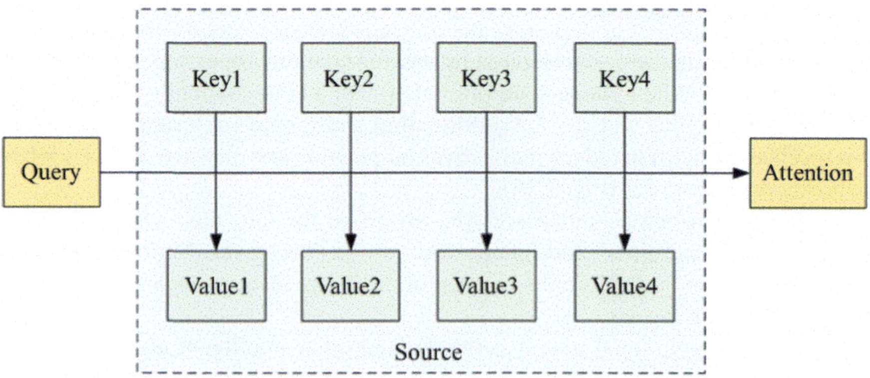

Fig. 7.16 Attention mechanism concept

$$f(Q, K) = \sum_{i=1}^{i=n} \text{similarity}\,(Q, K_i)$$

2. Normalize the weights using the Softmax function, which can further highlight the weights of important information and reduce the weights of irrelevant information, as shown in Eq. (3.7):

$$\alpha_i = \text{softmax}(f(Q, K)) = \frac{\exp(f(Q, K_i)}{\sum_j f(Q, K_j)}$$

3. Compute the weighted sum of the values (V) using the re-distributed weights, to obtain the feature vector A loaded with attention, as shown in Eq. (3.8):

$$A(Q, S) = \sum_{i=1}^{i=n} \alpha_i V_i$$

(2) **Normative Perception Network Model**

By introducing the concept of the attention mechanism, this paper designs a Normative Awareness Network (NAN). The NAN can excavate and quantify the importance of each work behavior in the sequence of the work behavior process for normative judgment. It assigns attention weights to the work behaviors based on their importance. This allows the extraction of key work behavior information and results in a more distinctive high-level feature vector. The model structure is shown in Fig. 7.17.

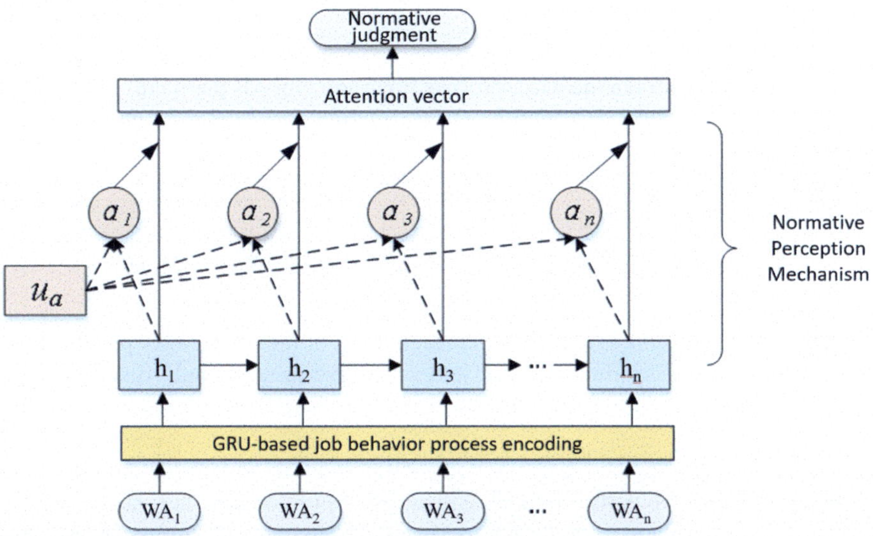

Fig. 7.17 Structure of the normative perceptual network

The normative awareness attention mechanism is designed to capture the importance information between the work behaviors in the process, i.e., the weights of the work behaviors. The method used in the processing is as follows.

1. First, for the state information obtained at each time step from the previous step of work behavior process encoding, it is processed through a simple MLP neural network to obtain a new latent representation. As shown in the following equation, where u_i is the latent representation of the processed information, and W_s and b_s are the training parameters.

$$u_i = \tanh(W_s h_2 + b_s)$$

2. Next, the Softmax function is used to allocate weights to the various latent representation information. Where α_i is the allocated weight, expressed in the form of a probability, which can measure the degree of association between a certain work behavior in the sequence and the normative nature of the current work behavior process. u_s is a randomly initialized weight vector that is jointly learned during training.

$$\alpha_i = \frac{\exp(u_i^\top u_s)}{\sum_i \exp(u_i^\top u_s)}$$

3. Finally, the obtained weights are weighted with the encoded state information correspondingly, which can then obtain the attention vector v.

$$v = \sum_i \alpha_i h_i$$

The normative perception network mining extracts the normative weights of the various work behaviors in the work process flow, and embeds them into the original feature set. This results in a high-level information representation with stronger discriminative power.

7.3.6 Normative Judgment Based on Logistic Regression

The attention vector v is a high-level representation of the entire work process flow in the normative discrimination task, and can serve as a strongly discriminative feature for normative judgment. It can be used to perform high-precision normative judgment.

In this study, the logistic regression algorithm is used, with the attention vector as the input feature, to perform the classification of normative or non-normative.

Logistic regression is a commonly used classification model. It uses the sigmoid function to process the input features and map the predicted values into the (0, 1)

Fig. 7.18 Sigmoid function of logistic regression

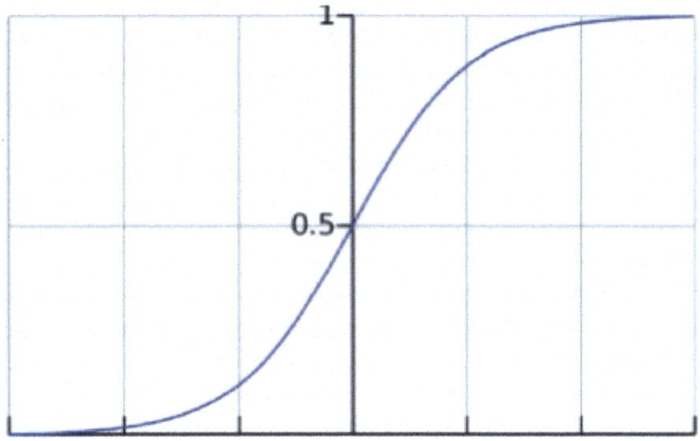

interval, resulting in an intuitive probability value for classification. The essence of the logistic regression algorithm makes it more suitable for binary classification problems [24–26]. The formula is as shown in Equation [X], and the curve is as shown in Fig. 7.18.

$$Y(x) = \frac{1}{1 + e^{-(\theta x + b)}}$$

The function Y(x) represents the predicted value, where x is the input feature and e is the natural logarithm base. The parameters θ and b are the coefficients of the linear transformation.

The predicted value Y(x) has a range between 0 and 1, and the curve is centrally symmetric. Based on these characteristics, the input features can be divided into positive and negative samples by using a threshold of 0.5 for the predicted value Y(x).

Specifically, the predicted class ŷ can be represented by the following equation:

$$\hat{y} = \begin{cases} 1 & Y(v) \geq 0.5 \\ 0 & Y(v) < 0.5 \end{cases}$$

Given that Y is a probability distribution function, the further a feature is from the central point, the greater the likelihood of it belonging to the corresponding class. The cost function C can be written as follows, where y represents the true normative class label (Ground Truth Label):

$$C = \begin{cases} -\log Y(v) & y = 1 \\ -\log(1 - Y(v)) & y = 0 \end{cases}$$

For input features x with a label value y of 1, the closer the predicted value is to 1, the smaller the cost. Conversely, the further the predicted value is from 1, the larger the cost. The same principle applies for label values of 0.

By summing the costs for all M samples and taking the average, we obtain the model's loss function L, which can be expressed as follows:

$$L = \frac{1}{M}\sum_{i=1}^{M} C_i$$

Given that the value of y can only be 0 or 1, combining Eqs. (3.14) and (3.15), we can obtain the complete expression for the loss function:

$$L = \sum_{i=1}^{M} y_i \log \hat{y}_i + (1 - y_i)\log(1 - \hat{y}_i)$$

By applying the gradient descent optimization method to the loss function L, we can solve for the optimal model parameters. Once the optimization process has converged, we will obtain the optimal model for determining the normative behavior flow of the task [27, 28].

7.4 Machine Collaborative Assembly Case Based on Operation Behavior Detection

In the human–robot collaborative assembly process, accurate recognition of human behaviors and movements is required to instruct the robot to make accurate response actions. This section uses human behavior detection in this process as a case study.

7.4.1 Case Introduction

This section utilizes the assembly case of a worm gear reducer to further demonstrate the feasibility of the aforementioned method. For a worm gear reducer, the main components include the worm gear, drive shaft, worm, bearings, upper cover, front cover, rear cover, and base.

To simplify the complexity of the part-behavior assembly tree during the experimental stage, the bearings are first fixed onto the worm to form a worm assembly, the drive shaft is fixed with the bearings onto the worm gear to form a worm gear assembly, and the rear cover is pre-attached to the base. After simplification, the basic assembly components include the worm gear assembly, worm assembly, front cover, rear cover and base (Fig. 7.19).

Based on the feasible assembly sequence of the worm gear reducer, an assembly part-behavior AND/OR graph can be constructed as shown in the figure above. This

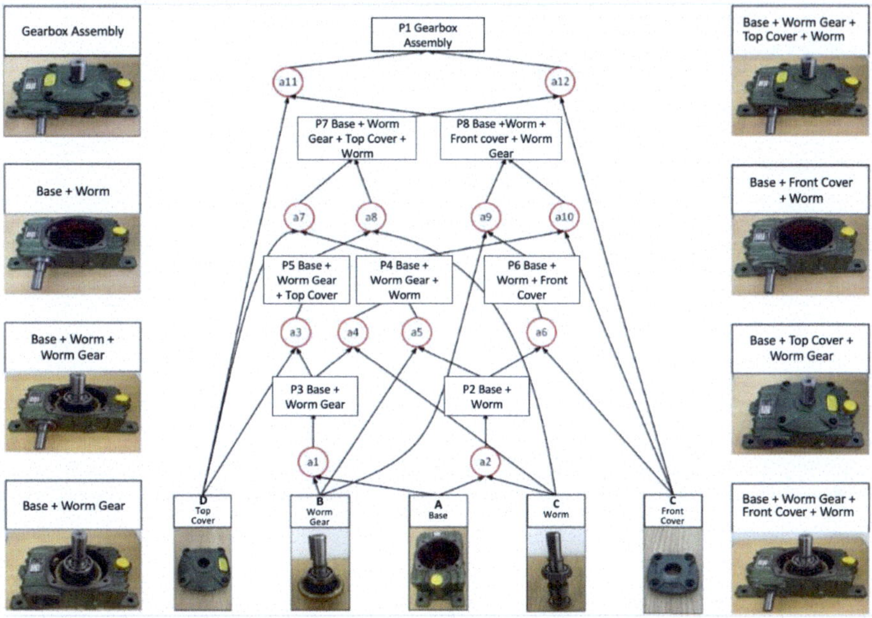

Fig. 7.19 Gearbox part—behavior and OR diagram

graph includes five basic parts, eight intermediate component states, and twelve assembly behaviors.

To address the possible part assembly states and assembly behaviors in the graph, we can construct parts and behavior tables as follows (Tables 7.1 and 7.2).

For the assembly case of the worm gear reducer, a dataset as shown in Fig. 7.24 was constructed (Fig. 7.20).

The fixed Realsense camera was positioned above and in front of the human worker, allowing for clear filming of the worker's assembly limb movements and the state of part assembly. The camera's resolution was set to 640×480, with a frame rate of 30 fps. Following to the part-behavior AND-OR graph sequence, manual assembly was performed, resulting in a total of 200 video segments, each lasting 16–24 s and containing 4 different assembly behaviors. Each segment may feature different assembly behaviors, with a total of 12 behavior distinct types.

As shown in Fig. 7.21, the method can be divided into the following three parts according to the data flow process:

1. **Assembly Behavior Feature Description and Extraction**

The most important feature of an assembly behavior is the worker's limb assembly behavior, so mathematical modeling of the worker's body is required to provide high-quality human body features.

Table 7.1 Turbine gearbox component list

Label	Quality	Parent node	Assembly	Remark
P1	0	− 1	a1, a2	Gearbox assembly
P2	0	a5, a6	a2	Base + worm
P3	0	a3, a4	a1	Base + worm gear
P4	0	a7, a10	a4, a5	Base + worm gear + worm
P5	0	a8	a3	Base + worm gear + top cover
P6	0	a9	a10	Base + worm + front cover
P7	0	a11	a9, a10	Base + worm gear + top cover + worm
P8	0	a12	a7, a8	Base + worm + front cover + worm gear
A	1	a1, a2	− 1	Base
B	1	a1, a5, a9	− 1	Worm gear assembly
C	1	a2, a5, a9	− 1	Worm assembly
D	1	a3, a7, a11	− 1	Top cover
E	1	a6, a10, a12	− 1	Front cover

Table 7.2 Turbine gearbox behavior list

Label	Frequency	Required parts	Produced parts
a1	0	A, B	P3
a2	0	A, C	P2
a3	0	P3, D	P5
a4	0	P4, E	P2
a5	0	P2, B	P4
a6	0	P2, E	P6
a7	0	P4, D	P7
a8	0	P5, C	P7
a9	0	P6, B	P8
a10	0	P4, E	P8
a11	0	P8, D	P1
a12	0	P7, E	P1

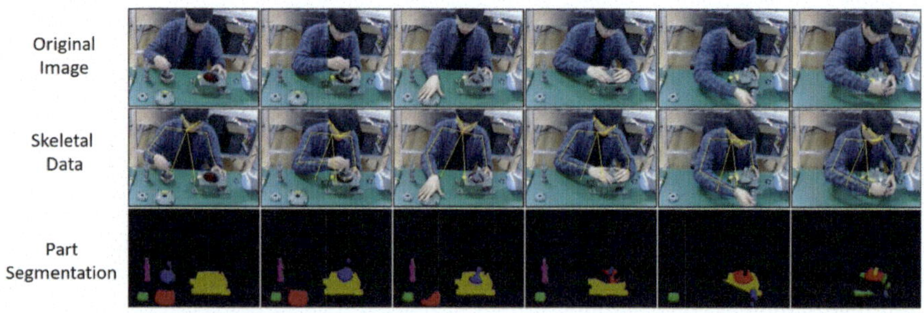

Fig. 7.20 Gearbox assembly behavior dataset

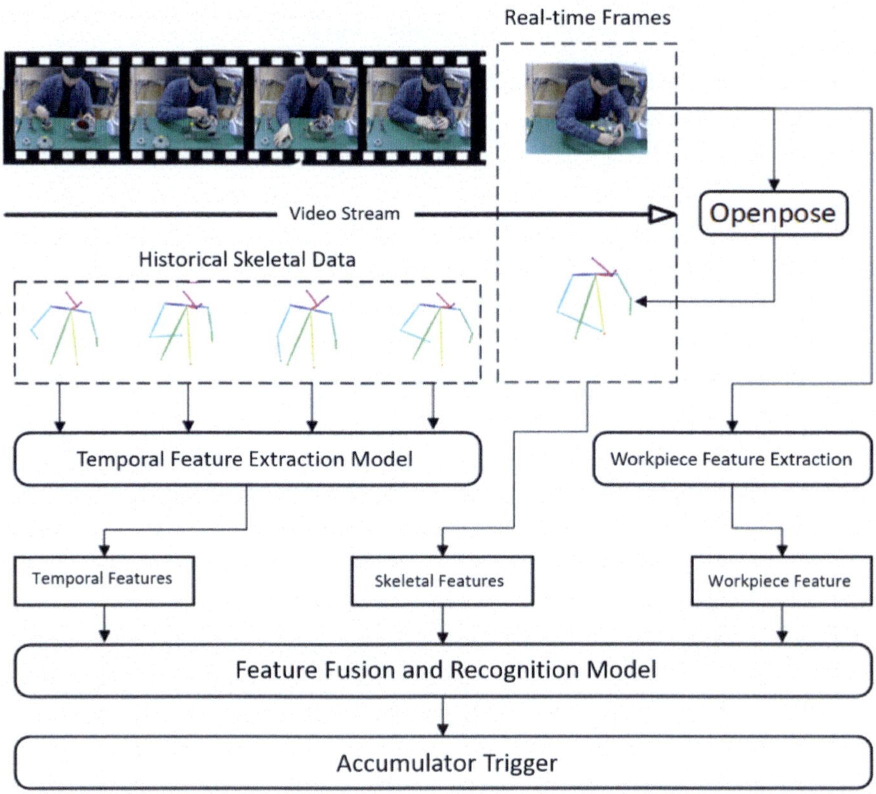

Fig. 7.21 Methodology workflow

The worker's operation objects have a guiding role in the recognition of the behavior, so the workpieces, tools and other assembly objects need to be recognized and expressed in the form of digital vectors.

Assembly behavior is a dynamic process, so it is necessary to analyze the time series information and use the self-attention model to extract the time series features.

2. **Feature Fusion and Frame-Level Assembly Behavior Classification**

Due to the online real-time requirements of human–machine collaboration, the classification of assembly behaviors cannot adopt the mode of behavior detection + behavior recognition. A self-attention model based on feature fusion is proposed to perform real-time classification of frames in the video stream.

Since the behavior is continuous and there are indistinguishable transition behaviors, the soft-label method is used to annotate the data to improve the model performance.

3. **Real-Time Fusion Processing of Classification Results**

The frame-level classification results cannot directly reflect the worker's assembly behavior, so the results need to be merged and post-processed, combining the continuous single-frame results into one assembly behavior, thereby triggering the robot's assistance behavior.

This paper adopts an accumulated threshold triggering mechanism to complete the behavior triggering, ensuring that only one continuous individual behavior is triggered, and smoothing the jittery recognition results.

7.4.2 Experimental Environment Setup

The experimental environment for the assembly behavior detection and recognition method (as shown in Fig. 7.22) consisted of a Windows-10 64-bit computer equipped with an Intel i7 8700 processor, an NVIDIA RTX2060 graphics card, 16 GB of RAM, and a RealSense D435 depth camera. The software environment was based on Python 3.6.

Two RealSense depth cameras were used to detect the worker's position and the part's position, respectively. A UR5 six-axis robotic arm was integrated to provide auxiliary operations, with communication between the computer and the robot handled via ROS (Robot Operating System). The spatial relationship between the depth cameras and the robot is depicted in the figure. The entire experimental platform was divided into two areas: the parts area and the assembly area. The parts area was used for part recognition and grasping, while the assembly area was used to complete the assembly and recognize the assembly behavior.

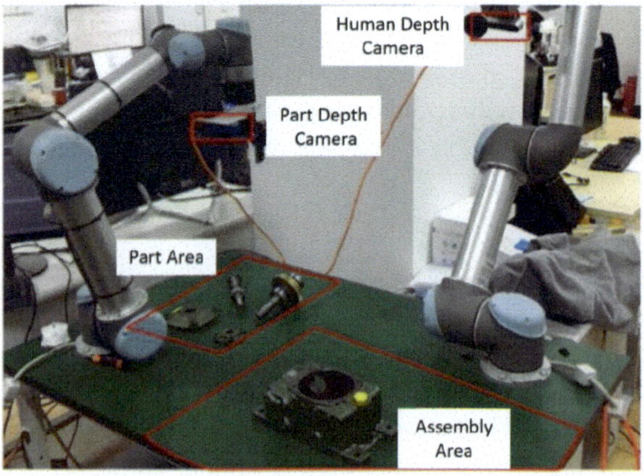

Fig. 7.22 Experimental environment

7.4.3 Functional Verification

(1) **Verification of Human–Machine Collaboration Scene Recognition Function**

After real-time processing of the color image obtained from the depth camera using OpenPose, the image coordinates of the key skeletal points of the human body can be determined. By registering the color image and the depth image, these key points can be mapped to the depth image, and the three-dimensional spatial coordinates of the key points in the camera coordinate system can be obtained through inverse operation, as shown in Fig. 7.23. The three-dimensional coordinates of the human skeletal structure can basically represent the positions of the various joints in the human body. The robot can use these skeletal coordinate positions to locate the worker for assistance or obstacle avoidance.

For the color image and depth image of the parts obtained by the depth camera, point cloud information can be generated. By inputting the point cloud into the pre-trained VoteNet [29], the three-dimensional bounding boxes and category information of the various parts can be obtained. The three-dimensional bounding boxes of the parts can basically represent the position and pose of the parts. Since the main focus of this project is not on the gripper pose for robotic grasping, a simple short-edge vertical grasping method is adopted to complete the robot's grasping task (Fig. 7.24).

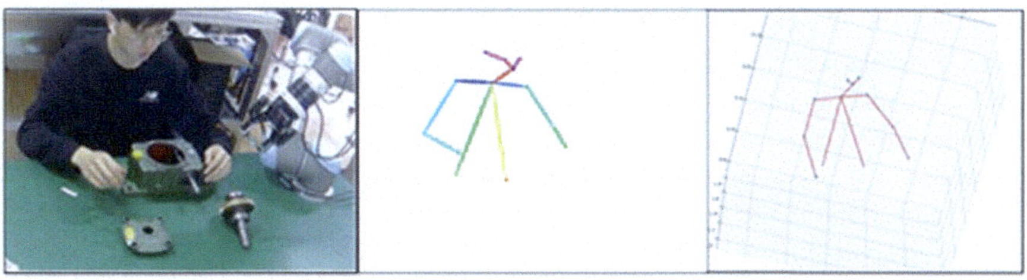

Fig. 7.23 Obtaining 3D human coordinates using a depth camera and OpenPose

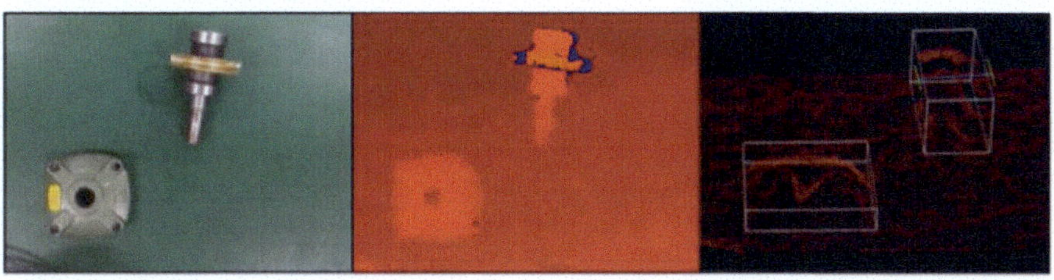

Fig. 7.24 Obtaining 3D part location information using a depth camera and VoteNet

The robot unifies the coordinates by mapping the human body skeletal key point coordinates and the part position bounding box coordinates to the robot base coordinate system. For a specific grasping task, the robot moves to the short edge of the part position and grasps the part, then moves the part to the position behind the wrist joint to complete the auxiliary grasping task.

(2) **Verification of Dynamic Human–Machine Collaborative Assembly Function**

The computer is responsible for understanding the scene in the human–robot collaboration through computer vision technology, including the worker's assembly behavior, the part's pose, the worker's pose, etc. The recognition results are dynamically updated on the behavior table and the part table to form the part-behavior assembly graph.

Based on the dynamic querying and updating method, the computer obtains the current assembly behavior and selects the target part for the next feasible assembly path. The part pose and worker position information in the scene are then converted to the robot coordinate system, and instructions are sent to the robot side. The robot then executes the corresponding auxiliary actions according to the instructions.

Figure 7.25 records two examples of human–robot collaboration in assembly. In the left case, the assembly is completed entirely according to the parts provided by the robot. In the right case, the worker exercises subjective choice. First, the worker completes the assembly of the base and the turbine component. According to the assembly OR graph rules, the robot then grasps the top cover and places it within the worker's reach. In this process, the worker can exercise subjective initiative to select another feasible path to install the worm gear component. The robot, by querying the dynamically updated assembly OR graph, abandons the top cover and instead grasps the front cover, which the worker then installs. Finally, the assembly task is completed.

This shows that in the human–robot collaboration process, the robot can provide guidance and parts for the worker, while the worker can subjectively choose to execute the next assembly step according to the robot's selection or their own initiative. The two work together, better utilizing their respective advantages.

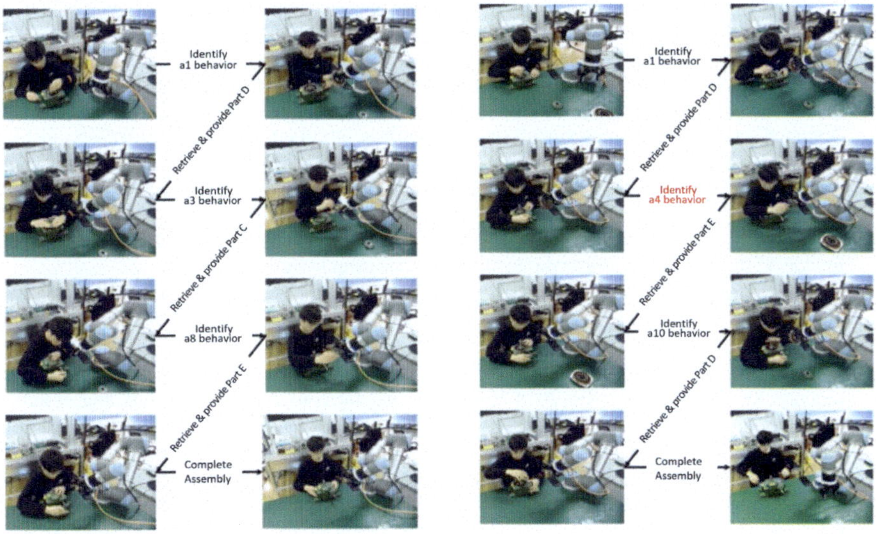

Fig. 7.25 Example of a human–robot collaboration process

References

1. Liu TY, Lu Z, Sun YF et al (2019) Workshop production behavior recognition method based on three-dimensional depth convolutional neural network. In: Computer integrated manufacturing systems, pp 1–20
2. Rude DJ, Adams S, Beling PA (2018) Task recognition from joint tracking data in an operational manufacturing cell. J Intell Manuf 29(6):1203–1217
3. Huang H (2018) Research on real-time monitoring system of motion fatigue. University of Electronic Science and Technology of China
4. Wang P, Liu H, Wang L et al (2018) Deep learning-based human motion recognition for predictive context-aware human-robot collaboration. CIRP Ann 67(1):17–20
5. Shotton J, Fitzgibbon A, Cook M et al (2013) Real-time human pose recognition in parts from single depth images. Springer, Berlin
6. Toshev A, Szegedy C (2013) DeepPose: human pose estimation via deep neural networks. IEEE
7. Tompson J, Jain A, Lecun Y et al (2014) Joint training of a convolutional network and a graphical model for human pose estimation. ePrint Arxiv: 1799–1807
8. de Gusmao Lafayette TB, Burle AQ, Almeida AA et al (2021) The virtual Kinect. In: Proceedings of the 23rd symposium on virtual and augmented reality, pp 111–119
9. Cao Z, Simon T, Wei SE et al (2017) Realtime multi-person 2d pose estimation using part affinity fields. In: Proceedings of the IEEE conference on computer vision and pattern recognition, pp 7291–7299
10. Han J, Shao L, Xu D et al (2013) Enhanced computer vision with microsoft kinect sensor: a review. IEEE Trans Cybernet 43(5):1318–1334
11. Li Z, Bao J, Liu T et al (2020) Judging the normativity of PAF based on TFN and NAN. J Shanghai Jiaotong Univ (Sci) 25:569–577
12. Heilbron FC, Niebles JC, Ghanem B (2016) Fast temporal activity proposals for efficient detection of human actions in untrimmed videos. In: Proceedings of the IEEE conference on computer vision and pattern recognition, pp 1914–1923

13. Jordan MI (1997) Serial order: a parallel distributed processing approach. Adv Psychol 121:471–495
14. Hochreiter S, Schmidhuber J (1997) Long short-term memory. Neural Comput 9(8):1735–1780
15. De Geest R, Tuytelaars T (2018) Modeling temporal structure with LSTM for online action detection. In: 2018 IEEE winter conference on applications of computer vision (WACV). IEEE, pp 1549–1557
16. Simonyan K, Zisserman A (2015) Two-stream convolutional networks for action recognition. In: Proceedings of the neural information processing systems (NIPS), p 11
17. Simonyan K, Zisserman A (2014) Two-stream convolutional networks for action recognition in videos. In: Advances in neural information processing systems, p 27
18. Feichtenhofer C, Pinz A, Zisserman A (2016) Convolutional two-stream network fusion for video action recognition. In: Proceedings of the IEEE conference on computer vision and pattern recognition. pp 1933–1941
19. Wang M, Lu S, Zhu D et al (2018) A high-speed and low-complexity architecture for softmax function in deep learning. In: 2018 IEEE Asia pacific conference on circuits and systems (APCCAS). IEEE, pp 223–226
20. Chung J, Gulcehre C, Cho KH et al (2014) Empirical evaluation of gated recurrent neural networks on sequence modeling. ePrint arXiv: https://doi.org/10.48550/arXiv.1412.3555
21. Cho K, van Merrienboer B, Gulcehre C et al (2014) Learning phrase representations using RNN encoder-decoder for statistical machine translation. arXiv Preprint. arXiv: 1406.1078
22. Vaswani A, Shazeer N, Parmar N et al (2017) Attention is all you need. arXiv: https://doi.org/10.48550/arXiv.1706.03762
23. Ghorbani MA, Zadeh HA, Isazadeh M et al (2016) A comparative study of artificial neural network (MLP, RBF) and support vector machine models for river flow prediction. Environ Earth Sci 75(6):1–14. https://doi.org/10.1007/s12665-015-5096-x
24. Peduzzi P, Concato J, Kemper E et al (1996) A simulation study of the number of events per variable in logistic regression analysis. J Clin Epidemiol 49(12):1373–1379. https://doi.org/10.1016/S0895-4356(96)00236-3
25. Friedman J, Hastie T, Tibshirani R (2000) Additive logistic regression: a statistical view of boosting (with discussion and a rejoinder by the authors). Ann Stat 28(2):337–374. https://doi.org/10.1214/aos/1016218223
26. Royston P, Altman DG (2010) Visualizing and assessing discrimination in the logistic regression model. Stat Med 29(24):2508–2520. https://doi.org/10.1002/sim.3994
27. Jain P, Kakade SM, Kidambi R et al (2016) Parallelizing stochastic gradient descent for least squares regression: mini-batching, averaging, and model misspecification. https://doi.org/10.48550/arXiv.1610.03774
28. Tao Y, Wenjie L (2013) Neural network model research based on gradient descent algorithm. In: Network security technology & application
29. Ding Z, Han X, Niethammer M. Votenet (2019) A deep learning label fusion method for multi-atlas segmentation. In: Medical image computing and computer assisted intervention—MICCAI 2019: 22nd international conference, Shenzhen, China, 13–17 Oct 2019, Proceedings, Part III 22. Springer International Publishing, pp 202–210

Chapter 8
Human–Robot Collaboration in Manufacturing Processes

Smart manufacturing does not aim for a fully unmanned approach. The role of humans in complex manufacturing processes may never be completely replaceable. The process of "AI + Manufacturing" differs fundamentally from the "automation" pursued in past manufacturing processes. "Automation" seeks to achieve robot-driven production, essentially replacing human labor, with an emphasis on large-scale mechanical manufacturing. In contrast, "Intelligence" pursues flexible production by robots, with the essence being the "collaboration between humans and robots." It emphasizes robots' ability to actively assist human work and adapt to environmental changes autonomously.

8.1 Introduction

"AI + Manufacturing" is not merely a straightforward replacement of workers with robots. It pulls the highly segmented assembly line work that has characterized industrial revolutions back to the "human-centered" organizational model. This approach enables both robots and humans to focus on tasks they excel at—robots undertake more repetitive, tedious, and hazardous tasks, while humans undertake more creative endeavors.

Manufacturing is an exceedingly complex industry where a single product may involve dozens of raw materials at the least, or millions of components at the most. Producing the same product, different enterprises employ varying production techniques, equipment, and component inputs. Divergent manufacturing processes pose significant challenges. Incompatible equipment interfaces complicate matters further. Differing data formats add to these issues. These factors make it hard to digitally connect supply chains, both upstream and downstream. As a result, each enterprise must undertake a unique and challenging digital transformation. Implementing fully

T. Liu et al., *Industrial Intelligence: Methods and Applications*, Springer Series in Advanced Manufacturing, https://doi.org/10.1007/978-3-031-81477-8_8

manual manufacturing approaches is difficult. Fully automated manufacturing also faces significant challenges.

8.1.1 *Machine Perception Assists Human–Robot Collaboration*

In the first step towards realizing human–robot collaboration, it is essential for robots to possess the capability of intelligently sensing their environment and tasks. Achieving intelligent perception of target workpieces is crucial. Robots need to acquire the pose and positional information of these workpieces. Only then can they accurately grasp them. This enables effective collaboration between robots and humans. In the context of collaborative manufacturing tasks, specialized sensing equipment is used. This equipment allows robots to perceive their environment. This perception is a prerequisite for effective human–robot collaboration. The robot's perceptual abilities are primarily realized through multi-sensor information fusion techniques. Specifically, this encompasses the employment of visual sensors, force sensors, as well as infrared distance sensors, among others.

Initially, robots use depth cameras to perceive objects in their environment. They perform precise localization of these objects. This enables the robots to actively assist human operators. They can help with tasks like grasping specific workpieces. For particular sub-tasks, analysis of the workpieces involved is required. Depth cameras are then used to capture the 3D point cloud data. This data pertains to the workpieces to be picked up in the working environment. Subsequently, appropriate neural network algorithms are applied. These algorithms segment and process the acquired point cloud information. The goal is to extract the characteristic point cloud features of the objects to be retrieved.

Before the start of the production manufacturing process, these workpieces are usually haphazardly stacked. They are often found in workpiece bins or on the work platform. Manual picking for installation, assembly, welding, and other tasks is inefficient, lacks precision, and does not ensure safety. Consequently, the use of robotic arms for workpiece handling becomes imperative.

In current industrial applications, the challenge of picking such scattered components is commonly categorized as the Bin-Picking problem, which revolves around how robots can identify and pick up disordered, stacked objects. Bin-Picking constitutes a systematic discipline, encompassing scene analysis, pose estimation, grasp planning, and motion planning, among others. Scene analysis involves identifying what targets are present in the image scene, understanding their spatial distribution, and discerning occlusions among the targets in the image. Pose estimation entails calculating the position and orientation of objects in the camera coordinate system. Grasp planning refers to the methodology employed to grasp objects in the scene, while motion planning concerns the robot's kinematic forward and inverse solutions, obstacle avoidance, and path planning.

In real-world scenarios, acquiring scattered parts often entails high data costs, cumbersome operations, and slow data acquisition. Conversely, employing a rapid virtual scene setup solution simulates the disorderly arrangement found in actual environments. By converting depth images captured by depth cameras into point cloud images, object localization is achieved through feature matching techniques. Moreover, under various constraints, a grasp objective function is constructed. This function determines the optimal sequence for grasping. This enhances the efficiency and precision of the process.

In industrial settings, this issue is predominantly categorized as the Bin-Picking problem, which revolves around how robots can identify and retrieve target objects from a disorganized, stacked environment. It encompasses key aspects including scene analysis, pose estimation, grasp planning, and motion planning.

In the identification of scattered components, machine learning methodologies typically employ classical supervised learning algorithms, combined with manually crafted descriptors to tackle point cloud segmentation. Given the vast amount of point cloud data, crucial point features must be extracted. This requires a balance between algorithm speed and classification accuracy. This balance often leads to poor generalization capabilities. Furthermore, when objects to be detected are in cluttered and overlapping scenes, feature extraction is time-consuming. This often results in inferior detection outcomes. The detections are prone to errors. Additionally, the algorithms show poor real-time performance.

Deep learning-based methods are generally classified into voxel-based approaches, point-based methods, unsupervised learning methods, and those based on projection and views. Among these, unsupervised learning leverages unlabeled data for clustering, enabling the bridging of different data types and reducing the need for dataset annotation. Projection and view-based methods, as the name suggests, involve projecting 3D point cloud coordinates onto a 2D plane, thereby reducing the dimensionality of the data. This is followed by applying 2D image processing techniques for further analysis.

Robot grasping can broadly be categorized into three main approaches: traditional robot grasping methods, deep learning-based grasping methods, and reinforcement learning-based grasping methods. Traditional grasping techniques primarily involve optimizing the end effector and selecting appropriate sensors. Depending on whether objects in the environment are known, robot grasping can be divided into grasping of known objects with predetermined locations and grasping of unknown objects. For known target objects, model-based methods are typically employed, utilizing a modeled object or point cloud model as a reference for grasping. In the case of grasping unknown objects, current practices often use deep learning. Sometimes, transfer learning is adopted to train models for grasping similar objects. Relevant thresholds are adjusted to adapt to grasping in unknown environments. This frequently involves on-the-fly training during the grasping process.

8.1.2 *Reinforcement Learning Enhances Collaborative Capabilities*

With the advancement of artificial intelligence, the rational and efficient utilization of AI stands to significantly enhance manufacturing efficiency and precision. Currently, methods such as supervised and unsupervised learning have already been implemented in the industrial sector. However, reinforcement learning, an AI methodology running parallel to these two, has yet to see widespread application. In practical manufacturing environments, annotating a large volume of workpiece data is time-consuming, labor-intensive, and challenging. Consequently, the integration of reinforcement learning into human–robot collaboration in manufacturing processes is being considered to enable agents to cleverly infer logic based on accumulated experience and knowledge. Human–robot collaboration represents a new manufacturing paradigm. It is distinct from traditional manual assembly. This new approach necessitates designers to prioritize operator safety. They also need to focus on manufacturing efficiency when devising systems.

Reinforcement Learning is a prominent research focus in machine learning. It has found broad application in several areas. These include industrial manufacturing, simulation, and robot control. It is also used in optimization and scheduling, as well as gaming. Its fundamental principle revolves around learning the optimal strategy to accomplish a goal by maximizing the cumulative reward an agent receives from the environment. Reinforcement Learning problems are typically modeled as Markov Decision Processes (MDPs). MDPs are encapsulated by a classic four-tuple $\langle S, A, P, R\rangle$. In this tuple, S represents the set of states. A represents the set of actions. P denotes the probability of transitioning from one state to another. R represents the rewards achievable by an agent in the environment. MDPs simulate stochastic policies in environments where the system states exhibit the Markov property. By simplifying the environment to exhibit Markovian characteristics, the next state of the agent is solely dependent on the current state, independent of previous states. Deep Reinforcement Learning, an extension of Reinforcement Learning, integrates deep neural networks to augment the perception capabilities of the original model. The manufacturing environment encompasses a myriad of manufacturing activities, including workpiece machining, welding, painting, and assembly, among others. Enhancing the cognitive reasoning abilities of robots during the manufacturing process significantly influences the performance of human–robot collaboration.

Typically, by treating the manufacturing environment as a setting where the reinforcement learning agent interacts, a wealth of interaction data is generated, such as the positional and pose data of various objects in the environment. Then, the robot's task planning or control data is used as action values. A reinforcement learning iterative framework is established. This framework is created between the robot and the manufacturing environment. Through extensive learning, the robot eventually acquires the optimal action strategy. Given that human operators are dynamic elements in collaborative manufacturing, they undergo performance and spatial positioning changes over time. This has several implications for the collaborative manufacturing process: (1) it leads to daily progress in collaborative manufacturing fluctuating with human performance changes; (2) it introduces errors in human–robot collaboration due to variations in human performance; (3) it causes discrepancies between planned and actual completion times for manufacturing tasks. To mitigate disparities introduced by variations in human collaborators' performance, adjustments must be made. These adjustments are to the robot collaborator's performance. They are based on the deviation values of human performance. This achieves a dynamic equilibrium in the collaborative performance between human and machine.

Certainly, in addition to compensating for variations in human collaborators' performance, it is necessary to enhance the robot's recognition of human behavior based on changes in the human collaborator's posture. Integrating a human action recognition network with the reinforcement learning framework achieves this objective. Two primary methods exist for recognizing human actions: (1) action recognition based on RGB video sequences, and (2) action recognition based on human skeletal data. The latter method, based on skeletal data, is less affected by spatial occlusion issues. Thus, it allows for inferring the type of task the human collaborator will undertake next based on the recognition of their current actions. The robot then takes preparatory actions in advance and assists the human operator in completing the corresponding manufacturing task accordingly. Skeleton information of the human body, obtained via a Kinect vision sensor, can serve as input to a trained action recognition network, yielding corresponding action labels. Using an action-task mapping table, the upcoming human–robot collaborative task information can then be inferred. Within the reinforcement learning framework, state values are defined according to the robot's collaborative task performance, with the robot's six-joint angle values representing action values. Following the input of the current state, the output is a strategy guiding the robot's movement. As the robot executes its action strategy, new environmental state values and a reward evaluating the strategy's efficacy are generated. Over time, the robot learns to optimize its action strategies for collaborative tasks. This process of reinforcement learning can be seen as the cognitive reasoning process for the robot. It promotes the robot's intelligent assistance capabilities.

Reinforcement learning empowers robots to self-enhance without external supervision, thereby elevating their intelligent levels based on autonomous reasoning and facilitating various supportive tasks alongside humans. Upon acquiring human behavioral data, robots leverage a task inference module to ascertain their subsequent collaborative tasks and subsequently plan their own movements and postures. Given the variability of human performance over time, robots adjust their work schedules. They also adjust their status to compensate for deviations. These deviations are caused by fluctuations in human performance. Ultimately, this drives human–robot collaboration toward a dynamic equilibrium. Hence, reinforcement learning holds direct and substantial applicability in fostering human–robot synergy within the manufacturing process.

8.1.3 Related Works

(1) Object Detection

Object detection technology has long been a focal point and challenge in computer vision research, typically requiring both object recognition and localization, with object recognition being a vital component of the technique. In recent years, rapid advancements in neural networks have led to the widespread application of object detection technologies. An increasing number of methodologies have been proposed. These include Support Vector Machines (SVM) [1]. Convolutional Neural Networks (CNN) [2] are also included. Artificial Neural Networks (ANN) [3] are part of these methodologies. Algorithms based on CNN principles have demonstrated exceptional performance in various recognition competitions.

Driven by the rapid progress in deep learning, the predominant approach to solving object detection problems is now based on CNN architectures. Virtually all current object recognition and detection algorithms are rooted in CNN principles, with various algorithms tailored to different application scenarios. Object detection algorithms can be largely classified into two categories. The first category includes those based on region proposals. Examples are RCNN [4], Fast RCNN [5], and Faster RCNN [6]. The second category includes regression-based detection algorithms. These include YOLO [7], SSD [8], and OverFeat [9].

The core idea behind object detection algorithms based on Region Proposals lies in the attention mechanism. The primary workflow of such methods is divided into feature extraction and classification. Feature extraction involves traversing the image using a search algorithm. This obtains a multitude of candidate regions. Then, CNNs are leveraged to extract features from these regions. These features are then used by classifiers as criteria for classification tasks.

Significant research has been conducted concerning regression methodologies. Redmon et al. [7] introduced the YOLO (You Only Look Once) approach, which deviated from the conventional practice of employing classifiers for detection tasks. YOLO reformulates object detection as a categorical probability regression problem, enabling direct construction of bounding boxes from images and classification via probabilities, thereby accomplishing object detection. By leveraging information from the entire image for predictions, YOLO achieves a remarkable detection speed. This speed is up to 150 frames per second. YOLO nearly doubles the classification success rate compared to Fast R-CNN. It also exhibits a robust model performance. However, YOLO is notably sensitive to variations in object sizes. OverFeat, proposed by Sermanet et al. [9], is a recognition algorithm built upon AlexNet, which optimally utilizes the feature extraction capabilities of Convolutional Neural Networks (CNNs). It repurposes features obtained from classification for both detection and localization tasks, facilitating versatility across diverse operational scenarios without necessitating extensive network re-tuning. This algorithm excels with its outstanding performance in both recognition and detection applications; nonetheless, its limitation lies in a relatively lower accuracy in identification.

Template matching stands as a prototypical image processing technique, applicable not only to pose estimation but predominantly utilized in object detection. Unlike deep learning, which requires training on vast datasets, this method merely necessitates a prepared library of standard templates. The algorithm operates in two primary phases: the OffLine phase and the OnLine phase. The OffLine phase is the preparatory stage. It involves acquiring an image set containing target objects. Training parameters are tuned based on the specifics of these images. Conversely, the OnLine phase serves as the detection stage, employing predetermined template matching regions to perform the matching. Successful matches result in the detection of target objects. A seminal algorithm within template matching is the Line-mod method. This method was proposed by Hinterstoisser et al. It is designed to address the identification and localization of 3D objects. These objects are found amidst cluttered, stacked scenarios. The method capitalizes on both RGB image data and Depth information. Core steps of the Line-mod algorithm encompass color image gradient computation, feature map divergence, and response map preprocessing. Characterized by its speed and superior detection performance, Line-mod stands out as the most real-time capable algorithm among its contemporaries.

(2) **Dispersed Object Grasping**

In general, the design of object grasping schemes in factory environments often involves objects being scattered and overlapping. In response to this challenge of picking up scattered objects, companies in the robotics field have conducted extensive research.

For instance, FANUC Corporation's robot pickup system in Japan [11] is capable of identifying and grasping scattered workpieces. The primary principles involved are as follows: Utilizing a 3D sensor, it projects infrared rays onto the surface of objects to construct a structural map of the current scene, processing visual information to acquire the pose data of the parts, thereby effectively identifying objects in the target scene. It also incorporates iRvision 2D vision technology to compensate for motion errors during the grasping process, ensuring accurate and error-free pickups every time.

ABB Robotics, based in Sweden, has developed a robot grasping system powered by their proprietary TrueView vision technology [12], which is capable of identifying and picking objects in disordered scenes. The in-house developed TrueView vision system is adaptable to various robots and cameras, demonstrating high versatility and meeting the requirements of most industrial settings.

TrueView integrates robotic kinematics algorithms, enabling it to perform motion planning for robots in diverse scenarios. It is also equipped with state-of-the-art vision processing algorithms that can handle both 2D image processing and 3D point cloud processing, ensuring the acquisition of object poses in the scene.

In addition to the top robot manufacturers, other robot or vision companies also develop their own solutions. Some well-known examples include [13]: (1) Universal Robots from Denmark, with their UR series of collaborative robot solutions. (2) Intel RealSense series' visual solutions from the United States. (3) Scape's bin-picking system from Denmark, among others.

Research on grasping randomly scattered objects has not only made strides in industrial settings but also finds widespread application in everyday life. Unlike industrial scenarios, object grasping in everyday environments poses unique challenges. There is a vast variety of objects. Backgrounds are complex. Items often have low grasping capability. These factors make the task more difficult. They also hinder generalizability [14–16]. To address these issues in the context of daily life, an increasing number of studies in China have focused on grasping scattered objects, with numerous companies and research institutes developing prototype systems tailored to this task. For instance, a team led by Dong [17] bases their approach on the three-dimensional dimensions of target objects to register and determine their pose, subsequently employing inverse kinematics to guide the robot in grasping. This algorithm is implemented using LabVIEW. Another example comes from Zuo Lizheng and his colleagues at Harbin Institute of Technology [18], who utilize a Kinect camera to acquire the pose of target objects, successfully achieving recognition and pose calculation for scattered components.

(3) **Human–Robot Collaboration Technology**

Robots lack the capability to independently handle complex and flexible assembly tasks. An effective approach to enhancing the flexibility of assembly systems is to combine the agility of humans with the efficiency of robots. Based on seamless human–robot interaction and intention recognition capabilities, human–robot collaboration (HRC) can reach an adaptive level. Liu et al. [19] modeled the assembly tasks of products as sequences of human motions, enabling the robot to be controlled and assist human workers in completing assembly tasks by predicting their motion intentions. Owing to the high autonomy and interoperability of Cyber-Physical Production Systems (CPPS), it is widely applied in the product manufacturing process. Yao et al. [20] introduced Functional Blocks (FBs) to enhance the efficiency of task planning, monitoring, and control. Darvish et al. [21] proposed a flexible HRC assembly framework that integrates sensing, expression, planning, and control. The application of an online reasoning module after recognizing human behaviors further improves the performance of HRC assembly. Considering the status quo of multimodal human–robot interaction, Wang et al. [22] outlined research directions for exploring human–robot collaboration in shared workspaces.

Digital Twin (DT) was initially proposed by Grieves et al. [23], representing a multidisciplinary, multi-physical, multiscale, and multi-probabilistic real-time simulation and optimization process. Building upon three-dimensional models, Tao et al. [24] introduced a five-dimensional model encompassing physical entities, virtual devices, services, data transmission, and connections. Characterized by comprehensive functionality, DT has been applied and studied in various contexts. Sierla et al. [25] extracted DTs from digital product descriptions, subsequently automating assembly planning and resource allocation processes. Sun et al. [26] investigated the assembly and debugging process of High Precision Products (HPPs), realizing the prediction of assembly feasibility and optimization based on a holistic assembly model. Guo et al. [27] proposed a Digital Twin Graduation Project Intelligent Manufacturing System (DT-GiMS) to tackle the assembly issues of fixed islands. This framework, integrating physical, digital, and service layers, provides visual information for managers. In human–robot collaboration assembly, the assembly process mirrors the sequence and methods of operations. Yi et al. [28] not only presented a decision support model for intelligent planning of the assembly process but also outlined an application framework based on a three-layer assembly supported by decision-making.

In human–robot collaboration, the uncertainty in robotic actions poses a challenge to ensuring human safety, rendering the description of variable spaces and human bodies crucial. Digital Twins (DTs) address this by capturing real-time data in the physical space and replicating it in a virtual space, thereby providing users with visual control functionalities. Droder et al. [29] explored the role of machine learning in controlling robot behavior within DTs, finding that robots can autonomously avoid obstacles using machine learning techniques. Oyekan et al. [30] established a DT workshop to analyze human responses to both predictable and unpredictable robot movements, revealing correlations between real-life human reactions to robot actions and those observed in virtual reality scenarios. This suggests that virtual reality technologies can inform safe human–robot interactions. Driven by DTs, human–robot interactions become smoother and more frequent. Ma et al. [31] utilized a five-dimensional DT model and designed a human–robot interface, facilitating efficient Human–robot Interaction (HMI). DTs provide high-fidelity virtual models, rich product lifecycle data, and services information, enhancing the effectiveness of Human–Robot Collaboration (HRC). Wang et al. [32] integrated Visual Question Answering (VQA) technology into DTs, enabling efficient human–robot collaboration. Compared to conventional methods based on static images, the proposed VQA approach better comprehends video information, making it more suitable for dynamic HRC scenarios.

Given DT's capabilities in monitoring, predicting, and optimizing, it can provide effective collaboration strategies for HRC assembly. Malik et al. [33] employed mirrored digital models of physical parts to simulate assembly plans. Their proposed distributed testing framework allows for online or offline experimentation, averting economic losses and personal injuries in actual production. Bilberg et al. [34] established a corresponding DT model for a flexible assembly unit, extending the use of simulation models to real-time control, task allocation, task sequencing, and program development, among other aspects. This framework prevents related economic losses and personal harm while enhancing the efficiency of human–robot collaboration. Kousi et al. [35] employed digital modeling techniques in production systems to enable system reconfiguration through shared environments and process awareness. Experimental results indicate that the application of DTs can adapt to different scenarios, thereby enhancing the flexibility of HRC assembly systems. Baskaran et al. [36] utilized the Siemens Tecnomatix suite to construct DT models for both humans and robots, exploring the relationship between ergonomics and assembly tasks. In the HRC assembly process, the virtual robot in the digital space physically assists the human and effectively evaluates the assembly outcomes.

8.2 Scene Understanding in Human–Robot Collaboration

With the transformation of manufacturing paradigms from mass production to small-batch customization, manufacturers require significant enhancements in both production flexibility and efficiency. Traditional manufacturing models, due to their poor adaptability, struggle to meet the demands of flexible manufacturing.

Human–robot collaboration (HRC) conducted in collaborative environments effectively enhances the flexibility of manufacturing systems, enabling adaptive production. To facilitate intelligent HRC, a reinforcement learning-driven HRC framework is proposed (as depicted in Fig. 8.1). This approach establishes a Markov Decision Process (MDP) based on the manufacturing process and employs reinforcement learning to enhance the robot's cognitive abilities through environmental interactions.

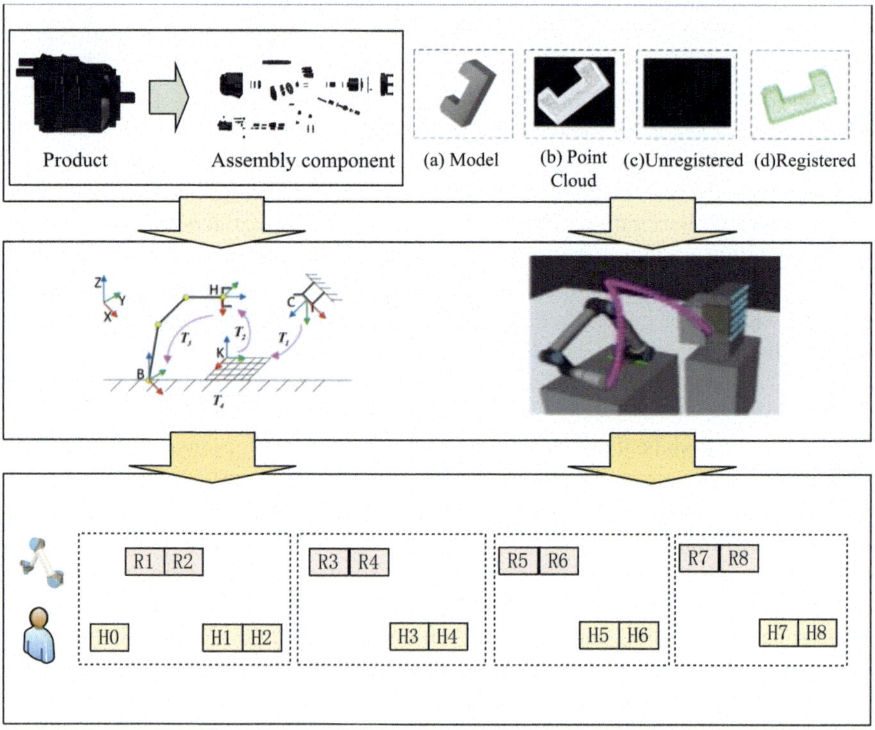

Fig. 8.1 Overall framework

8.2.1 Construction of Virtual Datasets for Cluttered Scenes

In real-world working environments, objects to be grasped often exist in disordered and scattered states, as illustrated in Fig. 8.2. For scenarios involving a large number of randomly stacked objects, training deep learning networks typically requires substantial datasets. Furthermore, stronger network models come with more hyper-parameters, necessitating even larger and higher-quality datasets. The proposed approach confronts the challenge of high costs associated with constructing real-world scenarios. It also addresses the time-consuming and labor-intensive processes for data collection and annotation. There is a common perception that uncontrollable factors during data capture may undermine dataset quality. This approach aims to overcome these issues. By adopting virtual physics engines for scene construction, one can leverage their strong randomness, high efficiency, and ease of annotation, thus overcoming these challenges.

When simulating scenes using the V-Rep simulation software, considering the low resolution of depth camera images and the loss of edge information pixels, it is necessary to perform data augmentation on the acquired image datasets. As shown in Fig. 8.3, two methods can be employed for depth image enhancement: local optimization and global optimization.

(1) **Construction of Virtual Datasets**

The design of virtual datasets primarily involves model creation, material rendering, and scene setup, accomplished using modeling, optimization, and simulation software such as Inventor, 3ds Max, and V-Rep. The main steps are as follows:

1. Utilize Inventor to construct standard structural component models at a 1:1 scale to real parts, resulting in STEP (Stp) 3D model files.

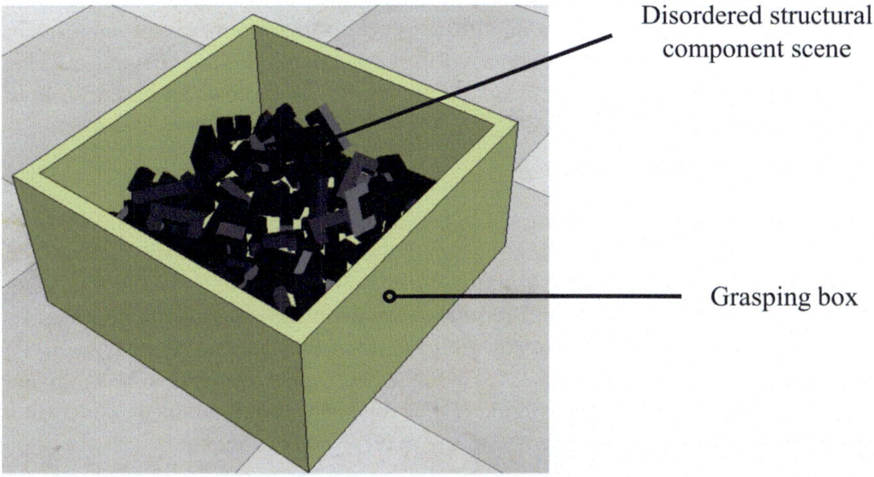

Fig. 8.2 Scatter scene with workpieces

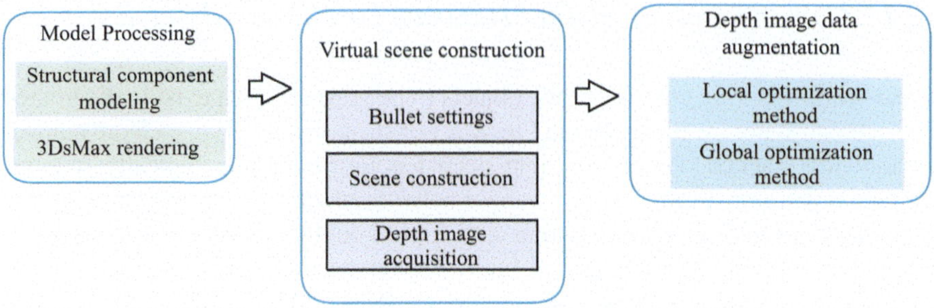

Fig. 8.3 Overall scheme for constructing virtual datasets

2. Apply 3ds Max to render materials and textures onto the standard structural component models, enhancing their resemblance to real-world components in terms of surface conditions. This yield saved OBJ model files, MTL material files, and TGA texture map files.
3. Import the OBJ model files into V-Rep and adjust V-Rep parameters according to the MTL and TGA files, configuring Bullet physics engine parameters to complete individual object parameter settings.
4. Replicate the import process for several structural components, positioning them at 200 mm along the Z-axis. Upon activation of the physics engine, the components undergo free-fall under gravity's influence, colliding and eventually forming a disordered, stacked scene.

(2) **Depth Image Data Augmentation**

Depth images captured by a Kinect in V-Rep exhibit characteristics such as low resolution and missing edge information. To make these depth images suitable for practical applications, it is necessary to enhance the depth images obtained from V-Rep, thereby improving their quality to meet the requirements of subsequent research. There are numerous methods for depth image data augmentation. This paper primarily investigates depth image enhancement approaches based on both local and global optimization techniques.

1. **Based on Local Optimization**

The principle of local optimization for depth images, as described in [37], is as follows: the characteristics of a certain pixel in the image are collectively determined by the features of points within a vicinity of that point. To repair and restore missing pixels in the image, let the region requiring repair in the image be denoted as Ω, with its boundary being $R(\Omega)$. For any arbitrary point $P(R)$ on this boundary, select a region centered at $H(P)$ with radius δ. Then, the pixel value of $P(R)$ can be estimated using the pixel points $Q(R)$ from region $H(P)$. When the radius parameter δ is chosen appropriately, the first-order estimate of the pixel $P(R)$ is given by:

$$I_P = I_Q + \nabla I_Q(P(R) - Q(R)) \tag{8.1}$$

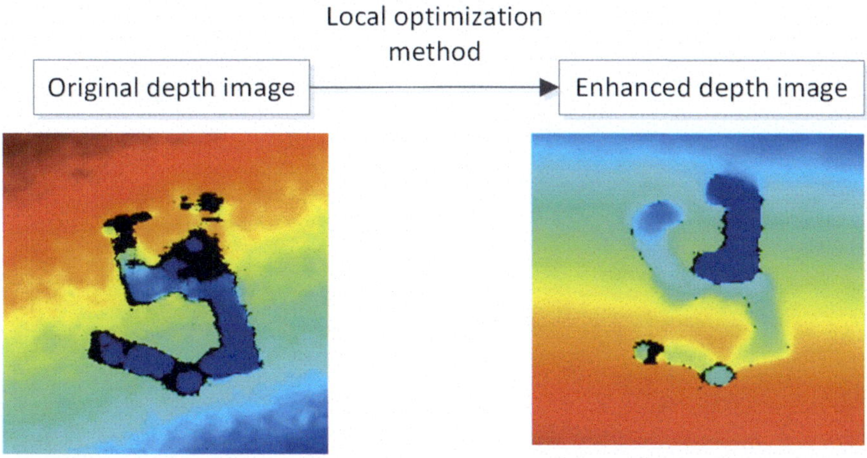

Fig. 8.4 Effect diagram of depth image local optimization

The pixel estimation formula for pixel $P(R)$ is as follows:

$$I_P = \frac{\sum_{Q \in R(\Omega)} k(P, Q)[I_Q + \nabla I_Q(P(R) - Q(R))]}{\sum_{Q \in R(\Omega)} k(P, Q)} \tag{8.2}$$

The effect of applying local optimization methods to the depth image dataset acquired from V-Rep is illustrated in Fig. 8.4. As shown, the original depth images obtained from V-Rep exhibit missing edge information of the structural components. After employing the local optimization method, the edge information of the components is successfully reconstructed, demonstrating excellent results suitable for subsequent point cloud conversion processes.

2. **Based on Global Optimization**

Depth image enhancement is not limited to local optimization techniques; employing a global method that adheres to specific constraints can also be a viable strategy for image augmentation. Among global image enhancement methodologies, methods based on undirected probabilistic graphs generally yield moderate improvements. This work primarily adopts the Markov Random Field (MRF) model [38], which combines low-resolution depth images with high-resolution RGB images to achieve enhanced depth images at a higher resolution. This approach serves as a preprocessing step for depth images in subsequent processing stages.

Assuming the depth measured by the Kinect in V-Rep is denoted as D, and the estimated depth is represented by Z, in accordance with Bayesian theory, when the measurement D remains constant, the posterior probability $P(Z|D)$ of the estimated depth $P(Z)$ can be obtained by combining the prior probability $P(D|Z)$ of the estimated depth with the maximum likelihood estimation probability, as illustrated by the following equation:

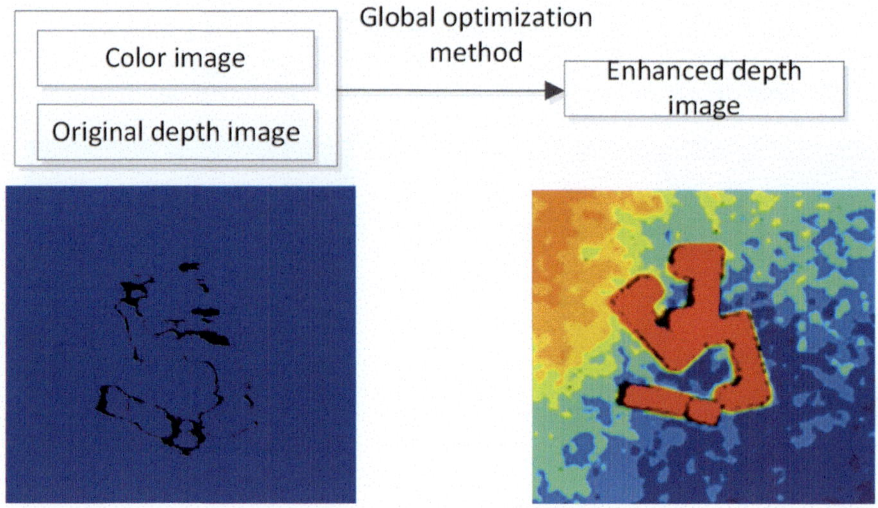

Fig. 8.5 Results using global optimization technique

$$P(Z|D) = \frac{P(D|Z) \cdot P(Z)}{P(D)} \tag{8.3}$$

where in $P(D) = \int_Z P(D|H) \cdot P(H)$, the formula after applying logarithmic operation becomes:

$$-\log P(Z|D) = -\log P(D|Z) - \log P(Z) + \log P(D) \tag{8.4}$$

Proceeding to solve for minimum energy:

$$E(Z|D) = E(D|Z) + E(Z) + E(D) \tag{8.5}$$

The effect of using global optimization is illustrated in Fig. 8.5, where the low-resolution depth image is globally optimized alongside the high-resolution color image. The shortcomings of the original point cloud diagram, such as the loss of edge information, are effectively addressed through the employment of global optimization techniques.

8.2.2 *Vision-Based Object Recognition and Localization*

After acquiring the depth image dataset, it undergoes point cloud image conversion and dataset partitioning. Subsequently, for structural component recognition, relevant contemporary network models are employed for model training individually, followed by performance evaluation to ascertain the optimal model. Furthermore, an object localization approach based on feature matching is adopted to locate and estimate the pose of structural components within the scene.

The virtual depth image dataset obtained in the previous section is transformed into point cloud images and then divided into training and validation sets. Notably, state-of-the-art models including PointNet++, ShapeNet, and 3DCNN are utilized for model training individually. Performance metrics are employed for model assessment, leading to experimental validation and analysis of structural component classification, ultimately selecting the model with the best recognition efficacy as the method adopted in this paper. Additionally, a feature matching-based object localization approach is proposed. This approach facilitates the localization and pose estimation of structural components within the scene. It provides precise pose data of these components. This data is used for the grasping study in Chap. 4. The research roadmap is depicted in Fig. 8.6.

(1) **Converting Depth Map to Point Cloud Image**

In the previous section, we obtained a dataset of depth images for structural components in a scattered state. Currently, point cloud data is commonly utilized as the dataset for deep learning applications. Consequently, to facilitate the training of subsequent models, it is necessary to preprocess this dataset and partition it into training and validation sets.

The principle behind depth imaging primarily involves capturing the distance of each point within the image, using this distance as pixel values to generate a distance image. Depth images can reveal contour information and positional information of objects within the image. Given that subsequent research is based on point clouds, this subsection will delve into the transformation relationship between depth images and point cloud images.

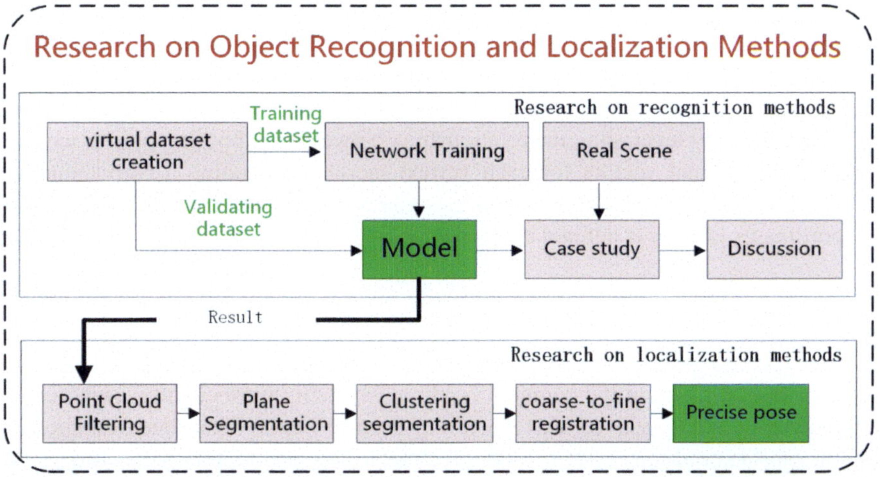

Fig. 8.6 Research framework for structural component recognition and localization

Typically, the movement of a camera can be viewed as a rigid body motion with six degrees of freedom, which can be represented by a transformation matrix P.

$$P = \begin{bmatrix} R_{3\times3} & t_{3\times1} \\ 0 & 1 \end{bmatrix} \tag{8.6}$$

where $R_{3\times3}$ represents the rotation matrix of the camera and $t_{3\times1}$ signifies the translation matrix.

Any point p_u in the camera's coordinate system can be transformed into a coordinate q_u in the world coordinate system through Eq. (8.7):

$$q_u = P \cdot p_u = R_{3\times3} \cdot p_u + t_{3\times1} \tag{8.7}$$

At each time instance, a depth image can be acquired using the camera, where the position of a pixel in the depth image can be represented using homogeneous coordinates indicating the depth value.

The representation of this point in the point cloud coordinate system is given by $P = (X, Y, Z, 1)^T$. Assuming the camera's intrinsic matrix is K, the projection equation for a pixel is formulated as follows:

$$n = u(P) = \frac{1}{D}[K, \vec{0}]P \tag{8.8}$$

To convert the depth image into a point cloud image, it is necessary to inversely solve Eq. (8.8). By back-solving the projection equation, we obtain the point cloud coordinates of the point in the camera coordinate system. The inverse solution formula is as follows:

$$m = u^{-1}(P) = D(V)K^{-1}n \tag{8.9}$$

For each pixel in the depth image, after performing a backward projection into the camera coordinate system, a corresponding 3D vertex map W_i is obtained. In this 3D vertex map, by leveraging the relationships between neighboring points, we can compute the normal vectors for each vertex, thereby generating a corresponding vertex normal map E_i. The relationship between the normal vector e_i of a vertex and its coordinates $w(x, y)$ is defined as follows:

$$e_i = (w_i(x+1, y) - w_i(x, y)) \times (w_i(x, y+1) - w_i(x, y)) \tag{8.10}$$

The vertex map and the normal vector map represent the spatial pose of pixels from the depth image in the camera coordinate system. Subsequently, each vertex is transformed along with its corresponding normal vector. This transformation is into the world coordinate system for point clouds. This accomplishes the conversion from the depth image to a point cloud image. As illustrated in Fig. 8.7, this process depicts the transformation of a depth image into a point cloud image.

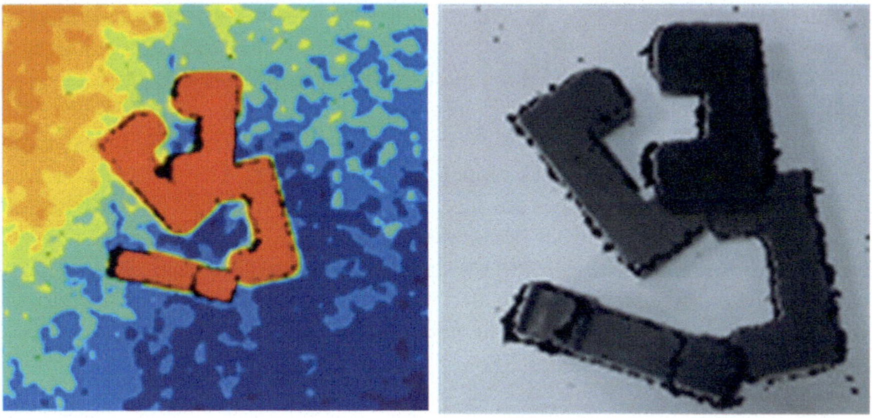

Fig. 8.7 Conversion of depth image to point cloud image

(2) **Deep Learning-Based Object Recognition Methods**

In recent years, deep learning has emerged as the most prominent research direction in computer vision, with extensive studies conducted on object pose estimation, object recognition, and target localization. This section will employ a virtual structured parts dataset to train three models: PointNet++, ShapeNet, and 3DCNN. The performance of these models will be evaluated through relevant metrics, allowing for the selection of the model with the optimal performance. Subsequently, the chosen model will be validated through experiments in real-world scenarios.

1. **Preprocessing and Parameter Configuration**

Currently, PointNet++, ShapeNet, and 3DCNN utilize common datasets. If you intend to train your own point cloud dataset, the following preprocessing steps are required.

Firstly, convert the point cloud format. Under the TensorFlow framework, the scene point cloud data obtained in PCD format needs to be converted into the HDF5 standard input format. Utilize the PCL library to batch process PCD files into PLY format, followed by converting the PLY files into H5 file format using the plyfile module in Python.

Secondly, given that more point cloud data during training incurs higher computation costs, adjustments need to be made to the NUM_POINT parameter in train.py. Depending on the characteristics of the textual dataset, subsampling is set up to downsample and update the H5 dataset in one pass. Typically, the scale space of point clouds encompasses all scales of the cloud, with information at these scales extractable through further processing to derive the primary features of the cloud. Analyzing all scales of the point cloud theoretically enhances model performance but necessitates a compromise in terms of time complexity, leading to unexpectedly high runtime and memory demands that can be counterproductive. Consequently, it is advisable to select a finite number of scales to expedite model training, optimizing the training scale for each of the three methods to speed up training times and enhance model efficiency. Scale parameters will vary according to the actual point cloud data, with the scale parameters for the three methods detailed in Table 8.1.

During the model training process, appropriate network parameters enhance the model's recognition capabilities, reduce training time, and can thereby improve the performance of the neural network model to a certain extent. Generally, the size of the batch (batch_size) determines the speed of gradient descent and convergence. A larger batch_size leads to a faster decline in the loss function and a quicker convergence rate; however, it also results in a substantial increase in GPU memory usage. An appropriate batch_size should be selected to ensure the sustainability of the current training environment. For example, the network parameter settings for PointNet++ during training are as illustrated in Table 8.2.

2. **Network Model**

PointNet++ was proposed to address the limitations of PointNet in complex scene recognition and segmentation tasks, where it exhibited insufficient generalization capability and failed to capture and process local features generated during training. Fundamentally, it adheres to the PointNet framework, commencing with hierarchical sampling of the scene point cloud. Post-sampling, point features bypass concatenation layers and directly enter segmentation layers. The output from this sampling serves as input to the next layer of the network. The primary structure is depicted in Fig. 8.8. The PointNet++ network model primarily resolves three issues: firstly, how to extract features within local regions; secondly, how these local regions are defined; and lastly, how to effectively extract local features under varying densities.

To address the first two issues regarding the generation and extraction of features within local regions, PointNet++ introduces a Set Abstraction (SA) layer to perform multi-layer segmentation of the point cloud and local feature extraction. This layer primarily consists of three network components: the sampling layer, grouping layer, and PointNet layer.

The sampling layer selects a subset of points to serve as centroids for the various local regions. Following this, the grouping layer performs a nearest neighbor clustering around these centroids, effectively segmenting the overall point cloud into

Table 8.1 Scale parameter settings

Model	PointNet++	ShapeNet	3DCNN
Scale setting	0.2	0.3	0.1

Table 8.2 Network parameter settings for PointNet++

Parameters		Parameters	
Optimizer	Adam	Dropout	0.4
Activation function	ReLu	Batch_size	32
Learning rate	0.002	Normalization	None

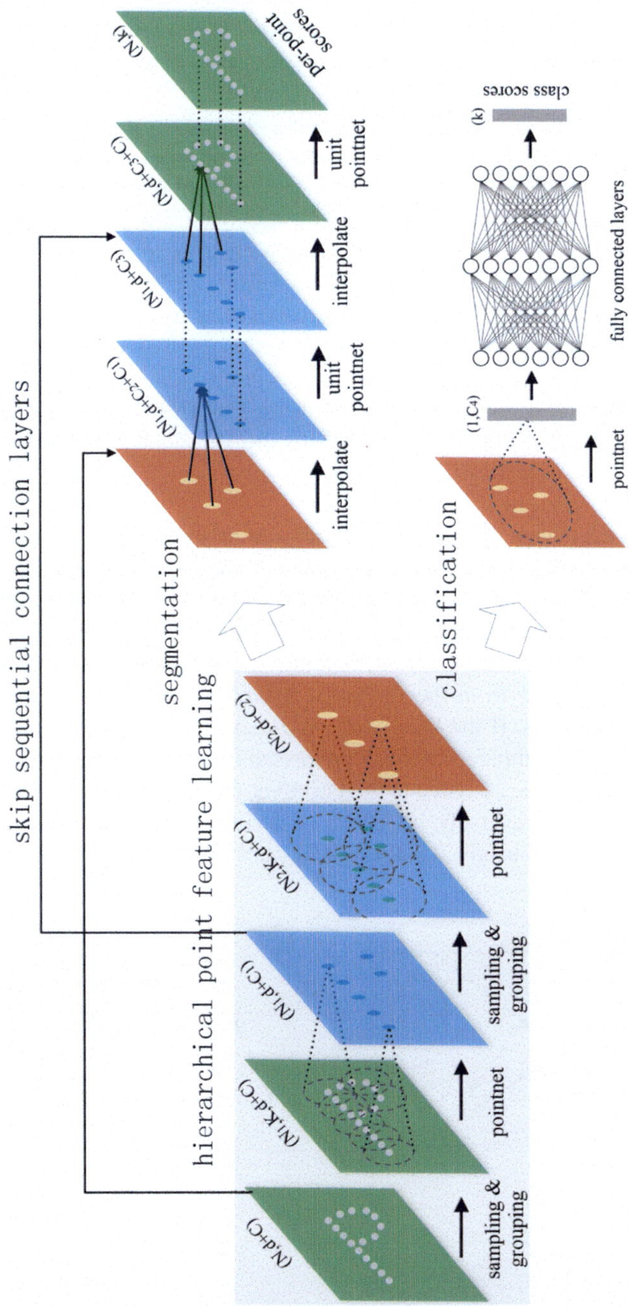

Fig. 8.8 Network architecture of PointNet ++ [39]

distinct local regions. The PointNet layer then extracts features from these localized regions. After the scene point cloud undergoes these three steps, it enables the extraction of local feature information between points in the scene.

Regarding the third issue, the challenge of extracting local features under varying densities, PointNet++ addresses the inherent non-uniformity of point cloud density. In the grouping layer, clusters with a greater number of points post-clustering indicate higher density, which in turn allows for the extraction of richer feature information. By incorporating multi-resolution and multi-scale combinations into the PointNet++ architecture, the network facilitates a clustering approach that effectively handles point clouds of different densities. This integration ensures comprehensive feature extraction across a wide range of point cloud densities.

As depicted in Fig. 8.9, the illustration showcases the concept of multi-scale and multi-resolution combinations. Multi-scale combination involves capturing spherical volumes at varying radii around key points in the point cloud. Features generated under these different radii are then extracted using PointNet layers. These multi-scale features are bundled together and fed into the subsequent network layer as input, with this process repeating iteratively. However, this method incurs significant computational costs and time consumption.

To enhance the efficiency of feature extraction, PointNet++ introduces the multi-resolution combination strategy. This approach extracts local features at different resolutions and merges high-level features with low-level ones, thereby enabling the algorithm to capture a broader range of local information. For example, in a specific layer of the network, when using the multi-resolution combination method, the local features of this layer come from the fusion of local features and those processed directly by PointNet layers. If the locally-derived features are less informative due to inadequate sampling compared to those processed directly by PointNet, a higher bias weight might need to be assigned to the PointNet-processed features for optimal combined performance. Conversely, if the locally-derived features are more informative, a higher weight should be given to them. A key advantage of the multi-resolution

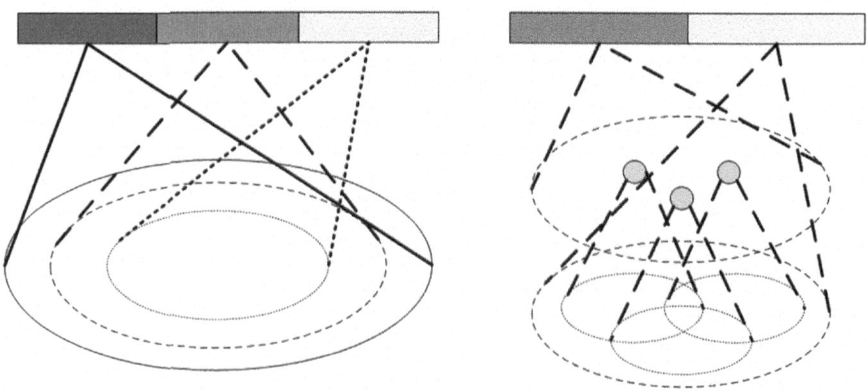

Fig. 8.9 Multi-scale and multi-resolution combinations

combination is its reduction in the need for extensive multi-scale feature sampling at lower network levels. It dynamically adapts to variations in point cloud density and proves computationally more efficient than the multi-scale approach.

3. **Performance Evaluation**

To validate the performance of the trained models, several parameters are typically analyzed.

Overall Accuracy (oAcc). Overall accuracy is the simplest metric used in deep learning, representing the ratio of correctly predicted instances to the total number of instances. Suppose we have a segmentation scenario with $(n + 1)$ classes. Let T_{ij} denote the number of points that actually belong to class i but are classified as class j. Similarly, T_{ji} represents the number of points correctly classified into class i. The formula can be expressed as follows:

$$oAcc = \frac{\sum_{i=0}^{n} T_{ii}}{\sum_{i=0}^{n} \sum_{j=0}^{n} T_{ij}} \tag{8.11}$$

Mean Intersection Over Union (mIoU). The mean Intersection over Union (mIoU) is defined as follows: Initially, the count of successfully predicted categories is determined. Then, the intersection between the predicted and ground truth values is calculated for each class. Finally, the average of these intersections over unions is computed across all classes. While mIoU is commonly used for the evaluation of image segmentation in 2D, its principle is equally applicable to the assessment of 3D point cloud scene segmentation, employing the same parameters as those used in overall accuracy. The formula is represented as follows:

$$mIoU = \frac{1}{n+1} \sum_{i=0}^{n} \frac{T_{ii}}{\sum_{j=0}^{n} T_{ij} + \sum_{j=0}^{n} T_{ji} - T_{ii}} \tag{8.12}$$

(3) **Object Localization Methods Based on Feature Matching**

PointNet++ can identify and categorize workpieces, as well as obtain a rough pose of structural components through relevant computations. However, the error in this rough pose tends to increase during coordinate system transformations, preventing the achievement of precise component poses.

In contrast, object localization methodologies based on multidimensional features initially entail preprocessing and scene segmentation of the post-recognition scene point cloud. Subsequently, the cosine of the angle is employed to secure matching point pairs, facilitating a coarse-to-fine registration of the point clouds. Ultimately, this leads to pose estimation and acquisition of the precise pose of the structural components. This approach is particularly suited for scenarios involving small sample sizes and low real-time requirements.

1. **Scene Point Cloud Preprocessing and Segmentation**

Given that the virtual dataset has undergone artificial noise addition, the properties of the target object's surface may alter. Moreover, due to line-of-sight occlusions, obstacles, and other factors, noise points and outliers inevitably emerge. Consequently, point cloud filtering is employed as the initial step in the preprocessing pipeline to eliminate noise points, outliers, or voids, thereby facilitating more effective point cloud segmentation and registration subsequently.

Common point cloud filtering techniques include median filtering, Gaussian filtering, and mean filtering. Additionally, in image processing, bilateral filtering is adopted, which performs a weighted average across the entire image, utilizing a weighted average of pixels within a certain range to represent the pixel value of a point within that range. Employing a Gaussian-weighted average achieves noise reduction.

Point cloud segmentation partitions the point cloud based on attributes such as shape and spatial distribution, ensuring that the segmented clusters exhibit similar characteristics. Segmentation is a prerequisite for point cloud registration, as clustering distinct categories of points within the scene point cloud is essential for extracting better features. Below, Random Sample Consensus (RANSAC) algorithm for plane segmentation will be introduced as an illustrative example.

A plane in point cloud can be defined by the equation $ax + by + cz + d = 0$. From the scene point cloud, points within a certain threshold range are segmented, transforming the scene point cloud into separate clusters of planar points and structural component points. The RANSAC algorithm, which iteratively estimates the parameters of the model (in this case, a plane) from random samples of the data, can be outlined by the following formulation:

$$\begin{bmatrix} x_1 & y_1 & z_1 & 1 \\ x_2 & y_2 & z_2 & 1 \\ x_3 & y_3 & z_3 & 1 \end{bmatrix} \begin{bmatrix} a \\ b \\ c \\ d \end{bmatrix} = \vec{0} \tag{8.13}$$

The steps of the RANSAC plane segmentation algorithm are as follows:

1. Initialize the plane parameters $a_0 = 0$, along with the number of inlier samples $\max(n) = 0$, and set an initial threshold value δ.
2. Randomly select three non-collinear inlier sample points, denoted as $P(x_i, y_i, z_i), i = 1, 2, 3$, and use these to form a plane model M. Solve for the plane's normal vector A' using formula (3–11).
3. For each data point $P\left(x_j, y_j, z_j\right), j = 1, 2, 3$ in the scene point cloud, calculate the distance $\left|P \cdot A'\right|$ from point P to the plane model M. If $\left|P \cdot A'\right| < \delta$, then point P is considered an inlier point of the plane; otherwise, it is classified as an outlier.
4. Repeat steps (2) and (3) until reaching the maximum number of iterations $k_{\max}$, at which point the segmented plane A is obtained.

Regarding the determination of parameter δ, tuning is necessary based on experimental results.

After segmenting the planar portion from the scene point cloud, resulting in a stack of scattered structural component point clouds, further segmentation of the structural component clusters is required. This is achieved using Euclidean clustering, which defines proximity or distance among points via the Euclidean distance. The main steps are as follows:

1. Obtain the curvature features of the point cloud, then extract the boundary lines of the target point cloud to acquire the cluster of points within the boundary, denoted as $P\{\emptyset\}$.
2. Construct a KD-tree to represent the relationships among the points in cluster $P\{\emptyset\}$.
3. Create a verification queue *Queue*, and initialize an empty set of clusters $S\{0\}$.
4. Enqueue any arbitrary point p_i from $P\{\emptyset\}$ into the verification queue *Queue*.
5. Search for points p_i within a radius r of from to form a neighborhood set $K\{\emptyset\}$. If a point p_{ik} in this neighborhood set $S\{\emptyset\}$ has already been processed in previous steps, add it to set; otherwise, place it back into *Queue*. This continues until all points in the verification queue have been processed and placed into a cluster in set $S\{\emptyset\}$, with the verification queue *Queue* reset to an empty set.
6. Repeat steps (3), (4), and (5) until all points in the cluster $P\{\emptyset\}$ have been processed.

2. **Construction of Matching Points and Key Point Sampling**

Following point cloud segmentation, the next step is to register the point clouds. The fundamental principle of point cloud registration involves determining a rotation matrix and a translation matrix that enable the transformation between two point clouds. Assume one point cloud is the target (reference) point cloud, while the other is the source (to be registered) point cloud. The algorithm computes these rotation and translation matrices to maximize the overlap between the two point clouds. The calculation formula can be outlined as follows:

$$\begin{bmatrix} `x_j \\ `y_j \\ `z_j \end{bmatrix} = R \begin{bmatrix} x_j \\ y_j \\ z_j \end{bmatrix} + T \tag{8.14}$$

where $R = \begin{bmatrix} r_{11} & r_{12} & r_{13} \\ r_{21} & r_{22} & r_{23} \\ r_{31} & r_{32} & r_{33} \end{bmatrix}$, $T = \begin{bmatrix} t_1 & t_2 & t_3 \end{bmatrix}^T$.

Based on two sets of point cloud data, computations can yield a rotation matrix R and a translation matrix T. However, when depth data is converted into a point cloud dataset, there exists inherent randomness in the depth distance information, leading to errors that prevent the point cloud data from matching accurately. Consequently, it is necessary to establish correspondence pairs between points. This paper will adopt the cosine of vector angles as the criterion for matching these points.

Using the cosine of vector angles as the matching condition enhances the precision and accuracy of matched points. The primary steps involved are as follows:

1. Initialize the set of source point clouds M and the target point cloud N;
2. For an arbitrary point p_i in the target point cloud N, obtain its multiple feature curvatures k_1 and k_2, and calculate the Gaussian curvature k_G and mean curvature k_a.
3. Compute the distance L of this point p_i to the centroid of the structural object, and the average angle θ_a of the normal vectors.
4. Construct the multiple feature vector $P_i = (p_{i1}, p_{i2}, p_{i3}, p_{i4}, p_{i5}, p_{i6})$ for this point, where $p_{i1} = k_1, p_{i2} = k_2, p_{i3} = k_G, p_{i1} = k_a, p_{i5} = L, p_{i6} = \theta_a$.
5. Identify the K nearest neighbor points q_j of point p_i in the source point cloud M, and similarly obtain their multiple feature vectors $Q_j = (q_{j1}, q_{j2}, q_{j3}, q_{j4}, q_{j5}, q_{j6})$. The similarity is calculated using the cosine formula, as shown by the equation:

$$F(p_i, q_j) = \frac{\sum_{k=1}^{n} (p_{ik}, q_{jk})}{\sqrt{\sum_{k=1}^{n} p_{ik}^2 \sqrt{\sum_{k=1}^{n} q_{jk}^2}}} \tag{8.15}$$

Match point p_i in the target point cloud with the corresponding point q_j that has the highest cosine feature similarity $F(p_i, q_j)$ within its neighborhood.

For key point sampling, the degree of undulation in local regions of a 3D point cloud is related to variations in the normal vectors near points in the scene point cloud. Local variation can be quantified using the cosine of the angles between normal vectors. This paper employs a multi-resolution key point sampling method to extract key points from the structural component point cloud.

In the process of point cloud registration, it is essential to select key points rich in feature information as the initial points for registration. The angle between K-nearest neighbor normals, θ_i can reflect the degree of undulation in the point cloud.

$$\theta_i = \frac{1}{k} \sum_{j=1}^{k} \arccos\left(\frac{\vec{n}_i \cdot \vec{n}_j}{|\vec{n}_i| \cdot |\vec{n}_j|}\right) \tag{8.16}$$

Using formula (8.16) to calculate the normal vector angles for all points, and dividing all points into n equal segments, the value range for the points in the ith segment is as indicated by the following formula.

$$F_i = \frac{\max(\theta) - \min(\theta)}{n} \tag{8.17}$$

where $\max(\theta)$ and $\min(\theta)$ represent the maximum and minimum values, respectively, of the angular deviations between normal vectors for all points in the point cloud.

During the registration process, alignment initially commences with lower-resolution data points. Sequentially, the registration resolution is increased, ultimately culminating in the completion of the registration task.

3. **ICP Registration Algorithm Based on Multi-resolution Sampling**

Point cloud registration can be seen as solving the mapping relationship between two coordinate systems. Currently, there are numerous algorithms for registration, including the classical Iterative Closest Point (ICP) registration algorithm [40], registration based on local descriptors (such as PFH, FPFH) [41], and registration based on probability distributions (Normal Distribution Transform, NDT) [42], among others. Building upon the traditional ICP registration method, this paper incorporates the multi-resolution sampling discussed in the previous section to enhance both the efficiency and convergence of the registration process. The steps for ICP registration based on multi-resolution sampling are outlined as follows:

1. Divide the target point cloud M and the point cloud N to be registered into n equal segments based on vector angles θ_i Initialize the maximum resolution for the iterative process as n, and perform point cloud data sampling according to the current resolution n_i. Additionally, initialize the registration threshold as δ
2. When $n_i < n/3$ this signifies the start of the coarse registration stage, where matching point sampling is conducted using cosine feature similarity $F(p_i, q_j)$.
3. When $n/3 < n_i < n$, this target the fine registration phase, where corresponding match points are obtained using k-nearest neighbors.
4. Perform Singular Value Decomposition (SVD) on the matched points to determine the transformation relationship between the point cloud M to be registered and the target point cloud N, thereby calculating the rotation matrix R and translation matrix τ.
5. Apply the rotation matrix R and translation matrix τ to transform the point cloud to be registered M yielding the set of points M
6. After the kth iteration, compute the mean squared error $E_k(R, T)$ for this iteration and compare it with the mean squared error $E_{k-1}(R, T)$ from the k-1-th iteration. If $|E_k(R, T) - E_{k-1}(R, T)| < \delta$, the loop terminates, completing the registration process; otherwise, the iteration continues.

4. **Estimation of the Target Object's Pose**

Each set of source point clouds, when registered with the target point cloud, yields a rotation matrix and a translation matrix. Consequently, this registration relationship needs to be mapped to the robot's end-effector coordinate system. This usually involves transformations across three frames: the point cloud coordinate system, the camera coordinate system, and the robot end-effector coordinate system.

To represent the object's orientation, rotation matrices are too man parameters, Euler angles suffer from gimbal lock issues; while quaternions avoid this problem, they can be difficult to understand and are abstract in representation. This paper opts to use rotation matrices for expressing the target object's pose due to their intuitive geometric interpretation despite their higher dimensionality.

8.2.3 *Sequence of Grasping for Robot Posture Adjustment*

For the optimal grasping sequence of disordered structural components, the analysis begins by examining four constraints during the grasping process. A grasping model is then constructed under multiple constraints, and an improved particle swarm optimization (PSO) algorithm is employed to optimize this model. This approach yields the optimal grasping sequence, with the grasping mode determined according to usage constraints of the structural components.

Initially, four constraints in the robot grasping process are analyzed: time constraints, physical constraints, workspace constraints, and incompleteness constraints. A grasping model is established under these multiple constraints, and the PSO is enhanced with Gaussian mutation operators and chaotic operators for optimization. By optimizing the objective function within the grasping scenario, the optimal grasping sequence is obtained, followed by validation of the algorithm's performance through experimental verification.

Through the use of an Iterative Closest Point (ICP) registration algorithm with multi-resolution sampling and pose estimation methods, the pose information of each structural component in the scene can be acquired. This pose information serves as the basis for the robot's grasping actions. Considering the final welding posture of the structural components, an adjustment algorithm based on pose matching is proposed, enabling the robot to adjust its posture to accommodate the grasping task. The methodological framework is illustrated in Fig. 8.10.

(1) **Particle Swarm Optimization-Based Method for Grasping Sequence**

When target workpieces are in a disorganized stacked state, the sequence in which a robot grasps these workpieces is influenced by multiple factors. These include

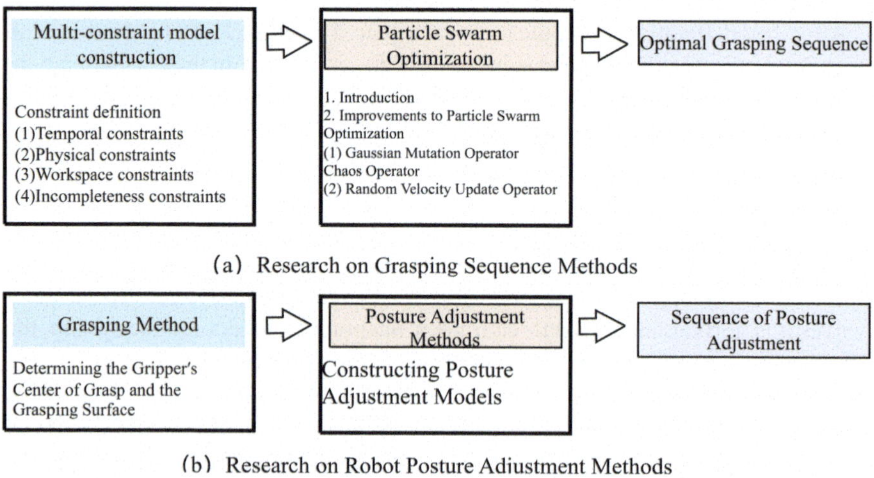

Fig. 8.10 Optimal grasping sequence and approach for posture adjustment

prioritizing the retrieval of top-layer objects, avoiding interference with other structures, adhering to joint limits during robot motion, among others. Consequently, the optimal grasping strategy can be viewed as a multi-constraint optimization problem.

The resolution of multi-constraint optimization problems typically involves employing an algorithm to find the best solution amidst multiple constraints. These constraints often interact, with effects that can be either synergistic or antagonistic. When a change in one constraint leads to a favorable improvement in another, it is referred to as a positive synergy relationship. Conversely, when a change in one constraint negatively impacts another, it is called a negative synergy relationship. In addressing multi-constraint optimizations, it's unrealistic to expect all constraints to reach their optimal states simultaneously. Instead, the goal is to find a dynamically balanced practical optimal solution. This solution aims to balance the optimization of target constraints while satisfactorily meeting all other constraints.

Generally, assuming the decision variables are of dimension N, an optimization model with M constraints can be defined as follows, where X and Y represent the decision and objective spaces, respectively:.

$$\min/\max \quad y = (f_1(x), f_2(x), \ldots, f_M(x)) \tag{8.18}$$

$$\textit{subject to} \quad x = (x_1, x_2, x_3, \ldots, x_N) \tag{8.19}$$

$$y = (y_1, y_2, y_3, \ldots, y_M) \tag{8.20}$$

During the robotic grasping process, there exists a relative positional relationship between the robot itself and the object to be grasped. With considerations for processing safety, reliability, and energy-efficient manufacturing, the following requirements must be met:

1. When identifying the grasping target, the robot end-effector's movement should minimize energy consumption and time when covering the same distance.
2. Objects must be within the robot's workspace, avoiding singular configurations to enhance overall maneuverability.
3. During grasping in layered or disorganized scenarios, the robot must avoid collisions with other objects to prevent secondary damage to both the grasped structure and the gripper.
4. The robot's acceleration and velocity must remain within predefined limits, without abrupt changes, to prevent damage to its components.

Current research also addresses the problem of finding optimal solutions under multiple constraints but struggles to satisfy these four points adequately. To obtain the optimal grasping sequence, this paper categorizes these requirements into four constraint conditions: temporal constraints, physical constraints, workspace constraints, and incompleteness constraints. Temporal constraints refer to the time and energy demands during actual grasping operations. Physical constraints pertain to limiting joint angles, velocities, and accelerations during the grasp to ensure task completion while prolonging the robot's lifespan, aligning with sustainability objectives. Workspace constraints involve assessing whether the robot can reach the object within the working environment and how to navigate obstacles to prevent collisions. Incompleteness constraints reflect the feasibility of grasping the object. This paper integrates these constraints and uses a high-performance particle swarm optimization (PSO) algorithm. The algorithm globally searches for the optimal grasping sequence between the objects and the robot. This approach leads to finding the best overall fitness solution.

1. **Temporal Constraints**

The most effective way to gauge efficiency is through time evaluation, with time metrics being a crucial indicator in robotic grasping tasks. The length of the cycle time when a robot grasps an object directly influence the processing efficiency. In this chapter, the grasping constraint is defined with the grasp and place positions predetermined. Given objects at different locations, the objective is to minimize the time taken by the robot to complete a grasp. Time assessment is formulated as Eq. (8.21).

$$t = \max_{i=1,\ldots,n} \frac{|l_{end} - l_{start}|}{v_{mi}} \tag{8.21}$$

By calculating the joint motion time for each joint operating at maximum velocity, along with the displacement of the nth joint $|l_{end} - l_{start}|$, the time taken by the slowest joint to move is determined, which in turn provides the overall motion time of the robot. Optimizing the speed of the slowest axis accelerates the overall movement time of the robot.

In addition to time constraints, the energy consumed during joint motions also significantly influences the robot's grasping efficiency. During the grasping process, robots typically operate in two modes: continuous motion, where not only the starting and ending points but also the path trajectory of the robot's motion are known—the energy consumed is related to the path taken and can be calculated accordingly; and intermittent motion, also referred to as point-to-point motion, where only the starting and ending points of the robot's motion are known. This paper defines energy consumption based on the torque exerted and the energy consumed during the robot's motion.

$$E = \int_0^j \left(\sum_{i=1}^{n} \left| T_i \cdot \dot{\theta}_i \right| \right) dt \tag{8.22}$$

where T_i is the torque of joint i, and i is the angular velocity θ_i of the ith joint.

2. **Physical Constraints**

Among the fundamental physical constraints of a robot, the primary one is the joint limit constraint. Due to factors such as the operating environment, drive motors, and joint structures encountered during the robot's manufacturing process, the range of motion for each degree of freedom must be restricted. Furthermore, each joint axis has a differing maximum attainable velocity and acceleration during its movement, necessitating separate limitations on both velocity and acceleration to prevent motor distortion. This distortion could lead to vibrations during the robot's grasping operation, affecting its lifespan. The physical constraints can be formulated as Eq. (8.23).

$$\begin{cases} \min(\theta_i) \leq \theta_i \leq \max(\theta_i) \\ |\omega_i| \leq \max(\omega_i) \\ |a_i| \leq \max(a_i);\ i = 1, 2, \ldots 6 \end{cases} \tag{8.23}$$

where $\min(\theta_i)$, $\max(\theta_i)$, $\max(\omega_i)$ and $\max(\alpha_i)$ represent the minimum and maximum joint angles, maximum angular velocity, and maximum angular acceleration respectively for the ith joint. In order to incorporate these constraints into the objective function, they need to be normalized onto a dimensionless scale spanning [0,1]. The normalized form of these constraints is given by Eq. (8.24).

$$\begin{cases} \vec{\theta}_i = \frac{|\theta_i - (\max(\theta_i) + \min(\theta_i))|}{(\max(\theta_i) - \min(\theta_i))} \leq 1 \\ \vec{\omega}_i = \frac{|\omega_i|}{\max(\omega_i)} \leq 1 \\ \vec{a}_i = \frac{|a_i|}{\max(a_i)} \leq 1;\ i = 1, 2, \ldots, 6 \end{cases} \tag{8.24}$$

3. **Working Space Constraints**

The workspace of a robot delineates its reachability in space, constituted by the set of points that the end effector can access, thereby reflecting the flexibility of the robot's motion. In reality, due to joint limits and mechanical structural constraints, the workspace of a robot is not a complete spherical structure. Thus, when attempting a grasp, it is essential to consider whether the object to be grasped lies within the robot's workspace.

Current research approaches include graphical, numerical, and analytical methods. However, graphical and analytical methods become increasingly complex as the degrees of freedom of the robot increase. Numerical methods, on the other hand, by selecting different sets of joint values and employing forward kinematics along with numerical integration, determine the workspace. Among numerical approaches, the Monte Carlo method stands out for its performance advantages. This method integrates statistical probability into the mathematical process, utilizing random numbers within a defined range to solve problems. This section employs the Monte Carlo method for analysis.

The application of the Monte Carlo method to solving the robot workspace problem comprises three primary steps: First, based on joint limitations, the range of motion for each joint is established [43]. Next, forward kinematics calculations are performed for each segment of the joint motion ranges, yielding a set of points representing the robot's possible positions. These points are then plotted to visually represent, and ultimately define, the robot's workspace.

The primary steps for solving the workspace of a typical serial robot are as follows. Perform a kinematic analysis of the robot, using forward kinematics calculations to sequentially determine the pose matrix of the robot's end effector. The pose vector can be represented as $[L_x, L_y, L_z]^T$.

Employ a random function $Rand(i), i = 1, \ldots, n$ to generate random values, which are then normalized to the range [0,1]. These normalized random numbers are further processed through multiplication with the respective minimum and maximum values of each joint's motion range to acquire random motion values for the joints:

$$\theta_i = \max(\theta_i) + (\max(\theta_i) - \min(\theta_i)) \times Rand(i) \tag{8.25}$$

Given the N random joint variable values obtained in the previous step, apply the inverse kinematics to solve for the pose matrices representing the robot's configurations. These calculated poses, through their distribution of N points, illustrate the authenticity of the represented workspace volume.

Finally, a point cloud diagram of the robot's workspace is generated, depicting the aggregation of all these calculated positions to visually represent the reachable areas in the workspace.

(2) **Particle Swarm Optimization Based on Gaussian Kernel Chaos Operator**

Traditional intelligent optimization algorithms are feasible for solving general optimization problems, but when it comes to tackling complex multi-constraint optimization issues, these algorithms typically handle all constraints by equivalently transforming them into a single constraint with additional weights assigned, a method that is heavily reliant on prior assumptions. Particle Swarm Optimization (PSO), one of the swarm intelligence algorithms, boasts advantages such as simple coding, fewer optimization parameters, and rapid convergence speed. It is a stochastic search algorithm. This section aims to employ PSO to optimize the grasping sequence of structural components. Addressing the challenge of PSO's tendency to get stuck in local optima, this section further integrates Gaussian mutation operators and chaotic operators to enhance the PSO.

Due to PSO's propensity to fall into local optima during iterative convergence, adjustments to the conventional PSO algorithm are necessitated, specifically through the incorporation of:

1. The Gaussian Mutation Operator [44]; in the process of PSO, the mutation operation is conducted utilizing a Gaussian probability, where the original particle is replaced by the mutated particle, formulated as:

$$Mutation(X) = X \cdot (1 + G(\delta)) \tag{8.26}$$

$$x_{ij} = z_{ij} \cdot (1 + G(\delta)) \tag{8.27}$$

where $G(\delta)$ represents the Gaussian mutation coefficient.

2. Chaotic Operator [45]; In the conventional PSO algorithm, the initialization of particles is typically random, which introduces considerable blindness. The randomness in the traditional PSO's initialization process entails a high degree of uncertainty. Conversely, chaotic state equations can address uncertainty issues by deriving stochastic particle motion states from a set of deterministic equations, a condition referred to as chaotic state. The ergodicity of chaotic equations can be utilized to tackle the randomness in the traditional PSO, thereby enhancing particle diversity and reinforcing the algorithm's exploration capability. The specific description is as follows:

$$z_{k+1} = \mu z_k(1 - z_k), \quad z_k \in (0, 1) \tag{8.28}$$

where z_k is a real sequence, and μ is the range parameter.

3. Velocity Operator: To enhance the optimization speed of the algorithm, an additional term incorporating a velocity value generated by a random variable is incorporated into the original velocity iteration formula, which is expressed as follows:

$$\begin{aligned} m_i(x+1) = \omega m_i(x) &+ p_1 r_{i1}(T_i^{tb}(x) - t_i(x)) + p_2 r_{i2}(T_i^{gb}(x) - t_i(x)) \\ &+ p_3(R_i(x) - t_i(x)) \end{aligned} \tag{8.29}$$

where $R_i(x)$ represents the point generated by the ith particle at the xth iteration, and p_3 is the diversity parameter.

4. Fitness Function: Initially, parameters such as population size and the elite solution set are determined within feasible ranges. The particle swarm is then initialized, with better particles selected to form the elite solution set. Particles in this elite set play a guiding role in the movement of other particles during the particle swarm optimization process. The fitness of the ith particle is defined as follows:

$$Y(i) = \begin{cases} 1 - \left(\frac{d(i,j)}{\sigma_s}\right)^{\alpha}, & d(i,j) < \sigma_s \\ 0, & others \end{cases} \tag{8.30}$$

where $d(i, j)$ denotes the Euclidean distance between particles i and j, and σ_s is generally a constant.

This algorithm employs chaotic state equations for particle initialization, thereby maintaining the chaotic sequential nature of particles while preserving a degree of randomness. After obtaining the initial particles, a Gaussian mutation operation is performed to enhance the diversity of the initial particles. During velocity updating, stochastic updates to the velocity operator are introduced to improve the global search capability of the particles. Probability-based mutation operations are applied to particles with similar fitness values to avert algorithm entrapment in local optima.

These improved strategies ensure both particle diversity and the convergence of the algorithm, with specific steps outlined as follows:

1. Initialization begins by generating an n-dimensional vector for the entire population, with each particle represented as $z_1 = (z_{1,1}, z_{1,2}, \ldots, z_{1,n}), z_{1,j} = rand()$, $j = 1, 2, \ldots, n$, utilizing formula (8.27) to derive initial vectors. The N initial particles are then computed using the Gaussian formula (8.26). Random numbers within a specified range generate initial velocities, followed by calculating the objective fitness function values for each particle according to formula (8.29). The best solution identified serves as the initial solution.
2. All particles' positions and velocities are iteratively calculated using formula (8.28), after which their fitness values are reassessed, and particles meeting the criteria for elite solutions are added to the elite set.
3. If the number of iterations is less than a predetermined threshold, steps (4) through (7) are repeated; otherwise, the algorithm concludes.
4. The fitness values of all particles in the elite set are obtained, and a global extremum calculation for these fitness values is conducted. Particles' positions and velocities are updated following formula (8.28), with subsequent computation of each particle's objective function value.
5. Proceeding to step (2), if the fitness value of a particle in the elite set falls below a certain threshold, it is removed from the elite set, ensuring the elite set remains within a manageable capacity.
6. If the position of the ith particle at the current iteration is superior to its previously best position, the best position $T_n^{tb}(x)$ is updated to the current position. If the difference between the current position and the previous best is within a predefined tolerance, $T_n^{tb}(x)$ may be replaced or retained based on a random replacement probability. Conversely, if the current particle's position is inferior to the optimal position, a Gaussian mutation with a 5% probability is initiated for reinitialization.
7. Ultimately, the solution identified within the elite set constitutes the optimal solution.

(3) **Construction of Objective Function**

When applying Particle Swarm Optimization (PSO) to determine the selection sequence of objects grasped by a robot, the problem can be translated into the trajectory of movements the robot needs to execute. Assume there are n paths to be traversed, with each path requiring the grasping of m objects. Let the objects be numbered from [1, m], implying that for each grasping instance, the order can be any permutation of the sequence [1, m]:

$$N = \{n_1, n_2, \ldots, n_i\} \tag{8.31}$$

$$\begin{cases} n_1 = \{4, 2, \ldots, m_i\} \\ \quad \cdots \\ n_i = \{7, 1, \ldots, m_i\} \end{cases} \tag{8.32}$$

Calculate the coordinates of the objects in the robot's base coordinate system along the grasping paths, and subsequently resolve these coordinates via inverse kinematics to obtain the rotation angles of each joint. The motion trajectories in joint space are then described using polynomial interpolation.

Considering the afore-mentioned analysis alongside various constraints, the objective function is defined as follows:

$$F = \begin{cases} k_3 \cdot cdt(3), cdt(3) \geq 1 \\ \max(k_1 \cdot cdt(1), k_2 \cdot cdt(2), k_3 \cdot cdt(3), k_4 \cdot cdt(4)), cdt(3) < 1 \end{cases} \tag{8.33}$$

where k_1, k_2 are weight factors, whose settings are illustrated in Eq. (8.34); $cdt(1)$ symbolizes the time constraint value, $cdt(2)$ represents the physical constraint value, $cdt(3)$ denotes the workspace constraint value, and $cdt(4)$ signifies the incompleteness value.

$$k_i = \begin{cases} 0 \quad Importance : cdt(i) > cdt(j) j \neq i \\ 0.5 \; Importance : cdt(i) = cdt(j) j \neq i \\ 1 \quad Importance : cdt(j) = cdt(i) j \neq i \end{cases} \tag{8.34}$$

Finding a solution that satisfies the established conditions represents the optimal feasible grasping solution, which is searched for using the enhanced particle swarm algorithm to locate the global optimum.

(4) **Method for Robot Posture Adjustment**

Taking the installation and welding of components on a spacecraft shell as an example, the final attitude and position for the weldment are depicted in Fig. 8.11. Practically, it is essential that the robotic end effector maintains a vertical orientation, as during subsequent welding operations, the welding plane must not coincide with the plane of the robotic end effector to prevent mis-welding occurrences.

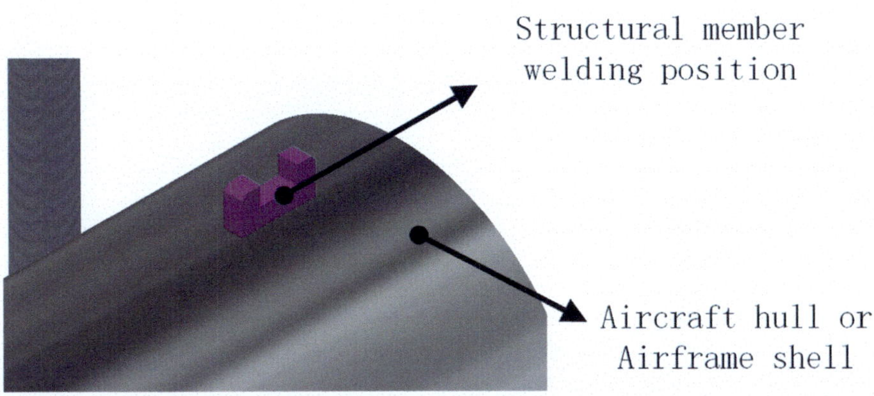

Fig. 8.11 Final pose configuration

Taking the Robotiq85 two-finger gripper as an example, when employing a two-finger gripper to grasp structural components, considerations extend beyond the component's pose and the sequence of grasping. It is also imperative to deliberate on the manner of grasping and how to adjust the robot's posture to accommodate this grasping approach.

Generally, there are two methods for grasping three-dimensional objects of regular shapes. One involves using the object's center of gravity as the grasping center point and clamping the shortest edge with the gripper. The other is to directly grasp the visible edges. In this study, the shortest edge of the structural component measures 20 mm, and its center of gravity is located on this edge. Given the Robotiq85's gripping range of [5, 85 mm], this paper adopts the approach of using the center of gravity of the structural component as the grasping center. As shown in Fig. 8.12, the purple circle represents the grasping center position, the red edge indicates the maximum gripper range, and the green color denotes the direction of the grasping surface. (2) Method for Grasping Posture Adjustment Based on Pose Matching.

When the robot grasps the structural component, it decomposes the component into a collection of feature surfaces. This collection results from affine transformations of the structure under different spatial orientations, essentially retaining principal features and their direction vectors. The component can be divided into four main feature faces $F1$, $F2$, $F3$, $F4$ and two virtual main feature faces ($F5$, $F6$), with each primary feature face corresponding to a feature surface set $\{\emptyset_1, \emptyset_2, \emptyset_3, \emptyset_4\}$. The structure has a total of twelve characteristic planes, as shown in Fig. 8.13.

1. Firstly, the least squares criterion is employed to construct the normal vectors $\vec{n}_i$ for all planes, where $i = 1, 2, \ldots, 12$. Subsequently, all these normal vectors are decomposed to derive the normal vectors of the four main feature faces $\{\vec{n}_{F1}, \vec{n}_{F2}, \vec{n}_{F3}, \vec{n}_{F4}\}$.

$$\vec{n}_{F1} = \vec{n}_1 \quad or \quad \vec{n}_2 \tag{8.35}$$

$$\vec{n}_{F2} = \vec{n}_5 + \vec{n}_9 \cdot \cos(\pi/4) \tag{8.36}$$

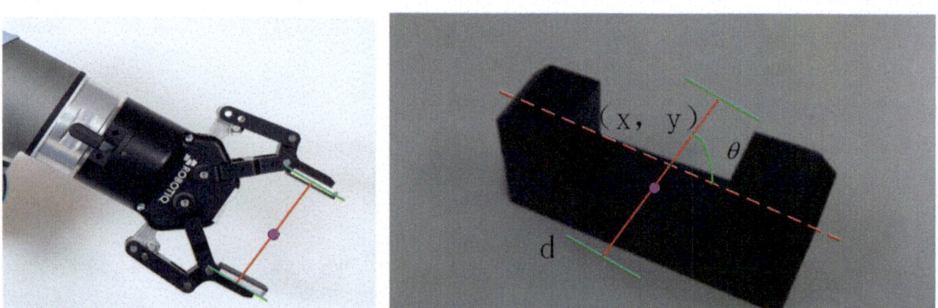

Fig. 8.12 Schematic diagram of structural component grasping

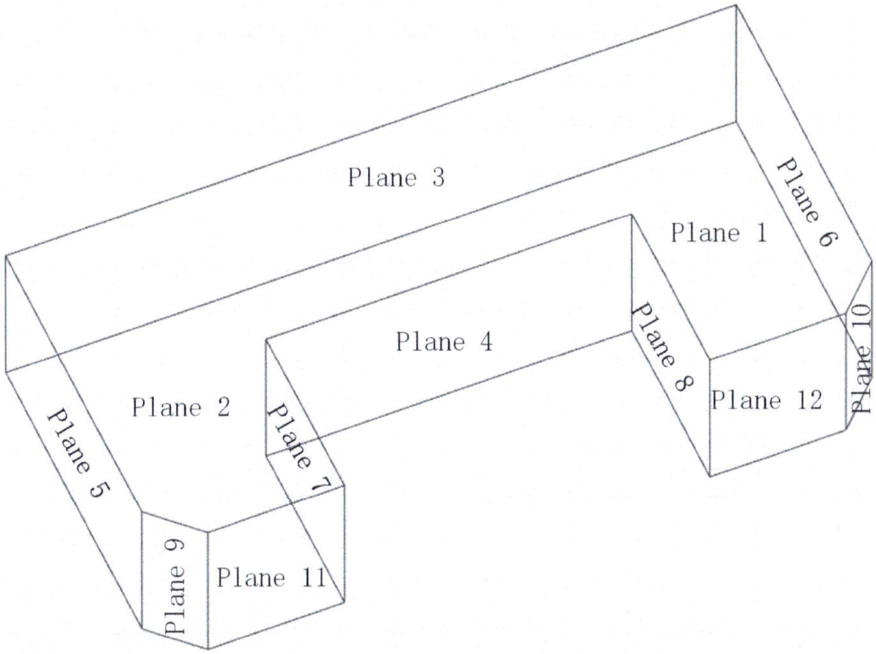

Fig. 8.13 Diagram of feature planes

$$\vec{n}_{F3} = \vec{n}_4 + \vec{n}_{11} + \vec{n}_{12} + 2\vec{n}_9 \sin(\pi/4) \tag{8.37}$$

$$\vec{n}_{F4} = \vec{n}_3 \tag{8.38}$$

2. The distances $\vec{x}_i, i = 1, 2, \ldots, 12$ from the structure's centroid (barycenter) to each feature plane $P(\vec{n}_i, d_i)_i, i = 1, 2, \ldots, 12$, where are computed. All directions are oriented from the centroid towards the respective feature planes.

$$b = \frac{\int_\Omega \rho(x) x d\sigma}{\int_\Omega \rho(x) d\sigma} \tag{8.39}$$

Let ρ represent the density function of the structural component, which varies depending on the material of the component.

For any point $p_{ij}, p_{ij} \in P(\vec{n}_i, d_i)_i$ chosen on the feature plane $|\vec{x}_i| = \left|\overrightarrow{bp_{ij}} \cdot \vec{n}_i\right| / |\vec{n}_i|$, the distance is calculated as follows:

$$\left|\vec{x}_i^*\right| = (|\vec{x}_i| - \mu)/\sigma \left|\vec{x}_i^*\right| \in (0, 1) \quad i = 1, 2, 3, 4 \tag{8.40}$$

where μ represents the mean of all distance data, and σ denotes the standard deviation of all distance data.

3. Finally, construct the feature weight vectors for the four main feature faces as follows:

$$\vec{m}_i = \vec{n}_{Fi} \cdot \vec{x}_i^*, \quad i = 1, 2, 3, 4 \tag{8.41}$$

The magnitude of the feature weight vector $\vec{m}_i$ reflects the proportion of the grasping weight allocated to that feature face. Quantifying the graspability of the four feature faces in this way facilitates subsequent experimental validation.

In the camera coordinate system, the feature weight vectors $\vec{m}_i$ are spatially decomposed into components, and the largest spatial component $\{\vec{m}_{ix}, \vec{m}_{iy}, \vec{m}_{iz}\}$ is identified. This component serves as the initial posture for the grasping set. If the third feature weight vector $\vec{m}_3$ has its spatial component as the maximum, then this first posture is considered the final posture.

As depicted in Fig. 8.14, upon acquiring the first posture of the structural component through the aforementioned method, this initial posture can serve as the first node in the trajectory for motion planning. However, the first posture is not necessarily the final one; further adjustments are required. To enable the robot to dynamically adjust its posture based on the current first posture autonomously, a posture matching library needs to be constructed.

The primary algorithmic process is outlined as follows:

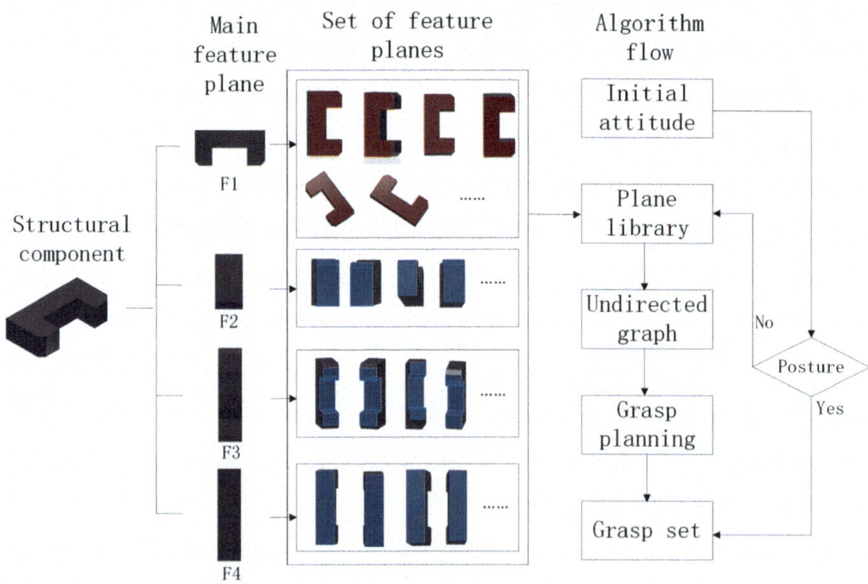

Fig. 8.14 Flowchart of pose matching process

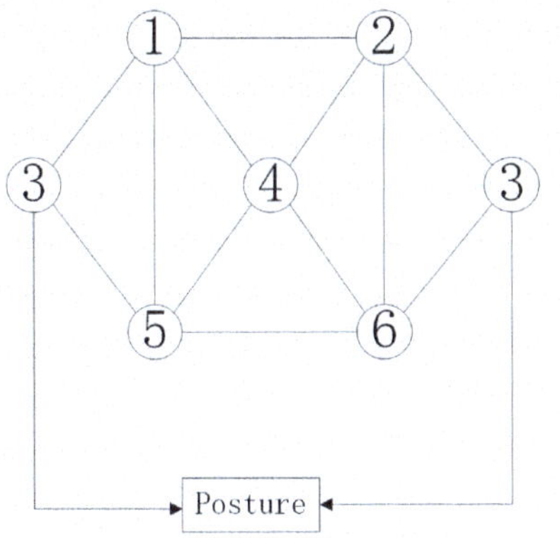

Adjacency matrix of the posture transformation graph

	1	2	3	4	5	6
1	0	1	1	1	1	0
2	1	0	1	1	0	1
3	1	1	0	0	1	1
4	1	1	0	0	1	1
5	1	0	1	1	0	1
6	0	1	1	1	1	0

Fig. 8.15 Undirected graph of posture transformation and its adjacency matrix

1. Among the above methods, the four principal feature planes (F1, F2, F3, F4) are considered equivalent due to the rotational symmetry of the structural component. However, the two virtual principal feature planes (F5, F6) are not taken into account. Nonetheless, in constructing the undirected graph $G = (V, E)$, all feature planes of the structural component must be considered. The graph for posture transformation is built based on the hexahedron rotating by 90° increments, along with its adjacency matrix for posture transitions, as depicted in Fig. 8.15.
2. This graph theory problem can be equivalently formulated as a special case solution of the minimum spanning tree method, where the undirected graph for posture transformation consists of n nodes (including one primary node and n-1 secondary nodes). Some nodes may form cycles, and cycle formation incurs a cost, with different cycles having varying costs. The question arises as to what is the minimal total cost required to connect these n nodes, such that any node can reach the primary node starting from any point. Let the weight of each cycle be denoted as, where w_{ij}, $i, j = 1, 2, \ldots, 6$. Each cycle's weight signifies the degree of difficulty in transforming from one feature plane to another and is represented by the inner product of the feature weight vectors $\vec{m}_i$.

$$w_{ij} = \vec{m}_i \cdot \vec{m}_j, i, j = 1, 2, \ldots, 6 \text{ and } i \neq j \tag{8.42}$$

3. Initially, a set U of grasp surfaces is defined, with the first posture a_{strat} designated as the first element of set U. Within the undirected graph, the node $a_{s\to 1}$ connected to the initial posture $U(a_{start}, a_{s\to 1})$ via the smallest weight edge is identified, thereby updating set. Subsequently, among the edges associated with both a_{start} and $a_{s\to 1}$, the one with the minimum weight is selected. Each time a new element is added to set, it is checked whether this addition constitutes the final position. If

so, set a_{end} is outputted as the grasp posture set. Otherwise, the process continues to search for the next posture.

4. Ultimately, a grasp configuration set U is derived based on the current orientation of the structural component. The robotic arm sequentially adjusts to each grasp posture within set U until the final grasping posture is attained.

8.3 Reinforcement Learning-Based HRC

Human–robot collaboration is a typical application scenario for reinforcement learning techniques. This subsection introduces the task decision-making method based on reinforcement learning on the basis of robot perception. Finally, human–robot collaborative manufacturing based on reinforcement learning is introduced.

8.3.1 Fundamentals of Robot Perception

Perception of the environment is fundamental to robots assisting humans in accomplishing various tasks. Traditional manufacturing processes are divided into multiple stages such as machining, welding, painting, and assembly, among others. Consequently, different manufacturing environments possess distinct characteristics. Taking collaborative assembly scenarios as an example, robots must perceive the location and structure of objects in the environment, as well as the current posture and behavioral intentions of the human collaborator. As illustrated in Fig. 8.16, by employing a Kinect vision camera, both RGB images and depth images of objects in the environment can be obtained. Positional and orientation data of assembly parts can be extracted through Convolutional Neural Network (CNN) methodologies. Moreover, structural data of the assembly parts, primarily manifested in assembly interface details, are also acquired to further constrain the generation of assembly sequences.

In addition to identifying parts in the assembly scenario, robots will also monitor human models in real-time. A digital model of the human form is created within a virtual environment, generating a 3D bounding box around the human model. This bounding box delineates the range of human activity, confining all actions within its interior. When planning the robot's motion trajectory, the production system regards the area of human movement as an obstacle region to optimize the trajectory accordingly. Beyond generating obstacle-free paths, robots also discern human action intentions and respond with appropriate assistance. The skeleton stream module of the Kinect camera facilitates the easy extraction and tracking of the human skeleton. By feeding this skeletal data into an action recognition network, the corresponding action categories of the human can be determined. For precise robotic assistance, estimation of human poses is crucial. Widely used open-source algorithms for human pose estimation, such as Openpose, enable efficient pose estimation processes. By

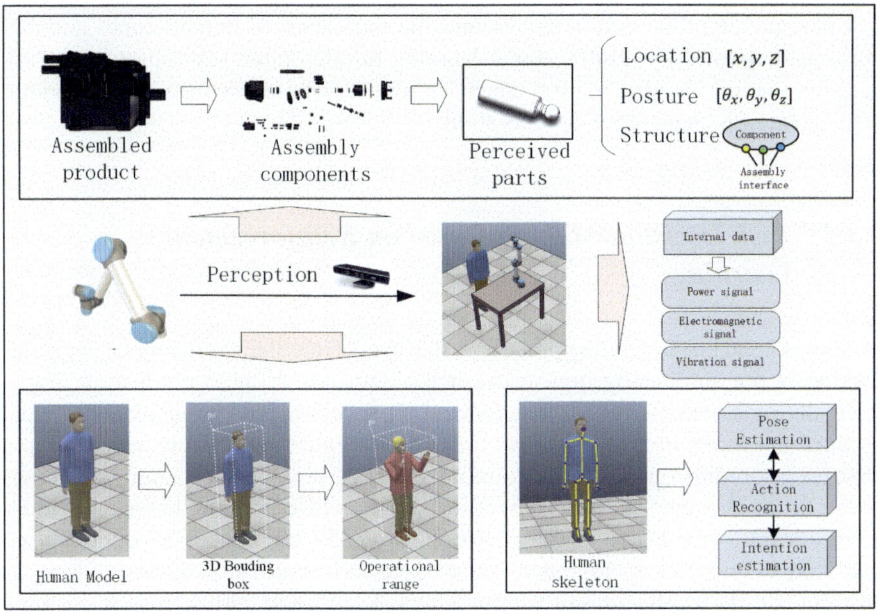

Fig. 8.16 Robot perception operations in human–robot collaboration environments

combining human action recognition with pose estimation techniques, robots can efficiently collaborate with human operators.

Naturally, during the operation of human–robot collaboration systems, a wealth of internal operational data is generated, including power signals, electromagnetic signals, and vibration signals, among others. Monitoring these internal data aids in the analysis and handling of the collaboration status. Should any internal data in the collaborative environment exceed predetermined thresholds or safety ranges, it would be regarded as a malfunction, necessitating prompt maintenance by technicians for normal operation to resume.

The manufacturing process, due to its complexity and dynamism, demands high flexibility and safety in the human–robot collaboration it entails. Reinforcement learning, through the iterative interaction between an agent and its environment, progressively enhances the agent's intelligence, thereby promoting adaptability in robots. Thus, the initiation of human–robot collaboration first necessitates the robot's recognition of its surroundings, followed by leveraging reinforcement learning to acquire skills pertinent to handling assigned tasks. In practical production scenarios, human work performance is subject to significant fluctuations; for instance, prolonged work leads to fatigue, which in turn decreases work efficiency and introduces more errors. To adequately compensate for the variations in human performance that impact overall productivity, robots must adjust their own work schedules to minimize discrepancies in the collaborative manufacturing process.

Consequently, this section delves into the intricacies of human–robot collaboration underpinned by reinforcement learning through three key aspects: (1) Task decision-making based on reinforcement learning; (2) Reinforcement learning within the context of human–robot collaboration.

8.3.2 Task Decision-Making Based on Reinforcement Learning

In conventional manufacturing processes, the level of intelligence in robotic decision-making is not sufficiently high to meet the demands of agile production. Especially during certain periods, when demand for specific products surges, traditional production models in manufacturing facilities struggle to effectively adapt to rapid changes in production planning. Reinforcement learning enables continuous interaction between agents and their environments, thereby enhancing the agents' intelligence levels. This holds significant implications for the human–robot collaborative production process. Taking a typical assembly scenario in manufacturing as an example, the application of reinforcement learning unfolds across three main stages: (1) Assembly sequence generation based on reinforcement learning; (2) Robot grasping path planning using reinforcement learning; (3) Human–robot collaborative assembly informed by reinforcement learning.

(1) **Assembly Sequence Generation Process Based on Reinforcement Learning**

Figure 8.17 illustrates the process of assembly sequence generation based on reinforcement learning. Given that human–robot collaborative assembly is a dynamically adjustable and optimizable process, a Markov Decision Process (MDP) can be constructed for the assembly scenario. This MDP encompasses four key elements: states, actions, transition probabilities, and rewards. The agent, perceiving the current state of the system, selects action strategies to alter the environment's state, thereby obtaining new environmental state values and reward values, with the accumulation of rewards over time referred to as the return. As depicted in Fig. 8.17, the robot, acting as the agent, adopts different assembly sequences (actions) to modify the current assembly environment. Since the product being assembled comprises multiple components with distinct structures, their assembly sequences will exhibit varying characteristics depending on the selection of base components. The assembly environment progresses through stages to generate assembly sequences, allocate tasks, and ultimately drive the entire human–robot collaborative assembly process. Environmental state data acquired is further classified and analyzed to form the various data included in the state space. The state space primarily encompasses three categories of data: assembly data, collaboration data, and product data. The efficiency, precision, and safety of the assembly process reflect the state of human–robot collaborative assembly under the current sequence. Additionally, the efficiency, precision, and safety of human–robot collaboration represent the relevant collaborative status.

Naturally, the state of the product assembled under this sequence is primarily characterized by its performance, such as: (1) functional effectiveness indicating the degree of functional restoration; (2) operational characteristics emphasizing the stability of product usage; and (3) ergonomic assessment from the perspective of human users analyzing comfort during product use. Based on the interaction between the action space and state space, reinforcement learning drives assembly performance towards improvement. The reasonable configuration of the reward function plays a pivotal role throughout the reinforcement learning process. In Fig. 8.17, the reward function is used to link assembly sequences with assembly outcomes. Under a specific assembly sequence, a reward value of + m is set when parts are successfully assembled; a larger reward value of + M is assigned upon the completion of the entire product; conversely, a penalty value of -N is set following product assembly failure. Moreover, to evaluate the assembly efficiency under a particular sequence, a continuous reward function f is separately established to reflect assembly speed. The basic framework of Actor-Critic [46] encounters difficulties in convergence; two optimization strategies can be adopted: (1) guiding training with prior knowledge; and (2) optimizing the structure, for example, using experience replay mechanism (DQN) [47], updating the main thread with multiple auxiliary threads (A3C) [48], or adopting a dual Actor and dual Critic network architecture (DDPG) [49]. Human intuitive judgments and experiential summaries can serve as crucial elements guiding the robot's action selection process. Original assembly constraints among components can also be employed as guiding conditions to restrain randomly generated assembly sequences, significantly enhancing assembly success rates. Ultimately, through reinforcement learning, the optimal assembly sequence for a specific product can be obtained.

(2) **Reinforcement Learning-Based Path Planning for Robot Grasping**

Figure 8.18 illustrates the process of generating robot grasping trajectories based on reinforcement learning. Numerous challenges arise in achieving adaptive generation of robot trajectories within complex assembly environments. For the agent, the environmental information obtained via a vision camera is relatively intricate, encompassing part location, pose information, as well as human motion data. Accurate data essential for robot grasping comprises primarily the positional and attitudinal information of objects in the environment. Utilizing a Kinect camera facilitates easy acquisition of RGB images and point cloud data from the scene. By employing Convolutional Neural Networks (CNNs) to process these RGB images, the 3D bounding box of objects in the image can be determined, which in turn enables the extraction of the centroid coordinates and rotation angles of this bounding box. To enhance the intelligence of robot part grasping, reinforcement learning is introduced as a data-driven methodology to improve grasping performance. Consequently, there exists substantial interaction data between the robot and the assembly environment to train the agent's problem-solving capabilities. Typically, the primary parameters controlling the robot are the joint angles of its six axes; defining these six values determines the position and orientation of the robot's end effector within the environment. In practical industrial applications, the TCP (Tool Center Point) coordinates are used to represent the position and orientation of the end effector, establishing a

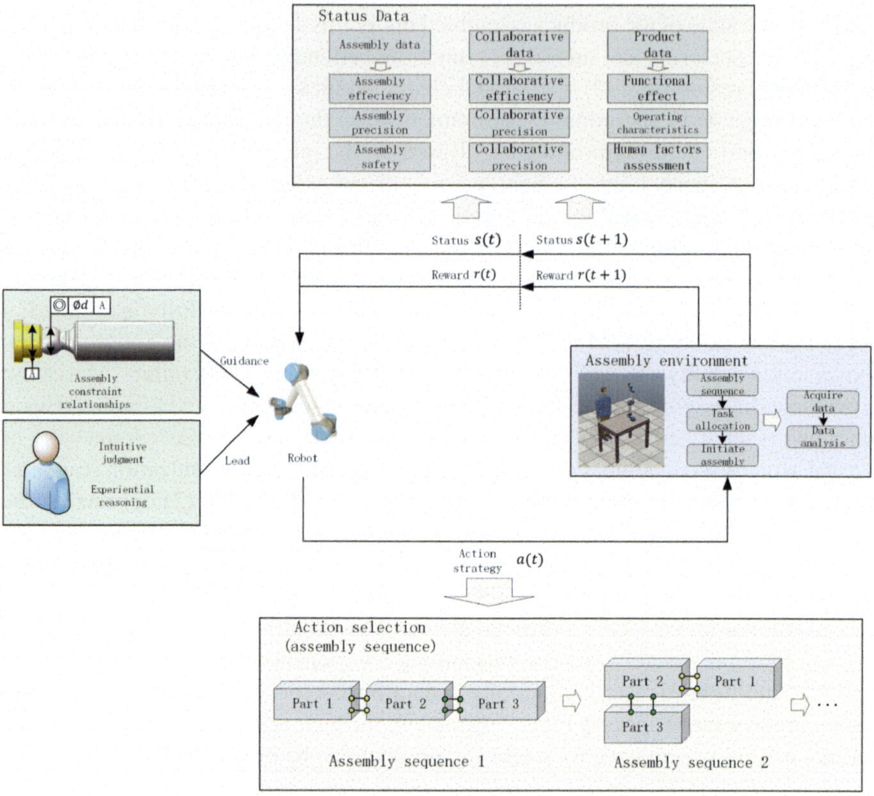

Fig. 8.17 Process of assembly sequence generation based on reinforcement learning

one-to-one correspondence between the robot's six joint angle values and the TCP coordinates.

To ensure safe and successful grasping of components in complex assembly environments, the robot must undergo an extensive trial-and-error process, involving random adjustments to its six joint angle values to create varied movements. These different actions serve as input commands transmitted to the robot controller to dictate specific behaviors. The state of the assembly environment changes to some degree with each action executed by the robot. For a clearer depiction of the grasping efficacy under the current action, the 6D pose difference between the robotic gripper's TCP point and the target part is utilized as the observed state. Alternatively, point cloud matching can be employed using the point cloud data obtained from the Kinect camera. The core idea behind point cloud matching is to assess the grasping status based on the spatial constraints formed by the point cloud, considering factors such as parallelism, contact angle, and grasp deviation.

Concurrently with the generation of new states, the system also produces reward values, which evaluate the performance of the current action strategy. When the

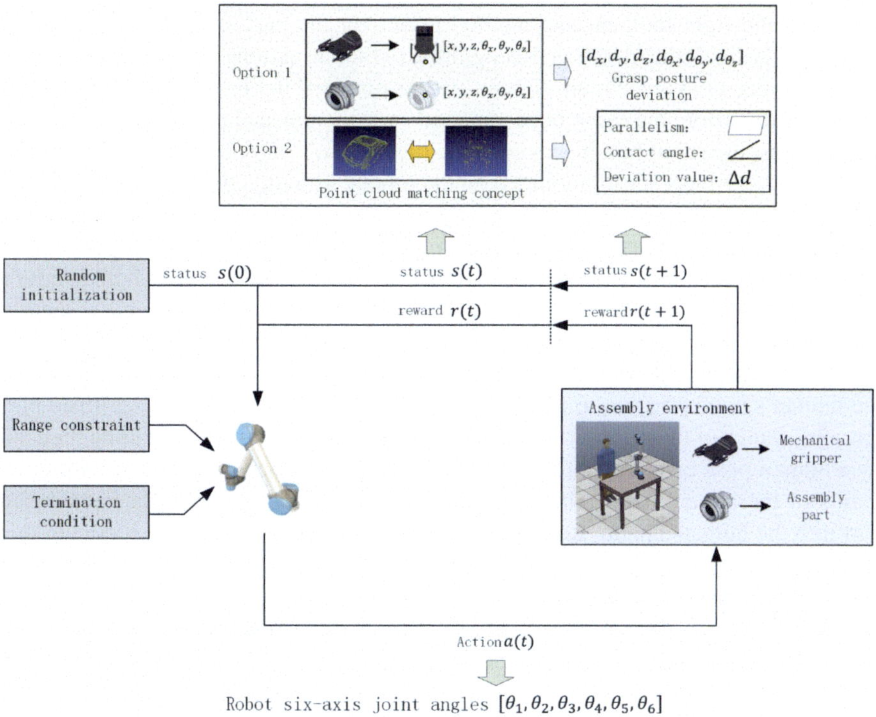

Fig. 8.18 Process of generating grasping trajectories for robots based on reinforcement learning

accumulated reward reaches its maximum, it corresponds to the formation of the optimal grasping trajectory. The reward function must reflect crucial aspects such as the efficiency of the grasp and the success of the attempt. If a single robot action constitutes a positive move, a small reward $+m$ is given; if the robot successfully grasps the part after a complete operation cycle, a larger reward $+M$ is awarded; if the robot fails to grasp the part but remains within an adjustable range, a reward $+N(N < M)$ is granted; if the robot fails to grasp the part and is in an unadjustable state, a significant penalty $-Q$ is imposed.

Random initialization of operations enhances the generalization capability of the robot's grasping actions, enabling it to adapt to more complex grasping scenarios. Moreover, implementing specific termination criteria governs the number of attempts the robot makes, ensuring that the grasping action is completed within a predefined number of attempts. Range constraints, on the other hand, confine the robot's trial-and-error scope within a pre-set limit, preventing the robot from operating outside the designated grasping zone and posing a safety threat to humans during its operation.

(3) **Collaborative Assembly Process Based on Reinforcement Learning**

Once the robot has completed part grasping, it must decide whether and how to transfer the part to a human collaborator, as well as assist in the assembly process.

Choosing the right moment and method to deliver the part to the human operator becomes a critical consideration. Recognition of human actions can further enable the robot to understand the human's intentions. As depicted in Fig. 8.19, during human–robot collaboration, the robot needs to hand over the grasped part to the human. To achieve this, the robot first has to perceive the human's current state and movement tendencies. A Kinect sensor mounted on the robotic arm directly captures the human body's skeleton data stream. Subsequently, an action recognition network composed of multiple LSTM units processes the input human skeleton data, outputting the type of action currently being performed by the human, such as raising both hands or extending a palm. When the human extends their palm, ready to receive the assembly part from the robot, the robot assesses the human's operational intent based on the recognized action. Upon receiving the assistance command, the robot actively locates the human's hand position to facilitate the controlled placement of the gripped part onto the center of the human's palm. This localization functionality can be realized by a human pose estimation algorithm, focusing particularly on the position of key points of the human hand. Thereafter, the robot is driven to gradually transfer the part into the human's palm. Reinforcement learning serves as the core model in this module, setting the positional and attitudinal difference between the human palm center and the robot's TCP as the observed state, while the robot's six joint coordinate values act as action signals, constructing a reinforcement learning framework for the environment shared by the robot and the human. Suitable reinforcement learning algorithms, such as DDPG or A3C, are employed for model training. Through the robot's random trials, it eventually succeeds in transferring the held part to the center of the human's palm.

As shown in Fig. 8.20, after the robot completes the part handover, it must collaborate with the human to fix the part onto the main component. Given the complex assembly constraints between the part and the main component, all assembly interfaces must be identified. Interfaces between parts typically exist in the form of points, lines, or surfaces, with various types of interfaces intertwined. The robot also employs an action recognition network to identify the human's current actions. Intelligent methods for human action recognition are primarily divided into two categories: (1) Action recognition based on RGB video sequences; (2) Action recognition based on human skeletons. RGB video sequences can suffer from occlusion issues, leading to incomplete representation of actions. Thus, action recognition efficiency based on RGB videos can be significantly disrupted. Conversely, human skeletons possess uniqueness in spatial and temporal dimensions, allowing for a clear expression of the current action. The action recognition network is primarily composed of LSTM units with strong temporal properties, capable of building models from both temporal and spatial perspectives. As depicted in Fig. 8.20, when the human aligns two parts with an assembly relationship, the system automatically identifies the action category and estimates the human's intention, such as deducing the control command—"Robot, assist in completing the assembly fixation process"—when the human aligns the parts. Upon receiving the control command, the system needs to locate the hole positions on the structured component with mounting holes. Solution one involves using conventional object detection algorithms to mark the positions of small holes on

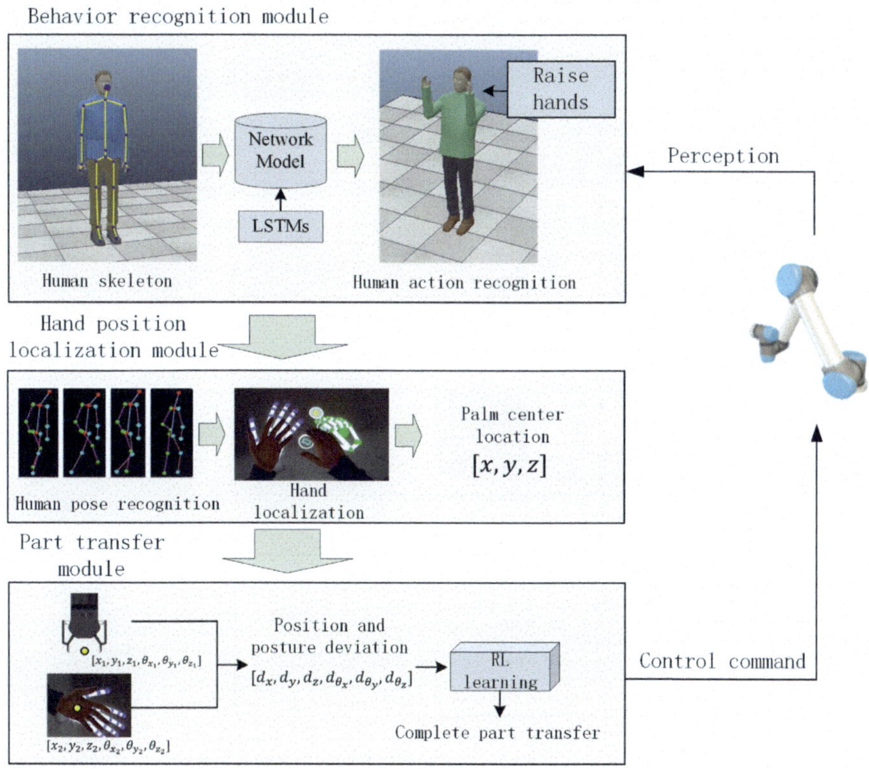

Fig. 8.19 Part handover operation in human–robot collaboration process

the structure. Commonly used neural network models include YOLO series object detection models and Efficient models, among others. Solution two leverages point cloud data captured by the Kinect camera to extract point cloud features. Networks like PointNet can effectively extract main features from point cloud images. Once the mounting hole positions are located, the robot, having picked up screws, must accurately fix them onto the mounting holes of the structured component. Reinforcement learning again serves as the key model for robot policy optimization, constructing an iterative graph between the robot and the assembly environment. The positional and attitudinal difference between the screw assembly interface and the target part interface is used as the observed state for the training process. Ultimately, the robot can fix the screws onto the structured component, thus completing the collaborative process.

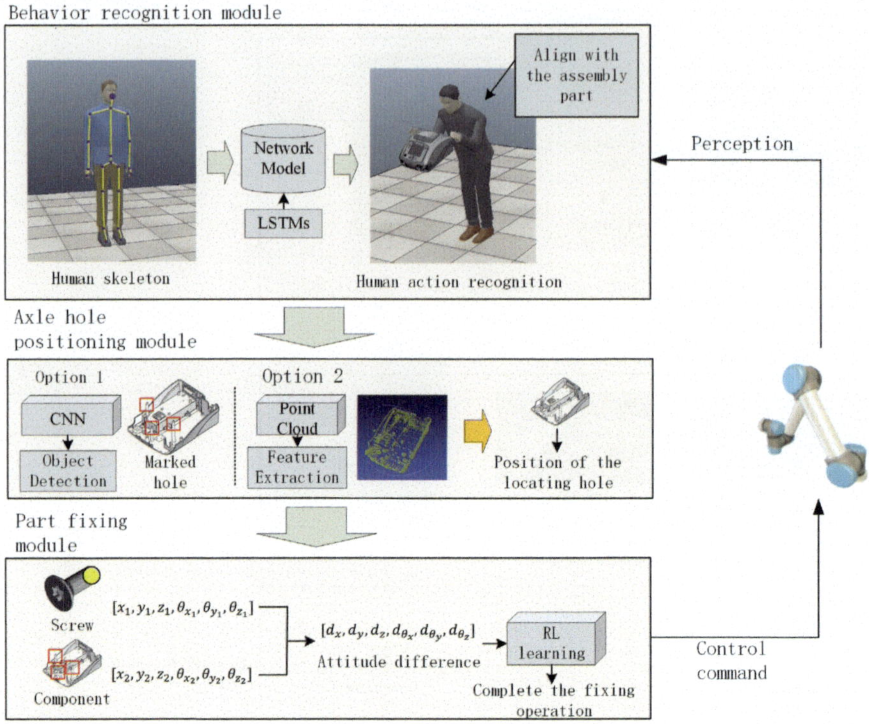

Fig. 8.20 Part fixing operation in human–robot collaboration process

8.3.3 RL Leaning on Human–Robot Collaboration

(1) **HRC Manufacturing Based on Reinforcement Learning**

The fundamental production process encompasses various stages such as blank manufacturing, component fabrication, and final assembly. In addition, manufacturing techniques involve diverse methodologies including machining, painting, welding, and assembly. To enhance the collaborative manufacturing capabilities of robots in these domains, it is imperative to consider improving the robots' adaptability to the manufacturing environment. During production, human performance is influenced by working hours; typically, after prolonged continuous work, human enthusiasm significantly wanes, accompanied by an increase in operational errors. Consequently, to maintain production output under extended working conditions, Harley Oliff and colleagues focused on studying the evolution of human performance over time and leveraged reinforcement learning (RL) to augment the robot's adaptive capacity [50]. Their RL methodology enables robots to proactively adjust their operational states to compensate for deviations resulting from variations in human coworkers' performance. Figure 8.21 illustrates the specific role of reinforcement learning in balancing human–robot collaborative manufacturing. Traditional manufacturing workflows encompass a sequence of operations such as product machining, welding, painting, and assembly. For each of these manufacturing stages, extended periods of manual labor affect worker performance, manifesting primarily in reduced efficiency and increased error rates. To better represent the collaborative manufacturing status within the manufacturing environment, a set of observables is employed to render the collaborative environment assessable. These metrics include: (1) robot cycle time; (2) collaborator cycle time; (3) remaining product count; (4) robot idle time; (5) collaborator idle time; (6) speed adjustment magnitude; and (7) buffer content, among others. Subsequently, these observables undergo normalization, confining values within the range of 0 to 1. Upon the robot adopting specific actions—corresponding to particular operating speeds and production schedules—the state information of the collaborative manufacturing environment is further altered. The resultant new state and reward feedback are then relayed to the agent to inform the next action selection by the robot. A tailored reward function is devised to guide the robot towards choices that minimize both robot and collaborator idle times in subsequent action selections (production planning). Ultimately, this achieves a substantial improvement in the performance of human–robot collaborative manufacturing. Human performance also fluctuates across different workdays, directly impacting the collaborative manufacturing process by reducing productivity. To mitigate variations in daily human–robot collaborative manufacturing capability, adjustments to the robot's production speed are made to offset the productivity decline caused by a decline in human worker performance.

(2) **HRC Manufacturing with an Enhanced Actor-Critic Framework**

The Actor-Critic framework, which combines value and policy-based reinforcement learning, excels at selecting appropriate actions in continuous action spaces and facilitates single-step updates. Given that a robot's motion process inherently involves continuous action sequences, the Actor-Critic framework is highly suitable for addressing robot control problems [51]. However, a notable drawback of the conventional Actor-Critic framework lies in its suboptimal convergence performance. Therefore, modifications to the original Actor-Critic structure are considered to enhance the convergence of the AC model.

Three primary methods for modification are:

1. Experience replay mechanism, where a replay buffer stores tuples of the form $\langle s, a, r, s' \rangle$;
2. Utilizing multiple threads to update the primary network parameters;
3. Adopting a dual Actor-dual Critic network architecture.

As illustrated in Fig. 8.22, incorporating the experience replay mechanism into the Actor-Critic framework allows the Actor, during policy updates, to not only consider temporally proximate state-action pairs but also to sample historical data stored in the experience replay buffer for training. This reduction in correlation between sampled

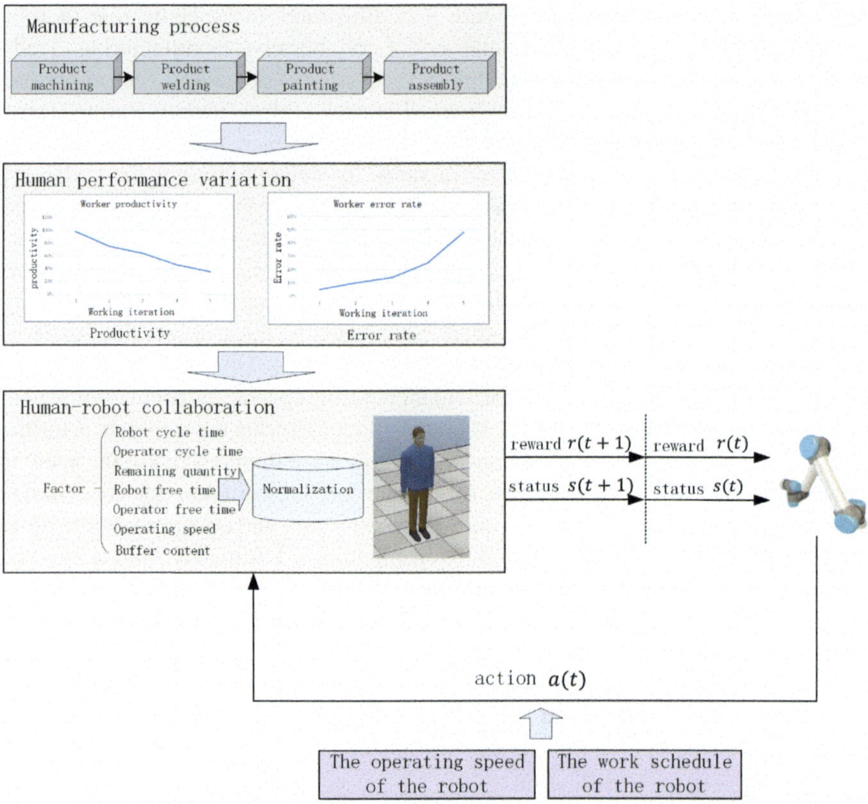

Fig. 8.21 Human–robot collaborative manufacturing based on reinforcement learning

data points facilitates a smoother convergence of the training process. A representative reinforcement learning algorithm that embodies this optimization approach is the DQN (Deep Q-Network) algorithm.

As depicted in Fig. 8.23, multiple reinforcement learning sub-modules, or workers, interact with the manufacturing environment. Simultaneously, a shared common component is established to facilitate connectivity between each subnetwork. Each neural network boasts an identical architecture, encompassing both an Actor and a Critic component. Each thread independently interacts with the manufacturing environment to gather pertinent experience data, without interfering with one another. Upon acquiring a sufficient amount of data through environmental interactions, each thread proceeds to compute the gradient of its local neural network loss function. The formulas for calculating the gradients of the local Actor and Critic are as follows:

$$\text{Actor}: \ d_\theta \leftarrow d_\theta + \nabla_{\theta'} log\pi_{\theta'}(s_i, a_i)\big(Q(s, i) - V\left(S_i, \omega'\right)\big) + c\nabla_{\theta'} H\left(\pi\left(s_i, \theta'\right)\right) \tag{8.43}$$

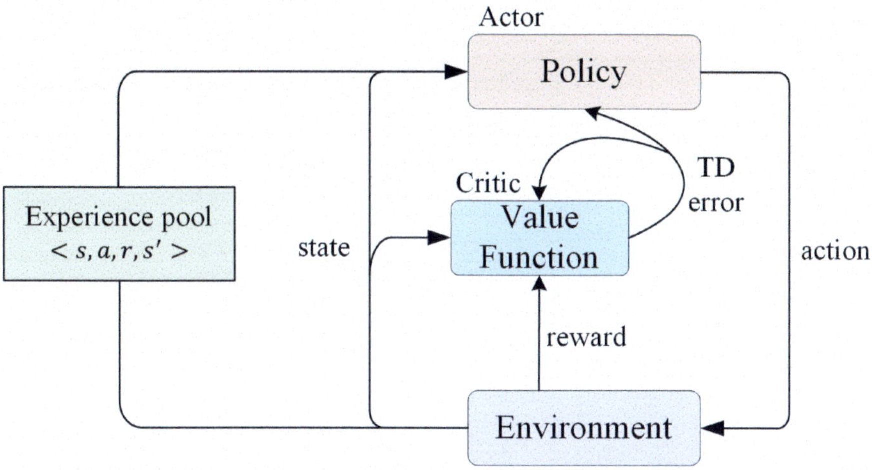

Fig. 8.22 Actor-critic framework with added experience replay mechanism

$$\text{Critic}: \ d_\omega \leftarrow d_\omega + \frac{\partial \left(Q(s,i) - V\left(S_i, \omega'\right)\right)^2}{\partial_{\omega'}} \tag{8.44}$$

Subsequently, the parameters of the global neural network are updated based on these gradients, with the update formula as follows:

$$\theta = \theta - \alpha d_\theta, \omega = \omega - \beta d_\omega \tag{8.45}$$

After backpropagating the gradients of the loss functions from all threads to the global network, updating its parameters accordingly, the parameters of the shared network, θ and ω, are outputted. All subsidiary thread networks then incorporate these shared parameters to update their own network parameters. This methodology is exemplified by the A3C (Asynchronous Advantage Actor-Critic) algorithm.

The third method employs a dual network architecture, where both the Actor and Critic networks consist of a current network and a target network. The Actor's current network manages the policy network, interacting with the environment and facilitating iterative updates to the parameters θ. Conversely, the Actor's target network selects the optimal action policy based on the next state sampled from the experience replay pool, with its parameters θ' derived from a copy of the current network's parameters θ. Meanwhile, the Critic's current network oversees the iterative updates of the value network parameters ω and computes the current Q-value. The Critic's target network is responsible for calculating the target Q-value Q', subsequently yielding the target Q-value, with its parameters ω' periodically replicated from ω. The loss functions for both the Actor and Critic networks are outlined as follows (Fig. 8.24):

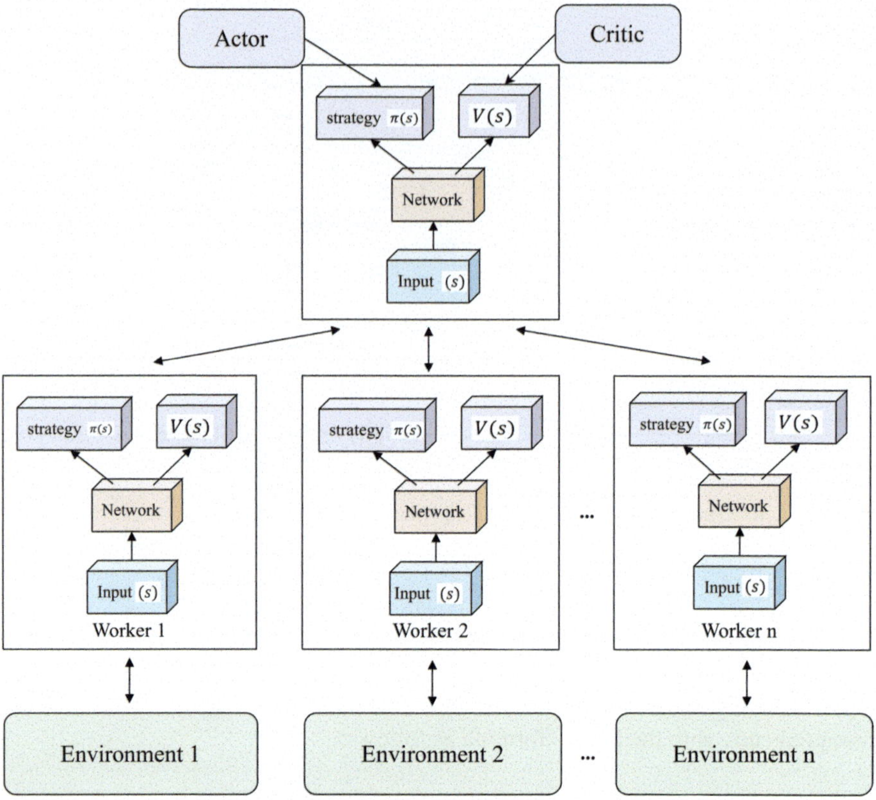

Fig. 8.23 Multi-threaded actor-critic framework

$$\text{Actor}: \ J(\omega) = \frac{1}{m}\sum_{j=1}^{m}\left(y_j - Q\left(\emptyset\left(S_j\right), A_j, \omega\right)\right)^2 \tag{8.46}$$

$$\text{Critic}: \ J(\theta) = -\frac{1}{m}\sum_{j=1}^{m} Q(s_i, a_i, \omega) \tag{8.47}$$

Employing the aforementioned enhanced Actor-Critic framework renders the reinforcement learning process in human–robot collaborative manufacturing more prone to convergence. By merely taking the collaborative states during the manufacturing phase as inputs and feeding them into the improved Actor-Critic framework, corresponding state values can be obtained. Given the need to evaluate multiple states within the human–robot collaboration process, the adoption of multi-threading accelerates the convergence of the overall reinforcement learning procedure. Hence, multi-threading is applied to DDPG to formulate a new reinforcement learning framework. Integrating DDPG's original experience replay mechanism, dual network

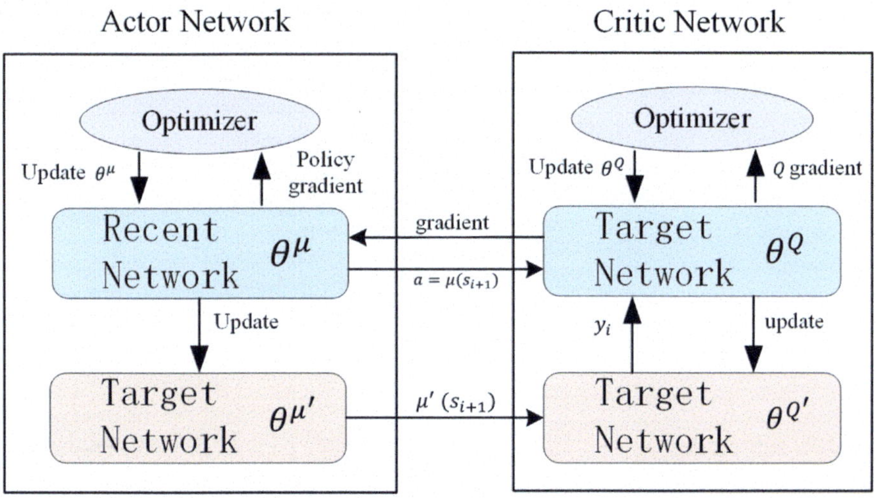

Fig. 8.24 Actor-critic framework based on dual networks

architecture, and A3C's multi-threading concept enhances both the learning convergence speed and outcome. The structure of this multi-threaded DDPG framework is depicted in Fig. 8.25. Each DDPG submodule interacts with the manufacturing environment to gather state information and refine the robot's action strategies. The policy gradients from each sub-network are backpropagated to the main network to update its parameters. Post the main network's update, its parameters are replicated to each DDPG submodule, enabling independent interactions of each sub-network with the environment.

In human–robot collaborative manufacturing, the state space encompasses several dimensions: (1) Robot cycle time (T_R); (2) Collaborator cycle time T_H); (3) Robot idle time (S_R); (4) Remaining product count (N_P); (5) Collaborator idle time (S_H); (6) Speed adjustment magnitude (N_s); (7) Buffer content (B_c). The primary control variable for the robot is the rate of its operational speed. Thus, the action space for the agent can be defined as $\langle V_R-, V_R+\rangle$, respectively.

Each observable in the state space serves as an evaluation metric to gauge the appropriateness of adjusting the robot's production speed under the current circumstances. Furthermore, multi-threading is employed to expedite the convergence of model training. The pseudo-code for reinforcement learning using multi-threaded DDPG is outlined as follows in Fig. 8.26. With the aid of updates from multiple threads, the main model ultimately acquires optimal parameters, thereby enhancing the environmental adaptability of the sub-networks. Implemented in a human–robot collaborative manufacturing environment, this algorithm facilitates a swift learning process for the robot and fosters robust adaptability. The robot can dynamically adjust its action strategies in real-time according to fluctuations in human performance, thereby minimizing discrepancies in collaborative performance.

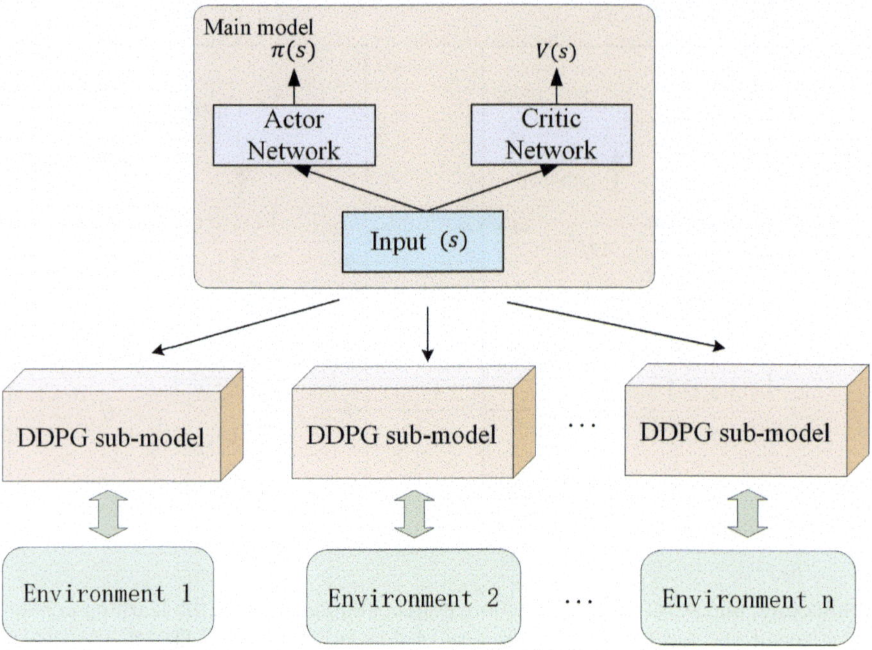

Fig. 8.25 Multi-threaded DDPG model framework

8.4 Case Study

Gripping and assembly are typical application scenarios for robots. This subsection illustrates the validation of human–robot co-manufacturing through 2 scenarios: loose parts recognition and generator assembly.

8.4.1 Identification and Pickup of Scattered Components

(1) **Experimental Environment**

Due to the need for robotic assistance in gripping and positioning during the welding process of structural components, the collaborative robot UR5 has been selected. Its high safety level, wide applicability in various environments, and rich programming interface make it an ideal choice. Considering the gripping method required for the structural components, the end effector Robotiq85 is selected. The relevant parameters of each axis of the UR5 robot are as shown in Table 8.3.

In industrial scenarios, cameras are generally required to have good environmental adaptability. For this purpose, the camera selected for this paper is the Intel Realsense D435 model. The D435 camera utilizes the latest USB 3.0 data transfer standard,

Multi-thread DDPG algorithm:
Randomly initialize all critic network $Q(s,a|\theta^Q)$ and actor $\mu(s|\theta^\mu)$ with weights θ^Q and θ^μ.
Initialize all target network Q' and μ' with weights $\theta^{Q'} \leftarrow \theta^Q, \theta^{\mu'} \leftarrow \theta^\mu$
Initialize all replay buffer R
For each thread do
  for episode=1, M do
    Initialize a random process $\mathcal{N}$ for action exploration
    Receive initial observation state s=$[T_R, T_H, S_R, S_H, N_P, B_C]$
    for t=1, T do
      Select action $a_t = \mu(s_t|\theta^\mu) + \mathcal{N}_t$ according to the current policy and exploration noise
      Execute action a_t and observe reward r_t and observe new state s_{t+1}
      Store transition (s_t, a_t, r_t, s_{t+1}) in R
      Sample a random minibatch of N transitions (s_i, a_i, r_i, s_{t+1}) from R
      Set $y_i = r_i + \gamma Q'(s_{t+1}, \mu'(s_{t+1}|\theta^{\mu'})|\theta^{Q'})$
      Update critic by minimizing the loss: $L = \frac{1}{N}\sum_i (y_i - Q(s_i, a_i|\theta^Q))^2$
      Update the actor policy using the sampled policy gradient:

$$\nabla_{\theta^\mu} J \approx \frac{1}{N}\sum_i \nabla_a Q(s,a|\theta^Q)\, |_{s=s_i, a=\mu(s_i)} \nabla_{\theta^\mu}\mu(s|\theta^\mu)|_{s_i}$$

      Update the target networks:

$$\theta^{Q'} \leftarrow \tau\theta^Q + (1-\tau)\theta^{Q'}$$
$$\theta^{\mu'} \leftarrow \tau\theta^\mu + (1-\tau)\theta^{\mu'}$$

    End for
  End for
End for
Perform asynchronous update of $\theta^{Q'}{}_{main}$ and $\theta^{\mu'}{}_{main}$ in main model:

$$\theta^{Q'}{}_{main} \leftarrow d\theta^{Q'}{}_1 + d\theta^{Q'}{}_2 + d\theta^{Q'}{}_3 + \cdots$$
$$\theta^{\mu'}{}_{main} \leftarrow d\theta^{\mu'}{}_1 + d\theta^{\mu'}{}_2 + d\theta^{\mu'}{}_3 + \cdots$$

Fig. 8.26 Pseudo-code for multi-threaded DDPG algorithm

Table 8.3 Parameters of the UR5 robot

Joint axis name	Motion upper limit	Motion lower limit	Maximum speed	Power (Watts)
Base	– 360°	360°	120°/s	55
Shoulder	– 85°	85°	240°/s	65
Elbow	– 170°	170°	240°/s	70
Wrist 1	– 170°	170°	240°/s	70
Wrist 2	– 360°	360°	240°/s	50
Wrist 3	– 360°	360°	300°/s	20
End effector	5 mm	85 mm	100 mm/s	30

enabling faster data transmission. It incorporates both a color module and a depth module, with the depth module belonging to the D400 series. It features a comprehensive optical solution and is accompanied by a well-developed SDK, making it highly suitable for prototyping and secondary development. The main parameters of the D435 are outlined in Table 8.4.

Table 8.4 Parameters of the D435 camera

Model	Realsense D435
Effective range	0.3–6 m
Operating voltage	16–32 V
Data interface	USB 3.0 Transfer protocol
Resolution	[1920,1080]

The real environment comprises the UR5 robotic arm, the D435 depth camera, and a pile of structural components. The structural components are fabricated using 3D printing, presenting as featureless black blocks, which aligns with an authentic welding scenario.

The robotic grasping experimental platform discussed in this paper is illustrated in Fig. 8.27. The scene primarily consists of the UR5 robot, Intel Realsense D435 camera, and a stack of structural components. The UR5 robot boasts six degrees of freedom with a positioning accuracy of 0.05 mm, and its base is securely fixed within the scene. Mounted at the robot's end is the Robotiq85 gripper, capable of executing two-finger grasps with a maximum finger opening distance of 85 mm and exerting a gripping force ranging from 10 to 110 Newtons. The D435 camera is situated to the front side of the robot, directly facing the grasping area. The grasping system encompasses Ubuntu 16.04, ROS-Kinetic control system, Rviz-Gazebo simulation environment, and V-Rep ROS simulation platform.

(2) **Experiment Design**

Both the structural component recognition and localization method and the grasp sequence method utilize point clouds that are processed in the camera coordinate

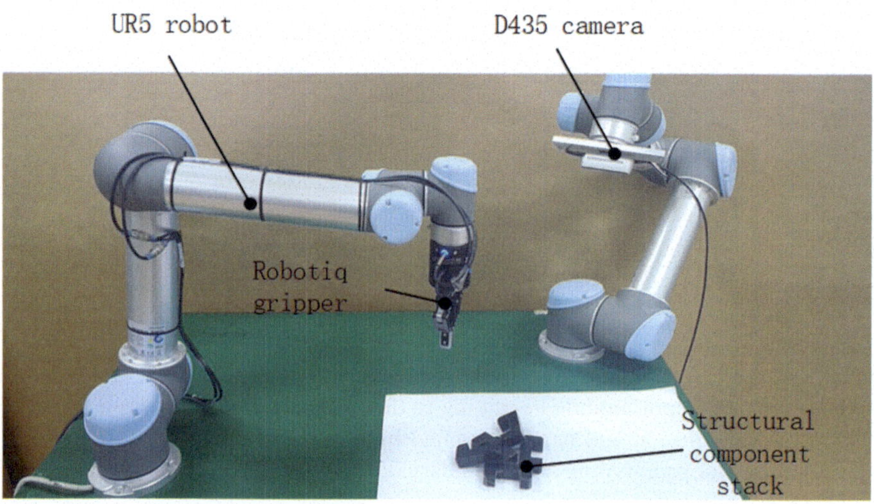

Fig. 8.27 Grasping experiment setup

system. To enable the robot to directly ascertain the pose of structural components in the scene, calibration of the camera as well as hand-eye calibration is imperative. Camera calibration involves determining the correspondence between a point in the camera and its coordinate system. This requires an analysis of the camera's imaging principles and experimental computations to obtain the camera's intrinsic and extrinsic parameters. Hand-eye calibration, on the other hand, aims to establish the transformation relationship between the camera coordinate system and the robot's end-effector coordinate system. Consequently, the robot can ascertain the precise pose of objects in its own coordinate system.

Upon completion of the calibration procedures on the established platform, actual robotic grasping experiments can proceed. Initially, the disorderedly stacked structural components in the scene are identified and localized to acquire their exact poses. Thereafter, the grasp sequence method is employed to derive a sequence for grasping these components. Once a component is selected for grasping, its pose is transformed via the hand-eye calibration to the pose in the robot's base coordinate system. Subsequently, based on a library matching the structure's orientation, a sequence for adjusting the robot's posture is determined. Finally, the grasping experiment is executed. The specific transformation relationships among various coordinate systems during the experiment are depicted in Fig. 8.28.

In the figure, the coordinate system $\{B\}$ represents the base coordinate system of the UR5, the coordinate system $\{C\}$ denotes the D435 camera coordinate system, and the coordinate system $\{F\}$ signifies the true coordinate system of the object. With respect to the positions of these three coordinate systems, the following relationships exist:

$$ {}_{C}^{B}T = {}_{F}^{B}T {}_{C}^{F}T \tag{8.48} $$

Thus, from Eq. (8.25), the relationship of the object $\{F\}$ with respect to the UR5 base coordinate system $\{B\}$ can be derived as follows:

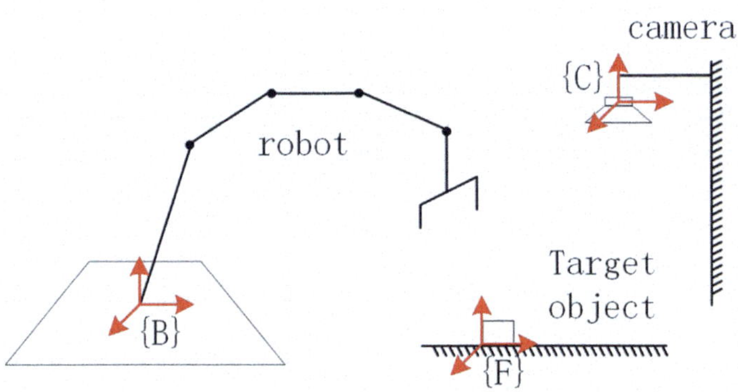

Fig. 8.28 Diagram of coordinate transformations

$$ {}^{B}_{F}T = {}^{B}_{C}T {}^{F}_{C}T^{-1} \tag{8.49} $$

The primary steps of the robotic grasping experiment are divided into the following.

1. Multiple structural components are stacked within the camera's field of view. After the camera captures a depth image of the structural component scene, it is transmitted to ROS-Rviz and converted into a point cloud image. A Python script is then executed to run the trained model, yielding classified point clouds of the structural component scene. Following this, a C + + script is used to apply bilateral filtering, RANSAC segmentation, Euclidean clustering, and multi-resolution sampling with ICP for precise registration, ultimately resulting in the precise pose of the structural components.
2. Based on the computed point cloud of the structural components, relevant constraints are calculated. Using a particle swarm optimization algorithm grounded in Gaussian and chaotic operators, the optimal grasping sequence for the current scene is obtained. Upon acquiring the target structural component and its pose, a sequence for adjusting the robot's posture is derived through computation.
3. This sequence for robot posture adjustment is fed into ROS-control, directing the robot to grasp the structural component and move it to the welding position. Post-welding, the robot returns to its initial position; the scene point cloud is updated. This process from steps (1) to (3) is iteratively repeated until no more structural components are left to be grasped in the scene, thus concluding the experiment. Given that the experiment involves an iterative grasping process, Fig. 8.29 illustrates the grasping procedure for a single structural component.

(3) **Experimental Results**

A repetitive grasping experiment was conducted on the stacked scene 50 times, with evaluations made on the grasp time, grasp precision, average grasp error, and grasp success rate. Grasp time refers to the cycle time from when the structural component is grasped until it reaches the final welding position and the welding is completed. Grasp error indicates the deviation of the structural component's center of gravity relative to the grasp center point. The grasp success rate represents the ratio of successful grasps of the structural components.

In this setup, the number of structural components in the stacked scene was five. The obtained experimental results are summarized in Table 8.5.

From the experimental results, it can be observed that the average grasp time for all five structural components in the 50 trials was within 20 s. Currently, manual assistance for welding takes at least 30 s, indicating that the use of robotic assistance has improved efficiency by over 30%. The average grasp error in repeated experiments was consistently below 3 mm, fully satisfying the requirement of a 5 mm threshold for final welding errors. Moreover, the average grasp success rate exceeded 98%, implying that out of 50 structural components, 49 could successfully undergo grasping for auxiliary welding. The component that was not grasped was

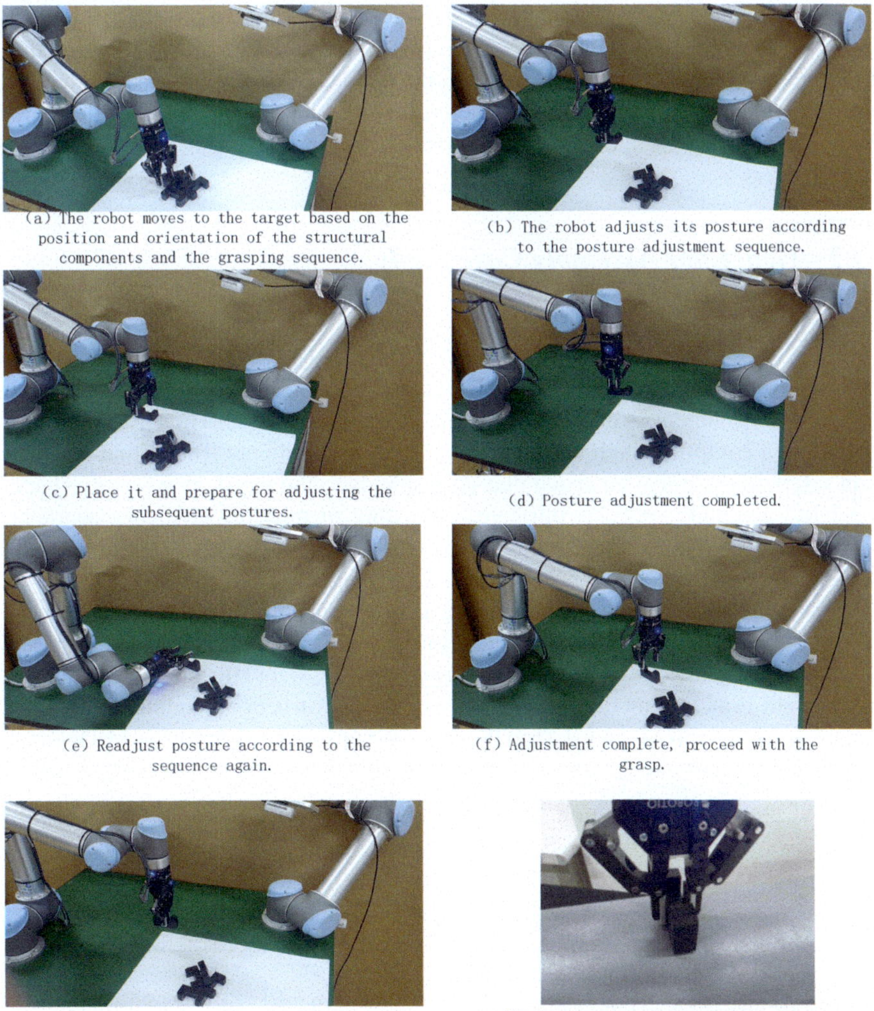

(a) The robot moves to the target based on the position and orientation of the structural components and the grasping sequence.

(b) The robot adjusts its posture according to the posture adjustment sequence.

(c) Place it and prepare for adjusting the subsequent postures.

(d) Posture adjustment completed.

(e) Readjust posture according to the sequence again.

(f) Adjustment complete, proceed with the grasp.

(g) Complete the grasp with the final posture.

(h) Move to the welding position for auxiliary welding.

Fig. 8.29 Schematic diagram of structural component grasping process

Table 8.5 Results of repetitive grasping experiments

	Component 1	Component 2	Component 3	Component 4	Component 5
Grasp time (s)	18.3	15.7	18.7	16.3	14.8
Grasp error (mm)	2.3	2.8	1.7	1.5	2.5
Success rate (%)	100	100	98	98	100

dropped during the posture adjustment due to inadequate clamping friction. This issue can be addressed in the future by replacing the gripper pads with ones that offer higher friction performance. Overall, the employment of robots in conjunction with computational vision processing methods for the auxiliary welding of structural components is feasible and can effectively replace manual labor.

8.4.2 *Generator Assembly Case Study*

(1) **Experimental Environment**

To verify the performance of human–robot collaboration (HRC) in assembly tasks, an assembly environment as depicted in Fig. 8.30 was constructed. The HRC assembly environment comprises both virtual and physical environments. The virtual environment refers to the mirrored space that corresponds to the physical assembly space. The HRC assembly space is intricate, consisting of multiple elements: workers, collaborative robots and control cabinets, visual or tactile sensors, assembly parts, assembly tools, workbenches, among others. Information from elements in the physical assembly environment is transmitted through sensors to a data management system, which in turn updates the status information of the virtual space. The left side of Fig. 8.30 showcases the digital twin system utilized for HRC assembly, encompassing four principal modules: (a) Environment Construction Module; (b) Virtual Assembly Module; (c) Control Module; and (d) Statistics and Analysis Module.

The functions of each module are as follows:

(a) Environment Construction Module: Allows users to add or modify assembly elements within the interface, creating a new assembly environment.
(b) Virtual Assembly Module: The HRC assembly process is carried out in a three-dimensional virtual space of the model. This simulation process serves to test the effectiveness of HRC, with its synchronization process with the physical scene enabling real-time interaction.
(c) Control Module: After confirming the safety of the assembly environment, the operator can control the assembly process.

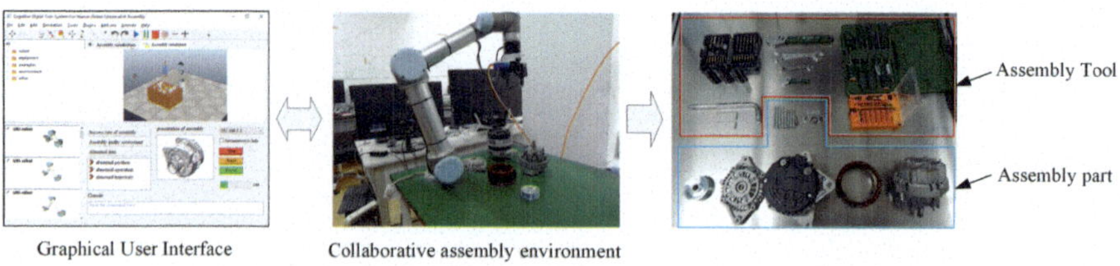

Fig. 8.30 HRC assembly scenario

(d) Statistics and Analysis Module: This module is designed to display and analyze assembly outcomes, such as success rates, assembly quality, and abnormal data, facilitating further optimization of the HRC assembly process.

(2) **Experiment Setup**

The HRC assembly experiment is structured around three aspects: human–robot collaborative action sequences, part grasping, and cooperative assembly. Two primary methodologies are employed in the experiments: (1) Assembly mode based on digital twins, and (2) Assembly mode based on fixed programming. In the first experiment, the generation of action sequences using a digital twin HRC model is investigated, compared to manually designed sequences. Experiments two and three utilize these two assembly modes for part manipulation and robotic assistance in fixing components. The key performance indicators observed during the experiments include operation time, success rate, and collision frequency, with the experiment designed across five groups, and results are reported as the average of all parallel runs. The object of assembly is a specific model of an AC generator composed of 16 distinct parts. According to the assembly constraint matrix, the HRC assembly process encompasses 15 assembly tasks. After task evaluation and initial allocation, the virtual space of the assembly system verifies the performance of the assignment. During task allocation, assembly tasks are categorized into robot-dominant and human-dominant tasks.

(3) **Experimental Result**

1. **Assembly Sequence**

As illustrated in Fig. 8.31, upon completing the HRC assembly, the following two conclusions are drawn.

Employing the assembly mode based on digital twins, the total HRC assembly time was recorded at 412 s, whereas under the fixed programming assembly mode, the total assembly time extended to 550 s.

Within the framework of fixed programming, tasks dominated by human intervention accounted for 70% of the entire assembly procedure, whereas in the DT (Digital Twin) assembly mode, this share decreased to 49.76%.

2. **Part Grasping**

During the part grasping phase, different planning methods within the data management system were compared: fixed programming, DDPG model, and D-DDPG model (incorporating two-layer reinforcement learning). Prior to training, both the DDPG and D-DDPG models were set with 1000 episodes, each episode consisting of 10 steps, a replay buffer capacity of 5000, and exploration noise initialized at 0.1. The parts to be grasped within the robot's capabilities were determined, and the test results represent the averages across all groups.

Table 8.6 presents the experimental outcomes for part grasping under various planning models. Notably, utilizing the D-DDPG model to grasp specific parts took approximately 14.832 s, whereas the DDPG model required 16.354 s. In contrast,

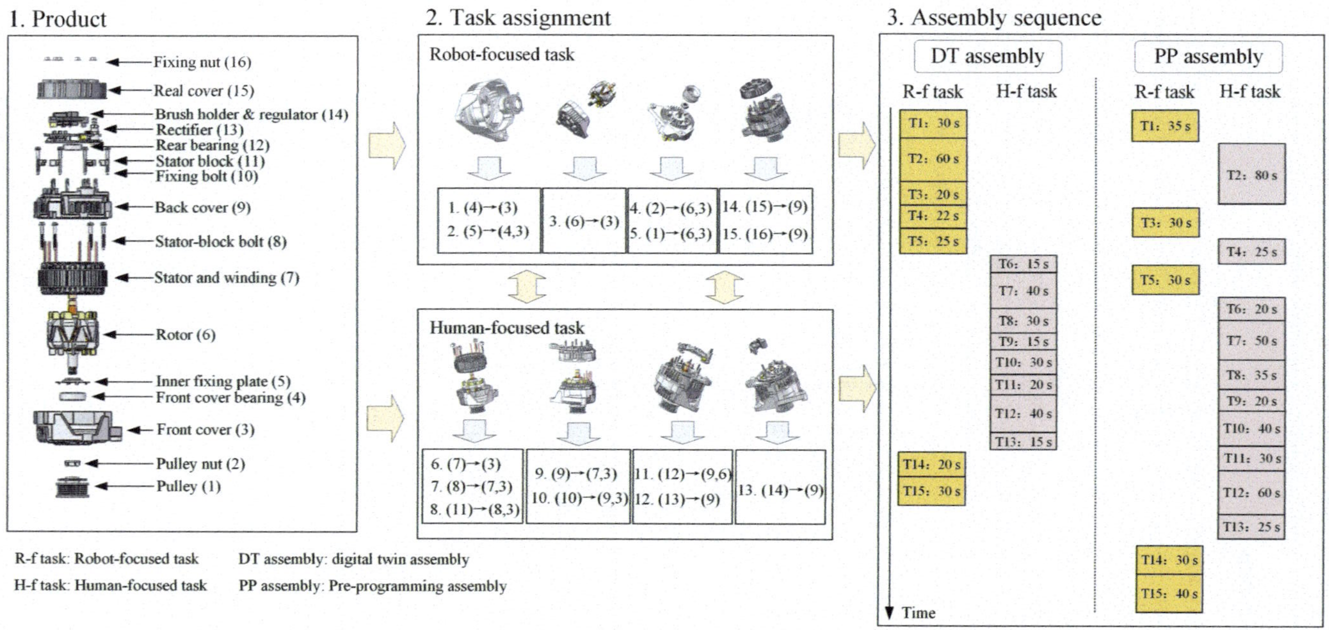

Fig. 8.31 Comparison of assembly performance under two different assembly modes

the pre-programmed grasping planning model consumed 25.645 s. This discrepancy arises because pre-programming may lead the robot to traverse redundant paths. In terms of collisions, the collision rate associated with the pre-programmed assembly mode reached 5.208%, significantly higher than that of the DDPG model at 1.563% and the D-DDPG model at 0.521%. Traditional methods are unable to provide safe trajectories for robots in dynamic environments, explaining this phenomenon. Furthermore, the grasping precision was assessed by the positional deviation between the TCP point and the part's center point. The pre-programmed method exhibited a grasping deviation of 2.447 mm, greater than that of the DDPG model (1.875 mm) and D-DDPG (0.895 mm). This is attributed to the inability of fixed programs to adjust grasping postures when object positions change dynamically.

3. **Collaborative Assembly**

During the collaborative assembly stage, comparisons were made between multi-LSTM layer action recognition models and spatiotemporal attention mechanisms. For specific assembly actions, the temporal attention mechanism assigns attention weights to different time steps. For instance, in Fig. 8.32a, the maximum attention weight is allocated at time step 8 for a given assembly action. Attention on key points of the human skeleton is also distributed according to their importance, with Fig. 8.32b illustrating that key point 11 receives the highest attention. Figure 8.32c depicts the accuracy of different action recognition models over time. Here, the accuracy of the recognition model improves as the training progresses. Moreover, there exists a correlation between the accuracy of the action recognition model and the number of LSTM layers. When the LSTM layer count reaches 4, the recognition accuracy peaks at 0.912453. However, this is notably lower than the accuracy of 0.966006 achieved by the spatiotemporal attention model. Figure 8.32d illustrates the average waiting time of the robot during collaborative tasks. Each of the different experimental groups underwent testing on ten assembly tasks. The robot's responsiveness was quantified by calculating the wait time after recognizing human actions. As shown in Fig. 8.32d, the robot's average waiting time during part transfer and collaborative assembly operations hovered around 4 s. These experimental findings suggest that the robot's waiting times indicate that the current high-performance robot assembly based on digital twins can meet the demands of high efficiency.

Table 8.7 provides the collaborative assembly test results from several perspectives, including assembly time, collision rate, and assembly precision. In terms of assembly time, the D-DDPG model completes a single assembly task in just 12.453 s, whereas the DDPG model requires 14.564 s for the same task. Compared to these

Table 8.6 Comparison of grasping performance under different planning models

Planning method	Grasping time (s)	Collision rate (%)	Grasping precision (Δd) (mm)
Pre-programming	25.645	5.208	2.447
DDPG	16.354	1.563	1.875
D-DDPG	14.832	0.521	0.895

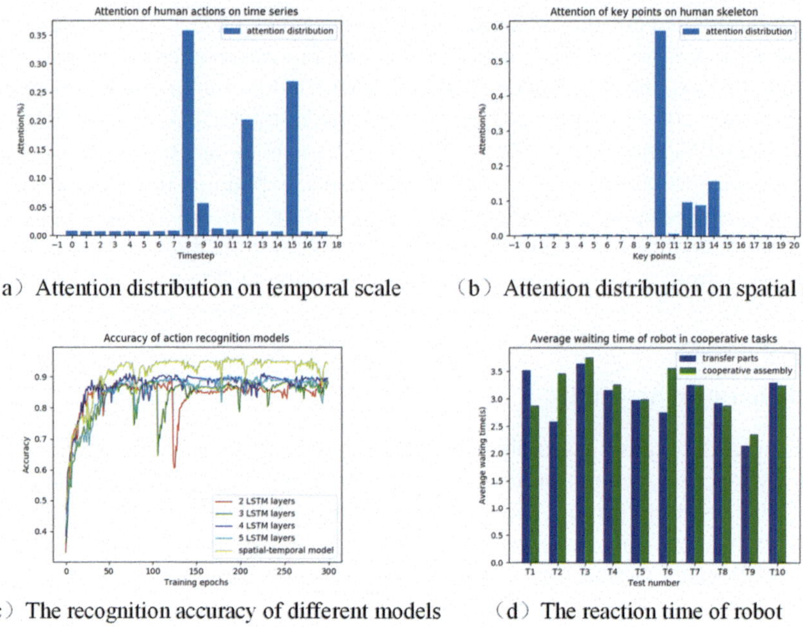

(a) Attention distribution on temporal scale (b) Attention distribution on spatial scale

(c) The recognition accuracy of different models (d) The reaction time of robot

Fig. 8.32 Results of HRC assembly experiments

Table 8.7 Comparison of collaborative assembly performance under different planning algorithms

Planning algorithm	Assembly time (s)	Collision rate (%)	Assembly precision (Δd) (mm)
Pre-programming	24.845	5.128	1.884
DDPG	14.564	1.538	0.964
D-DDPG	12.453	0.513	0.587

two models, the pre-programmed assembly mode takes 24.845 s to finish the task. This disparity arises because when the human's position changes, the robot follows a predetermined path without dynamic adjustment. From a collision standpoint, the pre-programmed assembly mode yields a significantly higher collision rate of 5.128%, far exceeding that of the DDPG model (1.538%) and the D-DDPG model (0.513%). The reason behind this lies in the robot's inability to adaptively adjust its own posture in response to changes in human body posture. There is a substantial correlation between assembly deviation and precision. The D-DDPG model reduces assembly deviation to 0.587 mm. Conversely, the DDPG model experiences the greatest deviation at 0.964 mm, and under the pre-programmed assembly mode, the deviation escalates to 1.884 mm. This phenomenon stems from the robot's incapacity to promptly respond to assembly tasks based on fixed programming.

References

1. Ou Y, Liu G, Xu T (2018) A learning-from-demonstration based framework for robotic manipulators sorting task. In: 2018 IEEE International conference on cyborg and bionic systems (CBS). IEEE, pp 42–47
2. Cruz JPN, Dimaala ML, Francisco LGL et al (2013) Object recognition and detection by shape and color pattern recognition utilizing artificial neural networks. In: 2013 International conference of information and communication technology (ICoICT). IEEE, pp 140–144
3. Qi J, Ma L, Cui Z et al (2024) Computer vision-based hand gesture recognition for human-robot interaction: a review. Complex Intell Syst 10(1):1581–1606
4. Girshick R, Donahue J, Darrell T et al (2014) Rich feature hierarchies for accurate object detection and semantic segmentation. In: Proceedings of the IEEE conference on computer vision and pattern recognition, pp 580–587
5. He K, Zhang X, Ren S et al (2015) Spatial pyramid pooling in deep convolutional networks for visual recognition. IEEE Trans Pattern Anal Mach Intell 37(9):1904–1916
6. Girshick R (2015) Fast R-CNN. In: Proceedings of the IEEE international conference on computer vision, pp 1440–1448
7. Redmon J, Divvala S, Girshick R et al (2016)You only look once: unified, real-time object detection. In: Proceedings of the IEEE conference on computer vision and pattern recognition, pp 779–788
8. Liu W, Anguelov D, Erhan D et al (2016) Ssd: single shot multibox detector. In: Computer vision–ECCV 2016: 14th European conference, Amsterdam, The Netherlands, 11–14 Oct 2016, Proceedings, Part I 14. Springer International Publishing, pp 21–37
9. Sermanet P, Eigen D, Zhang X et al (2013) Overfeat: Integrated recognition, localization and detection using convolutional networks. arXiv preprint arXiv:1312.6229
10. Hinterstoisser S, Cagniart C, Ilic S et al (2011) Gradient response maps for real-time detection of textureless objects. IEEE Trans Pattern Anal Mach Intell 34(5):876–888
11. Jun G, Xin W E I, Shiji Z et al (2022) Laser welding process of body galvanized sheet based on exhaust passage. Laser Technol 46(1)
12. Chen H, Zhang G, Zhang H et al (2007) Integrated robotic system for high precision assembly in a semi-structured environment. Assem Autom 27(3):247–252
13. Ran W, Liu S, Zhang Z (2020) A polling-based dynamic order-picking system considering priority orders. Complexity 2020(1):4595316
14. Chitta S, Jones E G, Ciocarlie M et al (2012) Mobile manipulation in unstructured environments: Perception, planning, and execution. IEEE Robot Autom Mag 19(2):58–71
15. Hornung A, Phillips M, Jones E G et al (2012) Navigation in three-dimensional cluttered environments for mobile manipulation. In: 2012 IEEE International conference on robotics and automation. IEEE, pp 423–429
16. Zaidi S, Maselli M, Laschi C et al (2021) Actuation technologies for soft robot grippers and manipulators: a review. Curr Robot Rep 2(3):355–369
17. Long Z, Jiang Q, Shuai T et al (2020) A systematic review and meta-analysis of robotic gripper. IOP Conf Ser Mater Sci Eng 782(4):042055
18. Wang X, Kang H, Zhou H et al (2023) Development and evaluation of a robust soft robotic gripper for apple harvesting. Comput Electron Agric 204:107552
19. Li Q, Zhang Z, You Y et al (2020) Data driven models for human motion prediction in human-robot collaboration. IEEE Access 8:227690–227702
20. Garcia CA, Montalvo-Lopez W, Garcia MV (2020) Human-robot collaboration based on cyber-physical production system and MQTT. Procedia manufacturing 42:315–321
21. Darvish K, Simetti E, Mastrogiovanni F et al (2020) A hierarchical architecture for human–robot cooperation processes. IEEE Trans Rob 37(2):567–586
22. Liu S, Wang L, Wang XV (2021) Function block-based multimodal control for symbiotic human–robot collaborative assembly. J Manuf Sci Eng 143(9):091001

23. Lattanzi L, Raffaeli R, Peruzzini M et al (2021) Digital twin for smart manufacturing: a review of concepts towards a practical industrial implementation. Int J Comput Integr Manuf 34(6):567–597
24. Tao F, Zhang M, Liu Y et al (2018) Digital twin driven prognostics and health management for complex equipment. CIRP Ann 67(1):169–172
25. Sierla S, Kyrki V, Aarnio P et al (2018) Automatic assembly planning based on digital product descriptions. Comput Ind 97:34–46
26. Sun X, Bao J, Li J et al (2020) A digital twin-driven approach for the assembly-commissioning of high precision products. Robot Comput Integr Manuf 61:101839
27. Guo D, Zhong RY, Lin P et al (2020) Digital twin-enabled graduation intelligent manufacturing system for fixed-position assembly islands. Robot Comput Integr Manuf 63:101917
28. Yi Y, Yan Y, Liu X et al (2021) Digital twin-based smart assembly process design and application framework for complex products and its case study. J Manuf Syst 58:94–107
29. Dröder K, Bobka P, Germann T et al (2018) A machine learning-enhanced digital twin approach for human-robot-collaboration. Procedia CIRP 76:187–192
30. Oyekan JO, Hutabarat W, Tiwari A et al (2019) The effectiveness of virtual environments in developing collaborative strategies between industrial robots and humans. Robot Comput Integr Manuf 55:41–54
31. Ma X, Tao F, Zhang M et al (2019) Digital twin enhanced human-machine interaction in product lifecycle. Procedia CIRP 83:789–793
32. Wang T, Li J, Kong Z et al (2021) Digital twin improved via visual question answering for vision-language interactive mode in human–machine collaboration. J Manuf Syst 58:261–269
33. Malik AA, Bilberg A (2018) Digital twins of human robot collaboration in a production setting. Procedia Manuf 17:278–285
34. Bilberg A, Malik AA (2019) Digital twin driven human–robot collaborative assembly. CIRP Ann 68(1):499–502
35. Kousi N, Gkournelos C, Aivaliotis S et al (2019) Digital twin for adaptation of robots' behavior in flexible robotic assembly lines. Procedia Manuf 28:121–126
36. Baskaran S, Niaki FA, Tomaszewski M et al (2019) Digital human and robot simulation in automotive assembly using Siemens process simulate: a feasibility study. Procedia Manuf 34:986–994
37. Telea A (2004) An image inpainting technique based on the fast marching method. J Graph Tools 9(1):23–34
38. Minaee S, Boykov Y, Porikli F et al (2021) Image segmentation using deep learning: a survey. IEEE Trans Pattern Anal Mach Intell 44(7):3523–3542
39. Qi CR, Yi L, Su H et al (2017) Pointnet++: deep hierarchical feature learning on point sets in a metric space. Adv Neural Inform Process Syst 30
40. Guo Y, Wang H, Hu Q et al (2020) Deep learning for 3d point clouds: a survey. IEEE Trans Patt Anal Mach Intell 43(12):4338–4364
41. Bello SA, Yu S, Wang C et al (2020) Deep learning on 3D point clouds. Remote Sens 12(11):1729
42. Yang J, Wang C, Luo W et al (2021) Research on point cloud registering method of tunneling roadway based on 3D NDT-ICP algorithm. Sensors 21(13):4448
43. de las Heras MJF, Manguán MC, Millán ER et al (2020)Determination of SEA loss factors by Monte Carlo Filtering. J Sound Vibr 479:115348
44. Jain M, Saihjpal V, Singh N et al (2022) An overview of variants and advancements of PSO algorithm. Appl Sci 12(17):8392
45. Periasamy M, Kaliannan T, Selvaraj S et al (2022) Various PSO methods investigation in renewable and nonrenewable sources. Int J Power Electron Drive Syst 13(4):2498
46. Zhang J, Koppel A, Bedi AS et al (2020) Variational policy gradient method for reinforcement learning with general utilities. Adv Neural Inf Process Syst 33:4572–4583
47. Arulkumaran K, Deisenroth MP, Brundage M et al (2017) Deep reinforcement learning: a brief survey. IEEE Signal Process Mag 34(6):26–38

48. Zhang W, Chen Q, Yan J et al (2021) A novel asynchronous deep reinforcement learning model with adaptive early forecasting method and reward incentive mechanism for short-term load forecasting. Energy 236:121492
49. Yarats D, Fergus R, Lazaric A et al (2021) Mastering visual continuous control: Improved data-augmented reinforcement learning. arXiv preprint arXiv:2107.09645
50. Li Y, Tee KP, Yan R et al (2020) Reinforcement learning for human-robot shared control. Assem Autom 40(1):105–117
51. Oliff H, Liu Y, Kumar M et al (2020) Reinforcement learning for facilitating human-robot-interaction in manufacturing. J Manuf Syst 56:326–340

Chapter 9
Intelligent Management and Control of Production Operation

The essence of production quality process control is to organize effective production, accurately identify and manage potential quality risks in the product manufacturing process. And constantly improve the driving force of self-improvement, which is a process of continuous improvement of the management system. However, due to the management guided by the production and manufacturing results. And the potential quality risks have not been unaware, which makes us ignore the control before and during the event, resulting in frequent production quality accidents. Managers cannot accurately judge the operation status of quality control, and the process improvement drive is weak. Artificial intelligence technology can bolster quality control in the production process. It effectively identifies potential risks that arise from various issues. These issues include inaccuracies in production organization and uncoordinated process sequences. It also addresses the risks from delayed information transmission and unclearly defined powers and responsibilities. Additionally, it deals with the use of unreasonable analysis tools. This technology helps to eliminate major and extraordinary production quality accidents. Ultimately, it contributes to achieving transparent production. This chapter introduces the knowledge management method of data organization in the production process and finally gives the intelligent control and energy consumption control methods for spinning production.

9.1 The Progress and Status of Intelligent Management and Control in Production Operations

The whole process of complex product production includes production demand, blank manufacturing, structural part processing, component assembly, etc. The whole production process faces the demands of high digitization levels, rapid response, and dynamic production configuration [1]. The entire machining production line is large in scale, with high equipment personnel allocation, quantity, and complexity,

T. Liu et al., *Industrial Intelligence: Methods and Applications*, Springer Series in Advanced Manufacturing, https://doi.org/10.1007/978-3-031-81477-8_9

resulting in a wide range of production data sources and dynamically changing production process states. Currently, the manufacturing industry faces challenges in the production and manufacturing process, mainly including knowledge management and reuse, collaborative management, quality improvement, and Six Sigma control. Many processes can be used to solve such problems. For instance, capturing process knowledge allows for its reuse through configurable process planning. This creates a superior platform for process planning. It also supports process verification and the preparation and management of process documentation. The platform aids in managing manufacturing processes and services. avoiding quality hazards and downstream errors in quality improvement, reducing waste caused by non-value-added activities, Continuously improving production efficiency, enhancing manufacturing agility and flexibility, and other measures. To better address the numerous issues faced in manufacturing production, the digitalized, intelligent, and informative production operation mode will bring greater production benefits to the production line.

The operational process management and control integrating intelligent manufacturing behaviors should at least include three dimensions of capabilities: the ability to understand, analyze, and solve problems, the ability to inductive and deductive reasoning, and the ability to adapt to the environment for survival and development. Intelligent production includes two aspects: the intelligentization of the production process preparation and the intelligentization of the machining production process. At the level of intelligentization of the production process preparation, various virtual simulations and advanced analysis software tools have been developed at home and abroad. The functions of modeling, simulation, analysis, data processing, and decision optimization are becoming increasingly complete. However, professional simulation software is not directly applicable to the production process preparation. Only customized products can meet the needs of automation and intelligence in the production process preparation. The intelligentization of the machining production process covers the entire product process. The production operation process of mechanical products is the process of converting various resources into final products. In the production line, it involves planning, scheduling, and execution behaviors, namely, combined scheduling behaviors. The part processing process is the execution of the production process, which is a general term for the process of converting raw materials and semi-finished products into target demands through certain procedures and methods. The main body is machine tools and other process equipment, and the core is the processing or operation process of the machine tools or process equipment. In the machining line, it is the autonomous execution behavior of the operation and condition handling of process equipment. Production process management and control is a general term for production operation management. In modern enterprises, it is a software engineering, management, and monitoring behavior, which can be summarized as perception and identification, data acquisition, and automatic reasoning. All these behaviors constitute the intelligent behavior of the automated machining line after being integrated through the production management and control system. Combined scheduling takes the machining line process as the main line, optimizes the scheduling of operation plans, material transportation, tool distribution, production operations, etc. According to the working status of the equipment and the principle of equipment load balancing, and performs combined scheduling and control based on balanced production methods and allowance design criteria. Autonomous execution arranges the scheduled job plans for the equipment in the form of action chains and executes them in the specified order. At the same time, it monitors the real-time status of the processed parts and each equipment and feeds back to the plan and scheduling for rescheduling to ensure the stable operation of the system. Perception and identification include comprehensive perception and monitoring of the real-time status and information feedback of material transportation, machine tool processing, production operation, and product conditions in the machining line. Data acquisition mainly obtains dynamic data through perception means, analyzes the obtained real-time operation status data in a timely and rapid manner, and then transforms it into dynamic knowledge. Automatic reasoning will make independent judgments and decisions based on the set rules and analysis results.

In the preparation of production processes and the application of various intelligent methods in manufacturing, the use of information technology greatly enhances the precision and efficiency of manufacturing equipment, leading to automation. Networked and flexible manufacturing is highlighted in the context of economic globalization, where network-based manufacturing technologies are extensively applied, emphasizing flexibility and reconfigurability as key features of twenty-first century. By leveraging IoT, big data collection, and other intelligent technologies, research is conducted on interconnecting machining resources, collecting real-time data on different production types in machine shops. And developing manufacturing process state awareness models for intelligent operations. Product intelligent manufacturing involves advanced information technology, electrical automation technology, and AI. Utilizing high-tech intelligent devices to build smart production lines, workshops, and factories, and implement intelligent management and services. Thereby achieving intelligent and innovative value chains within the manufacturing industry. Key technologies such as machine learning, deep learning, cognitive computing, knowledge graphs, and natural language processing are essential in the process of building intelligent manufacturing.

In the official Wikipedia entry, the knowledge graph is the knowledge base that Google uses to enhance the functionality of its search engine. Essentially, a knowledge graph is described as a semantic network that reveals relationships between entities, enabling the formal description of real-world entities and their interactions. Knowledge graphs are now commonly referred to as various large-scale knowledge bases, requiring support from various intelligent information processing technologies. Through knowledge extraction, entities, relationships, and attributes can be extracted from semi-structured or unstructured data. Knowledge fusion eliminates ambiguities between referring terms and factual objects, forming high-quality knowledge bases. Knowledge reasoning further explores implicit knowledge based on existing knowledge bases, enriching and expanding them. Distributed knowledge representations in comprehensive vectors are significant for the construction, inference, fusion, and application of knowledge bases.

Google's knowledge graph, proposed in 2012, symbolically processes big data, enabling machines to solve various production issues using knowledge graphs to attain cognitive intelligence like humans. Knowledge graph technology integrates information retrieval and natural language processing, playing a vital role in industrial settings with complex production processes and diverse manufacturing rules and databases. In an industrial context, with numerous production elements and complex manufacturing processes, different factories follow distinct rules stored in databases, process documents, and video materials. Constructing a knowledge graph for manufacturing resources in the industrial sector represents production elements. Facilitating resource prediction, self-organization, and enhancing the reuse of manufacturing resource knowledge to improve production efficiency. Knowledge graph technology offers a new feasible solution for handling complex data, organizing structured and unstructured data through semantic relationships, and emphasizing data governance and semantic connections for the utilization and migration of large-scale data. To better understand knowledge graph technology, it can be discussed from three perspectives: 1. Data Governance: Knowledge graphs alter traditional data storage methods by changing the form of data organization and enhancing industry data logic preservation. 2. Semantic Connections: Knowledge graphs bridge the gap between deep learning's understanding of underlying features and human comprehension of semantic spaces, aiding in converting human thinking into machine thought processes. Intelligent Applications: Knowledge graphs make knowledge accessible (searchable), query able (question-and-answer), and actionable (decision-making).

The optimization of production operation processes in the workshop mainly includes adjustments to the production process flow, workshop equipment planning, and the allocation of dedicated operational staff. It primarily encompasses two aspects: the optimization of non-optimal processes themselves, and the use of information tools. The tools are used to optimize operational processes based on the study of mechanical production requirements and environmental conditions based on product variety and batch status. The selection of unit-based operational process optimization adjustments is made to improve the production system, thereby adapting to the company's production and delivery needs. Based on intelligent process mining. The operational process is optimized mainly by discovering hidden unstable factors affecting workshop efficiency during work and taking a series of optimization measures to adjust the workshop process units [2].

Energy consumption control in the production process mainly involves various links and departments of production, and control of raw materials is essentially indicated by material, labor, and expense. Based on established drawings, process methods, and control standards for process consumption quotas, control of raw material consumption is carried out at various stages of the production process. Energy consumption constitutes the largest proportion when considering production costs. Different schemes are devised for energy consumption control based on the technical equipment level and processing level of the enterprise. This includes control of kinetic energy for various facilities, such as the use of mechanical, production building facilities, vehicles, lifting equipment, and electrical equipment. Energy consumption control also extends to thermal energy, which involves controlling the consumption of various production facilities, equipment, and processing technologies. Energy includes electricity, oil, coal, coke, wood, various chemical fuels, and combustion aids. Controlling production energy consumption enables rational resource allocation in mechanical processing.

In the field of textiles, intelligent management and control of textile production operations have developed alongside the world's new technological and industrial revolutions. In recent years, developed countries such as the United States, the European Union, and Japan have leveraged their comprehensive advantages in internet, computing, industrial big data to develop advanced textile manufacturing technology and are at the forefront. These countries have significantly upgraded and optimized textile manufacturing processes, establishing global networks. And integrating machinery, warehouse systems, and production facilities into CPS (Cyber-Physical Systems), fundamentally improving textile enterprises' manufacturing, engineering, material usage, supply chain, and lifecycle management. For instance, "Future TEX," one of Germany's Industry 4.0 projects, constructs a customer-centric flexible value chain and sets directions for future textile factories, digitalized manufacturing processes, large-scale customization, and new business models.

Advanced textile enterprises in Germany, the UK, Japan, and other countries have developed textile manufacturing data collection and information management systems. For instance, the UK's Crosrol company has developed a human–machine interface system for textile manufacturing. Germany's Trützschler company has created a data collection system for textile equipment. These systems have achieved interlocking control between textile machines to some extent. They have improved the operating efficiency of individual machines. The systems have significantly promoted the informatization of workshops and even entire enterprises. They have enhanced the core competitiveness of enterprises in the international market.

Most of the Fortune Global 500 textile enterprises have adopted ERP systems, enabling enterprises to rationally utilize resources, reduce inventory, lower costs, increase production efficiency, and market share. For example, the TEXIS software developed by the German company Orga System using the MEP principle and method. The method includes modules such as contract orders, production processes, finished raw material inventory, purchase plans, cost accounting, and finished product delivery [3].

Switzerland, Belgium, Germany, and other countries with advanced textile machinery technology have seen the adoption of information management modes in 60% of small enterprises, 80% of medium-sized enterprises, and over 90% of large enterprises. These countries' new intelligent textile equipment achieves interconnection, making information and data generated during equipment operation visual, transferable, and readable. MES systems are used to acquire and transmit equipment information and data to cloud platforms, from which SCM, APS, WMS, and other product modules can extract valuable data to support the overall efficient operation of the textile industry.

China's application of intelligent management and control in production operations has also begun. This is mainly embodied in firstly, the preliminary foundation of intelligent textile equipment and system integration technology, such as the fully automatic computerized flat knitting machine and flexible customized system of Ningbo Cixing, Changzhou Wuyang's intelligent high-speed double-needle bed knitting machine, and Jingwei Textile Machinery Company's E system covering the entire cotton opening, carding, combing, and spinning production process, Secondly, the demonstration pilot of intelligent manufacturing workshops covering the entire textile industry chain, such as the intelligent nylon production factory, intelligent spinning factory, and digital automatic dyeing production line, Thirdly, large-scale personalized customization services of garment enterprises have entered a growth stage, with leading companies such as Qingdao Kute, Zhejiang Baixi Niao, and Guangdong Aisida [4]. However, compared with the advanced level abroad, China still lacks standards in data interconnection, information systems for textile networked data collection, and textile manufacturing intelligent execution technology research, development, and application are still in the early stages, and there is a significant gap in the integration, informatization, and intelligence of textile equipment and electromechanical systems compared to advanced levels abroad, as well as the immaturity and imperfection of intelligent operation in textile enterprises.

9.2 Knowledge Graph-Based Intelligent Management and Control Model for Production Operations

Currently, there are a large amount of data resources in the industry, but the overall utilization rate of data resources is very low. This is mainly because the data accumulated by enterprises are multi-source and heterogeneous with complex relationships. There is a wealth of knowledge in many documents' information. At present, such information mainly relies on manual processing, which is cumbersome and has limited cognitive ability. In the long run, it is not conducive to promoting the development process of digitalization and intelligence of enterprises. Therefore, there is a higher demand for efficient processing and utilization of digital connections and knowledge management of the production process. It is necessary to improve the efficient utilization of production process information to guide and serve production, to improve production efficiency from the knowledge level, and to enhance the digitization level of workshops.

9.2.1 Model for Comprehensive Real-Time Monitoring of the Production Process

The large scale of complex mechanical product production lines leads to high levels of equipment personnel deployment, quantity, and complexity, resulting in a wide range of sources for production data and dynamic variability in production process states. By utilizing IoT and big data collection technologies, research has been conducted on methods for interconnected resources of complex mechanical products, aiming to achieve comprehensive perception, reliable transmission, and intelligent processing of mechanical product information. Addressing the characteristics of mechanical product production and the requirements of a secure network environment, various information sensors, RFID, infrared sensors, laser scanners, etc., are utilized in machining workshops to collect production data of different types (performance, size tolerances, surface quality, etc.). Subsequently, through networks such as 5G, Zigbee, Wi-Fi, Bluetooth, LoRa, Sigfox, eMTC, and NB-IoT, following specific communication and transfer protocols, the data is transmitted to be recognized as digital signals by computers, facilitating subsequent data processing and analysis [5]. Faced with vast, rapidly growing, and diverse big data, databases are employed to store and manage this data, forming databases for processes, tools, machine tools, etc. Data generated by machining workshop equipment is continuously updated and processed through IoT, various networks, and databases, enabling real-time online perception of the production status of mechanical product production lines. Figure 9.1 illustrates the equipment state perception throughout the entire process of mechanical product production.

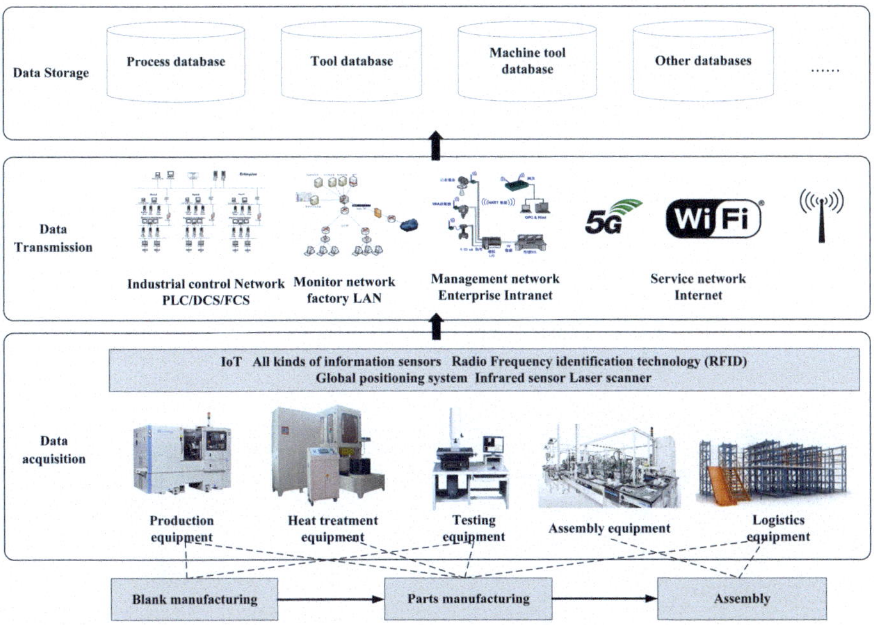

Fig. 9.1 Equipment state sensing in the whole production process of mechanical products

9.2.2 Data Modeling for the Production Process

The production process data serves as the basis for reflecting the essential characteristics and operational laws of the production system. Modeling and researching production process data can promote the development of emerging manufacturing modes, represented by intelligent production maintenance, and has garnered widespread attention in recent years [6]. In the context of manufacturing production processes, taking discrete manufacturing as an example, although the data input of manufacturing resources is intermittent during the production of individual projects, and the products are complex and diverse, the resultant volume of manufacturing process data is large and non-uniform. However, overall, the data can be divided into the following four levels: equipment data layer, process data layer, indicator data layer, and user data layer, as shown in Fig. 9.2.

The equipment data layer, as the bottom layer, collects millisecond-level sequential data on equipment states such as equipment vibration, power, cutting force, cutting parameters, etc. The second layer is the process data layer, which gathers relevant material, personnel operation, sensor control data. The third layer is the indicator layer, collecting online quality monitoring data, production efficiency, production

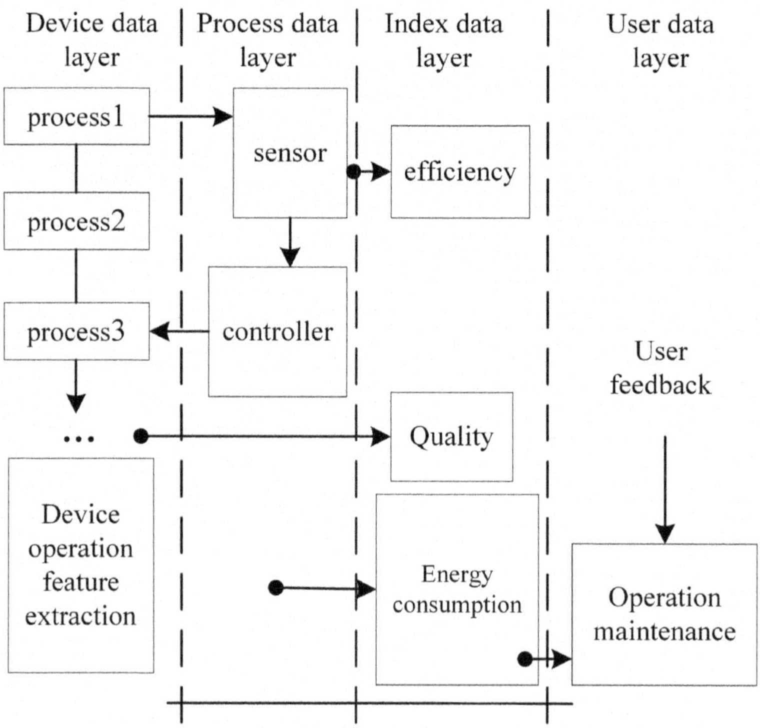

Fig. 9.2 Data types for the entire production process

progress, and manufacturing consumption data. The top layer consists of user feedback data on the product. The essence of modeling the entire production process data lies in the unified organization and management of the data, information, and knowledge generated during the process. By acquiring real-time manufacturing data and processing it into production information, further refining, and summarizing production knowledge, and applying knowledge engineering methods to solve production problems [7].

In the context of the entire production process, data is more dynamic, discrete, and time-sensitive. These data contain rich practical experience, processing rules, implementation methods, etc. In the modeling process of such data, it is necessary to transform data existing in individuals and scattered across different production stages into public, organized, and disseminated knowledge that can be shared and reused [8]. According to the production sequence, from the facility layout level, processing resource level, and organizational operation level to the technical implementation level, organizations can be modeled through the establishment of data, information, and knowledge spatiotemporal models at the equipment level, workstation level, production line level, workshop level, enterprise level, and cross-enterprise level. The process is illustrated in Fig. 9.3.

The data related to the four levels can be further classified into the following categories based on production characteristics and data sources.

1. Product basics and process data: This includes basic data such as product number, name, delivery time, as well as material, quantity, etc. Process data refers to the process requirements, machining drawings, quality requirements, processing procedures, and codes for the product.
2. Process logistics status data: This data pertains to the logistics of the production process, including workpiece flow data, workpiece transfer data from previous processes to the current one, start and end time data for workpiece clamping, machining and completion time data on machine tools, real-time location data of work-in-progress, real-time inventory data, etc.
3. Production process system operating condition data: This data represents the state of equipment during the machining process, such as material status data, tooling and cutting tool status data, personnel status data, production progress data, etc.
4. Workpiece quality characteristic evolution data: This data encompasses quality-related information during the machining process, including machining features, processing factors, benchmarks, and their evolution, as well as the relationship between them. It also includes data on process quality accuracy, surface finish, final value quality accuracy, surface finish, etc.

By utilizing the above-mentioned four categories of data, a multidimensional spatiotemporal representation of the production process can be constructed, enabling the correlation between production resources, data, information, and knowledge. This provides a fundamental data guarantee for information, knowledge, and knowledge engineering aspects.

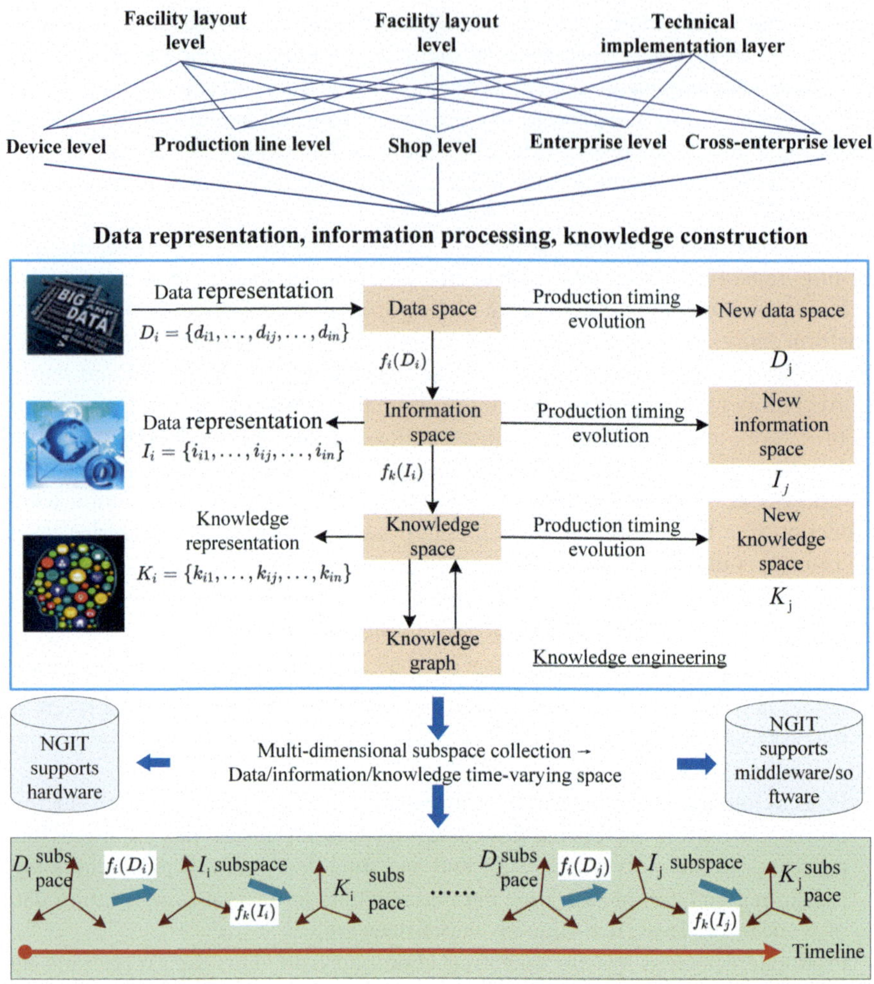

Fig. 9.3 Evolution of time-varying spatial information model for data oriented to the whole production process

Considering the diverse types and complex hierarchical relationships of production elements involved in the entire manufacturing process, it is essential to study the modeling and evolution process of product models based on knowledge graphs [9]. This primarily involves identifying and analyzing production element types in the processes of raw material manufacturing, structural component processing, and parts assembly.

Currently, there are limited modeling methods for manufacturing resources in the field of manufacturing. Ontology is the most popular knowledge modeling and representation approach, increasingly applied in various domains such as knowledge engineering and artificial intelligence. By combining ontology with knowledge

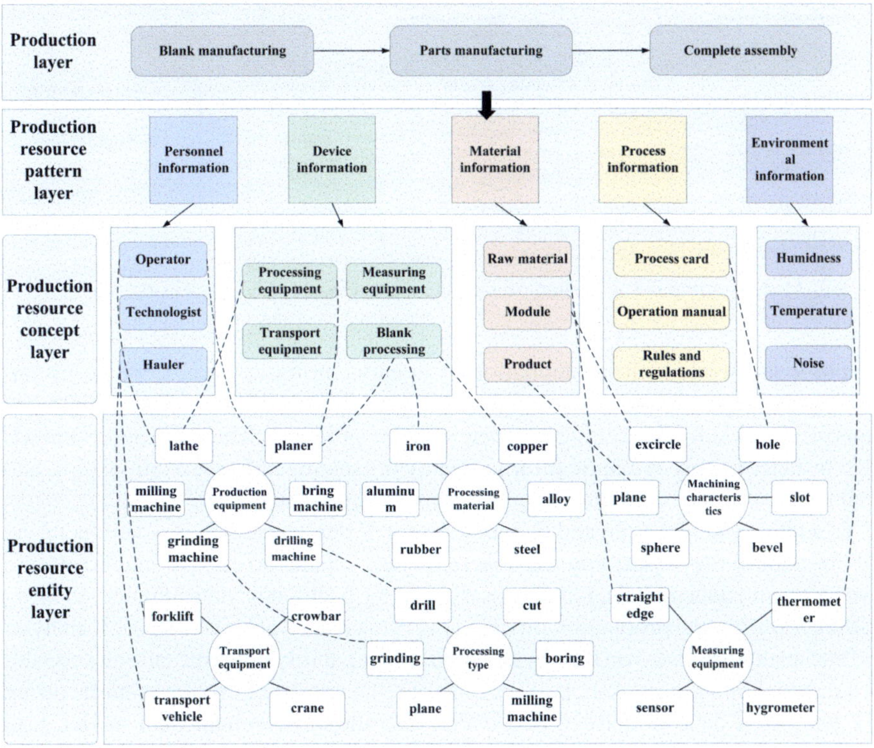

Fig. 9.4 Resource knowledge mapping model for the whole production process

graphs, hierarchical associations can be further explored to establish a static resource knowledge graph model for the entire production process as depicted in Fig. 9.4.

The establishment of the knowledge graph model for the entire production process reduces the confusion in the classification of production elements and helps in forming a comprehensive resource knowledge graph. Considering the "man–machine-material-method-environment" five factors that influence product quality in the theory of Total Quality Management, the model primarily consists of four layers: the production resource entity layer, the concept layer, the pattern layer, and the production process layer. This model is built top-down, corresponding to the transformation of data to information to knowledge. At the top layer, various structured data generated during the production process are stored in databases, including process documents, design drawings, and other semi-structured and unstructured data. These data are generally raw and unprocessed objective facts, lacking inherent meaning and value. The next layer consists of the five elements of the "man–machine-material-method-environment" ontology, which correspond to personnel information, equipment information, material information, process information, and environmental information. Through a series of analyses and processing, the data can be transformed into semantic information according to specified categories [10]. The

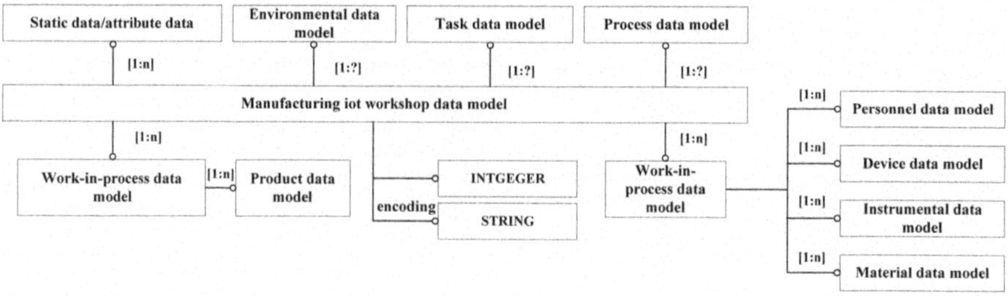

Fig. 9.5 Manufacturing IoT workshop overall data management model framework

bottom layer involves the instantiation of manufacturing resources, where instantiated knowledge serves as the template for constructing the knowledge graph and supports knowledge reasoning for new custom products. These resource knowledge instances apply semantic information by organizing entities, relationships, and attributes, providing theoretical support for subsequent inference.

To address the real-time and dynamic nature of data generated by manufacturing IoT, in addition to describing the inherent static characteristics of manufacturing resources with data models. It is necessary to describe the state data of manufacturing resources during the production process [11]. This can be achieved through analysis and encapsulation based on the physical domain, capability domain, and state domain of the data. Taking a workshop as an example, the construction of a manufacturing IoT workshop data model can be divided into the environment data model, task data model, process data model, work-in-progress data model, and manufacturing resource data model [12]. The overall data management model framework for a manufacturing IoT workshop is illustrated in Fig. 9.5.

(1) **Manufacturing Resources Data Model**

Manufacturing resources primarily include human, machine, and material. The characteristics of data modeling of manufacturing resources in terms of physical domain, capability domain, and status domain is illustrated in Fig. 9.6. The equipment data model focuses on static and dynamic structured descriptions through essential attributes, operational states, manufacturing capabilities, and maintenance records [13]. The personnel data model describes employees based on their basic information and functional skills and provides structured descriptions of task states. It is linked to task position equipment codes and work-in-progress codes, enabling real-time mapping between dynamic manufacturing data elements and personnel data, thus achieving dynamic descriptions of personnel data.

(2) **Process Task Data Modeling**

Structured definition of process tasks is carried out to establish a task data model (as shown in Fig. 9.7), providing a structural framework for system task modeling and enabling a unified description of task data. By defining task data and search

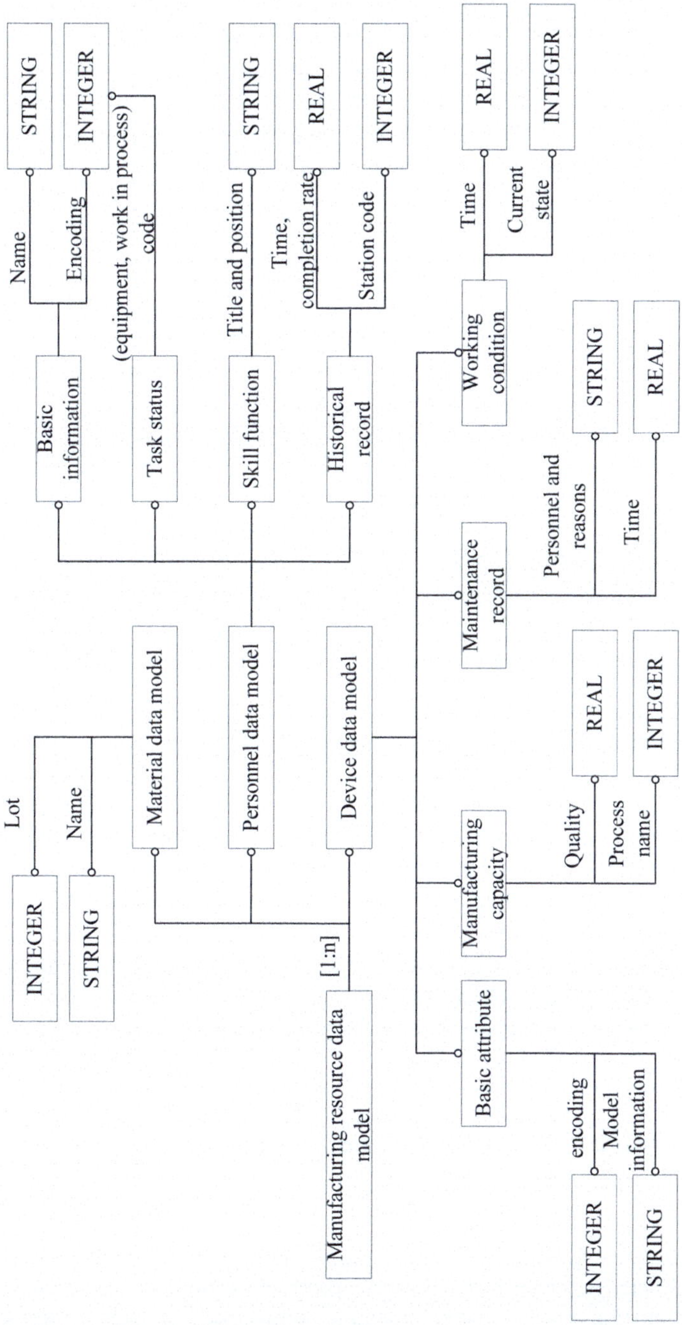

Fig. 9.6 Manufacturing resource data model

criteria for workstations, matching workstation tasks with process documents and quality standards, essential data conditions to support workstation task objectives are provided [14]. Task objectives encompass factors such as work efficiency, processing quality, target output, and delivery time, outlining the purpose of the tasks and linking workstation tasks to the overall workshop task data model. The status domain of the task data model describes the association of on-duty worker codes, equipment codes, processes, and work-in-progress codes, enabling real-time dynamic descriptions of status data elements.

(3) **Work-in-Progress Data Model**

The work-in-progress data model (as depicted in Fig. 9.8) encompasses the attribute parameters constituting the product, including raw materials, product names, batch information [15]. It provides a structured description of the status data related to work-in-progress, defining task status, process status, operation status, and quality inspection status during the production process. Furnishing a complete data structure required for real-time monitoring of work-in-progress status and quality traceability. Simultaneously, it involves the structural description of quality inspection standards, defining these standards, outlining inspection rules and non-conformance handling rules, and providing standardized descriptions of inspection results, thereby achieving a unified data description for non-conformance inspection handling.

(4) **Ontology Models for Manufacturing Process Characterization**

In production manufacturing, the production process of a product is completed in a specific technological sequence, exhibiting a structured and ordered characteristic. By arranging multiple processes, the planning and design of the manufacturing process are realized. To enhance the flexibility and scalability of expressing the structured and ordered characteristics and to quickly redesign when adding new processes or modifying equipment, manufacturing process knowledge and production equipment layout are expressed through the linkage relationships between ontology instances [16]. When constructing the ontology of process characteristics, only some elements directly related to the production manufacturing process are considered. Taking these elements as an example, the concepts and property relationships set in the ontology of process characteristics are shown in Fig. 9.9.

9.2.3 Knowledge Graph Construction for the Entire Production Process

Next, we will proceed with the corresponding introduction to data extraction, fusion, and knowledge generation from the production process, the construction of workshop-level knowledge graphs, as well as process monitoring and quality assessment methods.

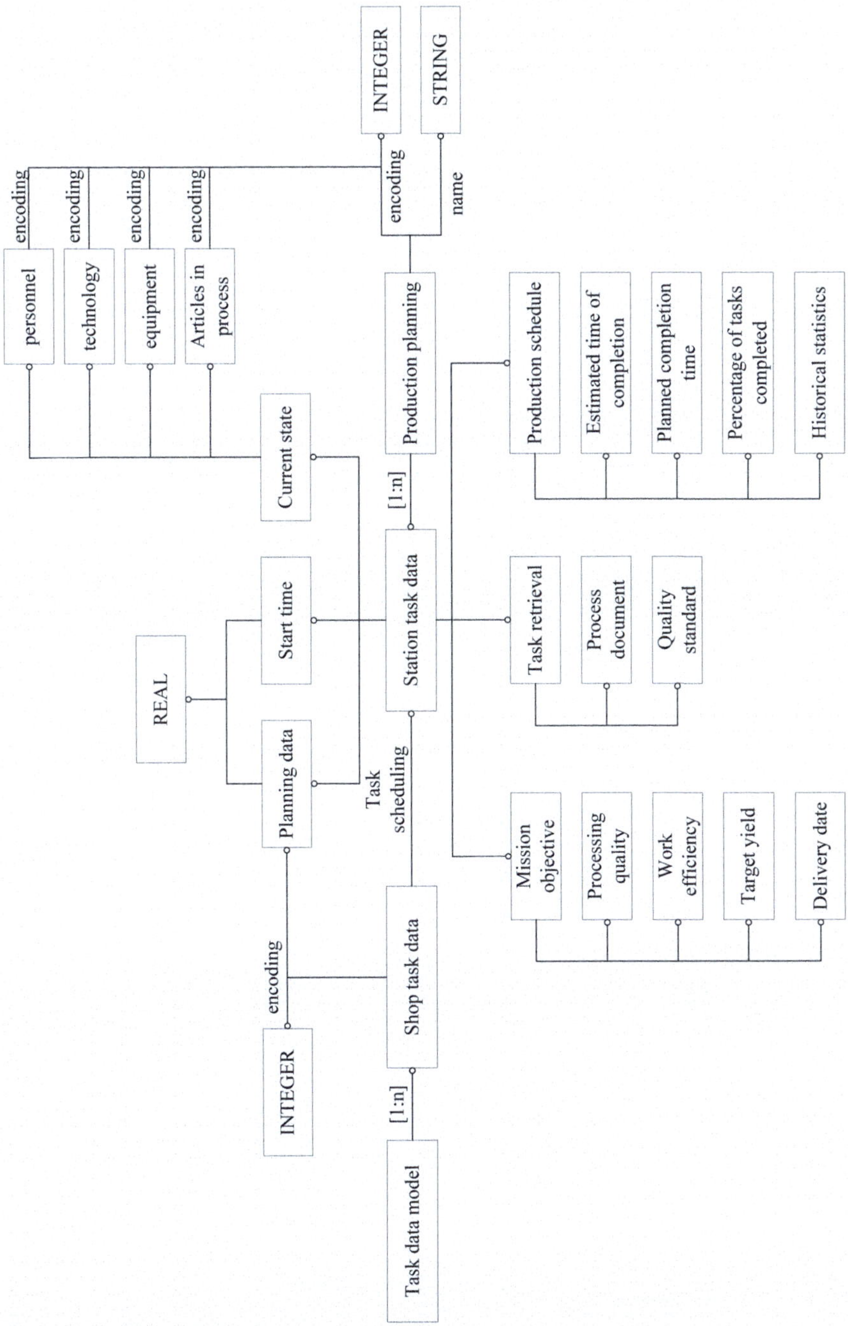

Fig. 9.7 Process task data model

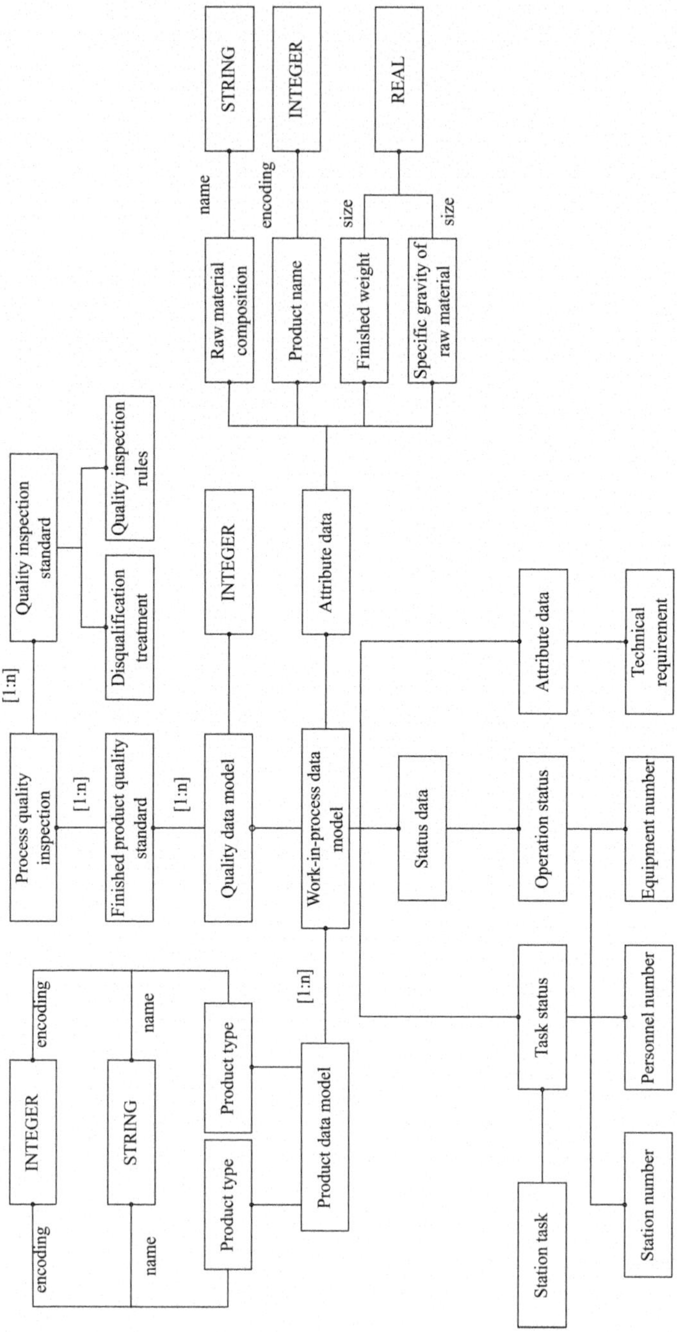

Fig. 9.8 Work-in-progress data model

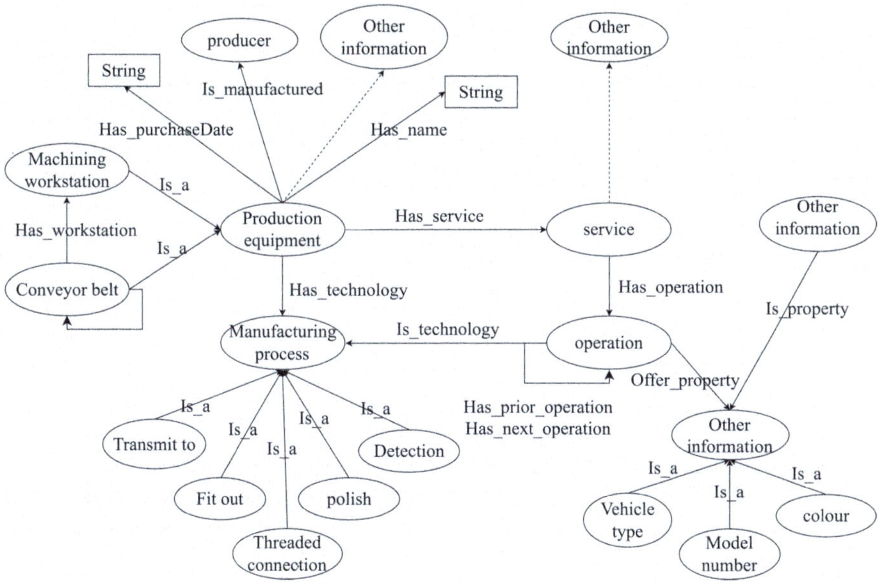

Fig. 9.9 Ontology relationship model for process characteristics

(1) Named Entity Identification

In manufacturing resources, there exist a large amount of semi-structured and unstructured data, such as process flow diagrams, video tutorials, etc., which are challenging to process. The quality of extracting knowledge from this information directly impacts the quality of the knowledge graph. By employing extraction rules that align with industrial scenarios and integrating them for application in resource entity identification, we aim to enhance the quality of knowledge extraction. Following the proposed human–machine-material-law-environment production process knowledge graph model, various elements in manufacturing resources are standardized. The five elements of human–machine-material-law-environment serve as the information foundation, gradually expanding the knowledge layer of manufacturing resources [17]. Addressing the issue of low efficiency in processing and utilizing unstructured data throughout the entire mechanical product production process, a modeling model based on the knowledge graph of the entire production process is utilized to standardize the production elements involved. Research is conducted on entity relationship extraction focusing on the entire production process, including the identification and extraction of production element types during the processes of raw material manufacturing, structural component processing, and component assembly. Natural language processing techniques are employed to handle data end-to-end, integrating data of entity relationships.

(2) Data Fusion and Knowledge Generation

The interaction of manufacturing data such as products, processes, equipment, and system operations in the mechanical manufacturing process results in complex operational characteristics within the workshop production process. Meanwhile, due to the lack of information exchange among various manufacturing units in the workshop, information cannot interact synergistically [18]. Utilizing knowledge graph technology, the extracted massive, multi-source, heterogeneous, and high-dimensional multi-scale data in the manufacturing process is classified and processed. The combination rules in D-S theory are used for the fusion of factual data, followed by the application of a credibility decay model based on time-domain evidence for process-oriented data fusion, achieving unified organization of all data for mutual collaborative use of information.

To address the low utilization of data in the entire process of structural component production and the increased management cost of data, research is conducted based on the linkage of entities, relationships, and attributes in knowledge graph generation. The results output by the fusion model are utilized to construct both a static manufacturing knowledge base for the workshop and a dynamic manufacturing temporal knowledge base. Subsequently, a complex network node association model and mapping rules between products, processes, equipment, systems, etc., are established based on the linkage of the two knowledge bases, forming a knowledge network. The static knowledge base model primarily establishes data associations among various resources between workshop units, creating structured, standardized entities, and relationship sets, facilitating data traceability and sharing. The dynamic knowledge base model establishes association relationships among manufacturing data such as process parameters, equipment parameters, and material flow during the manufacturing process. Finally, data link technology is used to utilize dynamic temporal data storage addresses as linked entities, achieving the linkage and interoperability of dynamic and static knowledge bases, completing the organization and fusion of the entire process data.

The implementation process of the two technologies combined is illustrated in Fig. 9.10.

(3) Knowledge Networking at the Shop Floor Level of the Production Process

The workshop knowledge graph is constructed using a "top-down" approach, mainly divided into pattern layer and data layer construction. The pattern layer includes ontology and concepts, primarily achieving the construction of the workshop ontology concept classification system through ontology construction methods. The data layer includes workshop entities and related attributes, realizing workshop knowledge extraction, knowledge fusion, and knowledge updating based on static resource data (S) in the basic data space and production process temporal data (D). Through S, personnel knowledge networks, machine knowledge networks, material knowledge networks, and method knowledge networks can be constructed, Through D, activity attributes in each knowledge network can be extracted, and a

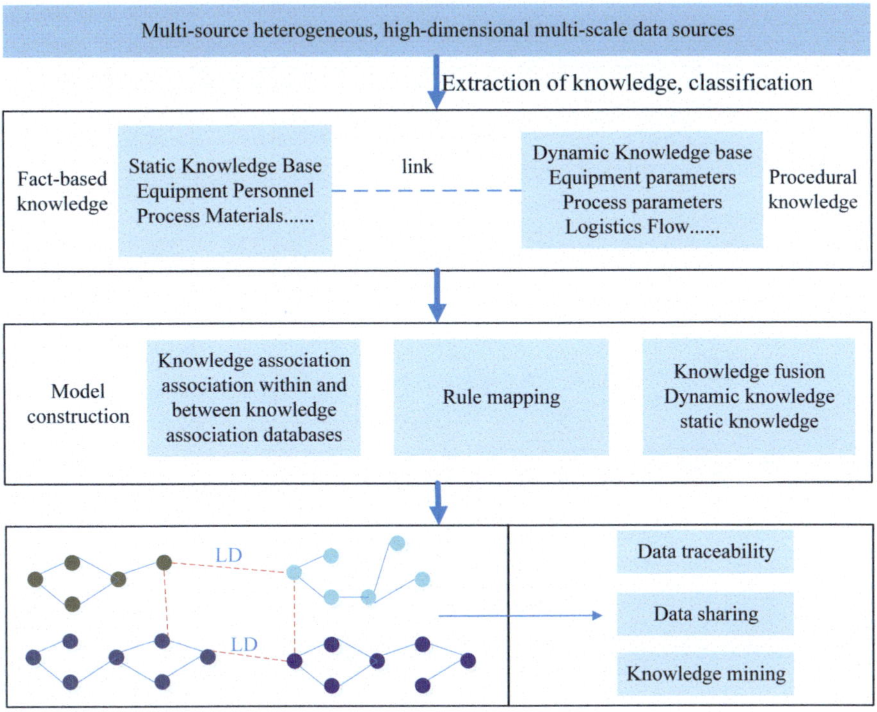

Fig. 9.10 Industrial spatio-temporal data fusion and knowledge generation model

four-dimensional spatiotemporal data space can be generated [19]. By leveraging the membership relationships of the four major knowledge network entities and the unique identification of activities. The four major knowledge networks and the four-dimensional spatiotemporal data space are virtually dynamically linked to form the workshop knowledge network. The main construction process is illustrated in Fig. 9.11.

By defining quadruples, the knowledge representation of the four major knowledge networks is conducted. For static resource data such as equipment and personnel in the workshop, automatic mapping is performed based on the established ontology data model. Since there are minimal changes in workshop processing personnel, the construction of personnel knowledge networks is relatively straightforward [20]. Entities are represented by personnel names in the database, with attribute lists determining static properties, while activity attributes focus on the personnel's availability status and task completion quantity. Figure 9.12 illustrates the mapping process of knowledge in the basic data space.

In this process, firstly, the data in the basic data space is mapped into knowledge networks according to the pattern layer. Simultaneously, a temporal space is created for personnel by associating entity-dynamic attributes in the knowledge network. The values of dynamic attributes of each entity are dynamically stored based on temporal

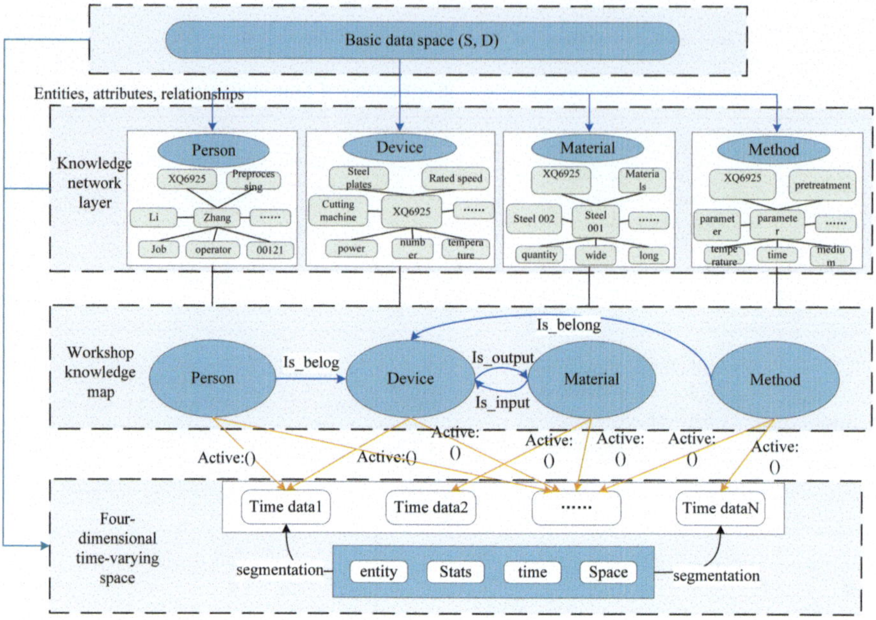

Fig. 9.11 Production process shop floor level knowledge network construction

granularity, and the activity attribute nodes in the knowledge network dynamically bind the address of data storage and the event function for data retrieval. In the personnel knowledge network, the relationships between entities are mainly categorized as the same workstation and hierarchical relationships between preceding and succeeding workstations. Using the same construction method, the knowledge networks for process methods and equipment are formed. The entities in these three types of knowledge networks are inherent resources of the workshop. The number of entities remains relatively constant. The attributes of interest being predefined, the updating function of the pattern layer is still retained. This allows for the dynamic addition and deletion of corresponding entities, attributes, and activities based on instances in the data layer to ensure the authenticity and variability of the knowledge network. Additionally, each entity's static attributes include workstation stations and operating equipment.

The material knowledge network differs significantly from the knowledge networks of equipment, personnel, and process methods. Due to variations in materials based on workstation requirements and product design in the manufacturing workshop, there may be an infinite number of entity nodes [21]. Therefore, extracting material entity names as nodes directly is not feasible. Entity nodes are divided into fixed entities and dynamic entities, where dynamic entities are continuously updated according to given rules, while fixed entities undergo minimal changes. The definition process is illustrated in Fig. 9.13.

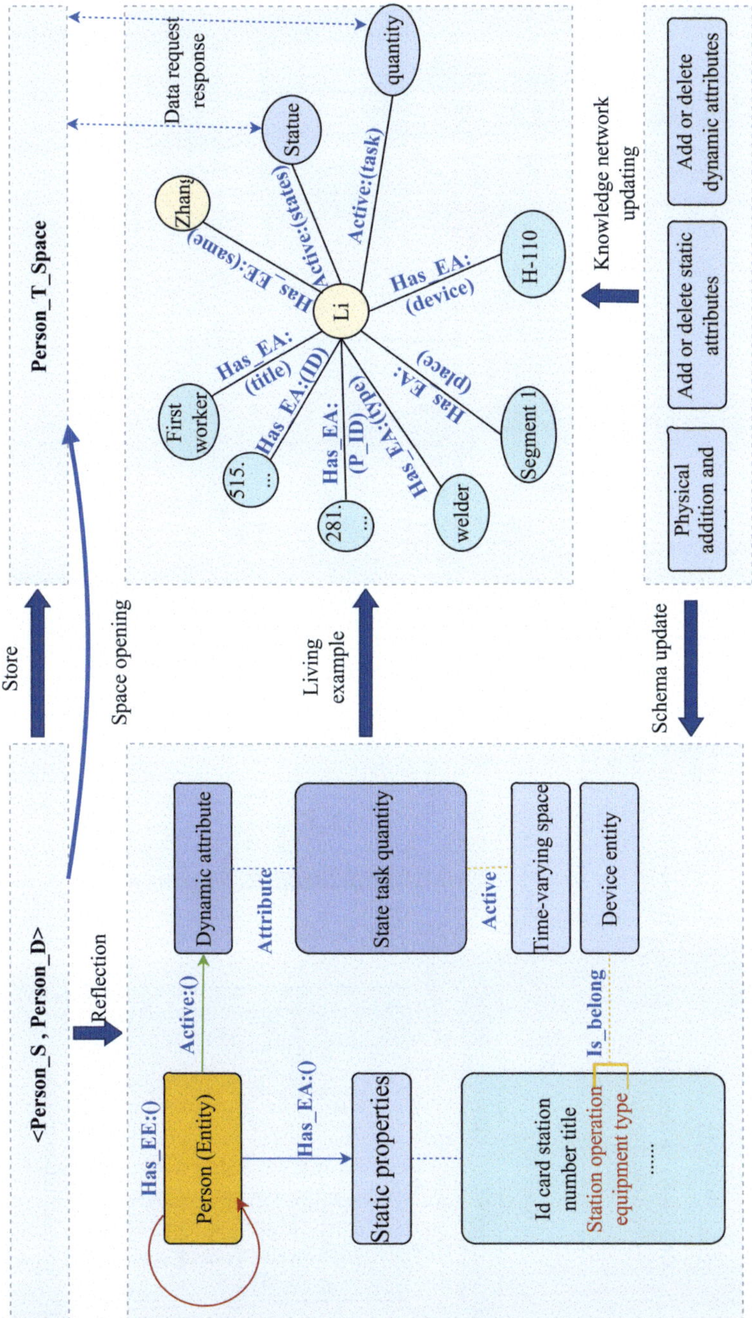

Fig. 9.12 The process of mapping knowledge in the underlying data space

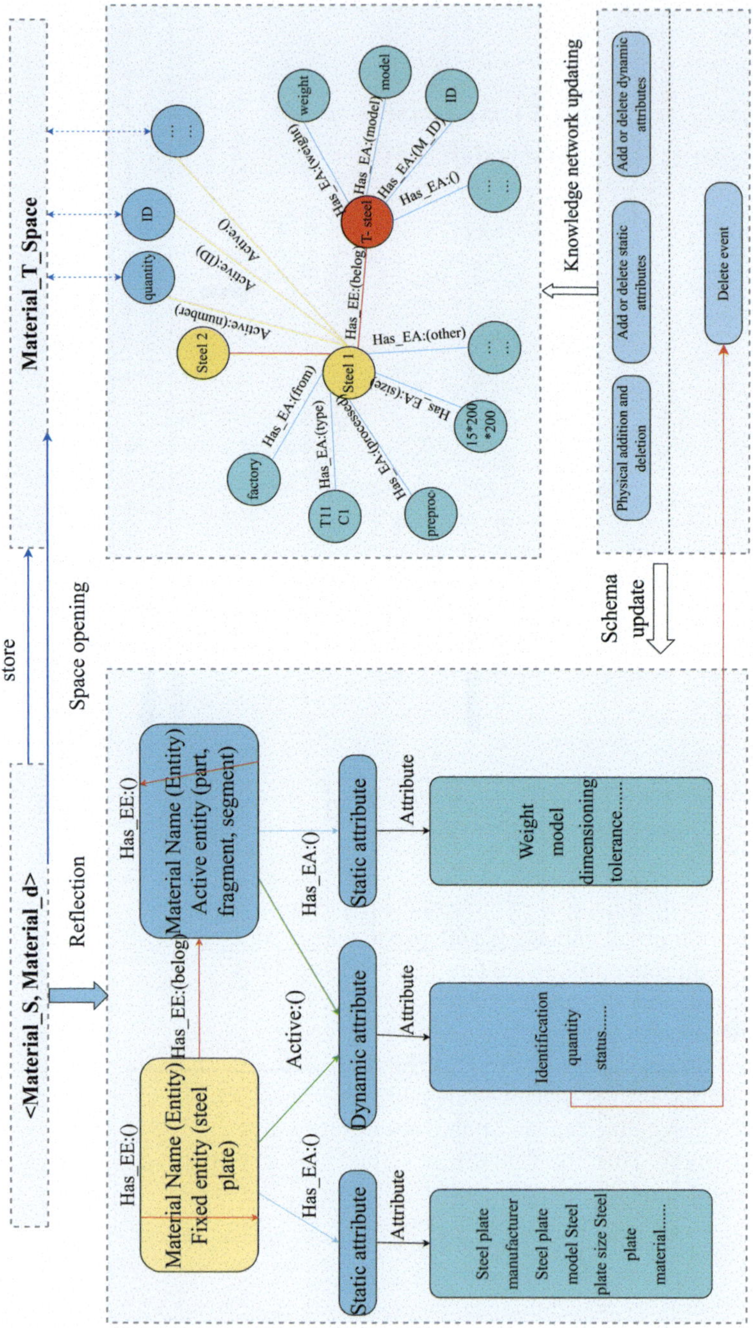

Fig. 9.13 Node definition process

In the construction process, activity entities are primarily obtained from the MBD segmented design model by reading the CATIA three-dimensional model structure tree. This involves retrieving the unique names of components, sub-components, and parts as well as their hierarchical relationships, which are then used to establish connections between activity entities. Subsequently, the attribute lists of each node in the structure tree are read to obtain the requirements and parameters of the model design, which serve as the static attributes of the activity entities. When the dynamic attribute state of a component or sub-component reaches completion, an automatic trigger initiates a deletion event, removing the entities of parts, components, and sub-components from the material knowledge network, while retaining the temporal data space. To generate the temporal data space, the four-dimensional temporal data space is expanded to include time, space, entities, and activity attributes. Entities and activity attributes are automatically extracted mainly from four knowledge network spaces. Spatial positions are primarily extracted from the equipment knowledge network space, and the time axis is dynamically extended based on the minimum temporal granularity. Data values are dynamically mapped from the basic data space D [22].

Within the context of activity attributes and the temporal plane, the data at each point is dynamically matched, mapped, and integrated from the basic data space. Entities and the spatial plane mainly record the positional information extracted from the equipment knowledge network, while entities and the attribute space primarily record the hierarchical relationships between entities and attributes. The space–time plane mainly records the trajectory information of each entity. Through the analysis of these four-dimensional planes, key information at any given space–time point can be pinpointed. The length of the time axis is mainly determined by the company's production cycle, and the removal of the tail end of the time axis is carried out dynamically to avoid the occurrence of super data bodies.

(4) **Process Monitoring and Quality Assessment**

To address the challenges of diverse data types at different stages of part production processes, the lack of information linkage among various resources. It also aims to overcome the difficulty of information traceability. The research is based on industrial knowledge to enhance the welding quality traceability process [23]. Starting from the relationships and entities in the welding industrial knowledge graph, graph theory techniques are utilized. When tracing a particular production resource entity, efficient querying of information related to that node is enabled. This includes providing attributes, equipment used, processing status, and other relevant information about the current node. Furthermore, it allows for in-depth exploration of entities with high relevance to the current entity, uncovering additional implicit associative information to achieve deep resource traceability and monitoring.

In response to the low correlation between spatial–temporal data on part welding and production resources, research is conducted on an industrial knowledge-driven welding quality assessment method. There is a lack of systematic accumulation of workshop-related knowledge for quality assessment. Research is carried out on an industrial knowledge-driven welding quality assessment method. Based on the quality-related knowledge graph involved in part welding, knowledge representation learning methods are employed to unearth implicit quality-related relationships between knowledge graphs, enrich the knowledge system of quality control, and enhance quality control mechanisms [24]. Additionally, machine learning techniques are utilized in conjunction with industrial knowledge graphs to establish a quality assessment optimization strategy model. By utilizing real-time traced production process data and a knowledge-driven model, real-time predictions of product quality during the production process are achieved, ultimately enhancing the quality of produced parts.

9.3 Data-Driven Intelligent Management and Control Optimization Technology for Production Operations

This subsection takes spinning production as an example, respectively, from spinning production process design and layout simulation optimization technology, spinning production process based on the industrial Internet platform, production process online detection and production status remote monitoring technology, based on the industrial big data platform for collaborative optimization of four aspects to illustrate the data-driven production operation intelligent control optimization technology.

9.3.1 Technology for the Design and Layout Simulation Optimization of Spinning Production Process

(1) **Spinning Production Process Planning**

Referring to the advanced domestic level of the Clearing, Scutching, and Lapping (CSL) process, this process consists of three stages: Opening and Loosening, Mixing and Opening, and Fine Opening. It progresses from coarse to fine, gradually loosening, removing impurities, and rationally blending the fibers. This process provides high-quality lint cotton, laying a solid foundation for the quality of subsequent semi-finished products. The design of the cotton carding machine is reasonable. After the lint cotton passes through the opening and cleaning zones of the carding rolls, it enters the Tin Plate Carding Zone [25]. Due to their optimal combination, the occurrence of cotton knots is reduced, micro-dust is removed, and the fibers are carded into a single fiber state. The dual fixed plate system before and after the Tin Plate Carding Zone can be flexibly adjusted according to different fiber characteristics, facilitating fiber carding and impurity removal.

The adoption of advanced combing preparation technology involves merging and coiling the slivers before feeding them into the combing machine. This results in good lateral evenness of the small coils, with no stripes and minimal lateral diffusion. The

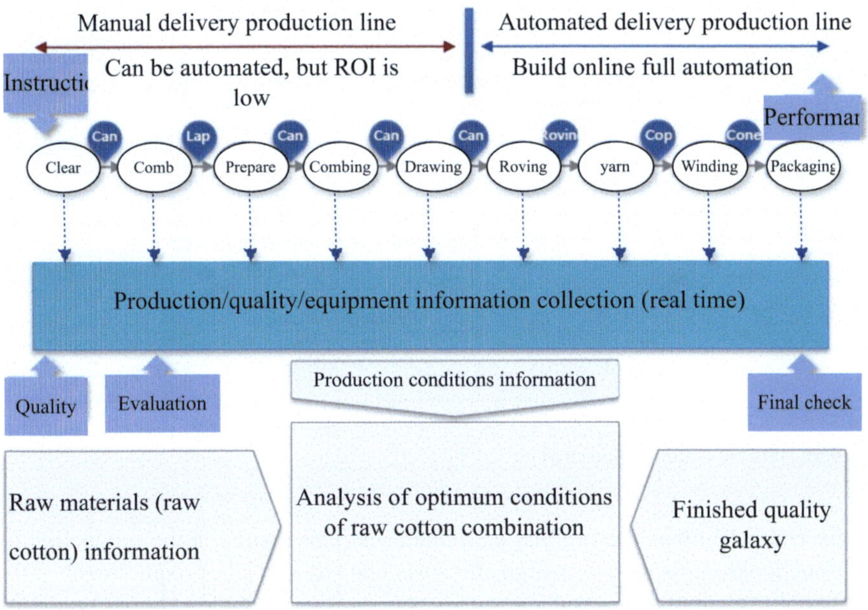

Fig. 9.14 Smart factory spinning process flow planning

quality of the formed coils is improved, leading to higher-quality combed yarn. This approach is conducive to saving cotton and reducing production costs (Fig. 9.14).

The roving frame is equipped with a four-roller drafting system, featuring a reasonable drafting process that is beneficial for fiber control. It can enhance fiber parallelism and straightness, laying a solid foundation for smooth drafting in subsequent processes. This equipment operates at high speed with high production capacity and excellent manufacturing precision, showcasing a high level of electromechanical integration.

The adoption of compact spinning technology involves incorporating negative pressure airflow and corresponding components in the spinning triangle zone to effectively control fiber bundles in this area. This results in improved yarn appearance and structure. By utilizing the principles of aerodynamics, a set of gathering elements in the spinning triangle zone, when loose fiber tufts pass over the surface of the negative pressure suction tube, uses air guidance to tightly bundle the fibers together [26]. These fibers are then twisted into yarn by a guiding apron roller, reducing the size of the spinning triangle zone during spinning and minimizing the number of floating exposed fibers (i.e., fly). This process enables the production of high-quality knotless yarn.

Electronic clearing, air twisting, and automatic winding machines are used to detect and correct defects. These defects include thick and thin sections, cotton knots, and irregularities. This process ensures that spliced joints have a tensile strength over 85% of the original yarn strength. The twisting diameter of the spliced joints does

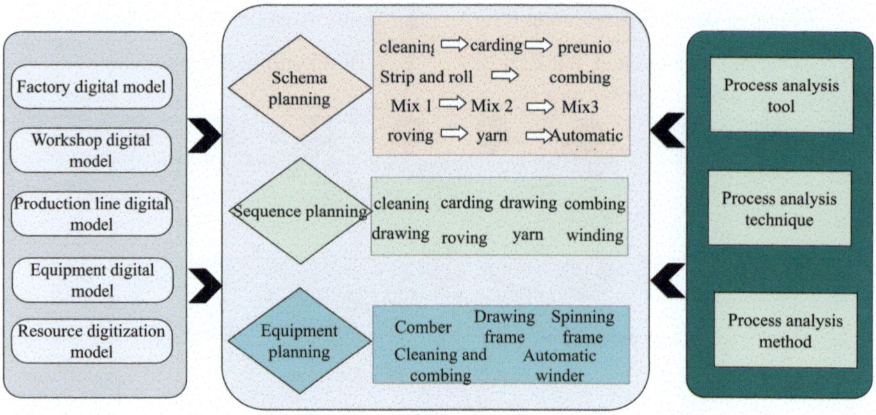

Fig. 9.15 Process planning flowchart

not exceed 1.15 times the original yarn diameter. This results in the production of high-quality knotless yarn with high efficiency and reduced labor requirements [10].

Digitally planning the production process involves creating a digital model of the factory, workshop, production line, production equipment, and manufacturing resources. With guidance from process analysis methods and technologies, supported by process analysis tools, a rational plan is developed for the product's process route, process arrangement, and manufacturing equipment. Through simulation analysis using digital models, the optimal process planning scheme is determined, ensuring that the factory's production scale is appropriate, meeting process requirements, and guaranteeing product quality [27] (Fig. 9.15).

The spinning process flow consists of carding, combing cotton, lap preparation, lap winding (cotton), combed yarn spinning (cotton), blending stage one, blending stage two, blending stage three, coarse yarn spinning, fine yarn spinning, automatic winding, automatic packaging, and automatic warehousing. The layout of the new workshop is arranged according to the process flow, where raw materials undergo carding, cotton combing, lap preparation, combed yarn spinning, lap preparation, coarse yarn spinning, fine yarn spinning, and winding processes before the finished products are stored in the warehouse. This ensures a smooth process without cross-transportation, meeting the normal production requirements. (Refer to Fig. 9.16 for the process flow).

(2) **Simulation Optimization of Spinning Production Layout**

An advanced smart spinning factory must have an integrated data system where relevant data is created, stored, and transformed throughout the entire manufacturing process, including product requirements, processes, and equipment data resources. The smart spinning factory serves as the foundation for establishing a digitized enterprise. A resource database is established based on the production needs of spinning,

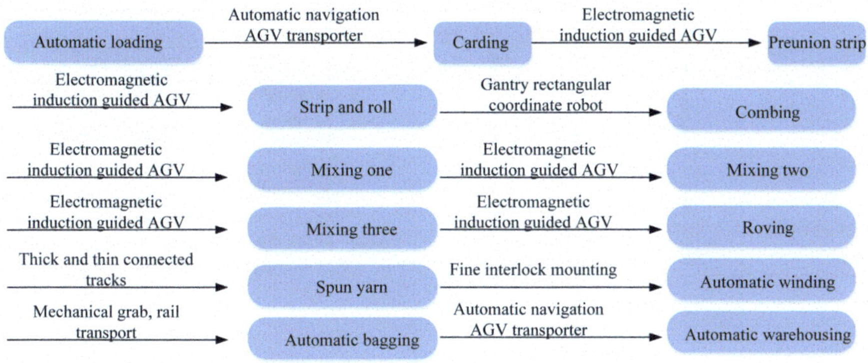

Fig. 9.16 Spinning production process flow

including knowledge repositories for processes, basic equipment information, equipment model libraries, instance libraries [28]. Process planning and equipment selection are performed according to production requirements. Layout design is carried out based on the actual conditions of the spinning workshop, and simulation software is used to simulate logistics based on the layout design.

The layout design of the spinning workshop is conducted in a virtual environment, and dynamic simulation analysis and evaluation are performed on the layout design process based on the information provided by PDM and ERP systems. The production layout is optimized and adjusted to meet the production requirements of the enterprise.

In the design process, Fig. 9.17 reflects the data relationships and data flow between the basic database, parametric model library, layout model library, and instance library. The basic data primarily serves the process planning, providing fundamental information for spinning, including basic information on spinning, process flow information, equipment information, etc. It also provides equipment modeling with the outer dimensions and interface data information, and through equipment modeling, it offers three-dimensional models of the equipment for the instance library. Upon completion of process planning, the process flow and equipment selection information are transmitted to the layout design system [29]. The layout design process requires the layout model library to provide simplified layout models of the equipment. Upon completion of the layout design, the results are associated with the simulation system, and the simulation system is used to simulate the layout design results. Based on the feedback from the simulation analysis, the layout design is modified. Subsequently, the layout design provides the connections and components of the equipment to the general assembly design. Upon completion of the general assembly design, the layout design of the production line is finalized, and the production line is then expanded into the instance library, becoming a new instance.

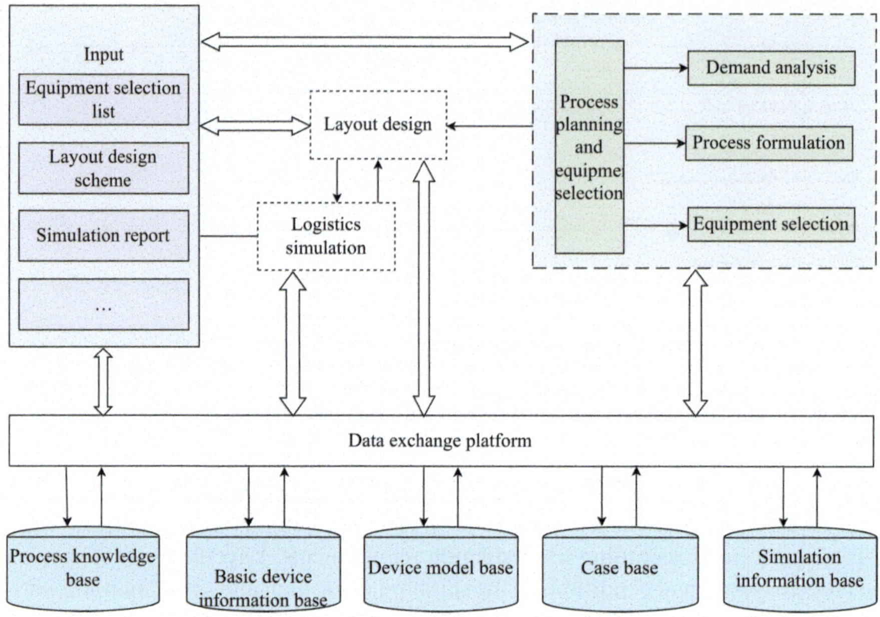

Fig. 9.17 Workshop layout and logistics simulation process modeling

9.3.2 The Complete Process of Spinning Production Based on an Industrial Internet Platform

The Industrial Internet platform is designed to meet the needs of digitalization, networking, and intelligence in the manufacturing industry. It constructs a service system based on the collection, aggregation, and analysis of massive data to support the industrial cloud platform. The platform enables ubiquitous connectivity of manufacturing resources, elastic supply, and efficient configuration. A precise, real-time, and efficient data collection and interconnection system is established. This system creates a development environment for industrial big data storage, integration, access, analysis, and management. It achieves the modeling, standardization, software-based representation, and reuse of industrial technology, experience, and knowledge. This process continuously optimizes the efficiency of resource allocation in research and development, design, production, manufacturing, and operation management. It forms a new ecosystem in the manufacturing industry that is characterized by resource enrichment. The ecosystem encourages multi-party participation and promotes win–win cooperation. It also fosters collaborative evolution within the manufacturing industry [30]. Figure 9.18 illustrates an Industrial Internet platform for interconnected and interoperable processes in spinning production.

By utilizing internet technology, wireless network technology, and mobile devices (tablets, smartphones, mobile PDAs), a collaborative sharing platform based on the mobile internet is constructed. This platform breaks down the information barriers

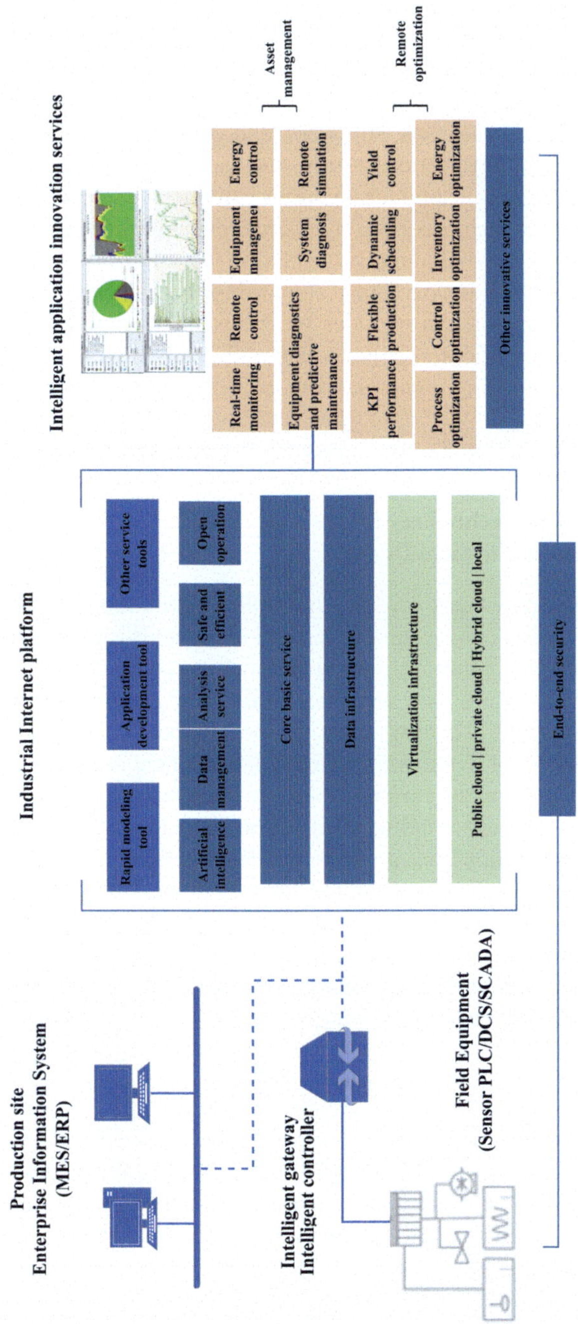

Fig. 9.18 Industrial internet platform for spinning production process connectivity

between spinning factories, suppliers, and distributors, and supports the integrated sharing of procurement information, quality information, inventory information. Financial information is integrated for better transparency and collaboration. The platform supports the exchange of technical information within the supply chain [31]. It facilitates the collaborative management of the entire supply chain business process between spinning factories, suppliers, and distributors, including demand planning, delivery, receipt, inspection, testing, and warehousing. The aim is to transform the relationship between spinning factories, suppliers, and distributors from a simple buyer–seller relationship to a closely interdependent relationship based on the industrial chain, ecological chain, and value chain. This transformation accelerates enterprise transformation and upgrading, supporting the achievement of strategic development goals. The platform includes functions such as model services, data management, analytical services, supplier collaboration, inventory collaboration, and mobile collaboration, enabling comprehensive coverage and information exchange within the enterprise.

Key steps in the construction of the Industrial Internet platform include:

(1) **Using Automation Technology for the Internet of Things (IoT) Enables the Interconnection of Physical Devices**

The traditional spinning process involves multiple discontinuous steps, making it extremely challenging to achieve automation and continuous production. Although recent advancements in continuous equipment such as comb-linking, coarse–fine linking, and fine-rove linking have automated tasks like cleaning, yarn dropping, packaging, and conveying. Thereby reducing the number of interruptions in the production line and freeing workers from heavy labor. There are still gaps between processes such as combing to carding, carding to drawing, drawing to roving, and roving frame to package winding [32]. To address this, the project refers to advanced spinning production lines domestically and internationally, upgrading and transforming spinning equipment including various process equipment, conveyor lines, inspection devices, mechanical hands (robots), and automated guided vehicles. Through various connecting and automated transport devices, the entire production line is interconnected, enabling a fully automated production line from raw material input to finished product warehousing.

(2) **Software Linkage Through Industrial IoT Technology**

Software linkage is established between devices and between devices and operators. Device systems with built-in or additional sensors are used to collect device status and production data information. Through the internet, industrial Ethernet, and Fieldbus, machine networking is achieved, allowing for rapid online data detection, adjustment of various process parameters, and remote fault diagnosis.

(3) **Proactive Operational Management Through Intelligent Technology**

Building upon big data, cloud computing, the internet, and other technologies, advanced sensor technology and communication technology. They are utilized to achieve real-time online monitoring of product characteristics and production equipment conditions throughout the entire spinning process. Intelligent control technology combined with control strategies is employed for SMS alerts and automatic adjustment of equipment operation and process conditions. Through intelligent analysis technology, real-time online analysis of equipment operation data, process quality data, and production energy consumption data is conducted. providing data support for production adjustment, management optimization, and energy consumption reduction. Thereby improving and perfecting traditional quality control methods in spinning production, stabilizing and enhancing product quality [33]. Additionally, real-time information on order progress and quality can be obtained through the internet. The system transmits, integrates, and analyzes real-time data, using these insights to guide production management in reverse. This creates a management platform that remotely monitors production status, automatically generates production reports, and tracks orders in real-time. The platform also monitors quality data in real-time, integrates seamlessly with ERP systems, and realizes networked, integrated production processes, thereby improving production efficiency and achieving manufacturing intelligence. Real-time data is fed back to the information system, which then communicates with the lower-level control system, such as PLC, to control various executive mechanisms [34]. By monitoring and managing factors such as raw materials, equipment, personnel, processes, and the environment and analyzing the relationships between parameters, proactive operational control is gradually achieved. Rapid estimation of expected product quality levels using raw material indicators enables precise adjustment of raw material proportions. It facilitates the exploration of raw material potential and helps in minimizing raw material costs. Linear analysis based on equipment load and equipment failure enables the identification of correlations between load rates and failure rates, facilitating the establishment of reasonable load settings.

9.3.3 Online Detection in Production Processes and Remote Monitoring of Production Status Technology

(1) **Online Detection Technology for the Spinning Production Process**

1. **Detection Method for Cotton Heterodimers Using Machine Vision Technology**

The detection method using machine vision technology typically employs various image acquisition devices to capture images in different spectral ranges. These images are then sent to processing devices for analysis and evaluation. The presence of heterodimers is determined, and if detected, corresponding removal devices are activated to execute the removal action. Currently, the main methods for image acquisition include visible light cameras, X-ray imaging, and infrared imaging. The processing is primarily carried out by computers or specialized image processing boards that execute algorithmic procedures (Fig. 9.19).

2. **High-Speed Cotton Flow Image Acquisition Based on Machine Vision**

In the automated detection technology for heterodimers in cotton flow based on machine vision, cameras, and lenses are the most critical hardware components, and selecting appropriate cameras and lenses is a key step. The choice of cameras and

Fig. 9.19 Typical machine vision application system

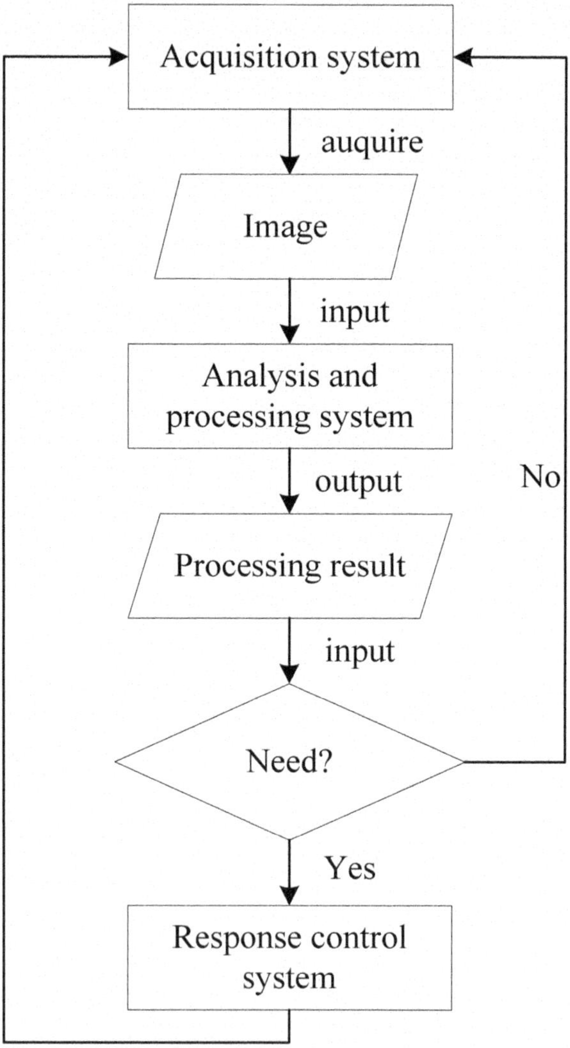

lenses determines factors such as the resolution and image quality of the captured heterodimer images, which are directly related to the subsequent image processing stage. Considering industrial application scenarios, cameras and lenses need to have high stability and anti-interference capabilities. Therefore, conventional consumer cameras cannot meet the requirements, and industrial-grade high-speed cameras need to be used [35]. When selecting high-speed industrial cameras, the following aspects are considered: the type of high-speed industrial camera, whether the acquisition rate of the high-speed industrial camera can keep up with the transmission speed of the high-speed cotton flow, whether the resolution of the high-speed industrial camera is sufficient to capture heterodimers, and whether the data transfer interface of the high-speed industrial camera can efficiently transmit the image data.

Industrial cameras can be divided into CCD (Charge Coupled Device) cameras and CMOS (Complementary Metal Oxide Semiconductor) cameras based on the type of photosensitive chip. CCD cameras have higher imaging resolution and lower noise. However, with the continuous development of CMOS technology in recent years, the gap between CCD and CMOS has become smaller. Moreover, CMOS cameras have advantages such as good integration, low power consumption, high-speed transmission, and wide dynamic range, making them widely used in the field of machine vision. According to the structural characteristics of the camera sensor, cameras can be further classified into line scan cameras and area scan cameras. Area scan cameras can capture two-dimensional images at once, providing intuitive measurement images [35]. However, they have a larger number of pixels. In contrast, line scan cameras only capture images line by line, ensuring high scan frequency and high resolution. In this paper, the application scenario involves high-speed cotton flow image acquisition on a fast production line, so a line scan CMOS camera is used.

The lens that matches the high-speed camera is also a crucial point that should not be overlooked. When selecting a lens, factors to consider include the sensor size of the CMOS, the field of view width of the target object, and the working distance of the high-speed camera. The lens should cover the entire surface of the camera's sensor and be compatible with the camera's interface. Assuming the field of view width of the target object is denoted as FOV, the sensor width as SW, the working distance of the high-speed camera as WD, and the focal length of the lens as F.

After selecting the high-speed camera, lens, and other hardware components based on the detection requirements, the final chosen hardware setup includes industrial cameras, lenses, acquisition cards, lighting systems, and background boards. After integrating all the selected hardware components, the overall structure is shown in Fig. 9.20.

3. **Automated Cotton Stream Foreign Fiber Detection Technology Based on Machine Vision**

The automated detection technology for heterodimers in cotton flow based on machine vision can be divided into three main modules: 1. the cotton heterodimer image acquisition module (image acquisition, lighting system), 2. the cotton heterodimer image processing and recognition algorithm module, and 3. the cotton

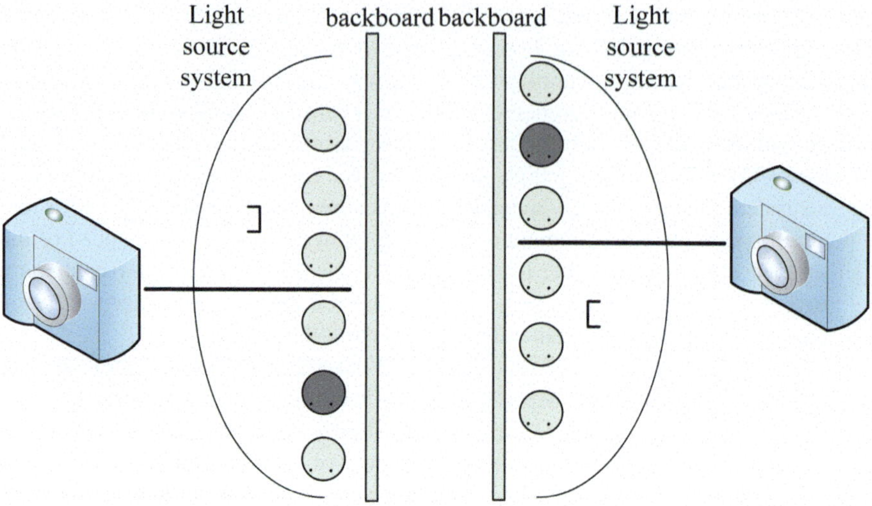

Fig. 9.20 Image acquisition hardware integration diagram

heterodimer removal module [36] (Fig. 9.21). The module invocation between the three modules is shown in Fig. 9.22.

The implementation of each module is designed based on the object-oriented approach to facilitate modification and reuse, reduce the overall complexity of the software platform, and shorten the execution time of algorithms.

In the detection and recognition module, for high-speed cotton flow heterodimer detection, it is necessary to capture images of both sides of the cotton flow simultaneously. This requires the parallel operation of cameras on both sides for image acquisition and recognition processes [37]. However, due to the presence of only one control mechanism, after identifying the heterodimers, the exclusion of heterodimers requires the serial operation of the two sides. Therefore, the execution flow of the entire software platform is shown in Fig. 9.23.

The overall design process of a system is as follows: first, based on the application requirements, calculate the required configuration parameters, and then search relevant literature to find suitable hardware based on the calculated parameters. Next, combine the hardware components for testing and use the hardware platform to collect image data once the requirements are met. Study the corresponding recognition algorithms and perform simulation experiments. Design the software architecture for these algorithms. Finally, integrate the software and hardware platform architectures to create a complete system for experimental testing. Any existing bugs are addressed, and the system is finalized. Following this approach, the final design of the automated cotton flow heterodimer detection technology based on machine vision is shown in Fig. 9.24.

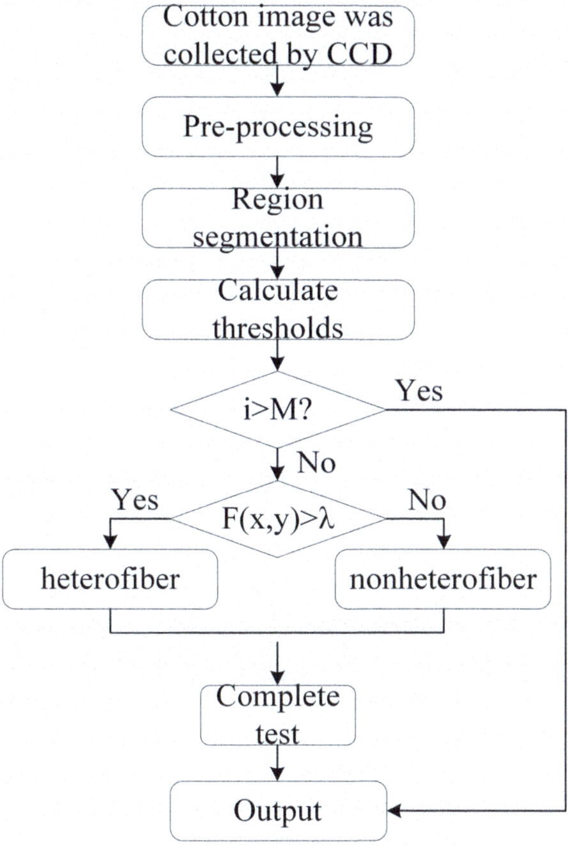

Fig. 9.21 Automated cotton flow foreign fiber detection technology

The overall workflow is as follows: Cotton fibers are processed through equipment such as a cotton opener, transforming the cotton into a snowflake-like state. With the help of a powerful air blower in the transmission channel, a thin layer of cotton is formed and rapidly transported at a speed of 10 m/s through the detection channel of the foreign fiber detection system [38]. The high-speed 3CMOS line scan camera scans the images of both sides of the cotton layer under the illumination of the light source system. The image data is then transmitted to the industrial control computer via the CamLink interface. The industrial control computer processes the image data using the foreign fiber identification algorithm. In a very short time, foreign fibers are identified, and the identification results are converted into response commands for electromagnetic valves. These commands are then sent to the control system of the electromagnetic valves through serial communication of the industrial control computer. Finally, the control system activates the corresponding electromagnetic valves to inject the detected foreign fibers, along with a certain amount of cotton, into the waste cotton box. The remaining cotton continues to be transported through the transmission channel and enters subsequent processing equipment.

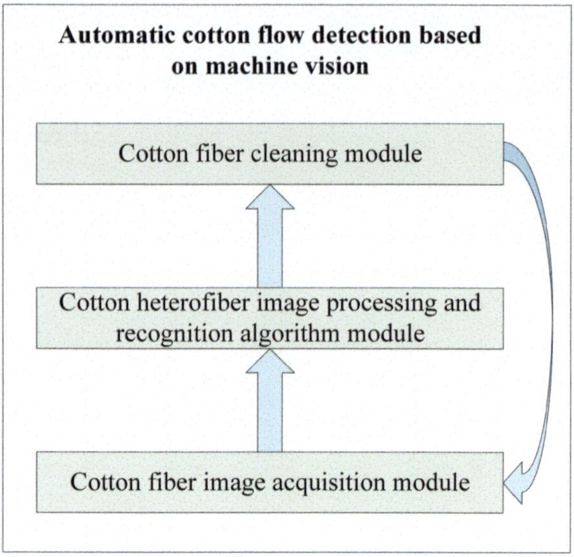

Fig. 9.22 Foreign fiber detection technology module calling diagram

(2) **Remote Monitoring Technology for Spinning Production Process**

Equipped with comprehensive online detection equipment for the spinning production process, establish a remote monitoring center for spinning production process based on the industrial internet. Develop data interfaces for spinning production equipment, which includes devices such as the cotton comber, drawing frame, roving machine, combing machine, coarse yarn machine, fine yarn machine, spinning machine, intelligent packaging and logistics equipment, and various intelligent auxiliary equipment. Construct a reliable data transmission network for these spinning equipment units. Additionally, configure advanced real-time data platforms to manage and utilize the transmitted data [39]. Through real-time data transmission and high-speed data acquisition, storage, and access, digitalize the spinning production process.

A centralized control room is set up for the entire system, where collection servers, application servers, and operator workstations can be installed according to requirements. A central data platform is also established to collect operational data, set data, quality data, energy consumption data, and yield data from devices, quality inspection systems, water and electricity flow meters, manual statistics, and detection systems through corresponding data interfaces. The collected data is then transferred to the central data platform in the control room. Computers perform real-time monitoring and archive key data. Operators and process engineers on the shop floor can directly receive production and process data sent by the central data platform through wireless smart terminals.

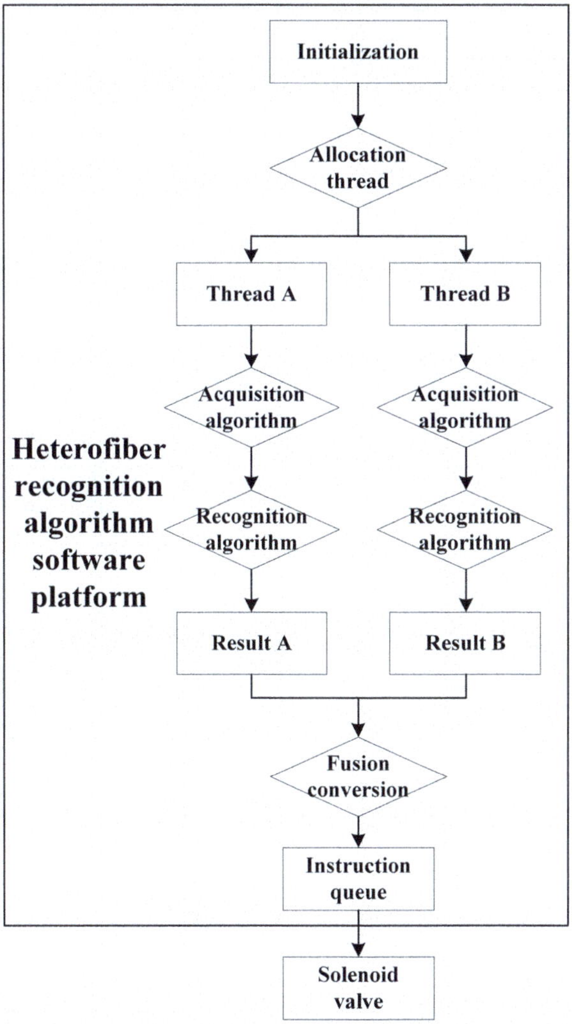

Fig. 9.23 Flow of the software platform for the foreign fiber recognition algorithm

9.3.4 Collaborative Optimization Based on Industrial Big Data Platform

Collaborative optimization of the spinning production process mainly involves providing support for enterprise production decisions through the analysis of historical data. It helps determine the demand for products, raw materials, and delivery deadlines by assisting in the execution of production plans, thereby selecting suitable production schemes to allocate production equipment and produce products in the required order. For example, based on the operation of production equipment, the quantity and types of products produced by the enterprise within the planning period

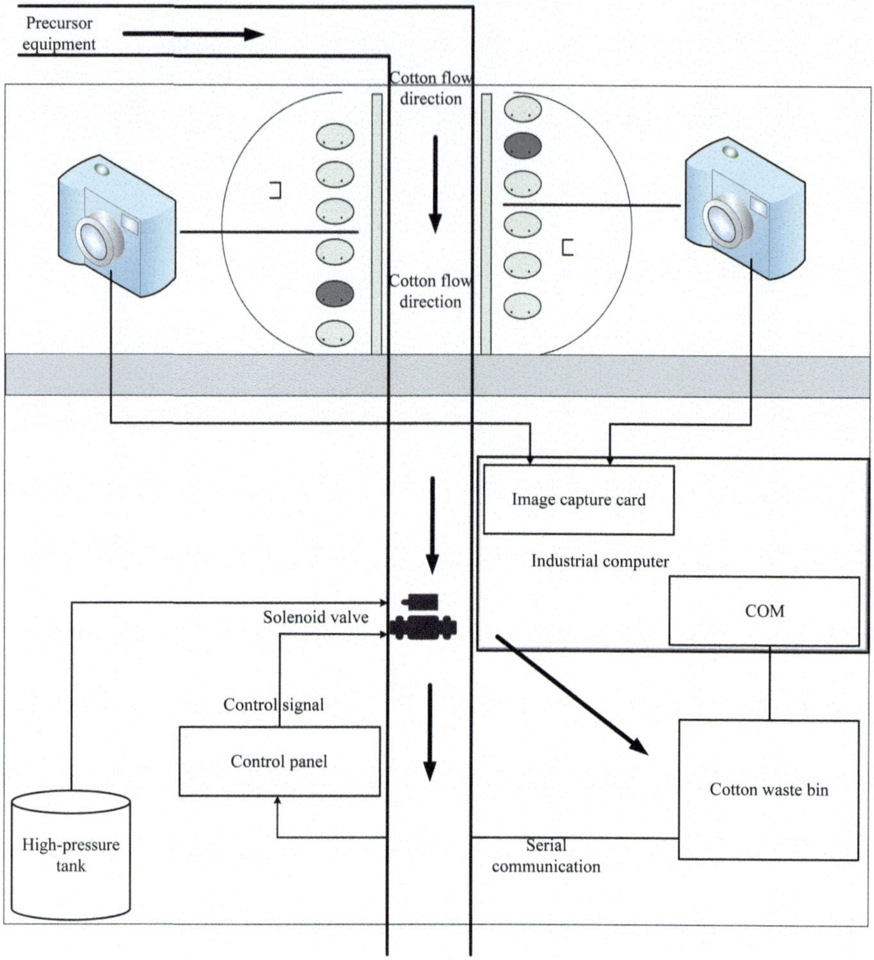

Fig. 9.24 Design of automated machine vision-based foreign fiber detection technology for cotton streams

can be arranged reasonably, making full use of enterprise resources for optimized scheduling. The goal is to minimize the total production time, produce high-quality products within the specified time, and maximize the overall efficiency of the enterprise. To address the massive structured and unstructured data generated in the operation of spinning enterprises, a big data analysis platform is established. It is used to achieve distributed storage, fast access, and in-depth analysis of data. This platform provides technical support for in-depth quantitative analysis of spinning production processes, knowledge mining for improving production efficiency, production capacity modeling, equipment state prediction and maintenance, and more.

The establishment of a big data platform greatly enhances the storage and processing capabilities of internal and external massive data for enterprises. It improves the integrated optimization capability of the spinning production process. And it enhances the ability of enterprises to perceive external environments and internal production processes and achieves the virtualization of physical production processes [40]. Thereby significantly improving the organization and management of production processes. The big data platform supports the development and application of the following technologies: 1. Quality analysis and early warning of spinning processes based on online detection data and manual inspection data to help identify uncontrolled factors affecting quality. 2. Based on historical production load data of equipment, failure records, and point inspections, analyze equipment status and drive preventive maintenance.

Using the big data platform, a systematic approach is applied to scientifically analyze and manage the production process. By analyzing online detection data of high-quality ring spinning intermediate processes, real-time operating parameter data of equipment, and laboratory inspection data, the controlled performance of the spinning production process can be analyzed to identify signs of process performance changes and potential issues with product quality. Event tracing and pattern analysis are performed to pinpoint potential root causes of failures. This assists in promptly addressing faults or implementing preventive measures to avert accidents during the production process. By doing so, it improves equipment operation and maintenance, boosts product quality, and reduces material consumption and quality degradation losses that result from accidents [41]. The use of the big data platform enables distributed storage, fast access, and in-depth analysis of data, helping enterprises conduct in-depth quantitative analysis of the spinning production process. This facilitates process improvement, operational optimization, quality analysis and control of the production process, equipment/production state prediction and fault diagnosis, and assists in management decision-making. Ultimately, the system constructs a knowledge automation system for high-quality ring spinning production processes, leveraging the big data platform. This system supports intelligent analysis and production decision-making in high-quality ring spinning production, thereby offering intelligent "services" to enhance enterprise operations.

(1) Real-Time Data Acquisition Technology for Spinning Production Process

1. Real-Time Data Acquisition Platform

The specific overall network design for real-time data acquisition is shown in Fig. 9.25.

2. Data Acquisition Architecture and Functionality

The construction of smart factories encompasses various production lines, warehouses, material transportation devices, and online inspection equipment in the production workshop. These equipment and production lines are provided by different suppliers. In this project, heterogeneous equipment data acquisition is achieved through the development of various standard interfaces (Fig. 9.26).

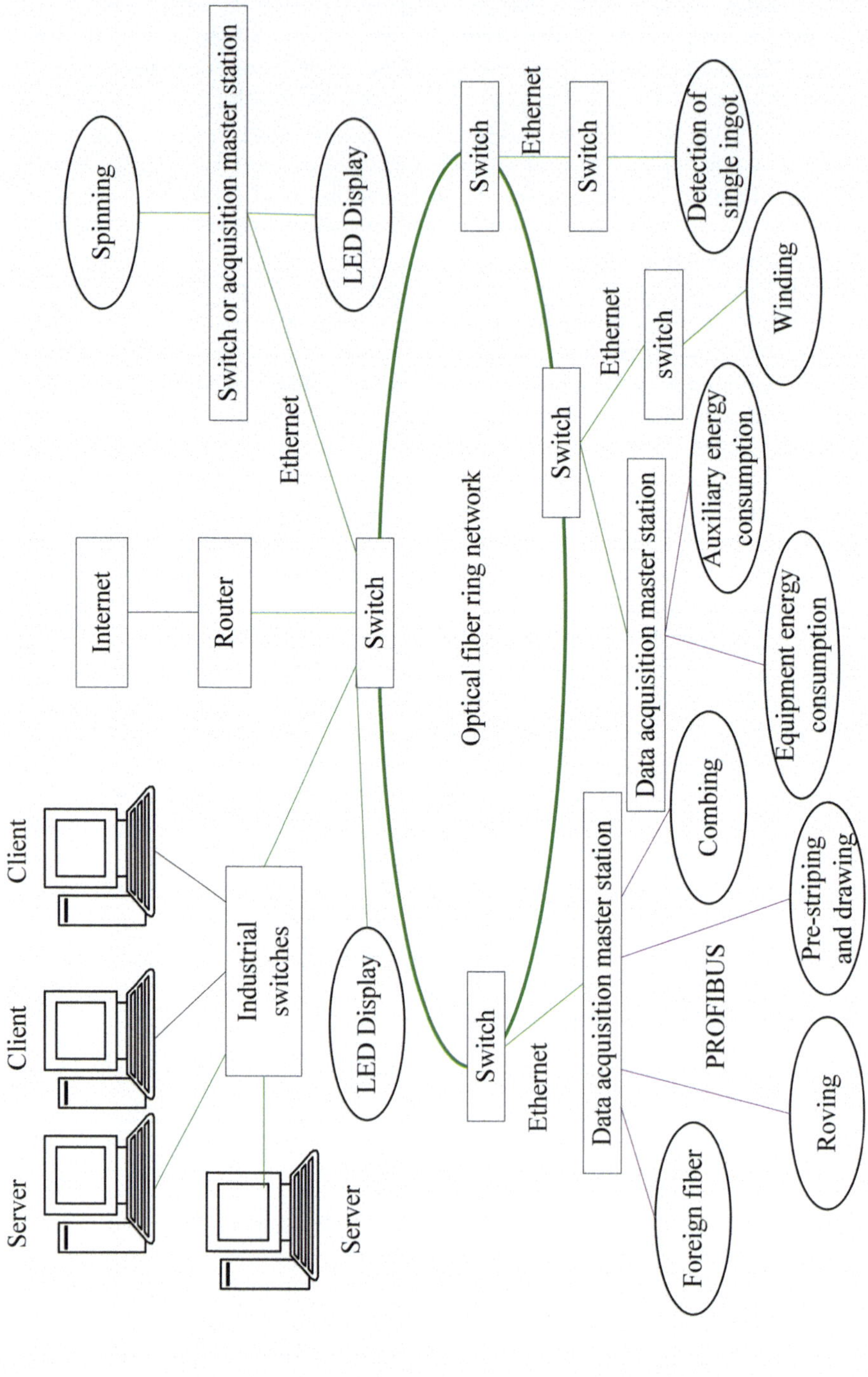

Fig. 9.25 System overall network architecture design

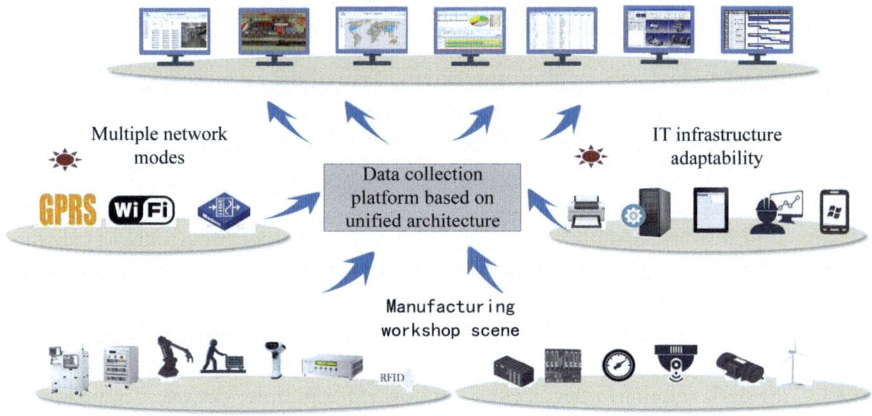

Fig. 9.26 Heterogeneous device interconnection and integration

By adopting a unified address space and object-oriented design pattern based on the OPC UA architecture, various production equipment from different manufacturers is transformed into a unified access service. This facilitates real-time monitoring services for systems at all levels to achieve information exchange [42]. Simultaneously, leveraging the security model and policies of OPC UA, complete authentication, authorization, and security auditing are provided to ensure the reliability of data exchange in production environments (Fig. 9.27).

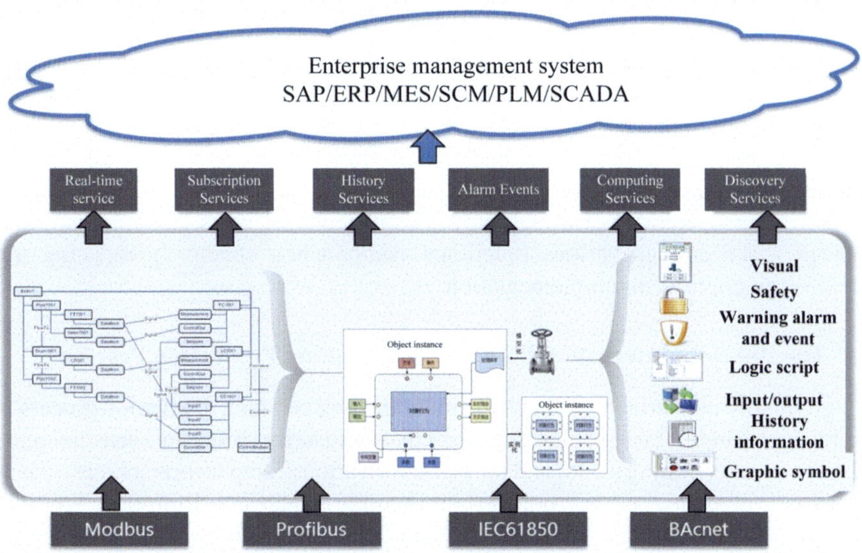

Fig. 9.27 Industrial data acquisition based on OPCUA unified architecture

3. **Real-Time Database Platform**

Distributed real-time database platforms are the core of data integration in the industrial Internet of Things, serving as a bridge between automation and information technology. Spanning from device communication at the lower end to interfaces with information systems at the upper end, data acquisition systems permeate the entire intelligent manufacturing system.

Using middleware equipped with an information model, distributed real-time databases integrate vertically integrated applications seamlessly via open interfaces and heterogeneous object models. This facilitates the management of data collection from the industrial Internet of Things to expansive industrial big data platforms.

Real-time database platforms collect real-time data from systems such as DCS and PLC through standard interfaces to support information systems. They support ODBC and OPC protocols, allowing for the real-time data collection from other systems and acting as a data source to support other systems. These platforms offer rich interface functionalities, enabling connections with various DCS, PLC, and other data acquisition systems from different manufacturers, supporting standard protocol interfaces for DCS, PLC, and data acquisition systems [43]. They provide graphical real-time database configuration management tools for setting up functions such as status queries, data simulation, point management, user management, and permission management. By efficiently compressing data, they manage the storage of long-term historical data. They offer analysis tools to support production managers in analyzing data at different levels, along with comprehensive and reliable security mechanisms that enable security management based on real-time tag numbers.

The architecture of real-time historical database products is illustrated in Fig. 9.28.

Considering the factory conditions and user requirements, the real-time historical data platform adopts a distributed architecture (as shown in Fig. 9.29). With distributed data collection, storage, and applications, the system is flexibly built and organized to alleviate performance pressure on the system, leading to a more secure and stable operation.

The database, data collectors, and various functional components all support remote management and configuration, eliminating the need for users to log in to servers for operations. Through network connectivity, users can configure and manage space and its various functional components, effectively reducing the workload of engineering implementation.

(2) **The Big Data Integration Platform for Spinning Production Process**

A high-quality industrial big data platform for the ring spinning production process is established using storage, networking, servers, and security virtualization techniques. This involves separating the logical properties of various computing resources—such as network domains, security policies, and routing strategies—from their physical attributes, like equipment types and physical locations, to create a virtual resource pool [44]. By combining various resource attributes, virtual resources in different domains are formed. By establishing an industrial big data platform, companies can

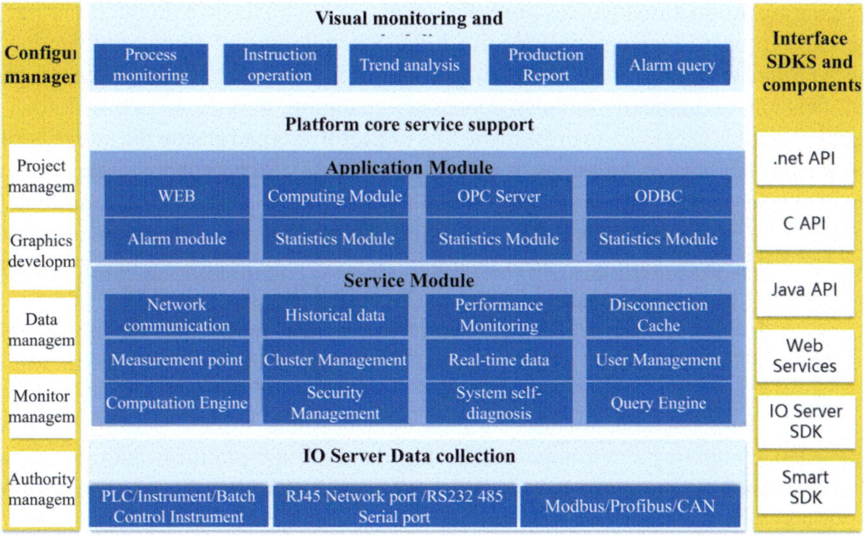

Fig. 9.28 Real-time history database structure

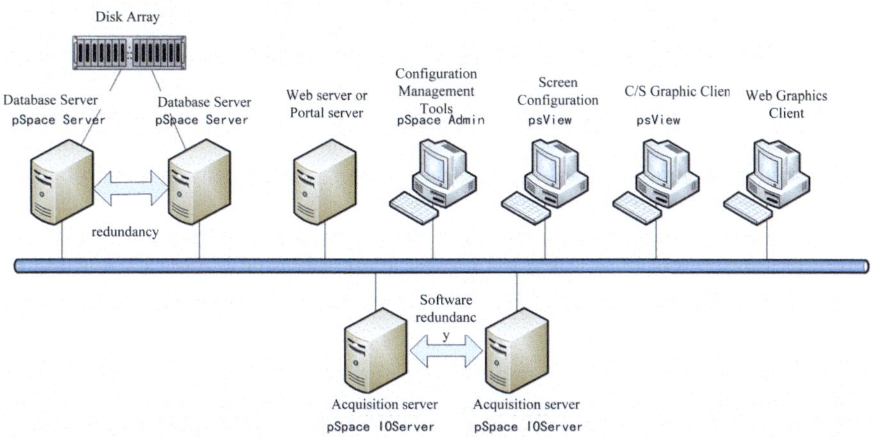

Fig. 9.29 Real-time database deployment

substantially enhance their capacity to store and process vast amounts of internal and external data. This platform boosts openness and the company's ability to understand both the external environment and internal production processes. Through collaborative optimization of production factors, the platform enables the virtualization of physical production processes, leading to a significant improvement in the organization and management of production processes.

Spinning production involves multiple stages, each with its own production rules. The entire process features multiple machines, operations, and continuous production. The data generated in the production process is not only large in volume and high in dimensional complexity but also constantly changing. In order to extract knowledge from it, it is necessary to prevent the loss of useful data and ensure the integrity of stored data. Traditional databases, both in terms of data structure and volume, are no longer able to meet the requirements of big data. Additionally, traditional centralized processing methods for massive data computation are relatively inefficient in terms of speed and efficiency, failing to meet the rapid and efficient processing requirements for historical production data. Considering Huaxing Group's spinning production process and data management model, an effective data processing technology is adopted to help companies store and manage production big data, as illustrated in the overall framework shown in Fig. 9.30.

The big data platform mainly consists of a big data storage platform and a distributed computing environment. The big data storage platform addresses the storage and management of production data collected by various intelligent devices in the textile production environment [45]. Due to the strong real-time nature of data in the spinning production process, which is constantly changing, it is necessary to display and analyze this on-site monitoring data in real-time. Moreover, with the accumulation of data over the years, the data volume has reached beyond the terabyte level. These production data reflect the entire production process and are crucial for the development of the enterprise, requiring preservation as historical records. Therefore, a new data storage strategy combining traditional databases with No SQL databases can be employed. Real-time data storage and querying can be achieved through the multi-table architecture of traditional databases, while historical data can be migrated to No SQL databases using data migration technology to store and manage massive production historical data.

The distributed computing environment primarily deals with the computational analysis of the enterprise's massive production of historical data, enabling distributed data processing. Textile enterprises generate a large amount of production data every day during the production process, and the accumulated historical data over the years is immense. Traditional centralized processing methods cannot achieve rapid data computation and analysis. By establishing a distributed computing environment, data can be partitioned across different servers, with each server only needing to compute a fixed amount of data [33]. Once the computation is complete, the results can be uploaded to the same server. This not only enhances computation speed but also mitigates the risk of system collapse due to the failure of a single node. The process of data processing based on the industrial big data platform is illustrated in Fig. 9.31.

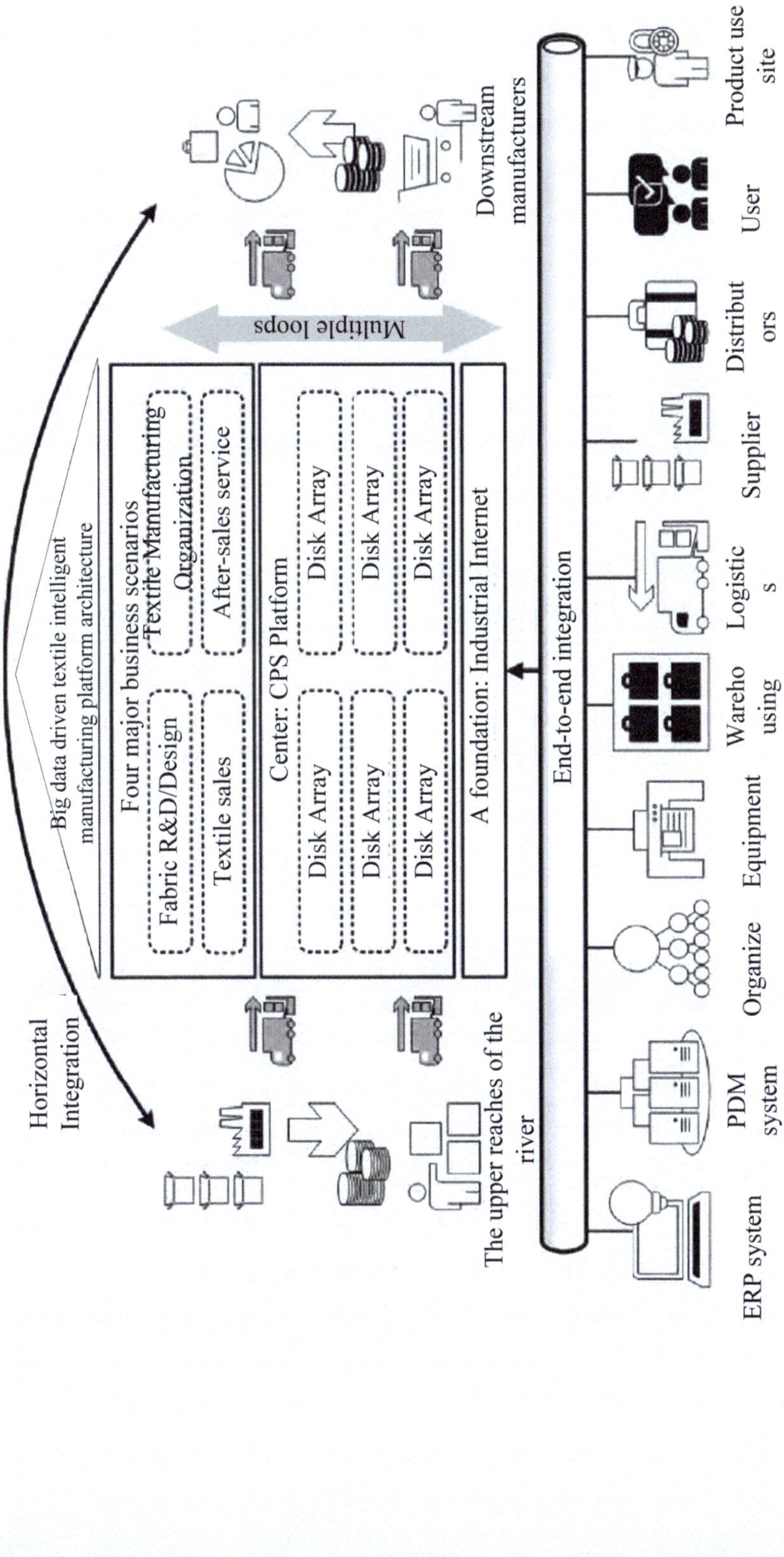

Fig. 9.30 Spinning smart factory industrial big data integration platform

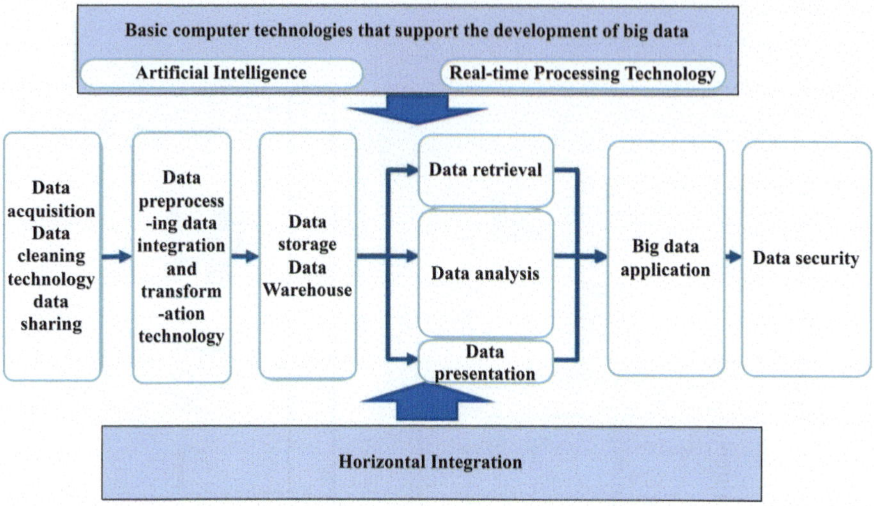

Fig. 9.31 Industrial big data processing flow

9.4 Case Study of Intelligent Management and Control in Production Operations

Taking the example of energy consumption control in ring spinning production, a full-process energy consumption control system based on artificial intelligence can be constructed. This system would entail real-time collection of production data, equipment operation data, real-time factory energy consumption data, and logistics data, and the establishment of a big data platform for energy consumption in spinning factories. Utilizing artificial intelligence techniques based on principal component analysis, real-time energy consumption analysis can be conducted to monitor the main influencing factors of energy consumption between processes.

Against the backdrop of the intelligent industrial internet platform and industrial big data platform for ring spinning, specific data sources such as IoT sensor nodes monitoring actual production equipment energy consumption data in the factory serve as the foundation. An innovative design concept for the overall architecture of the industrial energy consumption analysis system based on big data is proposed, adapting to the characteristics of industrial big data. A complete industrial big data processing system is designed and developed, and based on this system, energy-saving strategies for factory energy consumption big data analysis are put forward.

9.4.1 Design Scheme for an Intelligent Energy Consumption Management Platform Architecture in Ring Spinning

Energy consumption control in the spinning process refers to monitoring, analyzing, controlling, and optimizing the consumption of energy sources such as water, electricity, gas, heat, and raw materials during the manufacturing process. This aims to achieve refined management of energy consumption while effectively safeguarding the performance of the manufacturing system and the economic benefits of the enterprise. The goal is to achieve energy conservation, emission reduction, and cost reduction for manufacturing enterprises, and to maintain their competitiveness. The intelligent energy consumption control platform for spinning based on big data and artificial intelligence technologies is designed to function within the physical workshop. It utilizes various sensing technologies to perceive energy consumption information, production factor information, and production behavior status information. The virtual workshop, it reflects and simulates the production factors and behaviors of the physical workshop [46]. Through continuous interaction between the physical workshop and the virtual workshop during the actual production process, real-time control, and iterative optimization of manufacturing energy consumption in the physical workshop are achieved. The effectiveness of this approach is illustrated in Fig. 9.32.

Compared to traditional technologies and methods, energy consumption control in the spinning process based on big data and artificial intelligence possesses the following characteristics: 1. The data source has transitioned from single energy

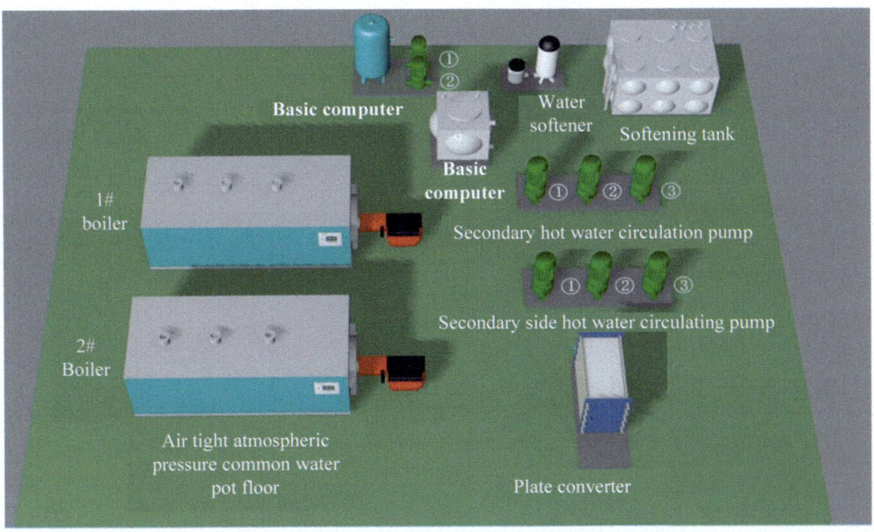

Fig. 9.32 Intelligent real-time control of spinning process energy consumption

consumption data to encompassing multiple types of equipment energy consumption, production factors, and production behavior data. The data sources not only include heterogeneous perception data from the physical workshop but also comprise simulated evolution data from the virtual digital workshop. 2. The interaction mode has shifted from traditional flat statistical chart display to immersive interaction based on virtual/augmented reality technology. 3. Efficient energy production process management has transformed from traditional experience-guided management to a physical-information fusion management driven by physical model-driven digital model knowledge evolution.

The architectural framework of the energy consumption control platform for the spinning process based on big data and artificial intelligence is illustrated in Fig. 9.33.

In the research and application of the energy consumption control model for the spinning process based on big data and artificial intelligence, the following challenges need to be addressed: 1. In the aspect of effective multi-state perception of energy, it is necessary to study intelligent perception devices and distributed perception networks that can adapt to harsh workshop conditions and provide accurate and reliable basic physical data acquisition. 2. In terms of effective iterative optimization of energy, during the simulation testing process of digital models, research is needed on self-organizing and self-learning optimization methods based on twin data to provide a basis for the system's optimized operation. Simultaneously, in the actual production

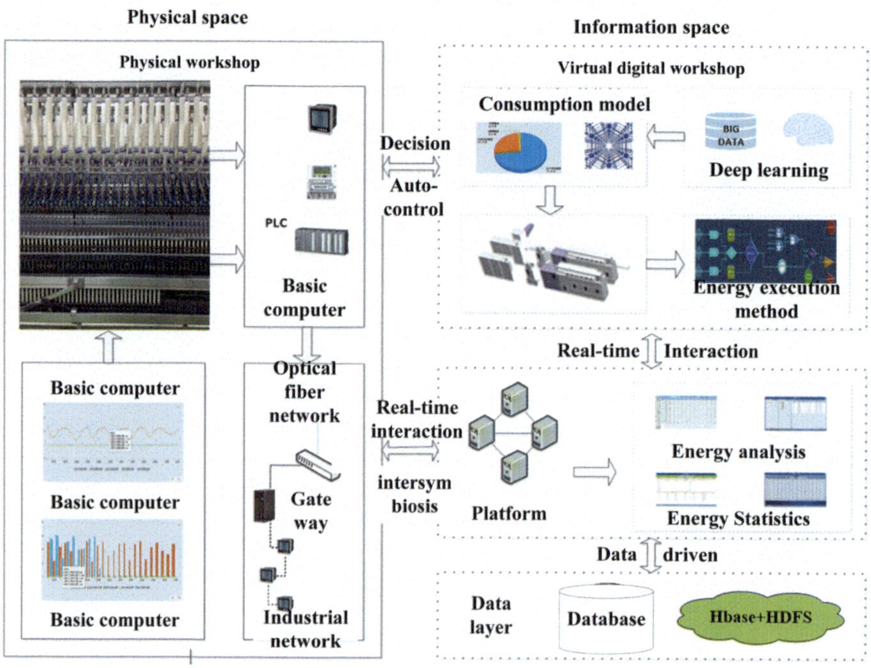

Fig. 9.33 Spinning process energy consumption intelligent control platform architecture

process within the physical workshop, efficient physical-information system iterative interaction mechanisms need to be studied to support the realization of dynamic environmental energy efficiency performance optimization.

These challenges are essential in advancing the research and practical application of energy consumption control in the spinning process using big data and artificial intelligence.

9.4.2 *Intelligent Monitoring and Early Warning of Energy Consumption in Spinning Based on Minimal Quantification Error*

In order to ensure safe and reliable operation of the production process, while meeting specified performance and quality indicators, it is necessary to minimize energy consumption in the spinning process by monitoring, diagnosing, and eliminating abnormal energy consumption conditions. The objective of energy consumption monitoring in the spinning process is to identify abnormal behavior by constructing energy consumption indicators, thereby ensuring planned process operation, minimizing energy consumption in the spinning process, maintaining equipment stability, and reducing production costs [47].

Additionally, with the advancement of computer technology, a large amount of process data is collected and stored. In a factory control center, there are usually thousands of types of analog and digital process data and hundreds of alarm messages displayed simultaneously. Operators often find it difficult to understand and judge the running status of the process based on such a large amount of data, and timely detect abnormal energy consumption, leading to a phenomenon known as "data-rich, information-poor". Therefore, how to fully utilize the deep-level information in these data and improve the energy consumption monitoring capability of the spinning process is a problem that needs to be addressed using artificial intelligence methods.

The statistical characteristics of energy consumption in the spinning process can theoretically be described using probability density functions, but in practical engineering, it is often difficult to determine a probability function and requires a large number of experiments to approximate its expression. To construct such a monitoring model, it is necessary to correctly collect data for various types of abnormalities, enabling the trained learner to have high generalization performance. However, in actual production, correctly collecting various abnormal data is quite difficult and may even be impossible, especially when the production system has just started and lacks sufficient diverse abnormal data. Training a supervised learning-based learner for ordinary users is also a great challenge because it requires users to have a deep understanding of the background related to the data and perform effective data processing and analysis. Therefore, constructing a learner with good performance is very difficult.

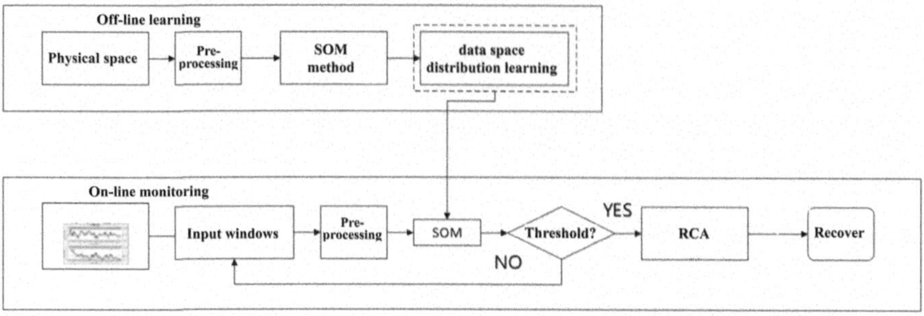

Fig. 9.34 SQM-based method for monitoring energy consumption in the spinning process

To address the energy consumption monitoring problem in the spinning process, the Minimum Quantity Error (MQE) control chart based on Self-Organization Mapping (SOM) can quickly identify subtle process anomalies and provide relevant process information accurately, thereby comprehensively displaying the process status. This enables the rapid identification and elimination of energy consumption abnormalities in the spinning process and the restoration of the process to a fully stable state. The flowchart of the energy consumption monitoring method based on SQM in the spinning process is shown in Fig. 9.34.

9.4.3 Intelligent Energy Consumption Control in Spinning Processes Driven by State Parameters

The high complexity and multivariable nature of the actual spinning process have a significant impact on energy consumption. Detecting and diagnosing the factors affecting energy consumption accurately and rapidly and providing timely energy adjustment solutions to address the root causes of the problems, can maximize energy savings, increase production stability, reduce production costs, and avoid greater resource losses for companies, resulting in significant economic benefits.

The intelligent energy consumption control strategy for the spinning process aims to automatically monitor energy consumption in real-time dynamic production environments. When energy consumption abnormalities occur, it first issues energy consumption abnormality warnings and determines whether to continue production based on the actual situation. It then automatically diagnoses the specific factors causing the abnormalities and evaluates and makes decisions regarding energy consumption abnormalities. Finally, it proposes adjustment plans to improve production and eliminate energy consumption anomalies.

Therefore, it is necessary to study and implement optimization and improvement strategies for energy efficiency in the dynamic environment based on accurate and reliable energy efficiency forecasting models for cotton spinning production processes. Unsupervised learning clustering algorithms can be used to analyze

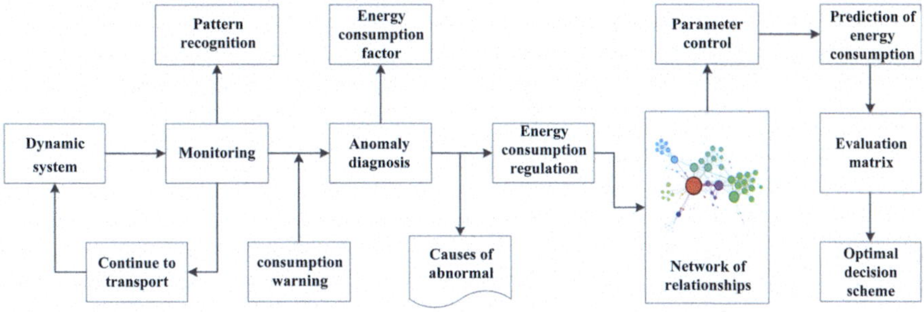

Fig. 9.35 State-driven method for intelligent regulation of energy consumption in the spinning process

the main factors affecting production energy consumption and propose energy efficiency optimization decision methods based on quantitative control mechanisms under the drive of big data. By using big data-driven approaches, the relationship network between state parameters (such as equipment operating parameters) and energy consumption indicators (such as electricity consumption per ton of yarn) in the spinning production process can be constructed. The evolution patterns of energy consumption in the spinning process can be explored, and intelligent decision-making methods for energy consumption in the spinning process based on state parameter control can be studied. Using a complex network model to describe the relationships among spinning manufacturing big data, a network of relationships between parameters can be established to determine control parameters and directions as inputs and search dimensions for energy consumption control in spinning workshops. Suitable control parameter values can be determined, and an evaluation matrix for candidate quality control plans can be established using intelligent workshop energy consumption prediction methods. Statistical analysis can then be conducted to select the optimal decision plan. The state-driven intelligent energy consumption control method for the spinning process is illustrated in Fig. 9.35.

9.4.4 *Energy Consumption Analysis and Prediction Based on Core Influencing Factors*

Energy consumption forecasting plays an important role in industries such as electricity and steel, with a long history of research and application both domestically and internationally. However, its application in cotton spinning enterprises is relatively limited. For cotton spinning enterprises, energy forecasting is crucial as it helps in the rational arrangement of energy production (including energy consumption planning, unit maintenance, equipment load adjustment, and economic unit combinations), improving energy conversion efficiency, reducing the production cost of secondary energy, optimizing energy allocation and scheduling, reducing waste of excess energy, and mitigating production fluctuations caused by inadequate energy supply [48]. Energy consumption forecasting is a vital component in formulating enterprise energy plans. By predicting energy consumption trends, waste can be minimized, production costs reduced, and significant improvements in market competitiveness, economic benefits, and the level of information management within the enterprise can be achieved.

The spinning production process is influenced by factors such as order scheduling, equipment operation planning, production scheduling, and climate variations, resulting in fluctuating demands for electricity, temperature, and humidity. As the central hub for energy conversion and supply in spinning workshops, significant energy waste occurs due to the conflicting supply and demand, necessitating efficient control of units to operate under optimal conditions. Therefore, leveraging vast amounts of actual production data and aiming for energy forecasting, data mining techniques are employed to extract valuable insights from production data to provide analysis and decision support for paper manufacturing enterprises, achieving a dynamic balance between energy production and consumption. The energy consumption forecasting method for the spinning process is illustrated in Fig. 9.36.

Energy consumption forecasting mainly involves the following steps:

1. **Goal Determination**

It is essential to clarify the significance of predicting the goal. By defining the forecasting objective, clear specifications should also be made regarding the forecasting subject, timeframe, region, and specific content.

2. **Object Analysis and Simplification**

The spinning production process is never a simple and straightforward process. In-depth analysis of the forecasting subject system, understanding its structure, characteristics, and dynamic patterns, can support the selection of appropriate forecasting methods and models, as well as provide essential reference information for evaluating forecasting results, greatly facilitating the establishment of forecasting models and analysis of forecasting results.

3. **Data Preparation**

Accurate and systematic data form the foundation for making forecasts. Based on the defined forecasting objective, relevant historical and current data must be collected, preprocessed, compressed, transformed, standardized, etc., to meet the requirements of forecasting.

4. **Selection of Forecasting Methods and Establishment of Forecasting Models**

There are various specific methods for forecasting. In practical applications, based on the forecasting purpose, data availability, and the characteristics of the forecasting

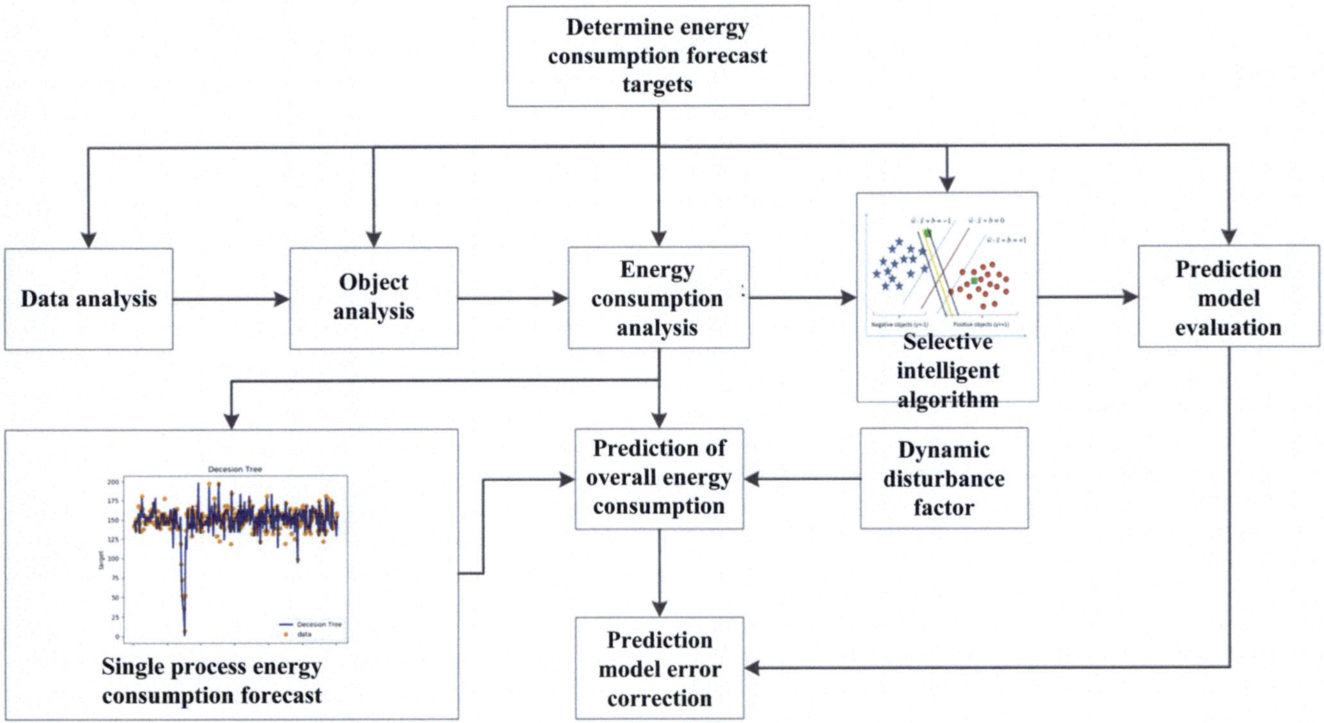

Fig. 9.36 Spinning process energy consumption prediction methods

subject itself. The correct selection and application of forecasting methods should be made. Sometimes, multiple forecasting methods need to be used simultaneously based on the available information, followed by a demonstration and evaluation of the forecasting results. After establishing the forecasting model, statistical, machine learning, and computer technology are utilized to calculate the model parameters.

The CART algorithm builds binary decision trees through recursion, segmenting the input or feature space into a finite number of regions. It calculates the probability distribution of predictions within these regions, which is the conditional probability distribution of the output based on the input conditions. This approach offers benefits like producing comprehensible rules and highlighting the impact of factors on energy consumption. Consequently, it is particularly well-suited for predicting energy usage in spinning processes.

5. **Result Analysis and Error Correction**

Thoroughly scrutinize the forecasting results, analyze forecast errors, assess the accuracy of the chosen forecasting model, propose correction measures, improve the forecasting model, and continuously enhance the quality and level of forecasting.

References

1. Chen X, Jia S, Xiang Y (2020) A review: knowledge reasoning over knowledge graph. Expert Syst Appl 141:112948
2. Chen Y, Li H, Li H et al (2022) An overview of knowledge graph reasoning: key technologies and applications. J Sens Actuator Netw 11(4):78
3. Lee MH, Kim S, Kim H et al (2022) Technology opportunity discovery using deep learning-based text mining and a knowledge graph. Technol Forecast Soc Chang 180:121718
4. Liu M, Li X, Li J et al (2022) A knowledge graph-based data representation approach for IIoT-enabled cognitive manufacturing. Adv Eng Inform 51:101515
5. Zhong RY, Xu X, Klotz E et al (2017) Intelligent manufacturing in the context of Industry 4.0: a review. Engineering 3:616–630
6. Paraschos PD, Koulinas GK, Koulouriotis DE (2020) Reinforcement learning for combined production-maintenance and quality control of a manufacturing system with deterioration failures. J Manuf Syst 56:470–483
7. Tah JHM, Abanda HF (2011) Sustainable building technology knowledge representation: using semantic web techniques. Adv Eng Inform 25(3):547–558
8. Chernenkiy V, Gapanyuk Y, Nardid A et al (2017) Using the metagraph approach for addressing RDF knowledge representation limitations. In: 2017 Internet technologies and applications (ITA)
9. Zarri GP, Zarri GP (2011) Knowledge representation and inference techniques to improve the management of gas and oil facilities. Knowl Based Syst (KNOSYS) 24:989–1003
10. Gayathri R, Uma V (2018) Ontology based knowledge representation technique, domain modeling languages and planners for robotic path planning: a survey. ICT Express S2405959518300985
11. Benavides C, García I, Alaiz H et al (2018) An ontology-based approach to knowledge representation for computer-aided control system design. Data Knowl Eng 118:107–125
12. Wu Z, Liao J, Song W et al (2018) Semantic hyper-graph-based knowledge representation architecture for complex product development. Comput Ind 100:43–56

13. Lin Y, Liu Z, Luan H et al (2015) Modeling relation paths for representation learning of knowledge bases. arXiv preprint arXiv:1506.00379
14. Chen WL, Xie SQS, Zeng FF et al (2011) A new process knowledge representation approach using parameter flow chart. Comput Ind 62(1):9–22
15. Zhang Y, Luo X, Zhang H et al (2014) A knowledge representation for unit manufacturing processes. Int J Adv Manuf Technol 73(5–8):1011–1031
16. Zhang C, Zhou G, Lu Q et al (2017) Graph-based knowledge reuse for supporting knowledge-driven decision-making in new product development. Int J Prod Res 55(23):7187–7203
17. Dai S, Zhao G, Yu Y et al (2021) Ontology-based information modeling method for digital twin creation of as-fabricated machining parts. Robot Comput Integr Manufact 72:102173
18. Yang T, Yi X, Lu S et al (2021) Intelligent manufacturing for the process industry driven by industrial artificial intelligence. Engineering 7(9):1224–1230
19. Khan MTH, Demoly F, Kim KY (2020) Constructing assembly design model capable of capturing and sharing semantic dynamic motion information in heterogeneous CAD systems. Int J Adv Manufact Technol 111(3):945–961
20. Arista R, Zheng X, Lu J et al (2023) An ontology-based engineering system to support aircraft manufacturing system design. J Manuf Syst 68:270–288
21. Guo Y (2023) Optimization of privacy-aware cloud crowdsourcing resource combinations for product development. Expert Syst Appl 227:120176
22. He Y, Hao C, Wang Y et al (2020) An ontology-based method of knowledge modelling for remanufacturing process planning. J Clean Product 258(10)
23. Zhou Q, Yan P, Xin Y (2017) Research on a knowledge modelling methodology for fault diagnosis of machine tools based on formal semantics. Elsevier Science Publishers B. V.
24. Oprea M (2018) A knowledge modelling framework for intelligent environmental decision support systems and its application to some environmental problems. Environ Model Softw 110:72–94
25. Wang Z, Yang W, Liu Q et al (2022) Data-driven modeling of process, structure and property in additive manufacturing: a review and future directions. J Manuf Process 77:13–31
26. Futia G, Vetrò A, Martin JCD, SeMi: a SEmantic modeling machine to build knowledge graphs with graph neural networks. SoftwareX 12
27. Kumar R, Kumar P, Kumar Y (2020) Time series data prediction using IoT and machine learning technique. Proc Comput 167:373–381
28. Cao P, Chen Y, Liu K et al (2018) Adversarial transfer learning for Chinese named entity recognition with self-attention mechanism. In: Proceedings of the 2018 conference on empirical methods in natural language processing
29. Zhu Y, Wang G, Karlsson BF (2019) CAN-NER: convolutional attention network for Chinese named entity recognition
30. Ristoski P, Gentile AL, Alba A et al (2019) Large-scale relation extraction from web documents and knowledge graphs with human-in-the-loop. J Web Semant 60:100546
31. Chen Y, Yang H, Liu K et al (2018) Collective event detection via a hierarchical and bias tagging networks with gated multi-level attention mechanisms. In: Proceedings of the 2018 conference on empirical methods in natural language processing
32. Smirnov A, Levashova T, Shilov N (2015) Patterns for context-based knowledge fusion in decision support systems. Inf Fusion 21:114–129
33. Yin S, Bao J, Zhang J et al (2020) Real-time task processing for spinning cyber-physical production systems based on edge computing. J Intell Manuf 31(8):2069–2087
34. Farooq B, Bao J, Raza H et al (2021) Flow-shop path planning for multi-automated guided vehicles in intelligent textile spinning cyber-physical production systems dynamic environment. J Manuf Syst 59:98–116
35. Shen X, Li X, Zhou B et al (2023) Dynamic knowledge modeling and fusion method for custom apparel production process based on knowledge graph. Adv Eng Inform 55:101880
36. Zhao S, Wang J, Zhang J et al (2020) Edge-cloud collaborative fabric defect detection based on industrial internet architecture. In: 2020 IEEE 18th international conference on industrial informatics (INDIN), vol 1. IEEE, pp 483–487

37. Tong X, Bao J, Tao F (2024) Co-evolutionary digital twins: a multidimensional dynamic approach to digital engineering. Adv Eng Inform 61:102554
38. Liu S, Lu Y, Li J et al (2023) A blockchain-based interactive approach between digital twin-based manufacturing systems. Comput Ind Eng 175:108827
39. Liu M, Zhou B, Li J et al (2023) A knowledge graph-based approach for assembly sequence recommendations for wind turbines. Machines 11(10):930
40. Bao J, Liu Q, Jin Z et al (2024) CarbonKG: industrial carbon emission knowledge graph-based modeling and application for carbon traceability of complex manufacturing process. J Comput Inf Sci Eng 24:081001–081011
41. Barker GL, Laird JW (1992) Temperature effects on cotton lint moisture regain rates. Trans Asae 35(5):1377–1380
42. Ozturk E, Cinperi NC, Kitis M (2020) Improving energy efficiency using the most appropriate techniques in an integrated Woolen textile facility. J Clean Prod 254:120145
43. Harrouz JP, Ghali K, Hmadeh M et al (2022) Design and control of MOFs-based indoor humidity pump integrated into the building's ventilated façade in hot and humid climates. Energy Convers Manage 268:115983
44. Wang J, Xu C, Zhang J et al (2022) Big data analytics for intelligent manufacturing systems: a review. J Manuf Syst 62:738–752
45. Zhao S, Zhong RY, Xu C et al (2024) A dynamic inference network (DI-Net) for online fabric defect detection in smart manufacturing. J Intell Manufact 1–16
46. Zhao S, Zhong RY, Wang J et al (2023) Unsupervised fabric defects detection based on spatial domain saliency and features clustering. Comput Ind Eng 185:109681
47. He Z, Xu J, Tran KP et al (2021) Modeling of textile manufacturing processes using intelligent techniques: a review. Int J Adv Manufact Technol 116(1):39–67
48. Lutz É, Coradi PC (2022) Applications of new technologies for monitoring and predicting grains quality stored: sensors, internet of things, and artificial intelligence. Measurement 188:110609

Chapter 10
Equipment Fault Diagnosis and Preventive Maintenance

The fault diagnosis of industrial equipment is the earliest application field of artificial intelligence technology. Pattern recognition and signal processing methods have been widely used, and excellent results have been achieved. However, the classical algorithm is also facing significant challenges: the data set is getting larger and larger, and the real-time requirement is increasing. It is one of the current hotspots for using the deep learning method to carry out this work.

10.1 Progress and Status of Equipment Fault Diagnosis and Preventive Maintenance

The purpose of fault diagnosis of industrial equipment is to enhance the stability of equipment operation and improve the life of equipment, which is an important feature related to system reliability and safety [1]. Using equipment fault diagnosis technology, the equipment's function and performance can be largely restored, extending the service life of core components and enhancing production quality. Therefore, fault diagnosis technology is the key to realizing the control system's health management and ensuring the equipment's safe and stable operation [2], which is of great significance to ensure the safe and stable operation.

As shown in Fig. 10.1, the fault diagnosis and health management of equipment are mainly divided into four levels.

1. Perception and analysis of control layer: collect data from the core component controller and embedded sensor for analysis.
2. Environment layer perception and analysis: collect working environment data, such as design data (e.g., the running program of the core component), process data, system integration control data and external PLC data for analysis.

T. Liu et al., *Industrial Intelligence: Methods and Applications*, Springer Series in Advanced Manufacturing, https://doi.org/10.1007/978-3-031-81477-8_10

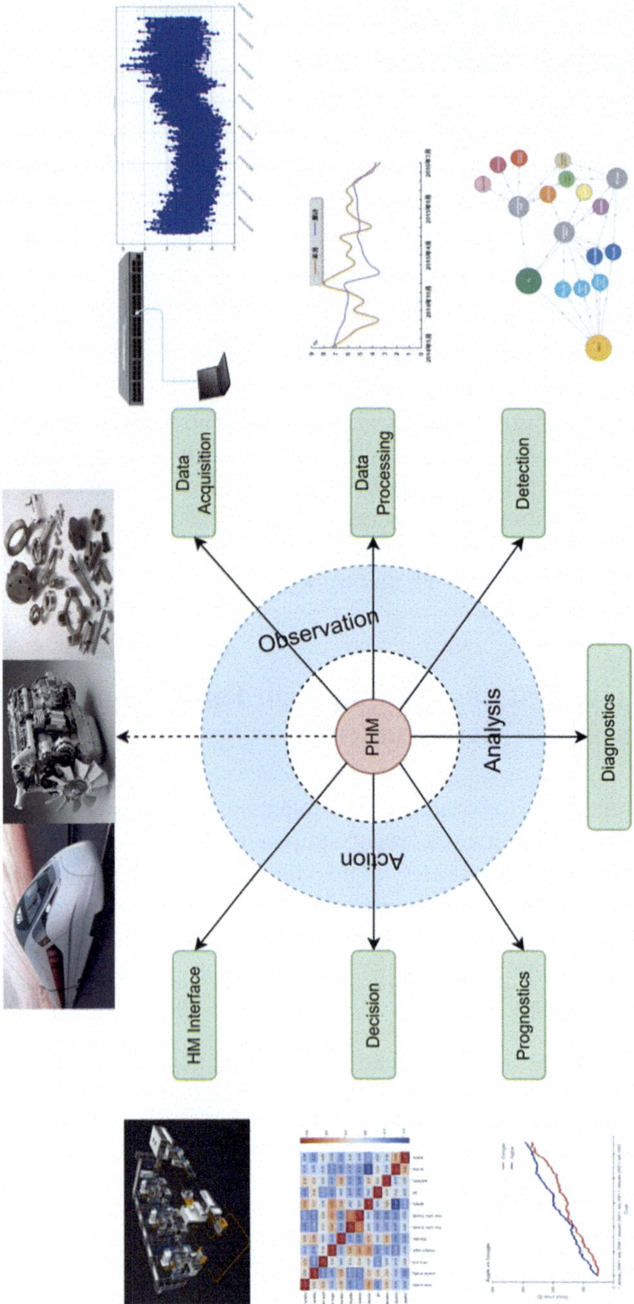

Fig. 10.1 Overall architecture of equipment fault diagnosis and health management

3. Additional layer perception and analysis: collect additional sensor data, such as the collection of additional force and torque sensors to obtain the impact of sub-component payload and tool installation.
4. Top-level perception and analysis: combined with vision for location recognition, considering the system architecture, system function and related parameters to evaluate the health status of the overall system.

With the change in information technology, equipment fault detection technology continues to carry out intelligent innovation and development. Fault diagnosis technology has experienced three development stages based on analytical models, signal processing, and artificial intelligence [3]. Among them, the analytical model [4] and signal processing [5] methods both promoted the development of equipment fault diagnosis technology at that time. With the improvement of the equipment structure and the complexity of the working environment, these two methods gradually cannot fully meet the requirements of current fault diagnosis. The fault diagnosis method based on artificial Intelligence (AI) is gradually favored by people and applied to fault diagnosis technology because of its strong calculation performance and high recognition accuracy.

Artificial intelligence methods for fault diagnosis include deep learning methods, reinforcement learning methods, and transfer learning. Deep learning methods for fault diagnosis include Convolutional Neural Networks (CNN), Deep Belief Networks (DBN), Stacked Autoencoder Networks (SAE), and Recurrent Neural Networks (RNN). The recognition accuracy and accuracy of these deep learning models are generally better than those of traditional machine learning models [6]. CNN is mainly used for feature extraction and fault diagnosis recognition. Literature [7] uses CNN to analyze the wavelet time-spectrum of fault signals for fault classification, which is superior to the traditional SVM algorithm in accuracy. In literature [8], the CNN algorithm is used to identify the fault of a high-voltage circuit breaker, and the accuracy is far better than the traditional fault diagnosis method.

DBN is developed from the research of biological neural networks and shallow neural networks. It is a unique form of Bayesian probabilistic generative model, and its data sample distribution can be inferred by Bayesian joint probability distribution. DBN comprises several layers of neurons composed of Restricted Boltzmann Machine (RBM). As a kind of neural perceptron, RBM comprises an explicit and hidden layer. Figure 10.2 is its structure diagram. The neurons between the two layers are bidirectional and fully connected, with the hidden and visible layers alternating. Each adjacent RBM forms the hidden and visible layers of different RBMs. The output of the last RBM is the input of the next RBM, and there is no connection between the same layer.

DBN can handle signal technology and diagnosis experience and is more adaptable and versatile. The comparison of DBN-based fault diagnosis and other fault diagnoses is shown in Fig. 10.3. In Ref. [9], DBN was used to extract original data features, and the fault recognition performance of the ensemble classifier was significantly improved. Literature [10] proposed a fault diagnosis method for gear transmission systems based on DL theory. Firstly, DBN was used to extract the vibration signal

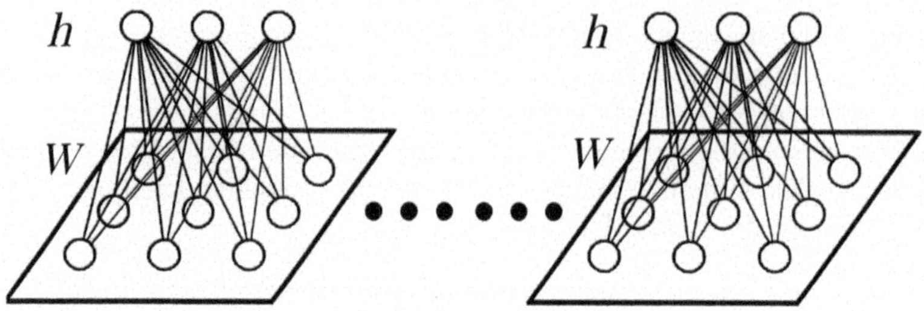

Fig. 10.2 RBM concatenated into DBN structure

of the gear transmission system, and then the fault signal was judged by the ability of DBN to make complex mapping representations.

The SAE method has a robust feature extraction ability and robustness and requires less sample data. Literature [11] uses SAE to identify dissolved gases in transformer oil and diagnose transformer fault types. In literature [12], the attenuation of DC line traveling wave shape with distance was used as a feature to establish the SAE regression model to realize end-to-end line fault location. Like DBN, SAE is composed of several units, but its structural units differ. The stacked autoencoder network is a multilayer perceptron neural network comprising multiple Autoencoder (AE) modules. The module transformation is shown in Fig. 10.4.

RNN considers the correlation between samples, compared with typical RNN networks such as LSTM, which can effectively extract the temporal and spatial

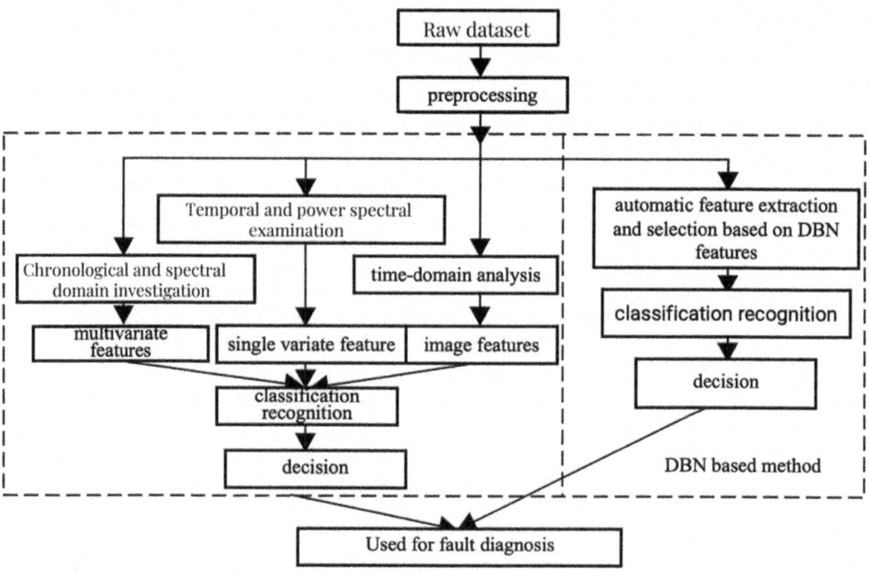

Fig. 10.3 Comparison of DBN and other fault diagnoses

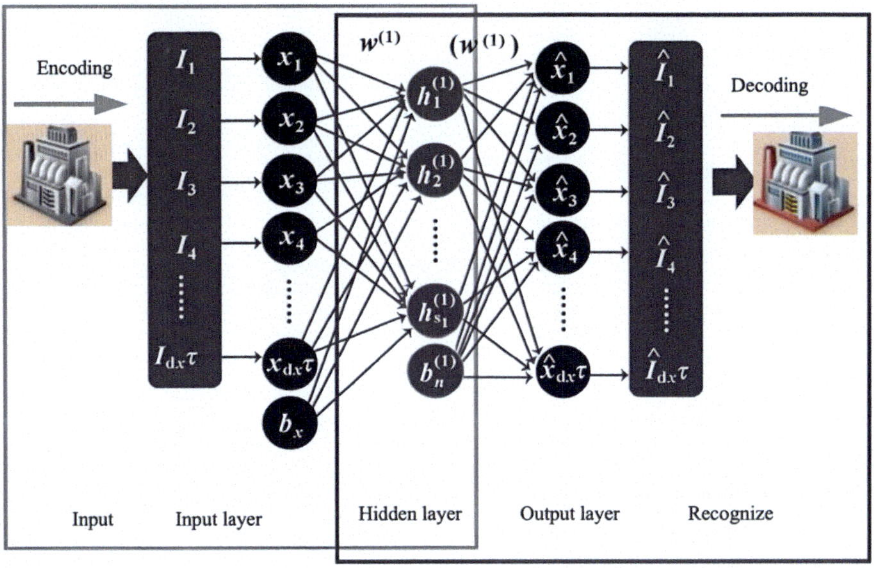

Fig. 10.4 SAE module transformation process

features of vibration signals. Literature [13] proposes an intelligent fault diagnosis method for medical equipment based on LSTM, which has higher fault diagnosis accuracy compared with other deep learning methods. Literature [14] proposed a batch standardized CNN-LSTM model, which can efficiently and accurately diagnose rolling bearings under various loads, various fault locations and various fault degrees. Different deep learning methods have different characteristics, and suitable methods can be selected according to the actual needs and scenarios of fault diagnosis.

Reinforcement learning methods are suitable for extracting fault features, or classifying fault features that have been obtained. Literature [15] proposed a three-state combination LSTM neural network based on reinforcement learning, which demonstrated high accuracy in predicting state degradation trends. In addition, Deep Reinforcement learning (DRL) provides new ideas for further improving the intelligence of fault diagnosis. Literature [16] proposed an improved DQN fault diagnosis method for rolling bearings based on deep reinforcement learning, and the proposed improved diagnosis model has good generalization. The fault diagnosis process based on reinforcement learning is shown in Fig. 10.5.

Transfer learning can be used to solve similar or similar problems, fine-tune the network parameters of machine learning models by means of transfer [17]. Literature [18] uses transfer learning in motor fault recognition to address the issue of poor model performance due to small sample sizes, enhancing the potential of AI methods in this field. Literature [19] established a data mining model based on transfer learning to solve the problem of less abnormal data in electric power. Different sample data

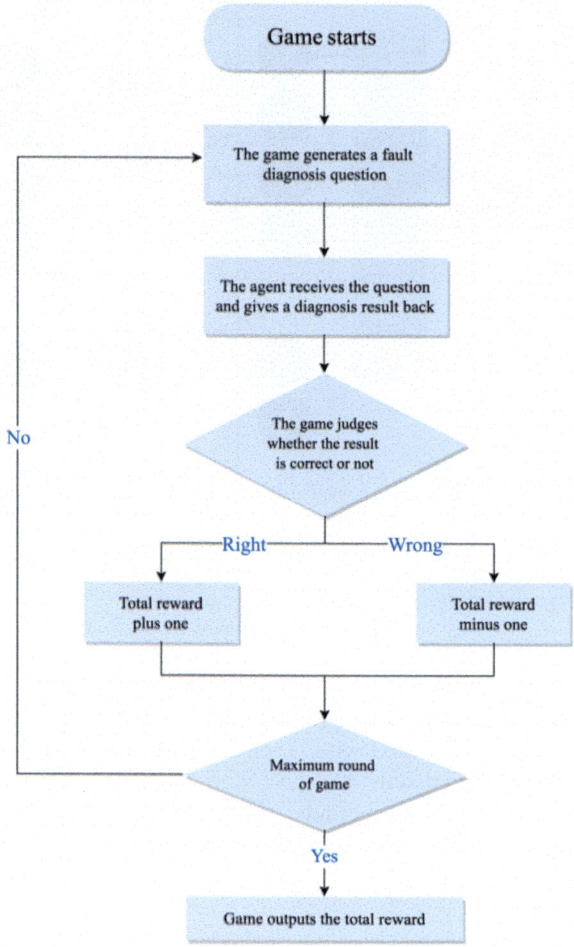

Fig. 10.5 Reinforcement learning-based fault diagnosis flow

were used to adjust the unified model accordingly to improve the utilization rate of fault data and the efficiency of training network. Therefore, transfer learning can enhance the computational efficiency of fault diagnosis and address the challenge of obtaining labeled fault data when there is insufficient labeled data available.

The preventive maintenance of equipment, like the fault diagnosis of equipment, is an effective measure to ensure the stable operation of equipment. The typical algorithms for preventive maintenance of equipment include real-time processing algorithm MRAM, off-line prediction algorithm MAOP and genetic algorithm. MRAM enables real-time evaluation by continuously monitoring key data from each maintenance unit and facilitates proactive maintenance by scheduling resources. MAOP is based on historical data stored in a distributed file system. Combined with the current equipment working status and historical information, the remaining effective working time of the current equipment is predicted. The individual evolutionary search of genetic algorithm can be connected with the objective function of maintenance strategy optimization to search and optimize the maintenance strategy. The Monte Carlo simulation module is nested in the genetic algorithm to solve the problem of individual fitness calculation.

Many researchers have taken equipment preventive maintenance as the focus of research. Literature [20] explored the optimization of preventive maintenance under non-uniform periods, establishing an equipment performance degradation model using a hidden semi-Markov model and multi-dimensional characteristic variables, offering a theoretical basis for preventive maintenance planning. Literature [21] developed a component preventive maintenance model that considers EMU failure risk, offering a more economical and safe approach, and providing theoretical support for EMU component maintenance decisions. Some researchers also focus on the dynamic adjustment method of preventive maintenance strategy. Literature [22] proposes a dynamic adjustment method for the preventive maintenance strategy of equipment affected by temporary shutdowns, providing a basis for reliability analysis during long-term shutdowns in a nuclear power plant. Literature [23] proposed a preventive group maintenance strategy for agricultural machinery equipment in service, which has good application in the service stage of agricultural machinery equipment. In addition, the equipment preventive maintenance method based on the new generation of information technology is also applied to the stable operation of equipment in various fields. Literature [24] puts forward the key technology system of digital twin satellite. Literature [25] constructs an F-15 digital twin, and uses aircraft life cycle data for fault prediction and health management, which effectively improves the accuracy of maintenance early warning and the reliability of maintenance plan.

The rapid development of technologies like big data, cloud computing, IoT, and 5G has created new opportunities for intelligent innovation in equipment fault diagnosis and preventive maintenance. Fault diagnosis and preventive maintenance are gradually integrated with the idea of intelligent manufacturing from system architecture to method model, reflecting the characteristics of industrial intelligence.

In the last two sections of this chapter, the intelligent models, methods and cases will be introduced in detail from two aspects of equipment fault diagnosis and equipment preventive maintenance.

10.2 Equipment Fault Diagnosis

This subsection firstly reviews and analyzes CNN based equipment fault diagnosis. Secondly, it analyzes the CNN based equipment fault diagnosis methods. Finally a case study of CNN based equipment fault diagnosis is given.

10.2.1 CNN-Based Equipment Fault Diagnosis

Reliability has always been an important aspect in the evaluation of industrial products and equipment. Good product design is of course essential for products with high reliability. But no matter how well designed products are, products will deteriorate over time because they are operated under certain pressures or loads in real environments, often involving randomness. Therefore, fault diagnosis and health maintenance have been introduced as an effective method to ensure a satisfactory level of reliability during the service life of products and equipment. Meanwhile, the rapid rise and widespread adoption of the Internet and IoT have led to an explosive growth in the data scale of industrial products and equipment, with increasingly diverse data types and more complex data structures.

In view of this significant change, people have begun to pay attention to the application of deep learning methods in this regard. Through the training of deep network with a large number of fault data, engineers can fully learn the potential characteristics of faults and realize the classification and prediction of faults. This not only provides a new fault diagnosis method for machinery with complex structure, shorts the time of fault diagnosis, but also ensures the safety of equipment and reduces the maintenance cost of equipment [26]. Convolutional Neural Networks (CNN) have been first applied in the recognition and classification of images, and have achieved remarkable success. Therefore, many researchers have also applied it to the field of mechanical fault diagnosis. When early researchers introduced the CNN model into rotating machinery fault diagnosis, they typically used the original CNN structure, constructing two-dimensional training samples, extracting features from two-dimensional data, and completing the diagnosis.

Most of the CNN architecture variants are very similar. Taking the LeNet-5 model proposed by LECUN as an example (the model is shown in Fig. 10.6) [27], the figure shows the classical CNN model and the main idea in the training process. Its main components include the convolutional layer, pooling layer, and fully connected layer. A typical feedforward neural network is built using multiple filters, which extract features from the input data. Convolution and pooling of all filtered data occur within the convolutional and pooling layers. After the recursion of the multi-layer neural network, the data input to the neural network will slowly hide its topology structure, and the data will finally be abstracted. The sign representation of the input data is invariant to translation, rotation and scaling to a certain extent, and the feature points are sparsely connected, weight sharing, and spatial or temporal downsampling data information. Specifically, the input feature map or original data map is first learned in a convolutional layer, which consists of multiple convolution kernels, and different feature maps are also calculated in this layer. Each neuron in the feature map is connected to the region of the neighboring neurons in the previous layer, and such a field is called the receptive field of a layer of neurons. The kernels of the first convolutional layer are designed to detect edge and curve low-level features, while the kernels of the higher convolutional layers are designed to encode more abstract features. This layer first convolutes the input with a learning kernel, and then

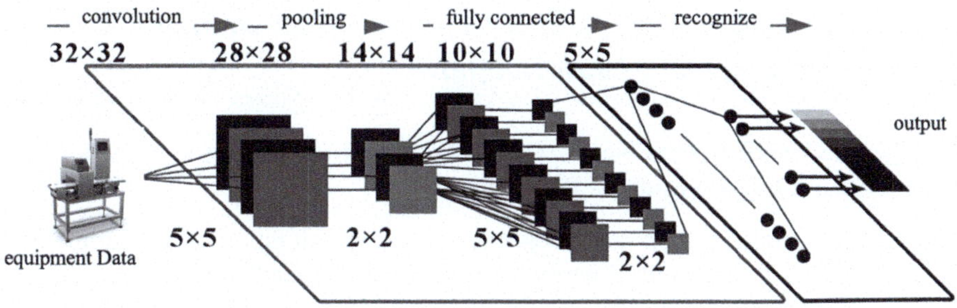

Fig. 10.6 CNN architecture

applies an elemental nonlinear activation function to the convolution result to obtain a complete feature map for several different special kernels. Finally, the purpose of this layer is achieved by reducing the resolution of the feature map in the pooling layer. Several convolutional layers and pooling layers are stacked continuously to gradually extract the useful representation of higher levels of features, and finally recognize and classify in the recognition area.

For fault diagnosis, CNN extracts effective features from input data, continuously abstracts the data through convolution and pooling, and refines local features into high-level features, thereby enabling fault diagnosis and recognition. At present, in most CNN application categories, feature extraction and recognition are two common fault diagnosis based on CNN. (1) When the rotating unit fails, its important feature is that the machinery shows abnormal vibration and noise. In 2016, Janssens et al. proposed a feature learning model for condition monitoring based on convolutional neural network to solve the problem that the vibration mode of rotating machinery is different due to faults or mechanical state changes. The model lets the data itself learn useful features for bearing fault detection, which makes the results significantly better than the feature learning method based on feature engineering [28]. (2) As a classifier, INCE et al. (2016) proposed a fast and accurate motor state monitoring and early fault detection system based on CNN [29], addressing the issue where feature extraction and classification are typically handled in separate blocks in conventional systems. The system has an inherent adaptive design, and integrates the feature extraction and classification stages of motor fault detection into the system in a learning body. The proposed method is directly applicable to raw data and therefore does not require a separate feature extraction algorithm, thus improving the speed and hardware efficiency of the system.

10.2.2 A Convolutional Neural Network-Based Method for Equipment Fault Diagnosis

(1) **Convolutional Neural Network-Based Method for Equipment Fault Diagnosis on a Fog Computing Platform**

In high data traffic environments, industrial systems often utilize cloud computing, which leverages internet cloud servers to provide on-demand, convenient access to shared computing resources, storage, applications, and other facilities. However, with the emergence of millions of IoT-connected devices generating vast amounts of data at high speeds, there is a data explosion. The sheer volume of data has put pressure on cloud storage for computing, storing, and managing this information. Cloud servers require more time to process data due to centralized storage and computation, and they are often far from IoT endpoints. This has led to the development of fog computing, which alleviates the burden on cloud computing services.

Fog computing is a decentralized computing architecture that processes, stores, and controls data intelligently around data devices. This flexible structure extends cloud computing services. Therefore, it shortens network distances, improves efficiency, and increases the volume of data transmitted to the cloud for processing, analysis, and storage [30].

Currently, the applied fog architecture consists of three layers: the terminal layer, the fog layer, and the cloud layer [31]. The terminal layer is close to the physical environment of IoT devices, such as sensors. The fog layer consists of Wi-Fi routers and gateways, typically at the edge of the sensor network. Gateways have computing capabilities that process most collected data with low latency, so only the most useful data is sent to the cloud layer for complex computations. In recent years, research on fog computing supporting manufacturing has increased. Wan et al. [32] designed a fog architecture to optimize energy consumption and workload of machines in manufacturing workshops. Wu et al. [33] proposed a fog computing-based framework optimized for manufacturing, integrating a random forest algorithm for tool wear prediction. O'Donovan et al. [34] developed a fog architecture integrated with machine learning algorithms, analyzing the execution times of fog and cloud layers, and demonstrating the advantages of designing the fog layer and implementing a particle swarm optimization algorithm within it.

Liang et al. [35] proposed a CNN (convolutional neural network) fault diagnosis model based on fog computing for CNC machining field. In CNC machining, deviations in dimensional accuracy and surface roughness can lead to product defects or waste. The root cause of the deviation may be due to poor tools or abnormal equipment condition, resulting in unsatisfactory accuracy, scratches on the processed parts, severe vibration or high temperature during processing. Therefore, designing predictive systems to effectively analyze state data from machines and processes is crucial for early fault detection, enabling intelligent diagnosis and adaptive optimization of machines. They used the CNN preprocessing mechanism to denoise the monitoring data to effectively improve the computational efficiency. On the other hand, they designed an appropriate CNN structure to improve the computational efficiency of

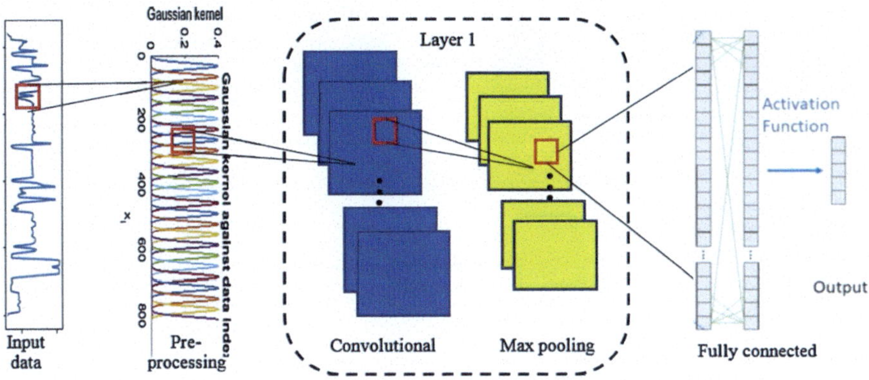

Fig. 10.7 Basic architecture of an improved CNN

prediction. At the same time, to optimize data transmission, the monitoring data collected on the device is processed on the fog layer in order to effectively identify faults using a locally deployed trained CNN. Intensive computations, such as training CNNs and performing rescheduling optimization to address anomalies, are conducted in the cloud. The improved CNN architecture, depicted in Fig. 10.7, is a lightweight deep network.

(2) Equipment Fault Diagnosis Method Combining Convolutional Neural Network and Attention Mechanism

As one of the key components of rotating machinery system, rolling bearing plays an important role in industrial production and application [36]. The health status of rolling bearings directly affects the safe and reliable operation of mechanical equipment. Therefore, timely and accurate detection of bearing faults is of great significance to ensure safe production, improve efficiency and eliminate hidden dangers.

Most of the traditional bearing fault diagnosis methods are based on manually extracted time domain, frequency domain or time–frequency domain fault features, which can only achieve good diagnosis results under specific defects and loads. Nowadays, deep learning methods have profoundly affected the field of fault diagnosis, and Convolutional Neural Network (CNN) with local connection and shared weight features has also been applied to the field of fault diagnosis [37].

On the other hand, the attention mechanism has garnered significant interest from researchers [38] due to its ability to selectively enhance useful information and its powerful parallel processing capabilities, which improve the model's performance and training speed. Wang et al. proposed a new data-driven fault diagnosis method for rolling bearing intelligent state recognition using multi-head attention and CNN [39]. The proposed method mainly consists of two steps: (1) by creating the corresponding fault data set directly from the raw vibration signal. Use simple signal-to-image transformation methods (2) Develop a new intelligent diagnosis model

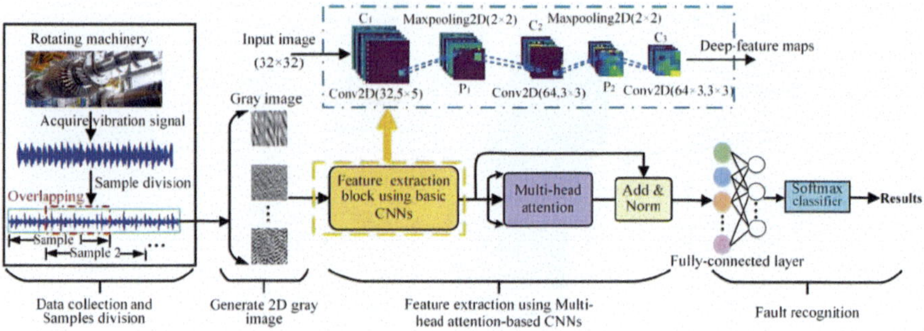

Fig. 10.8 Schematic diagram of the fault diagnosis

that combines multi-head attention with CNN to realize automatic feature extraction and fault pattern recognition (see Fig. 10.8). Compared to other methods, this signal-to-image conversion technique is simpler and has lower time complexity. By incorporating multi-head attention to optimize the fault recognition framework, the proposed approach achieves higher diagnostic accuracy and generalization ability with relatively few network parameters. The structure of a basic CNN. To enhance CNN model training and effectively learn abstract time-domain features of various bearing fault conditions, this method employs a novel data preprocessing technique that directly converts one-dimensional raw vibration signals into two-dimensional grayscale images [37], enabling end-to-end fault detection. Therefore, the proposed method is more suitable for on-line bearing fault diagnosis.

10.2.3 Equipment Fault Diagnosis Case Based on Convolutional Neural Network

(1) **Equipment Fault Diagnosis Case Based on Convolutional Neural Network Under Fog Computing Platform**

Liang et al. [35] proposed fog computing based methods to provide intelligent solutions for dynamic prediction and optimization of machining processes. The system consists of the following three layers:

1. End layer: Integration with physical devices via sensors, circuits, and routers. The power signal of the machine is continuously collected through the power sensor and transmitted to the fog layer for further processing.
2. Fog (computing) layer: The computing power and storage space of the fog layer are usually limited. A CNN (Convolutional Neural Network) is deployed in this layer. Fog computing is a concept derived from cloud computing, which is to move data processing and applications to devices at the edge of the network. The CNN is trained in the cloud layer and then deployed to the fog layer for detecting

abnormal conditions. This approach eliminates the need to transmit signal data to the cloud layer, ensuring that relevant data remains within the company. On the other hand, by leveraging the significant changes in the power signal during processing (where standby power is much lower than processing power), the power signal can be segmented into distinct cutting processes, which enhances the prediction accuracy of the CNN.

3. Cloud layer: The cloud layer has the advantages of computing power and storage space, so the reference signal is also stored in the database on the cloud. Since the scheduling optimization algorithm and CNN training usually require iterations to calculate the best results, they are also deployed in the cloud. At the beginning of each processing cycle, the optimal schedule is provided by using the Fruit Fly Optimization (FFO) algorithm to minimize energy consumption, manufacturing time, and machine utilization. The optimized schedule is transmitted to the terminal layer for production control. When the cloud layer receives a new reference signal, the database is updated, and this updated signal is used to retrain the CNN. This process allows the CNN to detect and identify new anomalies in the processing cycle. The updated CNN is then transmitted to the fog layer for anomaly detection, as illustrated in Fig. 10.9.

The system coordinator is arranged between the cloud layer and the fog layer, which plays the following four roles.

1. Update the knowledge database on the cloud layer to obtain the reference signal of the new machining part.
2. Retrain the CNN with the new reference signal.

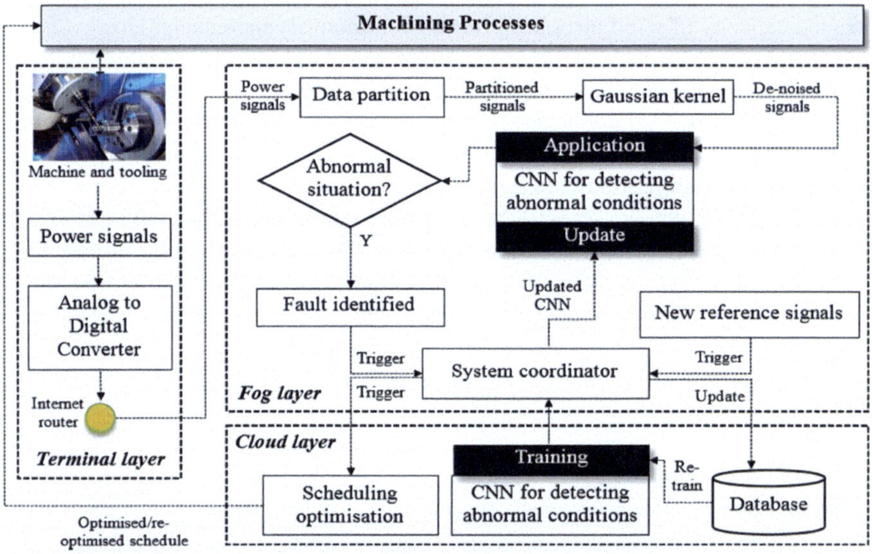

Fig. 10.9 Fault diagnosis system based on fog computing

3. Update the CNN on the fog layer.
4. Scheduling optimization is triggered for abnormal situations in the processing process, and the optimized schedule is sent to the terminal layer for production adjustment. When an abnormal condition is found, the machine is temporarily disabled. The disabling time is based on the problem complexity determined by the engineer. When the production schedule is rescheduled on the cloud layer, the disabled time is sent to the terminal layer. Fog nodes usually have limited computing power and storage space.

The system was successfully launched in a processing company in the UK. The total amount of data collected during an industrial trial was 12.98 GB. Due to the atomized design, only 3.36 GB of data is transferred to the cloud. As a result, 70.26% of the bandwidth is saved compared to the cloud solution. For 1 Mb of data, the transmission and processing of the monitoring signal to the fog node takes approximately 1.31 s, whereas transmitting and processing the same data in the cloud center takes about 3.96 s. Additionally, it has been observed that during periods of internet congestion, the data transmission speed for the cloud solution can be significantly slower. The time for data transmission in the fog system is reduced by 47.02%.

The fog design proposed in this study is further justified by the calculation time and storage situation. Table 10.1 shows the computation time and memory usage for all algorithm elements in the industrial trials. This indicates that preprocessing and CNN can be efficiently performed on the fog layer for prognosis based on capacity.

(2) **Equipment Fault Diagnosis Method Combining Convolutional Neural Network and Attention Mechanism**

Wind energy has shown strong competitiveness among all types of energy, so the installed capacity of wind turbines has been continuously increasing. However, due to the harsh, complex and variable working environment, the failure rate of components such as main bearings, gearboxes and generators of wind turbines is high, resulting in expensive maintenance and operation costs of wind turbines. Therefore, researching detection methods for the operational status of wind turbines is beneficial for identifying potential faults in a timely manner, formulating maintenance plans, and reducing economic losses [40]. Xiang et al. [41] combined CNN with LSTM and introduced an attention mechanism into the network to enhance useful information. CNNs are used to extract state-space features from wind turbines, while LSTM can better fuse temporal features of different part states. By giving different weights to

Table 10.1 Classification of transfer learning

Learning style		Ds & DT	Ts & TT
Traditional machine learning		Similar	Similar
Transfer learning	Inductive transfer learning	Similar/related	Related
	Unsupervised transfer learning	Related	Related
	Transductive transfer learning	Related	Similar

the input of model features, the attention mechanism highlights the influence of more critical factors and helps the model make more accurate judgments.

The model is applied to SCADA data from two wind farms. One set of data is SCADA data, which contains faults in wind farms in northern China from January 1 to July 14, 2015. The wind turbine was shut down on 14 July due to a gearbox failure. The data is first preprocessed and then abnormal data points such as stop points and bad points are removed. At the same time, the operating states of the wind turbine, such as the start-up of the wind turbine and the gearbox failure, are saved. Too many variables will lead to data redundancy and affect the accuracy of the prediction model. Therefore, through correlation analysis, variables with high correlation with output variables are selected as input variables. In this case, the selected target variable is the temperature of the gearbox bearing. Through correlation analysis, variables with high correlation with gearbox bearing temperature are selected as input variables, and 11 highly correlated variables are selected. Then, the model combining CNN with LSTM and attention mechanism is used to train preprocessed data. The prediction results after fault detection are shown in the root mean square error (RMSE) residuals in Fig. 10.10. It can be seen that the threshold is breached for the first time on day 128. Thereafter, the set threshold was repeatedly exceeded with large mutations, reaching a maximum of 4.48, before the outage on July 14. This change can indicate that the potential failure of the wind turbine has reached a certain level, which is in line with the expected change in the detection results before the failure. The predicted time is consistent with the actual failure time, so it can be determined that the wind turbine fails on the 128th day. The proposed model can effectively detect the faults of the wind turbine.

The other set of data is obtained from SCADA data, which contains faults in a wind farm in Zhejiang province from February 1, 2013 to May 2, 2014. Due to

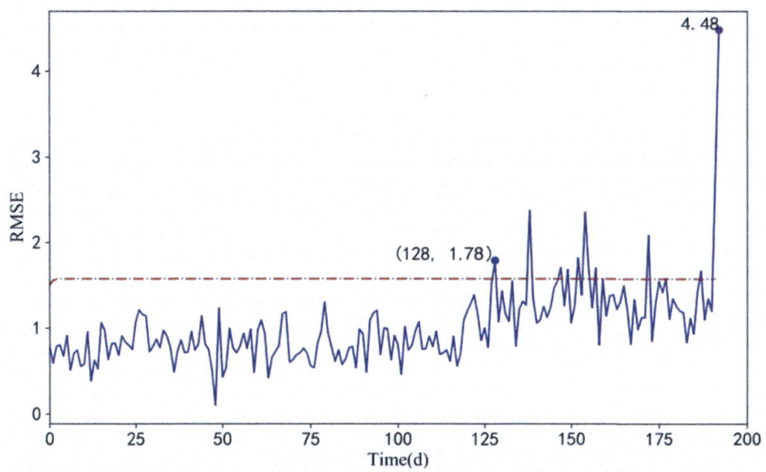

Fig. 10.10 Model predictions (root mean square error)

the generator failure, the wind turbine was repaired on June 4, 2014. The data are first preprocessed to remove abnormal data points such as stops and bad points. At the same time, the operating state of the wind turbine, such as the start-up of the wind turbine and the generator failure, are saved. Due to data redundancy, too many variables entered into the prediction model will reduce the prediction accuracy, so the generator bearing temperature is used as the output variable. Variables that are highly correlated with the temperature of the generator bearing will be selected as input variables. The 12 variables selected after correlation analysis, including the generator bearing temperature, average power, A-phase temperature, generator speed and A-phase current change trend.

In each phase of the wind turbine operation, each variable fluctuates within a certain range. Even in the case of faulty downtime, the trend does not change significantly, so the operating state of the wind turbine cannot be directly determined by the change in each variable condition. The residuals of the prediction results after fault detection are shown in Figs. 10.11. It can be seen that the threshold is violated for the first time on day 109 (Fig. 10.11).

Thereafter, the model repeatedly exceeded the set threshold and had large mutations before the outage on June 4, with a maximum value of 2.3, which was in line with the expected change in detection results before the failure. Therefore, it can be said that the detection method has detected the potential fault of the wind turbine on day 109. The prediction results (root mean square error) supported this conclusion.

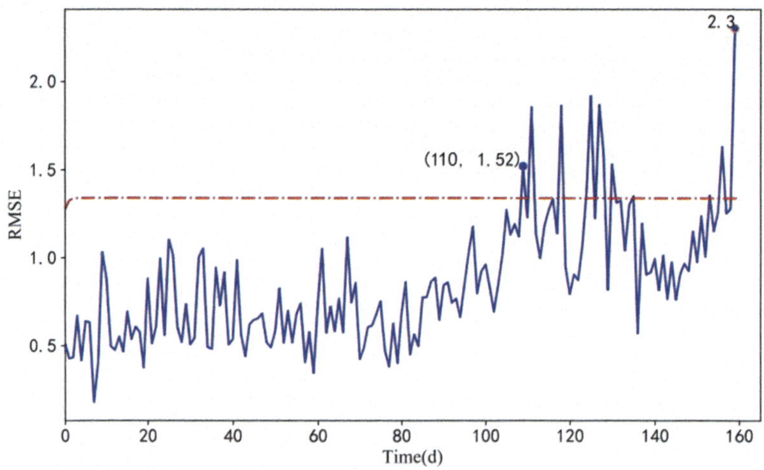

Fig. 10.11 Model prediction (root mean square error)

10.3 Preventive Maintenance of Equipment

Preventive maintenance of equipment can avoid failures and give recommendations for equipment maintenance in advance. This subsection first introduces the current state of research on transfer learning-based equipment preventive maintenance. Secondly, it introduces the method of equipment preventive maintenance based on transfer learning. Finally, bearing preventive maintenance is used as a case study.

10.3.1 Preventive Maintenance of Equipment with Transfer Learning

As one of the key innovation points of Industry 4.0, predictive maintenance technology is the key to ensure efficient and sustainable services in the future [42]. Current predictive maintenance technologies for mechanical equipment, centered on artificial intelligence, are predominantly based on data-driven models. These models assume that training and testing samples satisfy the condition of independent and identically distributed (i.i.d.) data. However, they face challenges such as limited target sample size and imbalanced datasets, which lead to a decline in model accuracy and reliability, and in severe cases, even model failure. Transfer learning, a new theory developed within the field of machine learning, effectively enables the sharing and transfer of information between similar domains or tasks. This transforms traditional learning from a ground-up approach into a cumulative process, enhancing the reusability of historical knowledge and information. It offers a novel approach to addressing the aforementioned challenges [43].

Transfer learning is a new machine learning method that uses existing knowledge to solve problems in different but related domains. The goal is to transfer the existing knowledge to solve the learning problems with few or even no labeled data in the target domain. Because the data set used to verify the effectiveness of various intelligent fault diagnosis methods satisfies the following two conditions: (1) Sufficient labeled data with fault information must be available. (2) The training and testing data are drawn from the same probability distribution.

However, in practical engineering, the success of machine intelligent fault diagnosis depends on these two conditions. For some machines, it is difficult to satisfy these two conditions due to the following problem. The main reasons are as follows. It is difficult to get labeled failure data from some machines. Specifically, there are two main reasons for the lack of labeled fault data. Firstly, machines may not be allowed to fail in operation, as unexpected failures often lead to machine failure or even catastrophic accidents. In this case, failure data is not available. Secondly, machines usually undergo a long-term degradation process from health to failure. This means that it is time-consuming and expensive to obtain the failure data of the machine.

An intelligent fault prediction method trained using labeled data acquired from one machine may fail to classify unlabeled data acquired from other machines. Although it is difficult for certain machines to obtain large amounts of labeled data, they can still be obtained from different but related machines. For example, it is difficult to obtain labeled data for railway locomotive bearings, while the labeled data for electric motor bearings can be obtained relatively easily. However, the probability distribution of the data acquired from different machines are different. Therefore, classification the performance of intelligent fault prediction methods degrade when training and testing datasets are obtained from different machines.

As shown in Fig. 10.12, transfer learning models trained using labeled data taken from one machine are able to generalize to unlabeled data taken from other machines. Transfer learning is able to use learned knowledge from the source domain to solve new but related tasks in the target domain. It is expected to solve problems where there are not enough labeled data to train on. An intuitive and commonly used approach in transfer learning is to obtain feature representations that are close across different domains while maintaining strong classification performance on the source data. The domain adaptation method is an important branch of transfer learning, and its network structure is shown in Fig. 10.13.

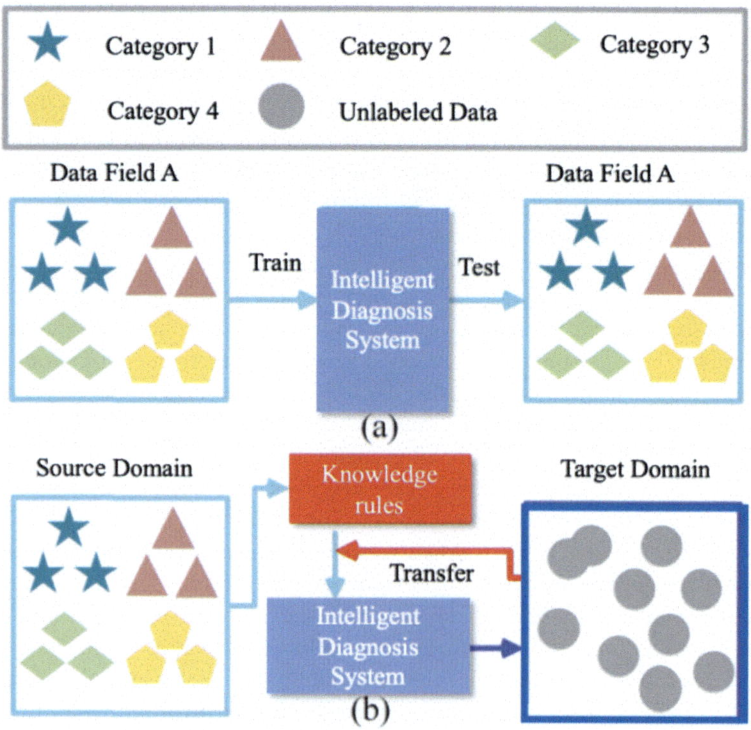

Fig. 10.12 Intelligent diagnosis logic with transfer learning

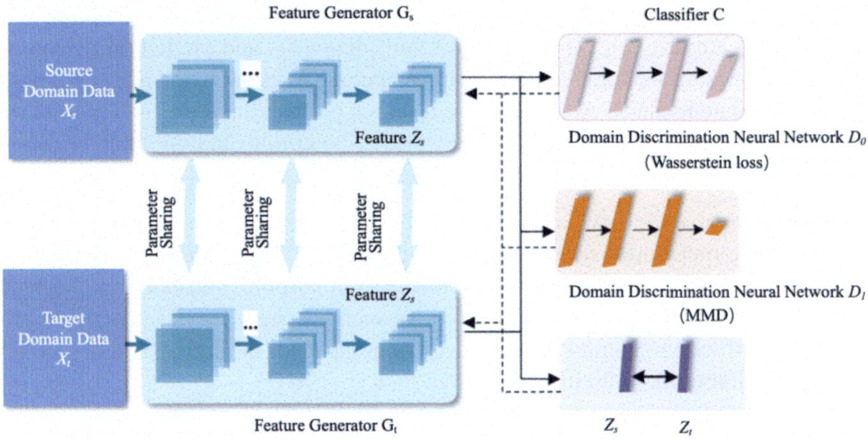

Fig. 10.13 Domain adaptive network architecture

The learned domain is usually called the source domain (Ds), the domain to be applied is called the target domain (DT), Ts represents the source task and TT represents the target task. According to the similarity between the source domain and the target domain, Pan et al. divided transfer learning into inductive, unsupervised and transductive transfer learning, as shown in Table 10.1 [44].

Transfer learning mainly solves the small sample size problem, the problem of different data distribution, and the problem of personalization. At present, the typical application research of transfer learning mainly includes natural language processing, sentiment classification, computer vision, medical health, artificial intelligence systems, etc. For example, Yang Hongfei proposed an audio scene recognition system based on sample balancing and transfer component analysis to improve the accuracy of audio scene recognition [45]. Zhu et al. studied a heterogeneous transfer learning method, which uses the tag information on the image as a bridge of knowledge transfer between the text and the image, so as to improve the classification effect on the image data [46].

With the continuous mining of the value of transfer learning, relevant scholars have also explored the application of transfer learning in the field of mechanical equipment predictive maintenance. Current research primarily focuses on transfer diagnosis and prediction for equipment under varying and complex operating conditions, as well as transfer diagnosis and prediction under small sample conditions. The main research fields are concentrated in the petroleum, chemical, electric power, aerospace and machinery manufacturing industries.

Large key mechanical equipment such as gearboxes and reciprocating compressors usually have complex structures, and often operate in harsh environments and variable working conditions, which lead to variable fault characteristics and difficulty in obtaining effective features. The samples no longer meet the independent and identical distribution, and the intelligent diagnosis model is difficult to build and has poor universality. Component failure is uncertain, and fault prediction and early warning are difficult. Therefore, the fault diagnosis and prediction of equipment under variable working conditions have become a research hotspot in the field of predictive maintenance. At present, the research on fault diagnosis and prediction of mechanical equipment under variable working conditions mainly includes three types of methods.

1. Pre-filtering technology, including discrete wavelet transform, kalman filter, adaptive kernel optimization method, etc., have the advantage of eliminating signal aliasing and interference under unsteady conditions, and extracting signals that are not affected by working conditions such as speed; The disadvantages are that it is difficult to extract the effective components of the signal, and the model optimization test is difficult.
2. Resampling techniques, including order ratio analysis, angular domain resampling, time domain resampling and frequency domain resampling methods, have the advantage of stable signal in angular domain (such as special domain); The disadvantage is that it is difficult to obtain the shaft rotation frequency under unknown faults, and resampling is not possible.
3. Data normalization technology, including regression difference analysis, subspace identification and other methods, has the advantage of eliminating environmental changes and working condition changes; The disadvantages are time-consuming and inefficient and difficult to determine parameters.

The diagnostic model established by the above methods for complex and variable working conditions still have the problems of requiring data to meet the requirements of independent and identical distribution, low accuracy and poor generalization ability. Relevant scholars have applied transfer learning to the fault diagnosis and prediction of equipment under variable working conditions, which greatly improves the accuracy and generalization ability of the model. Kang et al. addressed the challenge of low diagnostic accuracy in rolling bearings due to the difficulty or inability to obtain large amounts of labeled vibration data under variable operating conditions. They proposed a rolling bearing fault prediction method based on the combination of variational mode decomposition, multi-feature construction, and semi-supervised transfer component analysis [47]. Chen et al. addressed the challenge of target diagnostic data being inaccessible and the differing distribution characteristics between training and testing data due to complex operating conditions. They proposed a method combining quantitative analysis of transfer components with an improved LSSVM to tackle these issues [48]. Duan et al. addressed the issues of low reusability of gearbox monitoring data under multiple operating conditions, significant impact from complex conditions, and frequent failure of trained models. They proposed a transfer component analysis method based on auxiliary datasets from different operating conditions to improve equipment fault prediction [49]. Shen et al. have proposed a method for motor fault prediction under varying speed and load conditions. This approach combines feature extraction using singular value decomposition of the autocorrelation matrix with a transfer learning classifier. The method aims to enhance the accuracy of fault prediction in such dynamic operating environments [50].

Different data distributions are divided into two categories: different feature space distributions and different probability distributions. The aforementioned approaches leverage transfer learning algorithms to reduce discrepancies in data distribution by mapping samples from the source and target domains into a common feature space. Additionally, they introduce auxiliary data with different but related distribution characteristics into the target data. These methods, combined with iterative optimization, aim to improve the model's accuracy and generalization capabilities. In addition, transfer learning is often combined with data filtering and feature extraction algorithms to optimize the model.

The small sample size of fault data is one of the core difficulties in the field of predictive maintenance. In the field of predictive maintenance, the number of fault samples is scarce because the equipment is in fault state for less time.

When engineers prepare to collect fault sample data, the equipment may already be damaged and unable to operate normally. This results in a significantly lower number of collected fault samples compared to normal samples. Additionally, structural differences and varying operating conditions between similar and cross-category equipment make it difficult to apply fault samples from one piece of equipment to another. This challenge contributes to the scarcity of fault data within the industry. The research of relevant scholars on imbalanced data sets is essentially a small sample problem, which brings great challenges to the improvement of model accuracy and poor generalization.

Related scholars have introduced transfer learning to equipment diagnosis and prediction under the condition of small samples, which provides new ideas for solving problems. Gu et al. proposed an adaptive weighted interpolation-fitted transfer learning algorithm in the field of fault probability prediction for aviation maintenance support. This algorithm adaptively adjusts the proportions of interpolation, fitting, and transfer learning based on data volume and characteristics (such as distribution similarity and information entropy). This approach mitigates the risks associated with data sparsity and reduces the occurrence of negative transfer effects [51]. Given the challenge of missing labeled fault data in chemical industry scenarios, Hu utilized a stacked denoising autoencoder (SDAE) to achieve unsupervised abstract feature representation of chemical process data [52]. Ren et al. addressed the issue of low accuracy in assessing the effectiveness of weapon systems under conditions of high-dimensional noise and small sample data [53]. They proposed a hybrid model based on a stacked denoising autoencoder and support vector regression to improve learning and prediction accuracy on small sample noisy data. Zhuang et al. have proposed an analog circuit fault diagnosis method based on TL-LSSVM, which transfers the source domain data to the target fault training set, and significantly improved the diagnosis accuracy [54]. Yang et al. addressed the need for high computational speed in online fault diagnosis of distribution transformers. They also tackled the issue of insufficient variety in online monitoring data and the limited amount of routine test data for these transformers. To solve these challenges, they developed an online fault identification and diagnosis model for distribution transformers, utilizing parallel big data streams and transfer learning [55]. Zhang et al. addressed the issues of insufficient prediction accuracy and overfitting in the fault diagnosis of aviation engine bearings by proposing a semi-supervised ensemble learning model based on transfer learning [56]. This approach improved accuracy by over 9%.

The small sample condition of fault data, that is, too few training data, will lead to serious overfitting problems in traditional machine learning. Transfer learning can transfer samples with a certain degree of similarity from different target domains into the source domain for model training. Alternatively, it can effectively transfer source domain data related to the target into the target fault training set to assist in building a fault classifier. Additionally, transfer learning can leverage feature space mapping to train semi-supervised or unsupervised models. The use of transfer learning methods can effectively solve the problem of poor classification caused by small sample conditions and generalize the applicability of the model.

10.3.2 Equipment Preventive Maintenance Method Based on Transfer Learning

(1) **Preventive Maintenance Method Based on Transfer Learning Under Cloud/Edge Collaboration**

Collaborative computing is mainly divided into edge-cloud collaboration, edge-edge collaboration, edge-thing collaboration and cloud-edge-thing collaboration. Among the various forms of collaboration, edge-cloud synergy is the most explored by researchers and industry, representing a relatively mature technological stage. It is also the most widely applied collaborative model in fault diagnosis. In edge-cloud collaborative computing, there are three different ways for cloud and edge collaborative computing:

1. **Training-Prediction Edge-Cloud Collaboration**

In this case, the cloud designs and trains an intelligent model based on the data uploaded by the edge, and upgrades the model from time to time. The edge is responsible for collecting data and using the latest model to predict real-time data. This collaborative method is relatively mature and has been applied to many fields such as unmanned driving and video detection. Google's TensorFlow Lite framework is designed for this type of collaboration. It works by training a model with TensorFlow in the cloud and downloading it to the edge, where TensorFlow Lite loads the model and uses optimization techniques to complete and accelerate the prediction task [57, 58].

2. **Cloud-Guided Edge-Cloud Collaboration**

In this case, the cloud is responsible for some of the prediction as well as training the model. Specifically, the neural network model is partitioned, with the cloud handling the computation of the model's front end. The intermediate results are then transmitted to the edge, where the edge continues the prediction tasks to produce the final outcome. The key point of this collaborative method is to find the appropriate cut point and make a tradeoff between the amount of computation and communication. This collaborative method is still in the research stage, and its application in real scenarios is still relatively few.

3. **Edge-Guided Edge-Cloud Collaboration**

In this case, the cloud is only responsible for the initial training, and the model is downloaded to the edge after training. In addition to real-time prediction, the edge will also undertake the task of training, and the training data comes from the data of the edge itself. In this way, the final model can make better use of the locality of data and meet the needs of personalization. This collaborative approach is also in the research stage. The key technologies involved are as follows:

1. Transfer learning. Transfer learning technology has been widely used in traditional edge-cloud collaborative computing [59]. The initial purpose of transfer learning was to save time on manual labeling of samples, allowing the model to transfer from labeled data to unlabeled data, thereby training a model suitable for the target domain. In edge intelligence scenarios, however, it is often necessary to deploy the model, making it applicable to different environments. Taking face recognition applications as an example, different companies generally use the same model for face recognition access control. However, there are large differences between the original data set for training the model and the target datasets set of different companies. Therefore, transfer learning techniques can be employed to retain the original information of the model while incorporating a new training set for learning and updating. This results in a model that is well-suited to a specific edge scenario.
2. Neural network splitting. In the cloud-guided edge-cloud collaboration, the neural network model needs to be divided into two parts, one executed on the cloud, and the other executed on the edge. Therefore, it is necessary to identify an appropriate partition point, keeping computationally intensive tasks in the cloud as much as possible. The partition should be made where communication is minimized, allowing intermediate results to be transmitted to the edge. This approach balances the trade-off between computation and communication. Neurosurgeon [60] proposed by Kang et al. in 2017 is one of the representative studies. A regression-based method is employed to estimate the latency of each layer in the DNN (Deep Neural Network) model. Subsequently, the optimal partition point is determined to achieve desired latency or energy consumption objectives.
3. Neural network compression. In edge-guided edge-cloud collaboration, the computing power of the edge is relatively limited. In order to achieve effective training at the edge, reducing computing power is a key requirement. At present, many compression techniques for neural networks are used to solve this problem. In general, reducing insensitive parameters by means of parameter sharing and clipping can reduce storage and communication overhead, reduce network size, and reduce computing power requirements. Weights quantification and other work can also reduce the amount of network computation [61].

The fault prediction design framework based on transfer learning under cloud/edge collaboration is shown in Fig. 10.14. The cloud mainly stores the training sample data and trains the pervasive model. The training sample data consists of two parts: one is the public data set for fault diagnosis or data sets of related fields, and the other is the sample data set for device personalized diagnosis collected by the edge end. Subsequently, the ubiquitous fault diagnosis model for equipment is continuously trained and updated using the abundant training samples and computational resources available in the cloud. The resulting generalized training outcomes can serve as intermediate results, supporting various diagnostic scenarios.

The edge end collects the real-time status data of the equipment under personalized working conditions through the data sensing device, and transmits it to the cloud storage. Simultaneously, personalized training samples are generated at the edge based on specific operating conditions. These samples are used to customize the generalized diagnostic model transferred from the cloud, enhancing its applicability

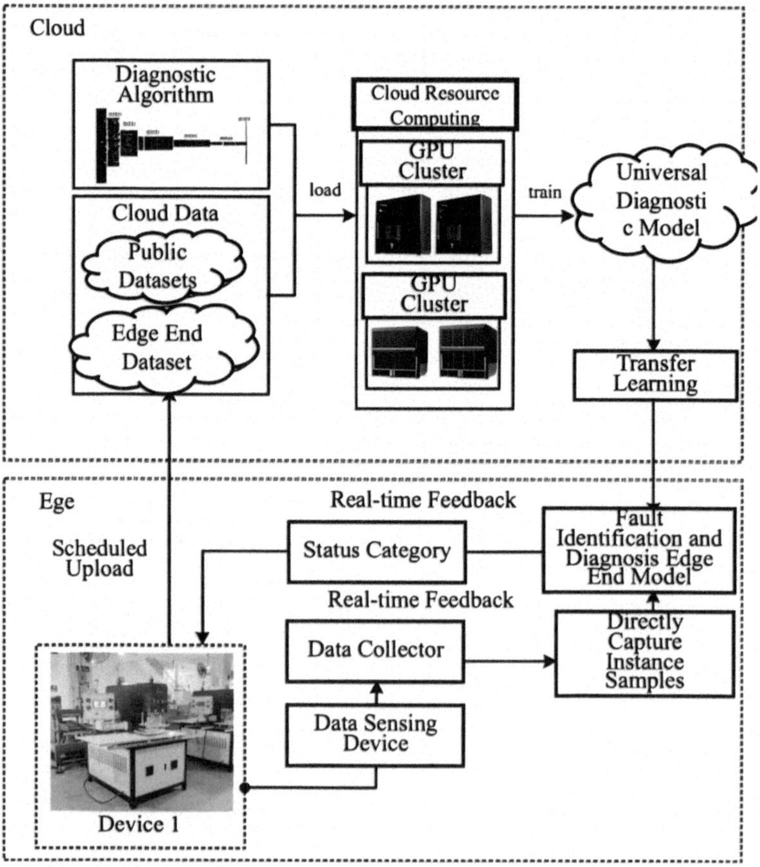

Fig. 10.14 Failure prediction design framework for cloud/edge collaboration

to specific diagnostic tasks. This process results in the development of personalized models for real-time prediction of equipment faults.

It can be seen from the figure that the data interaction between the cloud and the edge in the equipment fault prediction task mainly includes two ways. Firstly, the edge side will regularly upload the personalized samples stored in the local storage to the cloud storage to enrich the training set and validation set of the cloud. (2) As the cloud-based dataset is updated, the general diagnostic model parameters are periodically optimized through training. The optimized model parameters are then transferred to the edge for model refinement.

(2) **Preventive Maintenance Method Based on Digital Twin Model and Transfer Learning Fusion**

A single predictive maintenance approach based solely on either model-driven or data-driven methods fails to account for the consistency of the model lifecycle and the physical characteristics of the equipment within data-driven algorithms. This results in poor model method consistency and inadequate adaptability of the data-driven algorithms, ultimately leading to low prediction accuracy. Additionally, acquiring sensor data for current predictive maintenance of equipment is challenging, and calibrating the sensor data is time-consuming, which can disrupt production schedules and prove impractical. The Digital Twin model enables the injection of various faults and the simulation of sensor data, providing a data foundation for predictive algorithms. This approach addresses the challenges of data acquisition and improves the feasibility of predictions.

The technical roadmap of PHM for digital twin is shown in Fig. 10.15.

Based on the control layer perception and analysis method, and combined with the industrial Internet of things architecture, mainly includes the following (Fig. 10.16).

The diagnostic visual computing capabilities are deployed at the edge or in the cloud. If the diagnosis function is deployed at the edge, the edge side should respond to the update of the cloud model in time to ensure that the edge diagnosis is based on the latest cloud model. Cloud computing continuously trains models or builds a knowledge base using data reported from the edge. The edge then performs online diagnostics based on real-time controller parameters and the cloud model it has downloaded, as illustrated in Table 10.2.

Based on the above methods, the processing fault diagnosis and prediction of the core components are carried out, and the life is analyzed. This analysis is based on historical case records, as well as the simulation and prediction data from the Digital Twin. By combining these data sources with artificial intelligence methods, the root causes of historical faults in core components are thoroughly analyzed. At the same time, the improved processing technology of the core components is generated to ensure the quality improvement of the core components. Service life increases. This is shown in Fig. 10.17.

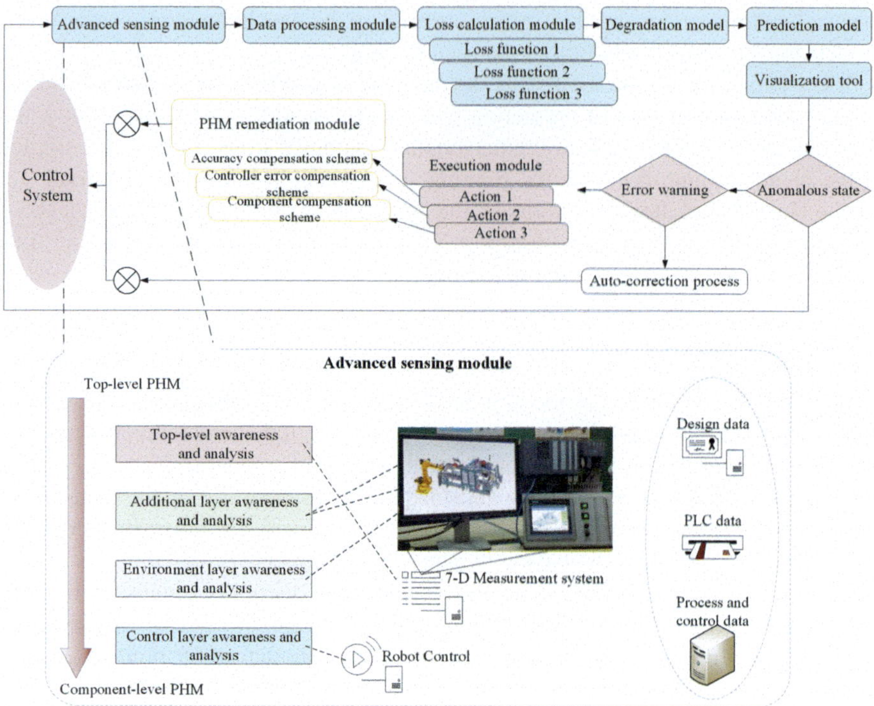

Fig. 10.15 Roadmap for digital twin based PHM

10.3.3 Equipment Preventive Maintenance Case Based on Transfer Learning

(1) Bearing Fault Prediction Case of Cloud/Edge Collaboration

Bearing vibration signals are easy to collect and effectively reflect the condition of the bearing. Therefore, the fault prediction algorithm utilizes these collected vibration signals as inputs to the fault prediction model. The output of the model is the classification of bearing fault types. After training the universal diagnosis model based on sample data in the cloud, the bearing fault prediction model for personalized working conditions is formed at the edge end by means of transfer learning, and the bearing fault is predicted in real time based on the bearing vibration data collected in real time.

The prediction algorithm in the above task scenarios requires comprehensive feature extraction of vibration signals, and the diagnostic model in the cloud needs to serve the personalized edge-side fault prediction task. At the same time, the edge-side needs to have the ability to accept the transfer of the cloud training model. Therefore, the construction of a neural network prediction and recognition model needs to meet the following criteria:

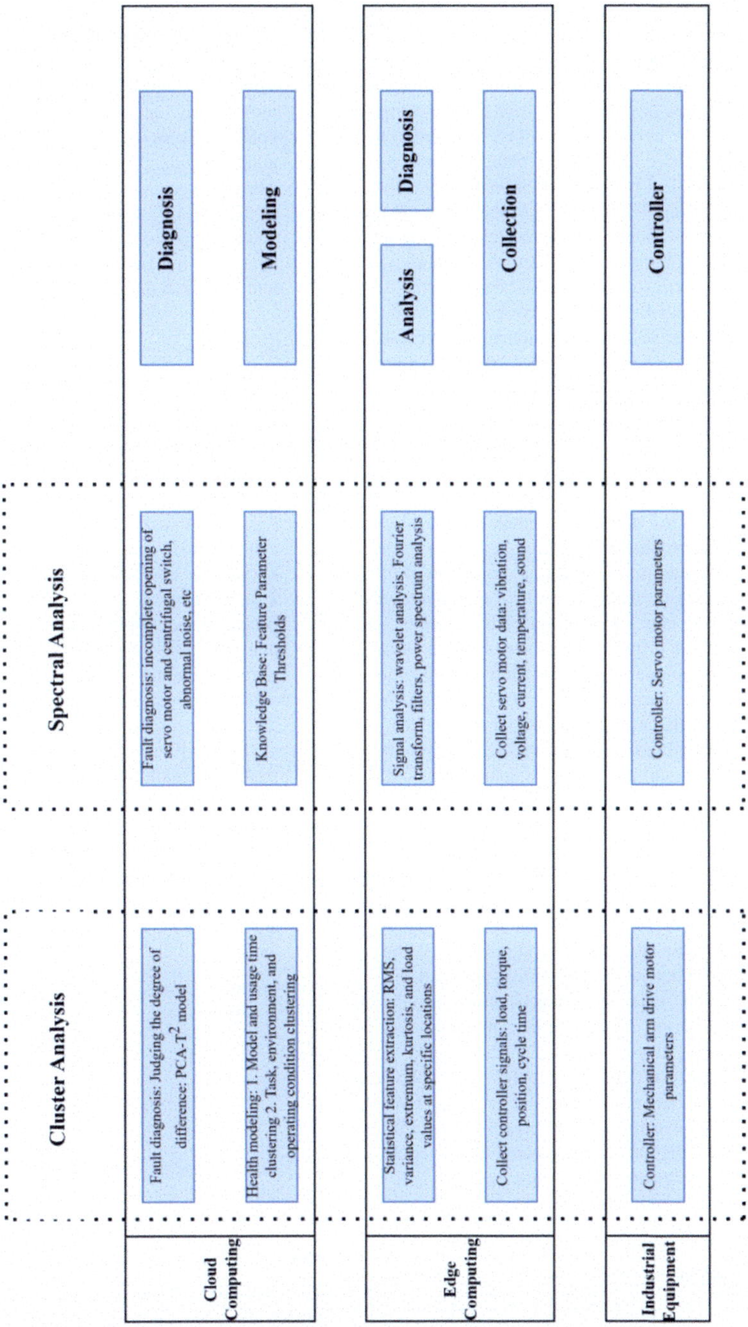

Fig. 10.16 Architecture for data acquisition, analysis, and diagnosis

Table 10.2 Diagnostic method

	Cluster analysis	Spectrum analysis
Edge computing	j collects the control signal When collecting the monitoring parameters in the controller, it is necessary to select the parameters of key parts for collection, such as load, torque, position, cycle time, robot type and other parameters in the drive motor of the manipulator k statistical feature extraction The sampling frequency of the signal obtained from the controller is low, so the feature extraction method for some high-frequency sampling or waveform signals will not be applicable. Instead, fixed statistical characteristics of the signal are extracted according to each action cycle, such as root mean square, variance, extreme value, kurtosis value and load value at a specific position	j collects the servo motor data The servo motor data including vibration, voltage, current, temperature and sound are collected from the controller k signal analysis Through wavelet analysis, Fourier transform, Butterworth Filter and power spectrum analysis View motor real-time and historical data
Cloud computing	j Health modeling Similarity clustering is performed on the devices. When clustering the equipment, the first round of clustering is performed according to the model and time of use of the equipment, followed by the second round of clustering according to the task, environment and operating condition of the equipment. In the analysis of the manipulator, the driving motor torque caused by different action cycles is different, and the maximum value, minimum value and average value of torque are selected here as the basis for clustering. When the manipulator performs similar actions, the distribution of the above features should be very similar, and clustering models such as DBSCAN can be used for automatic recognition k fault diagnosis There are many algorithms to judge the degree of difference, such as PCA-T2 model, Gaussian mixture model, self-organizing map, statistical pattern recognition and so on Using the analysis method of PCA-T2, the similarity degree between the monitoring parameter characteristics of each drive motor and the overall distribution of other devices in the unified cluster can be judged, and the T2 value is used as the final output result. T2 value represents the deviation degree between the current device and the cluster, and its distribution conforms to the characteristics of F-distribution. It is estimated that its control boundary can be determined according to the 90% or 95% confidence interval. When T2 value exceeds the control boundary and continues to increase, it means that the early fault is gradually developed and verified	j knowledge base establishment Select a certain section of the signal analysis data to calculate, set the data channel and fault type, calculate the threshold of characteristic parameters, as the knowledge base of fault type k fault diagnosis Based on the vibration signal analysis, supplemented by temperature, current and sound signals, the maximum amplitude and the corresponding frequency of the maximum amplitude on the motor vibration spectrum, the number of the wave peaks on the motor vibration spectrum, the sum of the amplitude of the wave peaks on the motor vibration spectrum, and the temperature and temperature rise of the motor are selected as fault features

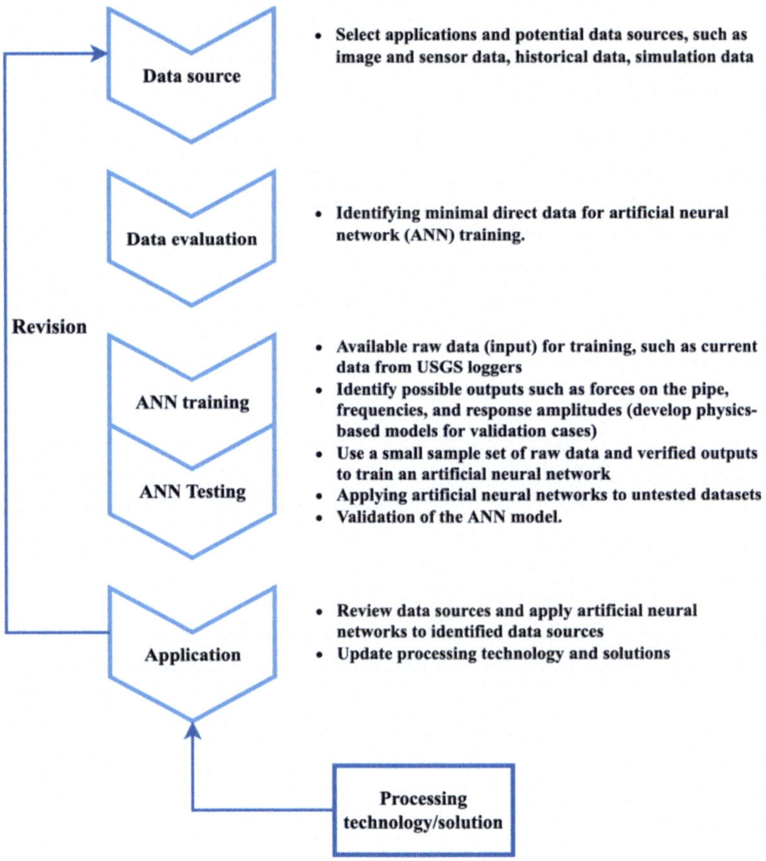

Fig. 10.17 Machining quality improvement strategy for core components

1. A large enough convolution kernel is needed to obtain the periodic transformation characteristics, and a small convolution kernel is needed to obtain the local features;
2. The depth of the neural network should be relatively deep to obtain high-quality feature representation more easily;
3. The problem of over-fitting should be fully considered in the process of constructing neural networks;
4. The overall parameters of the neural network should not be too many, which is convenient to be used in the edge end with insufficient resources;
5. The model framework of the edge end and the cloud end should be the same to meet the requirements of the transfer of the training results from the cloud to the edge end.

The DeepCNN with Wide first-layer Kernel (WDCNN) algorithm is in good agreement with the above design criteria. Based on this, the WDCNN algorithm is improved in the following two points:

1. The discarding method is used to suppress the overfitting of the neural network and improve the accuracy of the model.
2. The depth of the neural network is improved from 5 convolutional layers to 6 convolutional layers. Under the premise of ensuring that the total number of training parameters increases less, the overfitting phenomenon is avoided and the accuracy of fault diagnosis is improved. The improved algorithm structure is shown in Fig. 10.18, and the parameters of each layer are shown in Table 10.3. Figure 10.19 shows the training and updating process of the diagnostic algorithm built in the cloud.

In Fig. 10.19, in the early stage of the task, the public bearing data set is mainly used for universal model training. With the progress of the task, personalized samples generated by the edge end application will be added to the database for continuous update training. During training, the training samples are proportionally divided into a training set and a validation set. With each training iteration, the model parameters are updated through backpropagation. This approach leverages the extensive computational resources of the cloud to provide continuous and rapid updates of general model parameters for different task scenarios throughout the entire diagnostic service process.

The primary task at the edge is to perform real-time prediction of bearing faults. The algorithm follows the basic framework of the cloud-based algorithm, enabling the general model parameters from the cloud to be transferred to the edge through

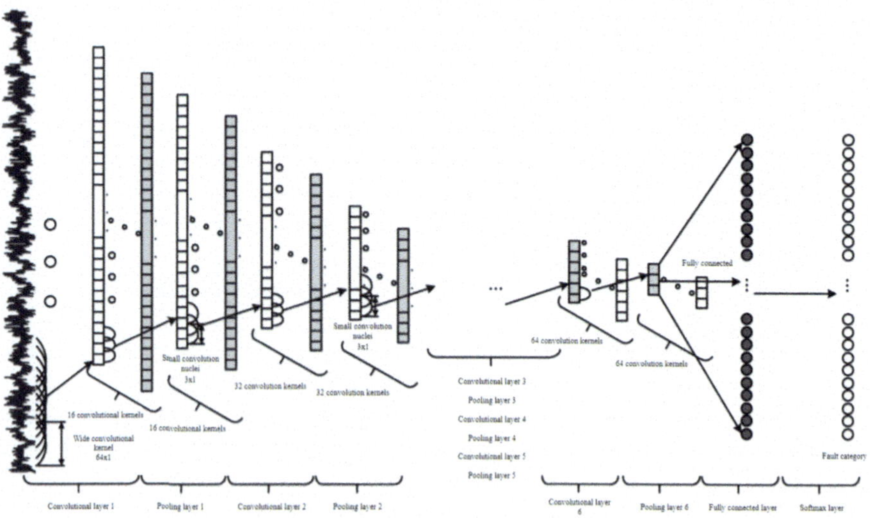

Fig. 10.18 Algorithm structure diagram

Table 10.3 Parameters of the improved CNN diagnostic model

ID number	Network layer	Convolution kernel size/stride	Number of convolution kernels	Padding
1	Convolution 1	64 × 1/8 × 1	16	Yes
2	Pooling 1	2 × 1/2 × 1	16	No
3	Convolution 2	3 × 1/1 × 1	32	Yes
4	Pooling 2	2 × 1/2 × 1	32	No
5	Convolution 3	3 × 1/1 × 1	64	Yes
6	Pooling 3	2 × 1/2 × 1	64	No
7	Convolution 4	3 × 1/1 × 1	64	Yes
8	Pooling 4	2 × 1/2 × 1	64	No
9	Convolution 5	3 × 1/1 × 1	64	Yes
10	Pooling 5	2 × 1/2 × 1	64	No
11	Convolution 6	3 × 1/1 × 1	64	No
12	Pooling 6	2 × 1/2 × 1	64	No
13	Fully Connected Layers	101	1	No

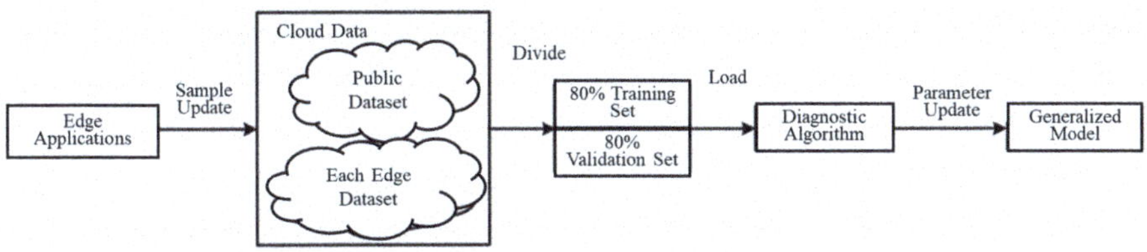

Fig. 10.19 Cloud prediction model training and updating process

transfer learning. But at the same time, it needs to be adjusted according to the characteristics of different edges. The main adjustment strategies are as follows:

1. The input data length is adjusted according to the different vibration signal sampling frequency of the edge end.
2. Select the matching neural network configuration according to the computing resources and local samples at the edge.

The update and application of the edge-side algorithm are shown in Fig. 10.20.

In Fig. 10.20, the prediction model initialized at the edge is obtained after parameter transfer from the cloud. Local samples are used to personalize the diagnosis model on the edge controller, and finally a personalized prediction model that can be applied to the edge is generated. At this stage, real-time vibration signals are collected and applied to the personalized prediction model, which generates predictive results in real-time. Appropriate responses are then made based on these predictive results.

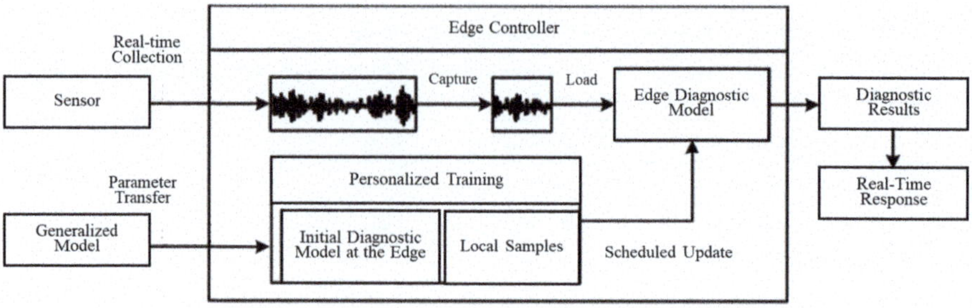

Fig. 10.20 Edge-end prediction algorithm update and prediction process

(2) A Predictive Maintenance Case of CNC Machine Tools Based on Digital Twin and Transfer Learning Fusion

In order to realize the predictive maintenance of model and data fusion, it is first necessary to build a data-driven algorithm model. The construction process of data-driven algorithm for predictive maintenance of CNC machine tools based on Digital Twin is shown in Fig. 10.21. Multi-type sensor data collected from machine tools need intelligent scene perception based on domain knowledge, including data noise reduction, preprocessing, feature extraction and feature selection, etc. Then the effective features obtained by analysis are stored in the database as health indicators of machine tool performance. Using the features obtained from scene perception, the classification algorithm and regression algorithm are constructed and trained respectively to realize fault diagnosis and prediction.

Developing predictive maintenance models based on scene sensing data needs to be based on correctly identifying the health status indicators of the system and building models to analyze and process these indicators. Identifying health condition

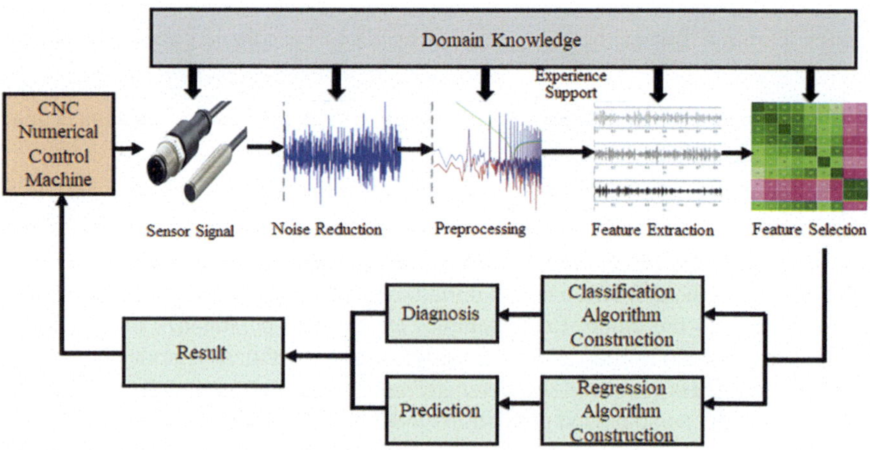

Fig. 10.21 Building a data-driven algorithm with digital twin

indicators for machine tools and training fault diagnosis and prediction models is an iterative process. It requires continuous refinement of the condition indicators and model parameters until the prediction or diagnosis results meet the expected outcomes.

There are many classification and regression algorithms for fault diagnosis and fault prediction, such as support vector machine, nearest neighbor, decision tree, random forest and Convolutional Neural Networks (CNN). Regression algorithms include linear regression, decision tree regression, random forest regression, and Long Short Term Memory (LSTM). Specifically, it can be selected according to the characteristics of predictive maintenance of machine tools.

As shown in the Fig. 10.22, it is difficult to obtain the fault and performance degradation data of CNC machine tools, and it is difficult to conduct data calibration in experiments. By obtaining fault and performance degradation data of CNC machine tools through Digital Twin simulation, a predictive maintenance approach can be achieved by integrating the Digital Twin model with data based on transfer learning.

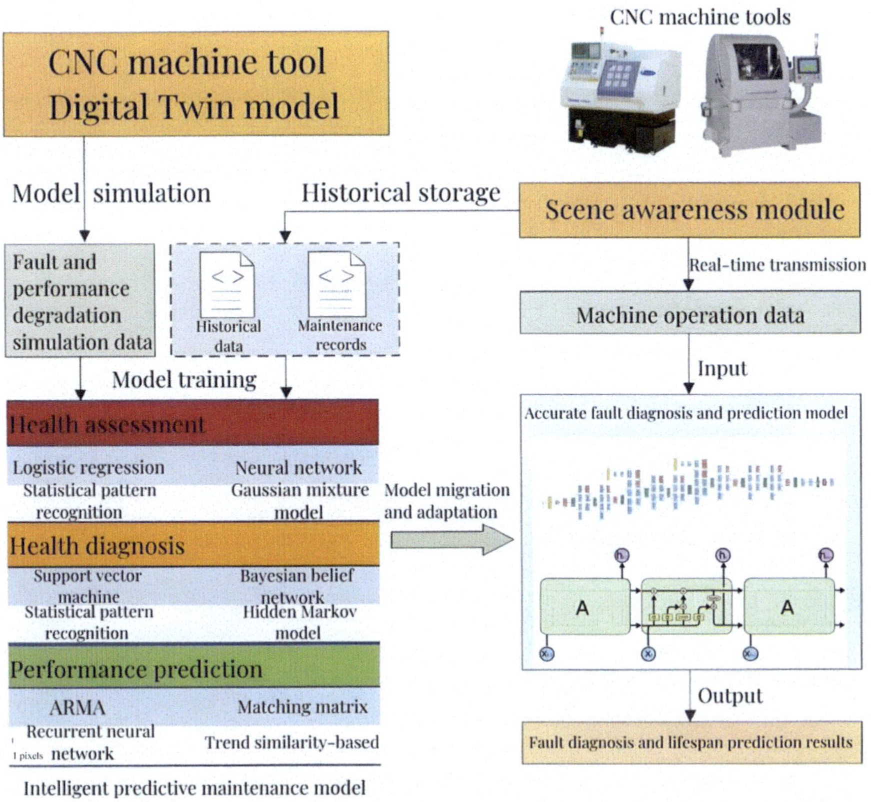

Fig. 10.22 Predictive maintenance of CNC machine tools based on fusion of digital twin and transfer learning

Firstly, the data generated by digital twin is used to train the predictive maintenance model, and then the trained model is transferred to the actual application scenario of CNC machine tools. Since the digital twin and the CNC machine tool can maintain consistency, the model trained by the digital twin data has the same model structure and parameters as the model trained by the actual machine tool sensor data. Therefore, the model can be smoothly transferred to the actual production and processing environment. The predictive maintenance of Digital Twin model and data fusion is realized.

The trained predictive maintenance model is transferred to the field equipment layer through transfer learning to realize the adaptation of the model. The real-time operation data from scene perception is fed into a transfer learning-based predictive maintenance model to locate faults or predict the machine tool's remaining useful life. It can solve the problem of insufficient labeled data of machine tools and insufficient computational resources for training.

References

1. Shen Y, Li L, Wang Z (2020) A review of fault diagnosis and fault—tolerant control techniques for spacecraft. J Astronaut 41(6):647–656
2. Li G, Qi Y, Li Y, Zhang J, Zhang L (2021) Research progress on fault diagnosis and state prediction of wind turbine. Automation of Electric Power Syst 45(04):180–191
3. Yuan L, Wang S (2021) A review on development of intelligent health management technology for spacecraft control systems. Acta Aeronautica et Astronautica Sinica 1–16
4. Frank PM (1990) Fault diagnosis in dynamic systems using analytical and knowledge-based redundancy a survey and some new results. Automatica 26(3):459–474
5. Patton RJ, Frank PM, Clark RN (2000) Issues of fault diagnosis for dynamic systems. Springer, London, pp 1–12
6. He J, Luo G, Cheng M, Liu Y, Tan Y, Li M (2020) A research review on application of artificial intelligence in power system fault analysis and location. Proc CSEE 40(17):5506–5516
7. Mirzaei M, Vahidi B, Hosseinian SH (2019) Accurate fault location and faulted section determination based on deep learning for a parallel-compensated three-terminal transmission line. IET Gener Transm Distrib 13(13):2770–2778
8. Huang X, Hu X, Zhu Y et al (2018) Pattern recognition of unknown PD types based on KELM-VPMCD. Electric Power Autom Equip 38(5):136–140, 147
9. Wang X (2019) Wind turbine fault prognostics based on deep learning. Chongqing Technology and Business University, Chongqing
10. Chen B, Liu H, Xu C et al (2019) Gear fault diagnosis based on DBNS. China Mech Eng 30(2):205–211
11. Shi X, Zhu Y, Sa C et al (2016) Power transformer fault classifying model based on deep belief network. Power Syst Protect Control 44(1):71–76
12. Luo G, Yao C, Liu Y et al (2018) Entropy SVM-based recognition of transient surges in HVDC transmissions. Entropy 20(6):421
13. Liu X, Zhang S, Xiao J, Fan L, Ma J, Zhong Y (2021) Intelligent fault diagnosis of medical equipment based on long short term memory network[J/OL]. J Biomed Eng 1–8
14. Shen T, Li S (2021) CNN-LSTM method with batch normalization for rolling bearing fault diagnosis. Comput Integr Manuf Syst 1–16
15. Feng L, Chen Y, Tang B, Wang J (2020) Reinforcement learning long and short time memory neural network for state prediction. J Vibr Measure Diagn 40(05):895–903+1021–1022

16. Kang S, Liu Z, Wang Y, Wang Q, Lan C (2021) A fault diagnosis method of rolling bearing based on the improved DQN network. Chin J Sci Instrum 1–13
17. Li X, Li B, Cai Y, Park J, Zhou Y, Li J, Chen S, Li H (2019) Artificial intelligence aids in the transformation and upgrading of smart power plants. Electric Power Equip Manage 12:23–24
18. Liang G, Yaguo L, Saibo X et al (2019) Deep convolutional transfer learning network: a new method for intelligent fault diagnosis of machines with unlabeled data. IEEE Trans Industr Electron 66(9):7316–7325
19. Wang Y (2019) Research on electric power data mining technology based on deep learning and transfer learning. Zhejiang University, Hangzhou
20. Su C, Li L (2021) Optimization of non-equal periodic preventive maintenance based on hidden semi-Markov degradation model. J Southeast Univ (Nat Sci Ed) 51(02):342–349
21. Wang H, Xiong L, Du W, Jiang Z (2016) Optimization of preventive maintenance for component of electric multiple unit with consideration of failure risk. J Mech Eng 52(24):123–129
22. Zhang J, Peng B, Liu X, Li G, Liu H (2021) Study on dynamic adjustment method of preventive maintenance strategy for affected equipment under long-term temporary outage of nuclear power units. Nuclear Power Eng 42(01):172–176
23. Liu J, Wei X, Ye J, Zeng B, Li M (2020) Preventive group maintenance strategy for in-service agricultural machinery equipment. J Agric Mach 51(S2):316–322+448
24. Liu W, Tao F, Cheng J, Zhang L, Yi W (2020) Digital twin satellite: concept, key technologies and applications. Comput Integr Manuf Syst 26(03):565–588
25. Glaessgen E, Stargel D (2012) The digital twin paradigm for future NASA and U.S. air force vehicles. In: Proceedings of the 53rd AIAA/ASME/ASCE/AHS/ASC structures, structural dynamics and materials conference. Rseton, Va, USA. AIAA, pp 7274–7260
26. Wu C, Feng F, Wu S et al (2019) Review of application research of deep learning in fault diagnosis of rotating machinery. Noise Vibr Control 39(5):1–7.
27. Lecun Y et al (1998) Gradient-based learning applied to document recognition. Proc IEEE 86(11):2278–2324
28. Janssens O, Slavkovikj V, Vervisch B, Stockman K, Loccufier M, Verstockt S et al (2016) Convolutional neural network based fault detection for rotating machinery. J Sound Vibr 377:331–345
29. Ince T, Kiranyaz S , Eren L et al (2016) Real-time motor fault detection by 1D convolutional neural networks. IEEE Trans Indus Electron 63(11)
30. Li Z. Research on key technologies of data security in fog computing environment. University of Science and Technology Beijing
31. Hu P, Dhelim S, Ning H, Qiu T (2017) Survey on fog computing: architecture, key technologies, applications and open issues. Netw Comput Appl 98:27–42
32. Wan J, Chen B, Wang S, Xia M, Li D, Liu C (2018) Fog computing for energy-aware load balancing and scheduling in smart factory. IEEE Trans Ind Inform (accepted and on-line early access).
33. Wu D, Liu S, Zhang L, Terpenny J, Gao R, Kurfess T et al (2017) A fog computing-based framework for process monitoring and prognosis in cyber-manufacturing. J Manuf Syst 43:25–34
34. O'Donovan P, Gallagher C, Bruton K, O'Sullivan D (2018) A fog computing industrial cyber-physical system for embedded low-latency machine learning Industry 4.0 applications. Manuf Lett 15:139–142
35. Liang YC, Li WD, Lu X et al (2019) Fog computing and convolutional neural network enabled prognosis for machining process optimization. J Manuf Syst 52:32–42
36. Chen Z, Li W (2017) Multi-sensor feature fusion for bearing fault diagnosis using sparse autoencoder and deep belief network. IEEE Trans Instrum Meas 66(7):1693–1702
37. Wen L, Li X, Gao L et al (2018) A new convolutional neural network-based data-driven fault diagnosis method. IEEE Trans Ind Electron 65(7):5990–5998
38. Vaswani A, Shazeer N, Parmar N et al (2017) Attention is all you need. arXiv:1706.03762 [cs.CL], Dec 2017

39. Wang H, Xu J, Yan R et al (2020) Intelligent bearing fault diagnosis using multi-head attention-based CNN. Procedia Manuf 49:112–118
40. Kumar A, Gandhi C, Zhou Y et al (2020) Latest developments in gear defect diagnosis and prognosis: a review. Measurement 158
41. Xiang L, Wang P, Yang X et al (2021) Fault detection of wind turbine based on SCADA data analysis using CNN and LSTM with attention mechanism. Measurement 175(8):109094
42. Predictive Maintenance: The Potential Outbreak Point in the Field of Industrial Digitization. China Ind Rev 2017(11):72–78
43. Wang K, Li Y (2019) Summary of application of transfer learning in predictive maintenance of machinery and equipment. China Instrument 12:64–68
44. Pan J, Yang Q (2010) A survey on transfer learning. IEEE Trans Knowl Data Eng 10:1345–1359
45. Yang H. Audio scene recognition based on sample balancing and transfer component analysis. Harbin Institute of Technology
46. Zhu Y, Chen Y, Lu Z, Pan SJ, Xue GR, Yu Y, Yang Q (2011) Heterogeneous transfer learning for image classification. In: Burgard W, Roth D (eds) Proceedings of the AAAI. AAAI Press, pp 1304–1309
47. Kang S, Hu M, Wang Y et al (2020) Fault diagnosis method of a rolling bearing under variable working conditions based on feature transfer learning. Proc CSEE 39(3):138–146+329
48. Chen C, Shen F, Yan R (2017) Enhanced least squares support vector machine-based transfer learning strategy for bearing fault diagnosis. Chin J Sci Instrum 38(1):33–40
49. Duan L, Xie J, Wang K et al (2017) Gearbox diagnosis based on auxiliary monitoring datasets of different working conditions. J Vibr Shock 10
50. Shen F, Chen C, Yan R (2017) Application of SVD and transfer learning strategy on motor fault diagnosis. J Vibr Eng 30(01):118–126
51. Gu T, Guo J, Li Z et al (2018) Airborne equipment fault probability prediction based on interpolation-fiting-transfer learning algorithm. Syst Eng Electron 40(1):114–118
52. Hu Z. Research on chemical engineering fault diagnosis methods based on deep learning. Hangzhou Dianzi University
53. Ren J, Hu X, Zhu F (2017) Effectiveness prediction of weapon equipment system-of-systems based on dep learning feature transfer. Syst Eng Electron 12:114–118
54. Zhuang C, Yi H, Zhang J, Liu S (2019) Analog circuit fault diagnosis using transfer learning and LSSVM. Chin J Electron Dev 42(03):668–673
55. Yang Z, Zhou R, Shen Y, Yang F, Lei Y, Yan F (2019) On-line fault identify and diagnosis model of distribution transformer based on parallel big data stream and transfer learning. High Voltage Eng 45(06):1697–1706
56. Zhang Z, Liu J, Huang L, Zhang X (2019) A bearing fault diagnosis method based on semi-supervised and transfer learning. J Beijing Univ Aeronaut Astronaut 1–15
57. Zhou Z, Chen X, Li E, Zeng L, Luo K, Zhang J (2019) Edge intelligence: paving the last mile of artificial intelligence with edge computing. Proc IEEE 107(8):1738–1762
58. Chen J, Ran X (2019) Deep learning with edge computing: a review. Proc IEEE 107(8):1655–1674
59. Pan SJ, Yang Q (2009) A survey on transfer learning. IEEE Trans Knowl Data Eng 22(10):1345–1359
60. Kang Y, Hauswald J, Gao C et al (2017) Neurosurgeon: collaborative intelligence between the cloud and mobile edge. ACM SIGARCH Comput Architect News AC 45(1):615–629
61. Han S, Mao H, Dally WJ (2015) Deep compression: compressing deep neural networks with pruning, trained quantization and Huffman coding. arXivpreprint. arXiv:1510.00149

Chapter 11
Digital Twin and Industrial Intelligence

With the introduction of Germany's Industry 4.0 and the United States' Industrial Internet strategy, manufacturing enterprises in various countries have gradually been developing towards digitalization. Achieving digital transformation is not only the inevitable path for the future development of China's manufacturing industry but also an important way to overtake industrial powers. In recent years, digital twin technology, as a key enabling technology to solve the difficulty of information-physical integration in intelligent manufacturing and to practice the concept and goals of intelligent manufacturing, has been deeply researched and promoted. The continuous deepening of the integration of digital twin technology and the national manufacturing industry has strongly promoted the digitalization, networking, and intelligent development process of the manufacturing industry, becoming a powerful driving force for the development and transformation of China's manufacturing industry.

11.1 Digital Twin Progress and Current Status

This section firstly introduces the development history of digital twins, secondly introduces digital twin related research work, and finally compares and analyzes typical definitions and reference models of digital twins.

11.1.1 History of Digital Twin Development

The concept of "twins" in the manufacturing field can be traced back to the Apollo project of the National Aeronautics and Space Administration (NASA) in the United States [1]. In that project, NASA needed to manufacture two identical spacecraft, with the one remaining on Earth called the twin, used to reflect (or mirror) the state

T. Liu et al., *Industrial Intelligence: Methods and Applications*, Springer Series in Advanced Manufacturing, https://doi.org/10.1007/978-3-031-81477-8_11

of the spacecraft performing the mission. During flight preparation, the spacecraft called the twin was widely used for training. During the mission execution, simulation experiments were carried out in the precise space-like model on the ground using this twin, reflecting and predicting the state of the spacecraft performing the mission as accurately as possible, thus assisting the astronauts in orbit to make the most correct decisions in emergency situations.

In 2003, Professor Michael Grieves proposed the concept of "virtual digital expression equivalent to physical products" in the Product Lifecycle Management course at the University of Michigan and gave a definition: a digital replica of one or a set of specific devices that can abstractly express real devices and can be used as a basis for testing under real or simulated conditions, as shown in Fig. 11.1. However, when this concept was proposed in 2003, it did not attract the attention of scholars at home and abroad, mainly because: 1. at that time, the technical means for collecting product-related information in the production process were limited; 2. the digital description of physical products was not yet mature; 3. the computer performance and algorithms at that time were difficult to achieve real-time processing of big data.

After 2011, digital twin ushered in new development opportunities. In 2011, the digital twin was proposed by the U.S. Air Force Research Laboratory and further developed to solve the maintenance and life prediction problems of aircraft under future complex service environments [3]. They plan to deliver a new model of spacecraft in 2025 along with a digital model corresponding to the physical product, i.e., a digital twin, which has super-realism in two aspects: 1. it contains all geometric data; 2. it contains all material data. In 2012, the U.S. Air Force Research Laboratory proposed the concept of "airframe digital twin" [4]: the airframe digital twin, as a

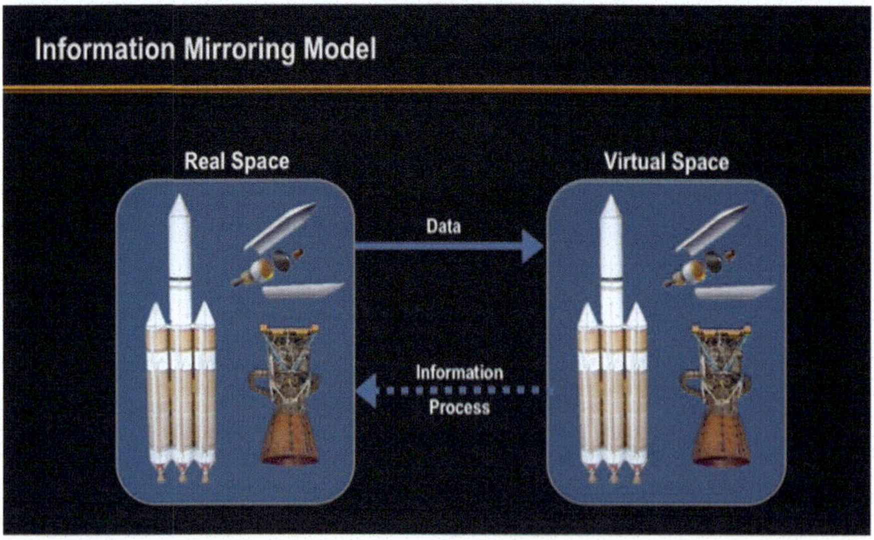

Fig. 11.1 Three-dimensional reference model of digital twin [2]

super-realistic model of the airframe being manufactured and maintained, can be used to simulate and determine whether the airframe meets mission conditions. At this point, digital twin has moved from the conceptual model stage to the initial planning and implementation stage, and the description and research on its connotation and properties have become more in-depth, specifically manifested in: 1. highlighting the hierarchy and integration of digital twin; 2. highlighting the super-realism of digital twin; 3. highlighting the universality of digital twin; 4. highlighting the consistency of digital twin throughout the product lifecycle; 5. highlighting the computability of digital twin.

Facing the future requirements of lightweight, high-load aircraft and longer service life in more extreme environments, NASA and the U.S. Air Force Research Laboratory collaborated and jointly proposed the digital twin paradigm for future aircraft. For aircraft, flight systems, or launch vehicles, they define a digital twin as an integrated multi-physics, multi-scale, probabilistic simulation model that utilizes the best available physical models, updated sensor data, and historical data to reflect the state of the corresponding physical entity [5]. In the same year, the digital twin was officially brought into public view in NASA's released roadmap for "Modeling, Simulation, Information Technology and Processing" [6]. In 2015, General Electric planned to implement real-time monitoring, timely inspection, and predictive maintenance of engines based on digital twins and through its self-built cloud service platform Predix, using advanced technologies such as big data and the Internet of Things [7]. In 2017, to achieve the interactive integration of the physical and information worlds in manufacturing workshops, Fei et al. [8] proposed an implementation model for digital twin workshops, clarifying their system composition, operating mechanisms, characteristics, and key technologies, providing theoretical and methodological references for the realization of cyber-physical systems in manufacturing workshops. From 2018 to the present, digital twin technology has achieved rapid development in both theoretical and application levels, with its application scope covering the entire lifecycle from product design [9] to product manufacturing [10, 11] and operation and maintenance services [12].

11.1.2 Related Works

The author has undertaken the task of developing a lunar rover test platform since 2005, and is responsible for the simulation of the virtual lunar rover, including kinematics, dynamics and control system, real-time communication with the ground test prototype, and synchronized display of the data of the simulation and control on a large-scale virtual reality platform, as shown in Fig. 11.2, which is also the prototype of the digital twin.

Due to the limitations of information technology conditions, the digital representation of real physical products is a novel but immature concept, the authors in the early stage of the research (1999), known as virtual products, virtual prototypes, digital prototypes, virtual functional prototypes, etc. Professor Yan Junqi and other

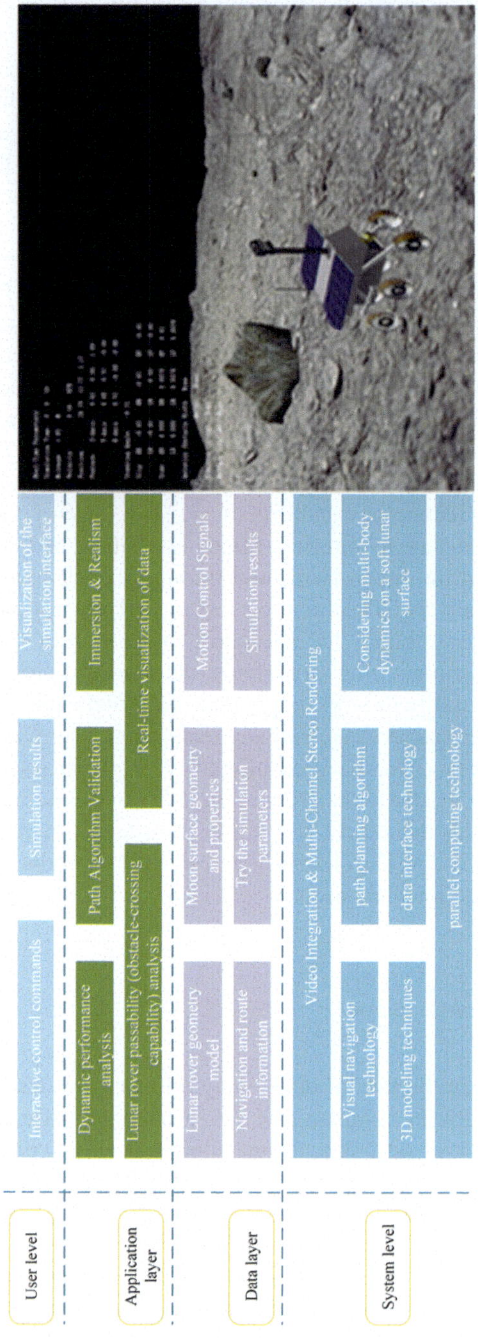

Fig. 11.2 China's lunar rover mobile subsystem virtual testbed

key projects undertaken by the NSFC, "Theoretical and technological foundations of virtual manufacturing ", the first proposed environment model. It is pointed out that the functions of digital prototypes are demonstrated in different working condition environments, which is much the same as the current mapping model of digital twin. It was not until 2011, when NASA and the U.S. Air Force Research Laboratory applied digital twin technology to the health maintenance and life prediction of space vehicles, that digital twins gradually attracted extensive attention from domestic and foreign academics and industries, while the conceptual model of digital twins has been continuously developed and improved.

11.1.3 Digital Twin Definition and Reference Model

In recent years, scholars at home and abroad have carried out in-depth research on the concept, connotation, practice and application of digital twins, however, there is still a lack of accurate and unified definitions and descriptions of digital twins, and the main definitions of digital twins are shown in Table 11.1.

As a result, there are two interpretations of "digital twin": one defines digital twin as a technology or method; the other defines digital twin as a virtual representation, data object or model.

Table 11.1 Definition and description of digital twins

Define	Descriptive	Source (of information etc.)
Integration of multi-disciplinary, multi-physical quantity, multi-scale, and multi-probability attribute simulation processes by fully utilizing heterogeneous data from multiple sources	Process approach	NASA
A digital replica of physical assets, processes and systems that can be used for different purposes	Model object	GE
Virtual representations of physical products or processes that are used to analyze and predict the performance characteristics of physical objects	Virtual representation data object	SIEMENS
Virtualized, digital representation of physical entities or systems	Virtual expression	IBM
Creating virtual models of physical entities digitally and simulating the behavior of physical entities in real environments with the help of data	Technical approach	Literature "Exploring digital twins and their applications"

Digital twins can be regarded as a technique for realizing CPS, and there are several popular international digital twin reference models that are worth introducing here.

(1) **GE Digital Twin**

GE was an early proponent of the digital twin, based on the Predix system for data analytics, and at one point became the face of the Industrial Internet. Figure 11.3 shows the GE digital twin power plant, which features a data analytics engine as the core of the digital twin.

(2) **Siemens Digital Twin Reference Model**

Figure 11.4 shows the Siemens spiral structure digital twin reference model, the perspective is placed on the whole life cycle, by building a digital twin model of the production system that integrates the manufacturing process, it can realize the digitalization of the whole process from product design, production planning to manufacturing execution, which is divided into 1. Digital twin-based product development, 2. Digital twin-based manufacturing, 3. Digital twin-based product performance testing and operation and maintenance through the collaboration platform Teamcenter for data integration.

(3) **Deloitte Digital Twin Reference Model**

The digital twin architecture presents the process of moving from the physical world to the digital world, and from the digital world back to the physical world. This "physical-digital-physical" process or cycle forms the basis of Deloitte's Industry 4.0 research, which broadly describes a digital manufacturing environment in which advanced production technologies are combined with the Internet of Things to enable manufacturing companies to be interconnected while at the same time transmitting and analyzing activities and using the information to take smarter practical actions. Figure 11.5 shows Deloitte's digital twin hexagonal model, with two main areas of focus in creating a digital twin: designing the digital twin's processes and information requirements for the product lifecycle-from the design of the asset to the asset's on-site use and maintenance in the real world; and creating the enabling technologies-the -integrating real assets and their digital twins to enable the real-time flow of sensor data with operational and transactional information in core enterprise systems.

(4) **Boeing Digital Twin Rhombus Model**

Figure 11.6 shows the Boeing digital twin model.

(5) **SAP**

Figure 11.7 shows the square model proposed by SAP, where each side is a twin interface, Twin-to-Device, Twin-to-SoI, Twin-to-SoR, and Twin-to-Twin, oriented towards the three dimensions of Industry 4.0 to unfold the integration.

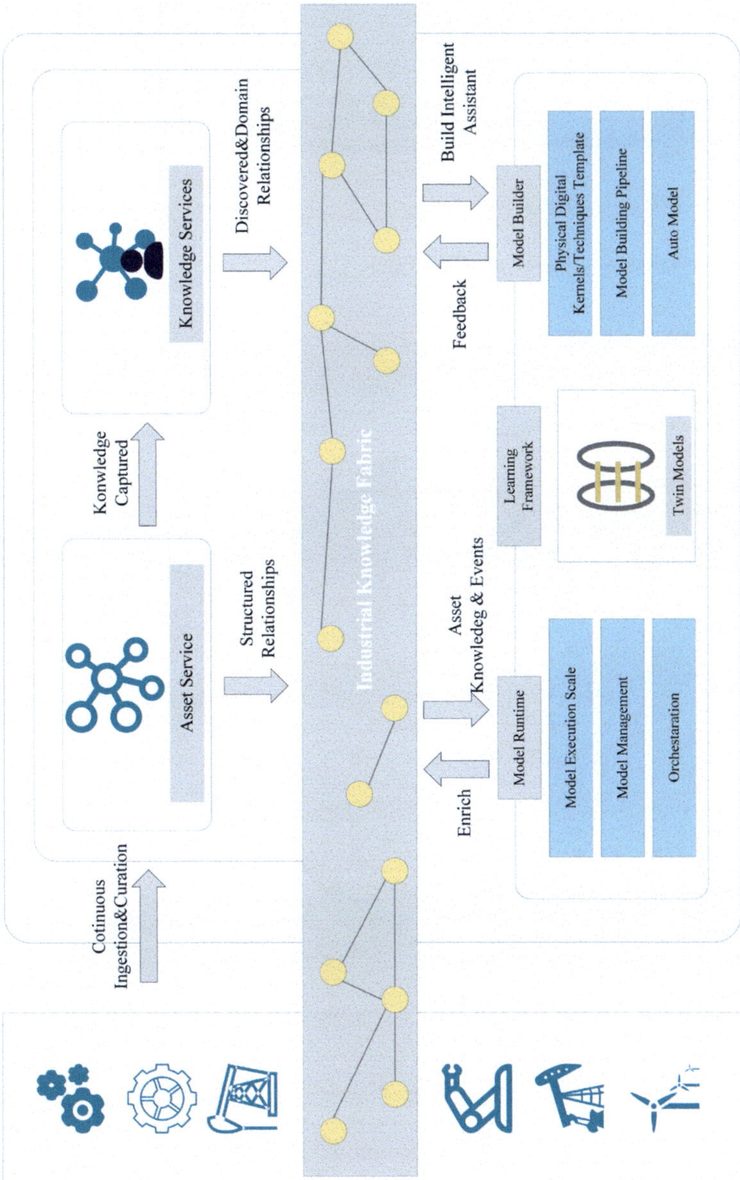

Fig. 11.3 GE digital twin power plant [13]

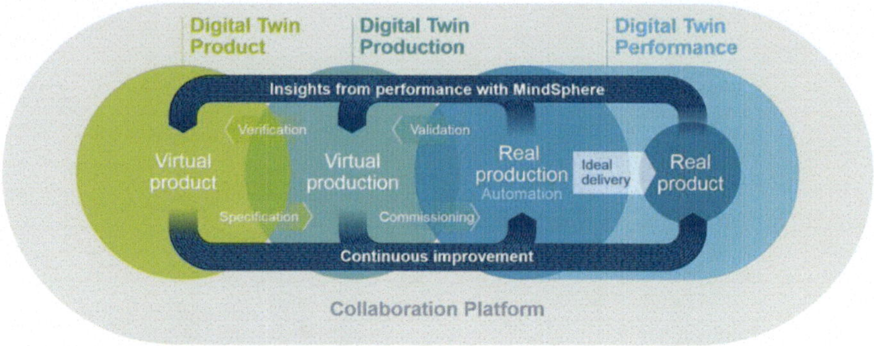

Fig. 11.4 Siemens digital twin reference model [14]

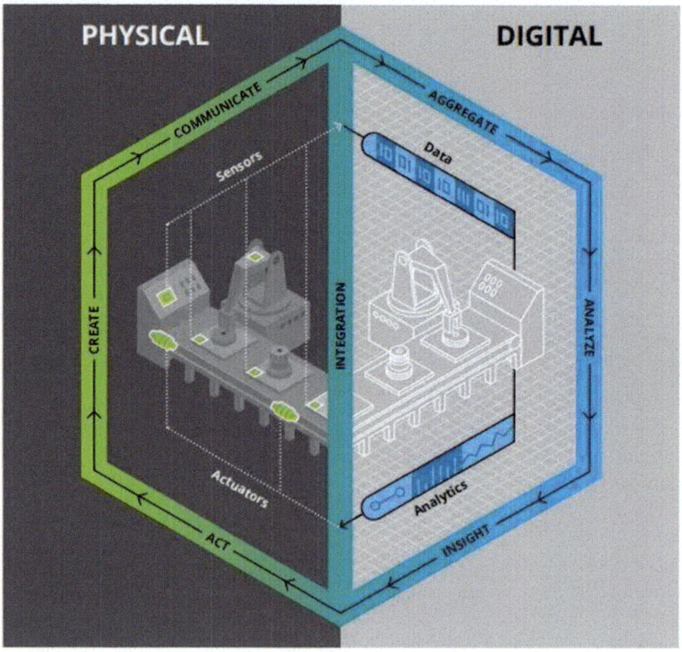

Fig. 11.5 Deloitte digital twin model of production processes [15]

(6) **PTC**

Figure 11.8 shows the Digital Twin reference model proposed by PTC in 2018, and it looks like this is just like a yin and yang bagua diagram. When the product life-cycle management (PLM) process can be extended to the site of product application and back to the next design cycle, a closed-loop product design system closed-loop process is established, and predictive maintenance can be realized before the product

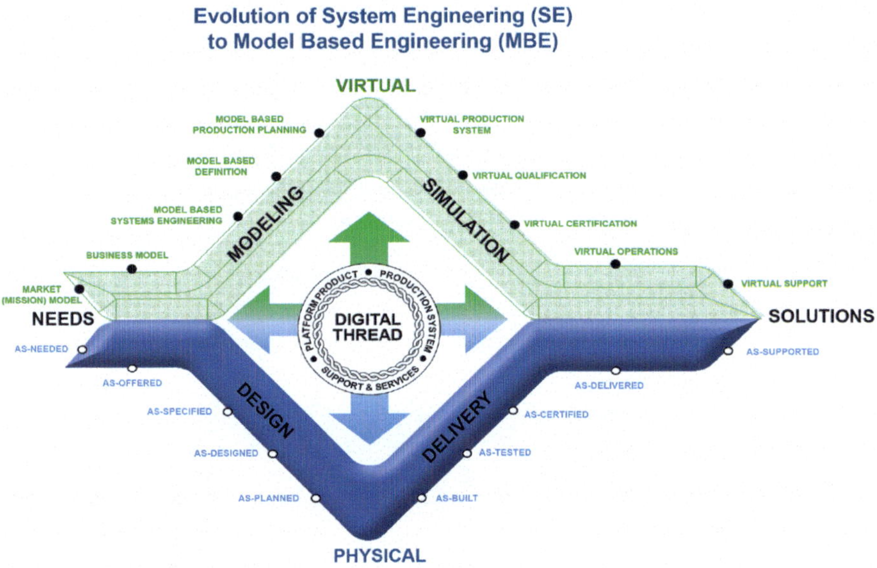

Fig. 11.6 Boeing digital twin modeling [16]

fails. Every action of the smart product is brought back to the designer's desktop, thus realizing real-time feedback, with AR as the main enabling technology.

(7) ANSYS

ANSYS stands for product design analysis and proposes a product digital twin (As Designed) as shown in Fig. 11.9, and integrates with Predix, GE's industrial data and analytics cloud platform, and with Thing Worx®, an IoT platform that creates a gateway between remote sensors and simulation software, lending machine learning and augmented reality to collect from the IoT the critical information and connecting the data to ANSYS analytical models.

(8) Microsoft Corporation

Figure 11.10 shows the Microsoft digital twin framework.

(9) Oracle

As shown in Fig. 11.11, Oracle IoT Cloud delivers digital twins through: Virtual Twin Go beyond simple JSON documents to enumerate observations and expectations through device virtualization Predictive Twin Problem solve through models built using a variety of technologies to simulate the complexity of real products Twin Projections Project the insights generated by digital twins onto your back-end business applications, making IoT an integral part of your business infrastructure.

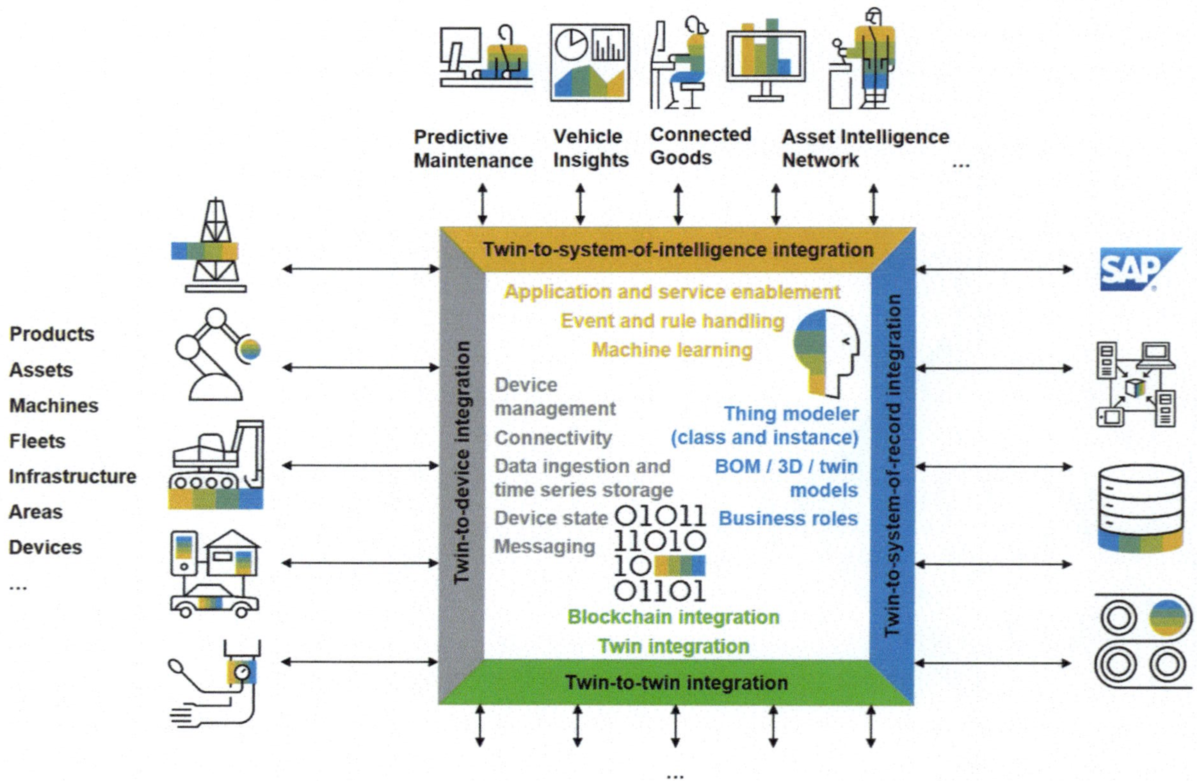

Fig. 11.7 SAP digital twin model [17]

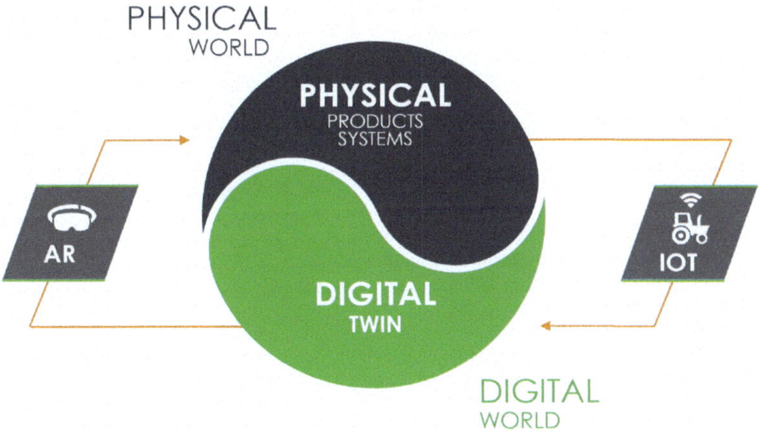

Fig. 11.8 PTC digital twin reference model [18]

(10) **The Fraunhofer Digital Twin Reference Model**

Figure 11.12 show the **Fraunhofer** digital twin reference model.

(11) **Beijing University of Aeronautics and Astronautics**

Figure 11.13 show the five-dimensional model of digital twin proposed by Tao Fei's team.

Overall, all require the creation of a highly complex virtual model that is an exact replica (or twin) of a physical object. In this paper, digital twin is defined as a mapping model of physical objects in virtual space, which is a virtual model expressed digitally, emphasizing the one-to-one correspondence between physical objects and virtual models. And digital twin technology refers to the technology and method to realize the interaction and integration between virtual and real, emphasizing the integration process between physical space and virtual space. The concept of digital twin is realized by using the "soft body" constructed by 3D graphics software to map the objects in reality. This mapping is usually a multi-dimensional dynamic digital mapping that relies on sensors or analog data mounted on the object to provide insight and rendering of the object's real-time state, as well as feeding back data carrying commands to the object that result in a change of state. The digital twin is a bridge between the real world and the digital virtual world. The digital twin embodies the mechanism of software, hardware, and IoT feedback. The data of the running entity is the nutrient delivery line for the digital twin. In turn, much analog or command information can be channeled from the digital twin to the entity for diagnostic or preventive purposes in a two-way evolutionary process.

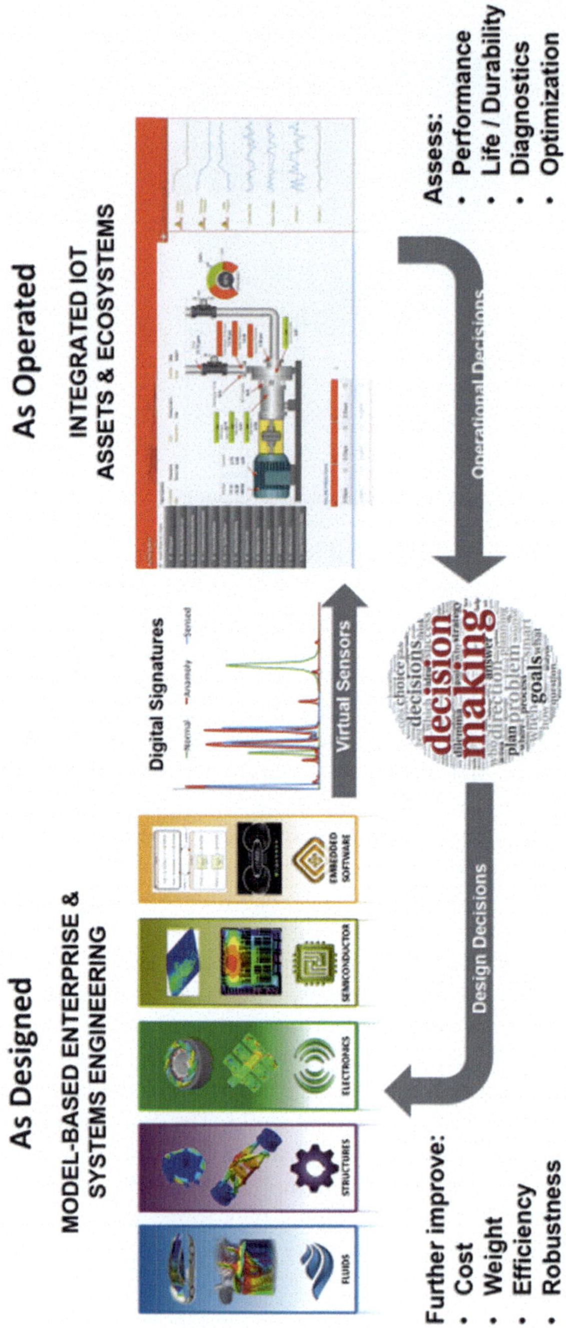

Fig. 11.9 ANSYS digital twin framework [19]

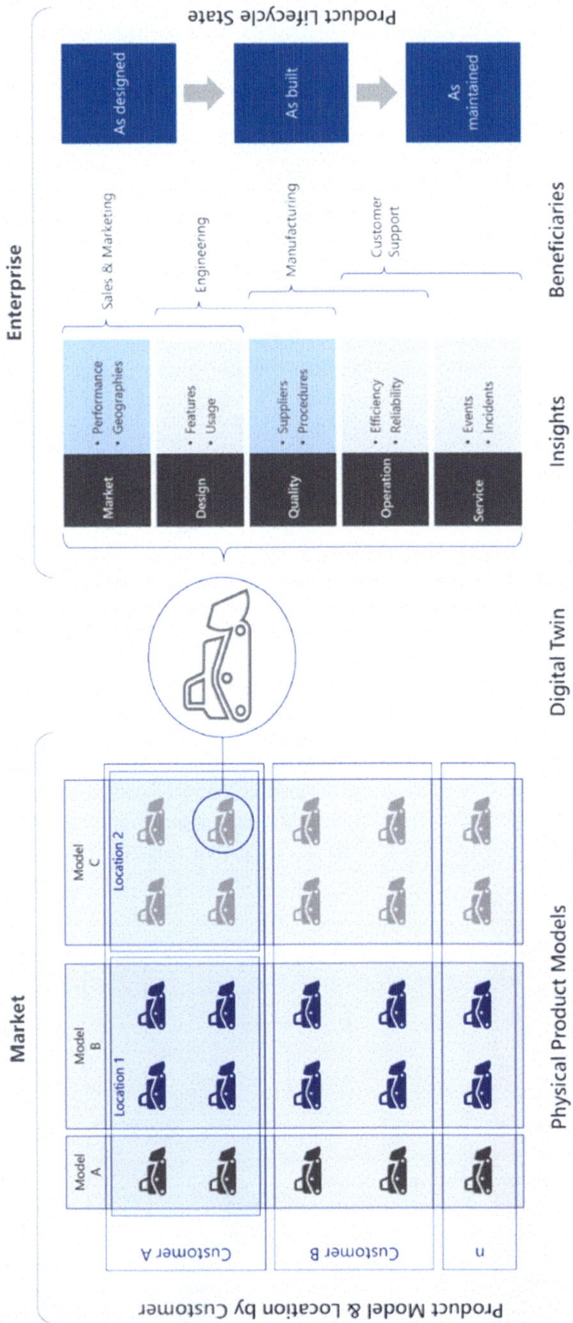

Fig. 11.10 Microsoft digital twin framework [20]

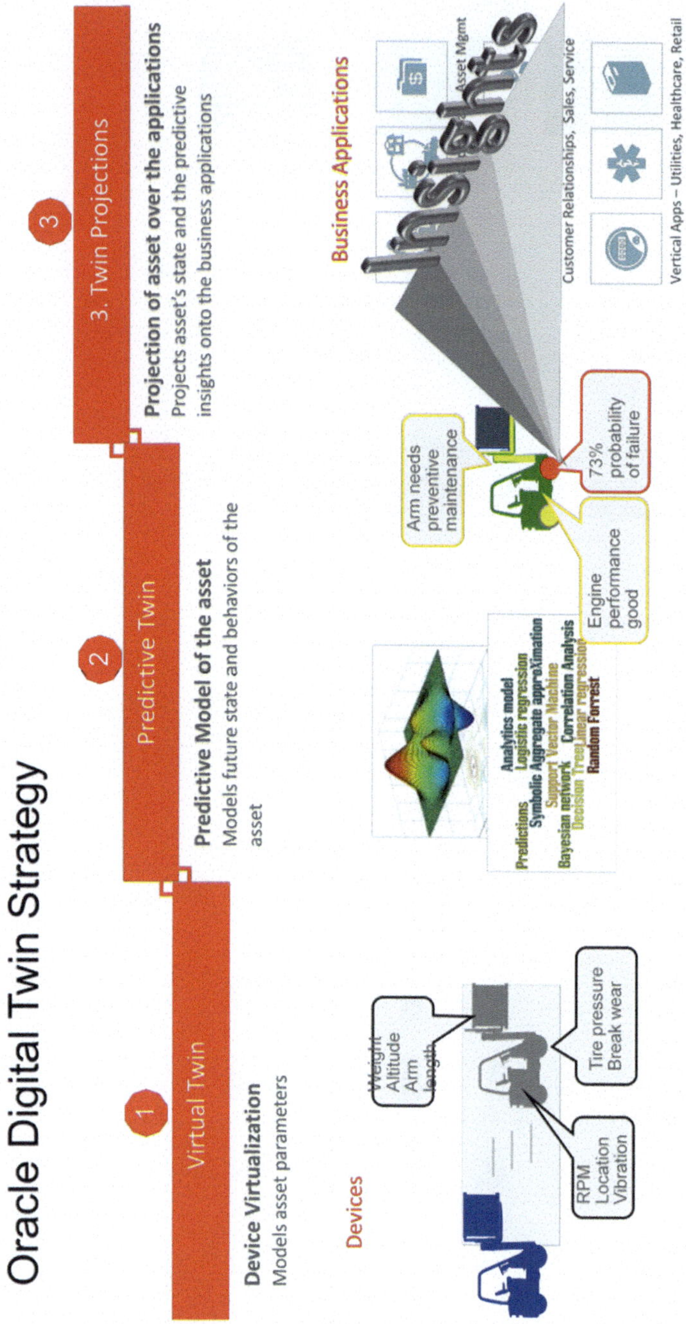

Fig. 11.11 Oracle digital twin model [21]

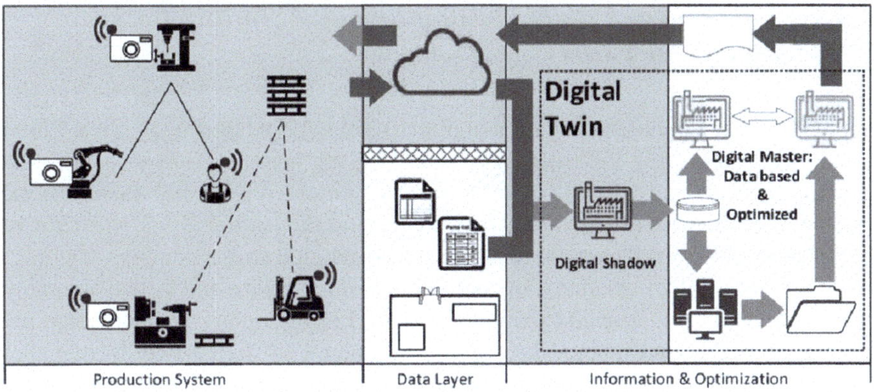

Fig. 11.12 The Fraunhofer digital twin reference model [22]

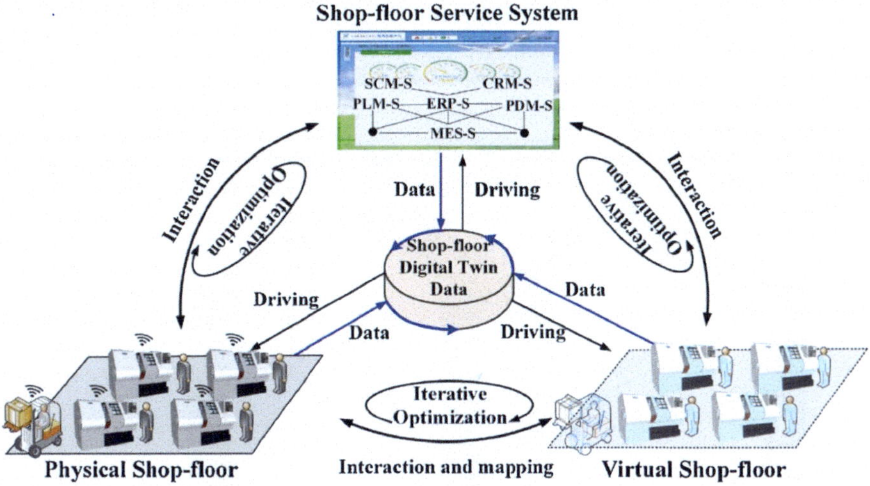

Fig. 11.13 Five-dimensional model of the digital twin [23]

11.2 Digital Twin Assembly

Assembly is an important application object of digital twin technology. This section firstly describes the theoretical system of assembly and tuning based on digital twin, secondly introduces the theoretical system method of assembly and tuning with digital twin, and finally takes the assembly of high-precision electro-hydraulic servo valves as an example to illustrate.

11.2.1 Theoretical System of Mounting and Shifting Based on Digital Twins

Digital twin assembly technology is a digital assembly based on digital twins. Based on the virtual assembly information model, with the help of the twin data in the assembly context, and with the quantitative calculation of assembly quality as the core, it realizes the virtual-real interaction, data fusion, decision analysis, and iterative optimization of the whole assembly process. In the digital twin assembly system for high-precision product assembly, the physical assembly space and virtual assembly space establish a virtual-reality mapping system. For the tightly coupled assembly needs of assembly and tuning, the scientific problems of digital twin in twinning and twinned are extracted. As shown in Fig. 11.14, the main research of digital twin-based assembly theory system contains four parts: 1. digital twin assembly theory for tightly coupled assembly, forming a multidisciplinary, multidisciplinary, multistage integrated theory system; 2. digital twin model construction of assembly elements; 3. data fusion and interoperability of assembly digital twins; and 4. digital twin data based on twinning for assemblability prediction, process optimization, and digital twin assembly operation.

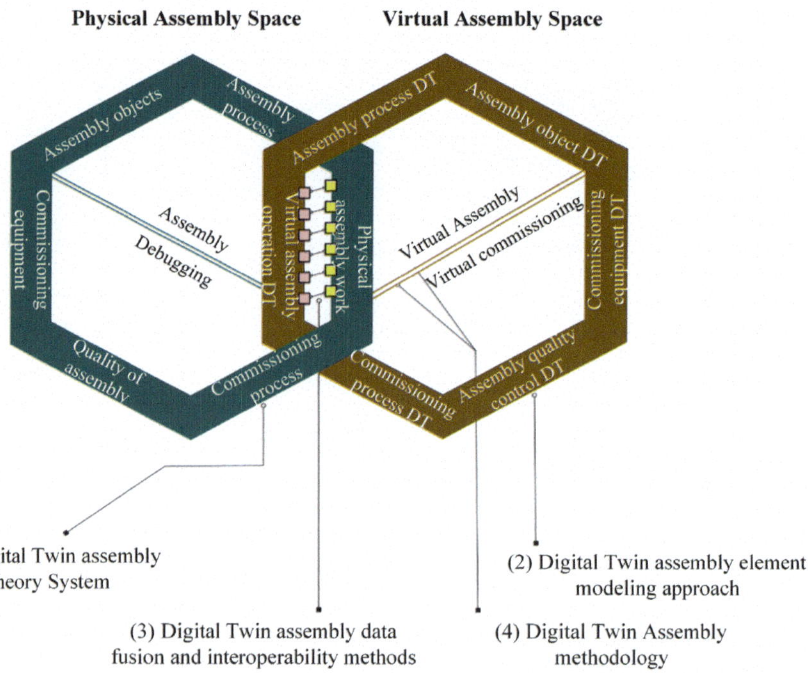

Fig. 11.14 Digital twin based fitting theory system

11.2.2 Systematic Approach

In the digital twin based mounting theory system, the following methods need to be implemented:

(1) **Digital Twin Assembly Element Modeling Methods**

In the digital twin assembly element modeling method, the main consideration is the zero-component reverse modeling method considering the key assembly features and the digital twin model construction method based on the virtual-real fusion of the whole element of the assembly.

1. Component inverse modeling method considering key assembly features. The method of constructing digital twin model is proposed with key assembly features as the alignment point. The assembly data fusion and deviation calculation with fusion of manufacturing deviation is shown in Fig. 11.15.

① Determine the assembly features of assembly components, reference features, boundary features, etc.; ② According to the geometric shape of the parts, key assembly features, determine the movement trajectory in the laser scanning system of high-precision assembly components, rotary table rotation, the scanning frequency of the special measuring equipment, to obtain high-precision measurements of three-dimensional features and shape of the point cloud; ③ Through the special high-precision indirect measurement equipment, to obtain the cavity internal confinement characteristics of the shape of the shape of; ④ For the indirect data measured through mathematical modeling analysis to generate confinement profile data information, and ultimately fused into the point cloud data. The measured indirect data are analyzed by mathematical modeling to generate the information of confined contour data, which is finally fused into the point cloud data.

For an assembly part, the traditional point cloud scanning method is to scan the surface of the measured workpiece without any difference, without considering the key parts such as assembly features, datum features, and so on, thus leading to the problem of extremely large amount of data and data redundancy. This project proposes to adopt the component inverse modeling technology that considers the key assembly features, analyzes the high-precision coupling relationship for the problem of multiple assembled parts matching features, and researches the reverse modeling technology of virtual-real mapping of the components on the basis of the cooperation between the performance measuring instrument and the dimensional measuring instrument. The specific process is shown in Fig. 11.16.

2. The construction method of the assembly all-element digital twin model based on virtual-real fusion. Analyze the mapping relationship between the integrated model and the entity model, and construct the assembly all-element digital twin model based on virtual-real fusion. The construction of the digital twin model of each assembly element is shown in Fig. 11.17, which mainly consists of five parts,

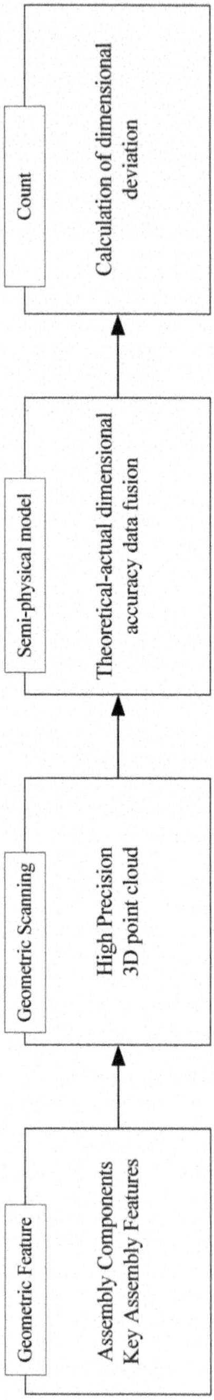

Fig. 11.15 Assembly data fusion and deviation calculation for fused manufacturing deviations

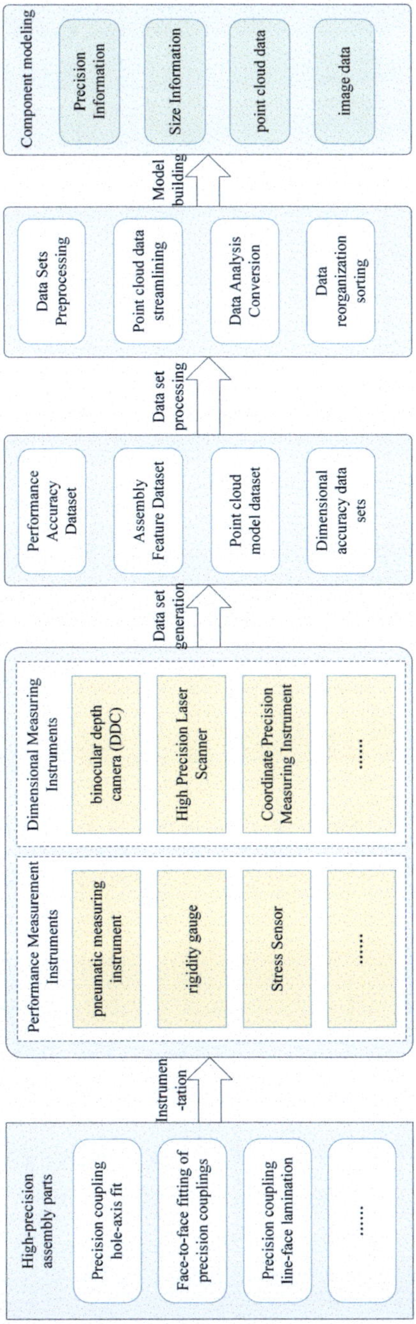

Fig. 11.16 Component reverse modeling process

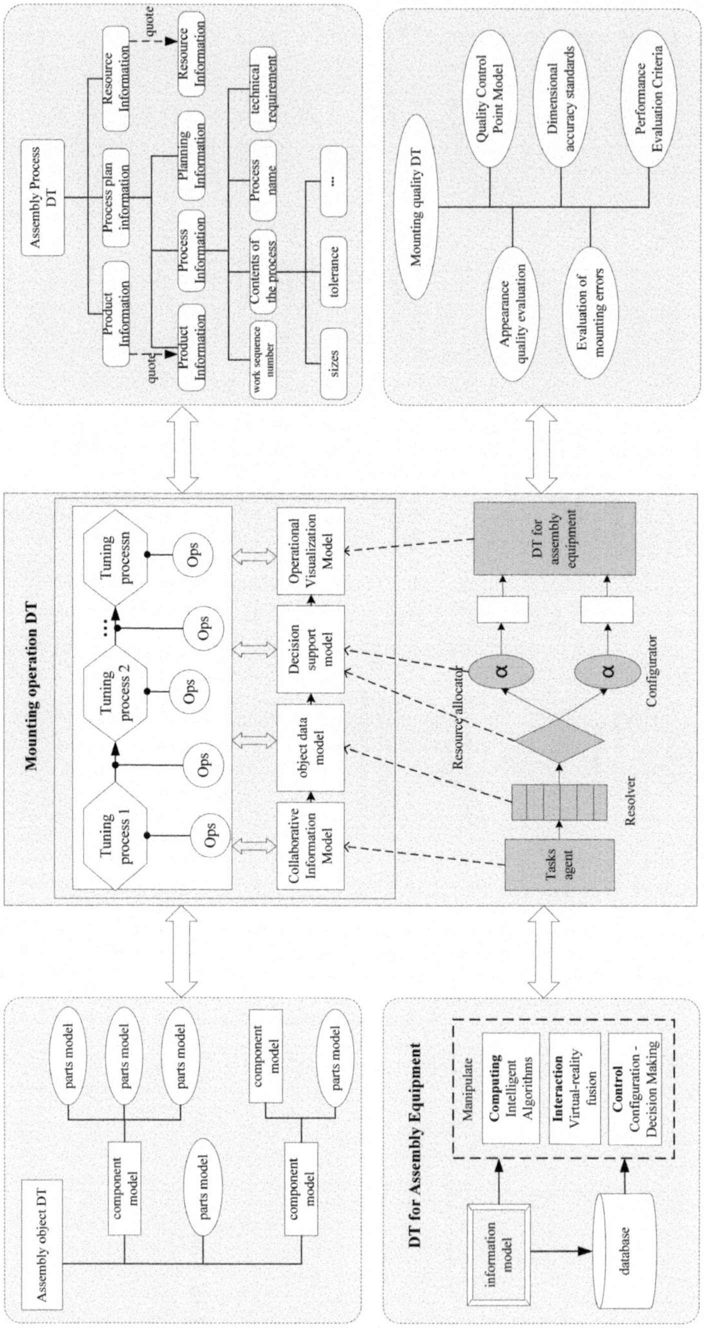

Fig. 11.17 Digital twin model construction for assembly elements

assembly object, assembly process, assembly equipment, assembly quality and assembly operation digital twin model, the remaining four models are mainly for the assembly operation process to provide services, to facilitate the entire assembly process to continuously iterate, fine tuning, to ensure that the assembly of high quality.

The relationship between the digital twin model of assembly elements and the fitting process is shown in Fig. 11.18. Obtaining data from the actual assembly fitting process, processing the data, constructing the assembly element digital twin model, including the model layer and the operation layer, and providing guidance for the fitting process through the relevant applications in the virtual space.

(2) **Data Fusion and Interoperability Method for Assembly Digital Twin**

The technical route of the data fusion and interoperability method of assembly digital twin is shown in Fig. 11.19. The top part of the figure shows the whole process of electro-hydraulic servo control valve assembly, the middle part is one of the assembly processes to develop the data fusion and interoperability of the assembly digital twin, and the bottom part is the whole process of electro-hydraulic servo control valve commissioning.

Data fusion methods between assembly digital twins: Electro-hydraulic servo control valves contain multiple assembly processes and commissioning processes in a complex assembly process. Among them, the physical assembly process contains parts size data (shape, geometry), assembly features, assembly location, assembly constraints (line fit, face line alignment) and other information through the scanning and sensor equipment will be transmitted to the virtual assembly process DT information, after the virtual simulation to generate a large number of interactive datasets, access to this information exists in a multidisciplinary multi-discipline and multi-professional structured, unstructured data (surface defects) There are multi-disciplinary and multi-specialty structured and unstructured data (surface defects, rotation detection noise, 3D measurement point cloud, etc.), sequential and non-sequential data (superposition of data from multiple lines, superposition of data from multi-assembly coordinated operation), and so on. For the unstructured data, CNN image semantic processing model, RNN-based noise anomaly signal extraction model, and PCNNs point cloud feature model are used to process and transform into structured information. For the non-sequential data, it is transformed into sequential data through data alignment, data reorganization, and data sorting technology processing. The obtained fragmented structured and sequential data are processed through knowledge fusion and knowledge processing to construct the electro-hydraulic servo control valve total assembly and tuning relationship mapping. The information interaction with other digital twins is realized through the interaction and interoperability interface based on the knowledge graph (using Web Service, with data types of unified attributes, nodes and edges).

Interoperability method between assembly digital twins: information interaction between assembly digital twins is carried out through knowledge graph-based interaction and interoperability interfaces. Its information interaction process is shown

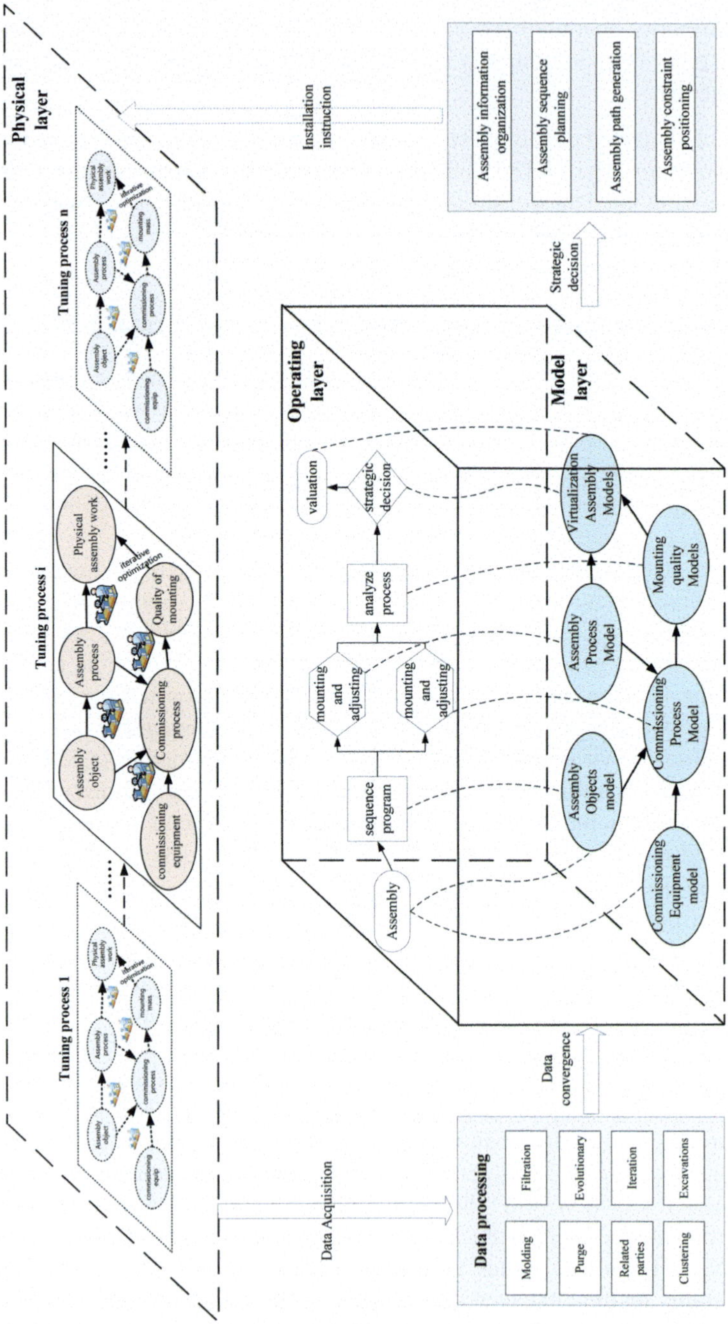

Fig. 11.18 Digital twin model relationship diagram for assembly elements

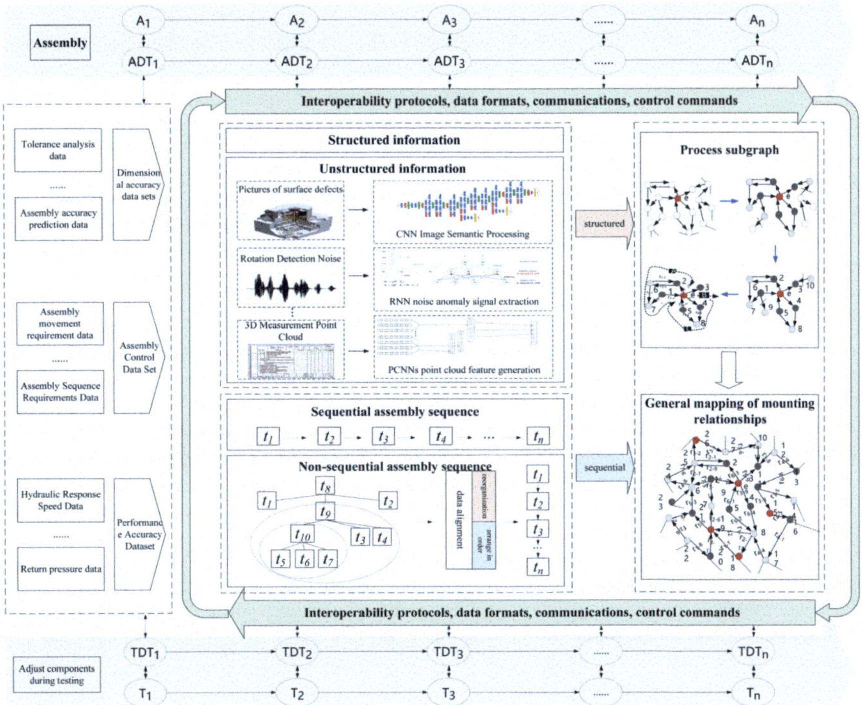

Fig. 11.19 Technology roadmap for data fusion and interoperability methods for assembling digital twins

in Fig. 11.20. The interoperability of assembly is mainly realized through the physical system of assembly and the information system of assembly. In the sequential assembly and adjustment process, after the assembly of the assembly digital twin ADT1 is completed, the assembly dimensional accuracy is evaluated based on the dimensional fine reading data set (tolerance analysis data, assembly dimensional chain generation data, assembly gap data, assembly accuracy prediction data, etc.), and after the debugging is carried out by the debugging equipment TDT1, the assembly dimensional accuracy is evaluated based on the performance accuracy data set (hydraulic response speed data, electromagnetic moment data, output flow rate data, return oil pressure data, etc.). After debugging-by-debugging equipment TDT1, debugging dimensional performance assessment is performed based on the performance accuracy data set (hydraulic response speed data, electromagnetic torque data, output flow data, return oil pressure data, etc.), followed by assembly of digital twin ADT2 assembly and debugging of debugging equipment TDT2 debugging, and so on in an iterative cycle of mounting. Among them, when the dimensional accuracy and dimensional performance of the tuning failed to meet the requirements, based on

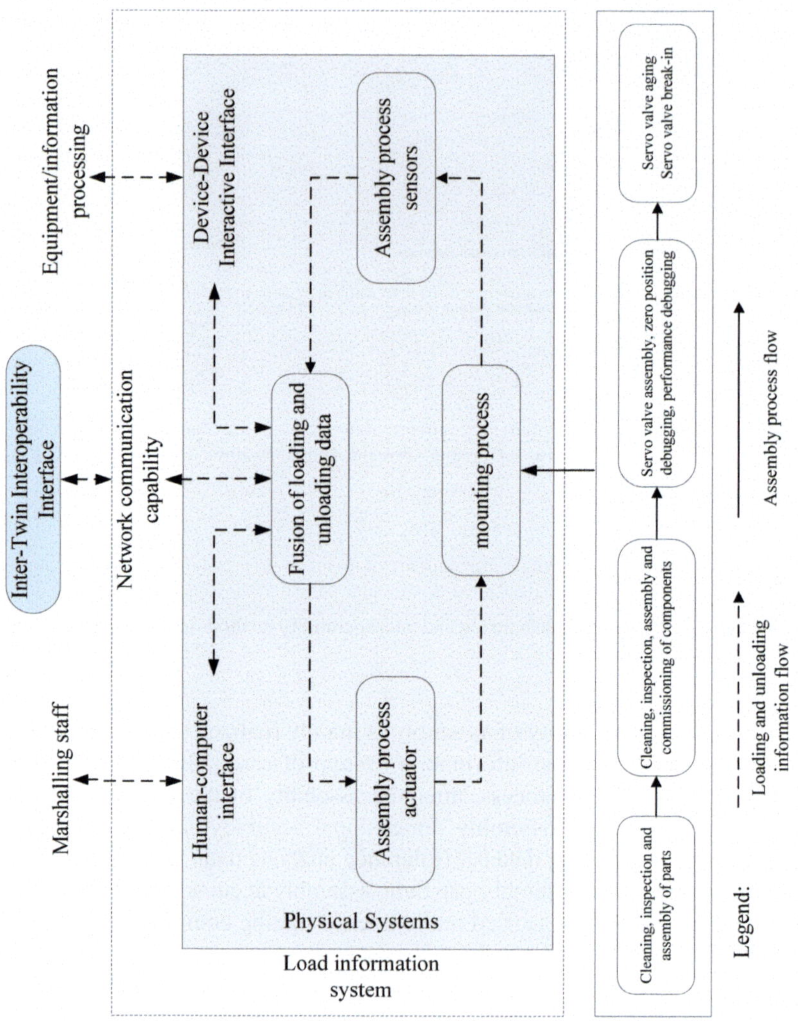

Fig. 11.20 Interaction process of loading and tuning information within the twin ADTi

the assembly control data set (assembly action requirements data, assembly sequence requirements data) for re-tuning, after the tuning optimization iteration, and finally complete the tuning of the servo valve.

(3) Digital Twin Assembly Operation Technology

The complex assembly of assembly tightly coupled products, its requirements while assembling and debugging, the need for assembly process with a high degree of real-time, assembly quality elements constraints need to be highly coupled. The actual state of the process, subject to the constraints of the manufacturing process and measurement technology, the actual value of the characteristic parameters of precision product parts and the ideal value of the error, directly affecting the quality of the assembled products and characteristics; at the same time, for the complex changes in the conditions of the assembly and the assembly of the object, the universal design process cannot meet the processing needs, should be adaptive changes and adjustments. To this end, on the basis of highly integrated twin data, data mining and modeling of twin data is required to analyze and deduce the measured data, performance data, simulation data, emulation data, and deviation data (e.g., shape deviation, dimensional deviation) in the assembly process contained in the twin data. Before the implementation of assembly activities, according to the specific working conditions, the actual characteristics of the parts, the corresponding processing technology, predict the quality characteristics of the finished product to avoid the emergence of ineffective assembly; in the processing process, according to the corresponding working conditions, the actual characteristics of the parts, to achieve the dynamic adjustment and real-time optimization of the assembly process. The two are highly integrated to collaboratively ensure the quality of product assembly, as shown in Fig. 11.21.

1. Twin data-driven assemblability prediction method: Aiming at the current assembly technology in the assembly process error accumulation, parts manufacturing errors on the assembly process lack of analysis and predictability of the problem, the use of deep neural network prediction model, to achieve based on the virtual and real fusion of the twin data of assemblability prediction. Based on the fusion of physical measurement data and virtual simulation data to form twin data in the data fusion method, the latest inspection and measurement data, progress data, performance data, and measured values of assembly process state parameters contained in the data fusion method are associated and mapped to the digital twin model, and based on the established integration model, theoretical values of key technical state parameters, and the prediction and analysis model, we can predict and analyze in real time the assembly progress, accuracy, and performance of the physical products, specifically, the method of predicting and analyzing the assemblability of physical products. progress, accuracy and performance of physical product assembly in real time based on the established integration model, theoretical values of key technical state parameters and predictive analysis model, the specific method is shown in Fig. 11.22.

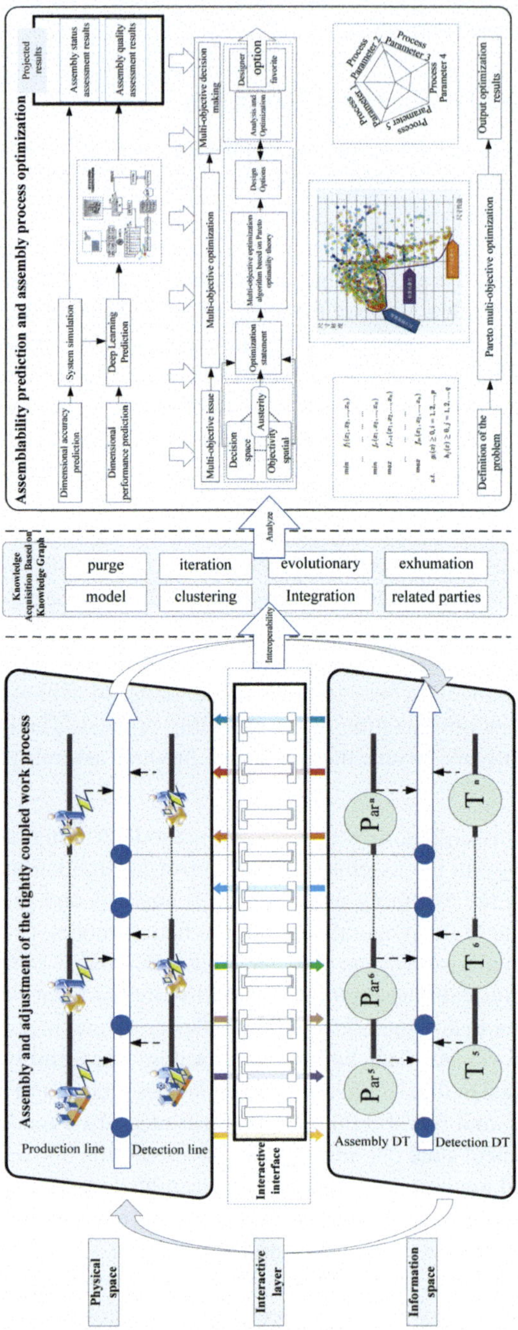

Fig. 11.21 Technology roadmap for digital twin assembly manipulation technology

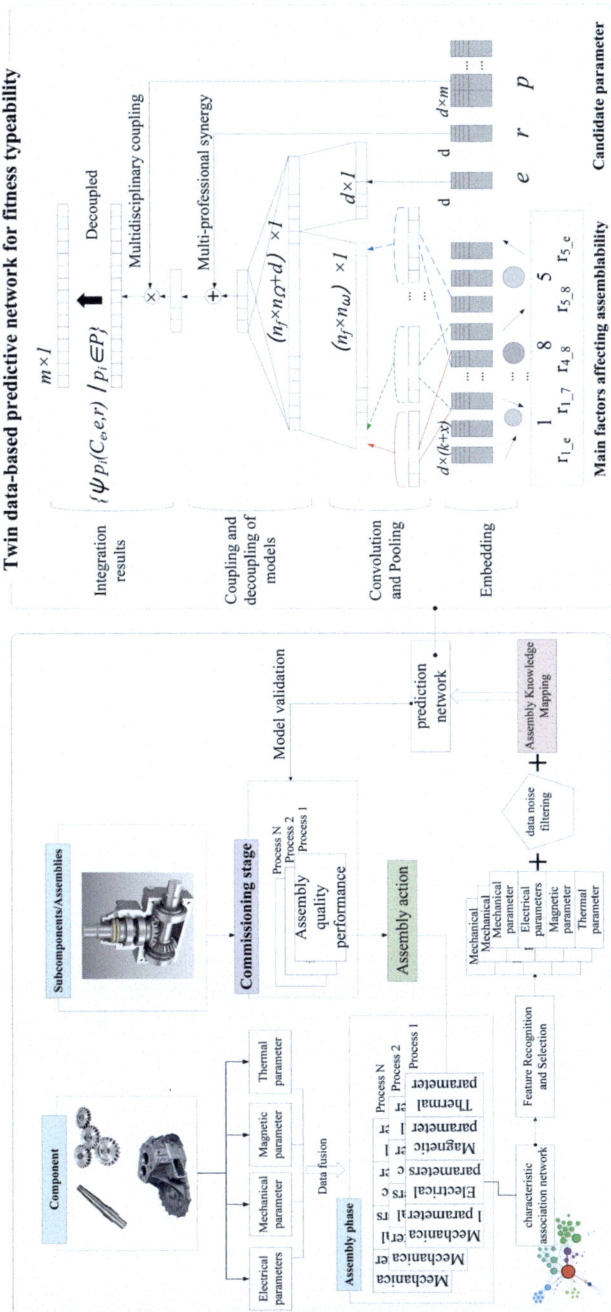

Fig. 11.22 Twin data-based assemblability prediction network

2. Assembly process optimization method based on Pareto Optimality Theory: By analyzing the typical assembly process defects of high-precision assembly parts, and targeting the key issues of parts grouping and selection, tolerance analysis, assembly dimensional chain generation, assembly path generation, etc., AI technology is introduced to realize the process knowledge modeling, knowledge refinement, and knowledge optimization based on data analysis, as well as autonomous process design, process optimization and process decision making based on autonomous learning. optimization and process decision-making based on autonomous learning. In general, the assembly process optimization problem is a multi-objective optimization problem.

Generally speaking, a multi-objective optimization problem consists of multiple objective functions with some related equations and inequality constraints, which can be described as follows from the mathematical point of view:

$$\begin{aligned} &\min f_1(x_1, x_2, \ldots, x_n) \\ &\ldots \\ &\min f_r(x_1, x_2, \ldots, x_n) \\ &\max f_{r+1}(x_1, x_2, \ldots, x_n) \\ &\ldots \\ &\max f_m(x_1, x_2, \ldots, x_n) \\ &s.t.\quad g_i(x) \geq 0, \quad i = 1, 2, \ldots, p \\ &s.t.\quad h_j(x) \geq 0, \quad j = 1, 2, \ldots, q \end{aligned}$$

Examine two decision vectors a, b ∈ X. a Pareto Dominate (Pareto Dominate) b, denoted as a > b, if and only if:

$$\{\forall i \in \{1, 2, \ldots, n\} f_i(a) \leq f_i(b)\} \wedge \left\{\exists j \in \{1, 2, \ldots, n\} f_j(a) < f_j(b)\right\}$$

If in the entire parameter space does not exist any decision vector Pareto prevails a decision vector, then the decision vector is said to be both Pareto optimal solution, all Pareto optimal solution composed of Pareto optimal solution set. Description of the process optimization method based on Pareto optimal theory: Under the loading and shifting condition, for the loading and shifting processing unit, multi-dimensional execution process, multi-parameter coupling influence can form the Pareto feasible solution space under the effect of weights.

11.2.3 Application Cases

High-precision electro-hydraulic servo valves are widely used in military weapons and are key components therein. The attitude control element of most weapon systems uses electro-hydraulic servo valves, which convert small power electrical signals into high power hydraulic outputs to realize the automatic control of mechanical equipment through hydraulic actuators. Electro-hydraulic servo valve has the advantages of small volume, high power amplification, good linearity, small dead zone, fast response, smooth and reliable movement, can adapt to the analog and digital modulation, etc., in a variety of missile weapons electro-hydraulic servo system has been widely used, compared to the servo control system of the "heart".

Aiming at the assembly process of high-precision electromechanical and hydraulic products, the research object is a high-precision electro-hydraulic servo valve, which is a core component of a certain type of weapon, and the research on intelligent assembly technology is carried out based on digital twin technology. High-precision electro-hydraulic servo valves are characterized by complex structure, multidisciplinary assembly and adjustment process, high precision, and great difficulty in assembly and adjustment. High-precision electro-hydraulic servo valve internal structure is compact and complex: the key components can be up to more than 100, representative products are nozzle baffle type, jet tube type, deflection plate jet type, moving circle slide valve type and other types of servo valves. High-precision electro-hydraulic servo valve assembly and adjustment process is complex: involves a multi-physical, multi-disciplinary, multi-level problem, in which the electromagnetic, mechanical, hydraulic and structural features interact with each other, mechanical, electrical, hydraulic and other interdisciplinary cross, process quality constraints of many levels, mutual constraints.

For this reason, in order to practice the digital twin assembly theory of high-precision electro-hydraulic servo valves, an assembly simulation prototype system based on digital twins was built, as shown in Fig. 11.23.

The digital twin-based assembly simulation prototype system is divided into fast reverse modeling module, integrated model building module, system interface module, 3D visualization kernel module, digital twin model, assemblability prediction module, process optimization module, assembly operation tuning equipment data acquisition module, each module relies on the interface to establish a connection to form a set of digital twin-based assembly simulation prototype system.

11.3 Digital Twin Processing

Digital twin technology can provide the basis for providing product quality prediction, monitoring, and optimization for machining processes. This section first introduces the digital twin model for machining, second introduces the intelligent methods for machining processes, and finally shows a manufacturing digital twin case study given by NIST.

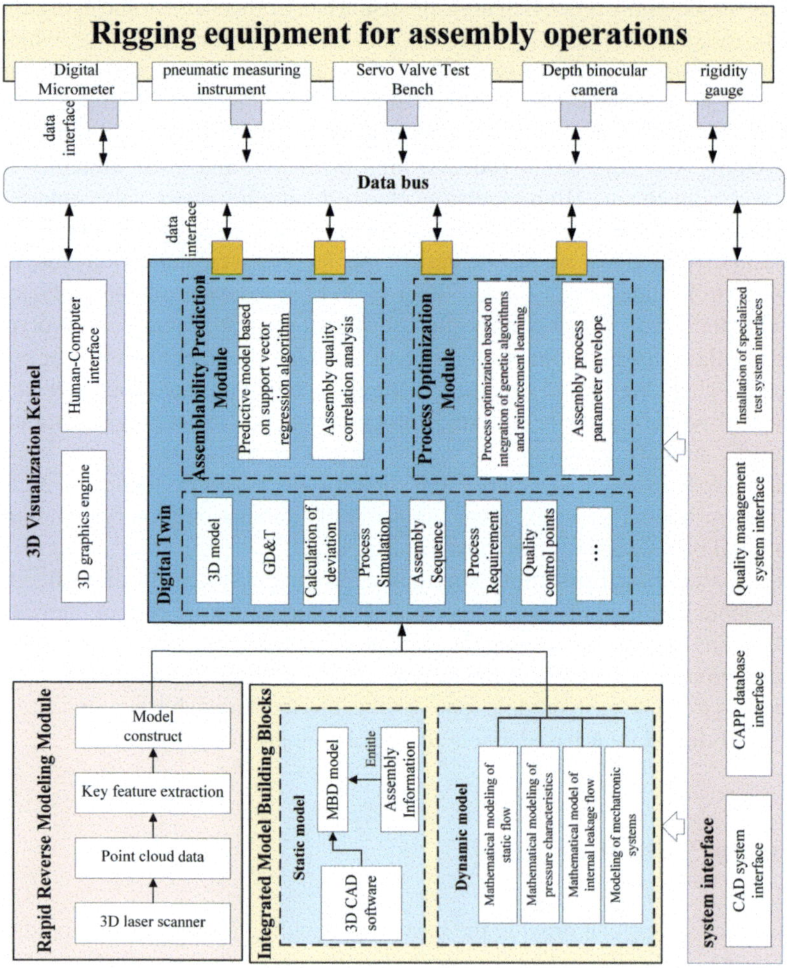

Fig. 11.23 Digital twin-based assembly analog simulation prototyping system

11.3.1 Process-Oriented Digital Twin Modeling

Figure 11.24 shows the processing-oriented digital twin model, in the manufacturing stage, the physical reality of the world will be the product of the production of measured data (such as processing progress data, processing equipment data, quality of work in progress data) to the virtual world of the virtual product and real-time display, to achieve the product model based on the monitoring of the production of measured data and monitoring of the production process (including the comparison between the designed value and the measured value, the comparison between the characteristics of the actual materials used and the characteristics of the designed materials, and the comparison between the planned completion progress and the actual completion progress, etc.). In addition, based on the production measurement data, through the intelligent prediction and analysis of logistics and progress, it realizes the prediction and analysis of quality, manufacturing resources, and production progress; at the same time, the intelligent decision-making module formulates the corresponding solutions based on the results of prediction and analysis and feeds them back to the physical products, so as to realize the dynamic control and optimization of the physical products, and to achieve the purpose of integrating the virtual with reality and controlling the real with the virtual.

How to realize the complex dynamic entity space of multi-source heterogeneous data real-time accurate collection, effective information extraction and reliable transmission is a prerequisite for the realization of digital twin. In recent years, the rapid development of the Internet of Things, sensor network, industrial Internet, semantic analysis and recognition and other technologies provide a set of practical solutions for this. The rapid development of artificial intelligence, machine learning, data mining, high-performance computing and other technologies provides important technical support for the role of digital twins in the integrated display of product data, product production progress monitoring, product quality monitoring, intelligent analysis and decision-making (e.g., product quality analysis and prediction, dynamic scheduling and optimization).

The core of the digital twin system is the product digital twin model, which is a dynamic integration of model and data [23–25]. The product digital twin model for CNC machining should meet the data requirements of uniqueness of the manufacturing process, and MBD technology provides a good technical path for the establishment of the model of digital twin. Considering the characteristics of CNC machining, the author, based on the MBD connotation and ideas, uses the integrated 3D solid model to completely express the product definition information, which contains the change situation of the product oriented to the process, and specifies in detail the expression of geometric, physical and process in the 3D solid model [26].

The MBD model is extended with behavioral and contextual information of the product [27]. Behavioral information is the functional state of the product that the physical model responds to the external environment and stimuli; contextual information, i.e., a complete description of the key elements of the processing process, including the product process and the actual processing information in each process,

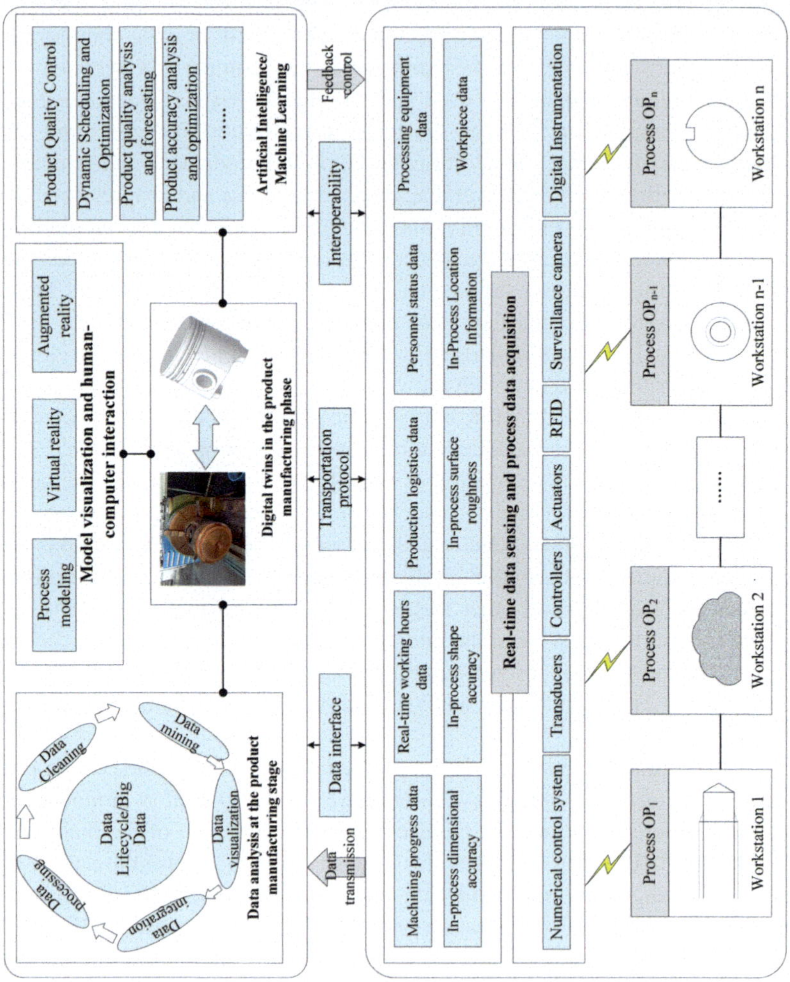

Fig. 11.24 Processing-oriented digital twin model

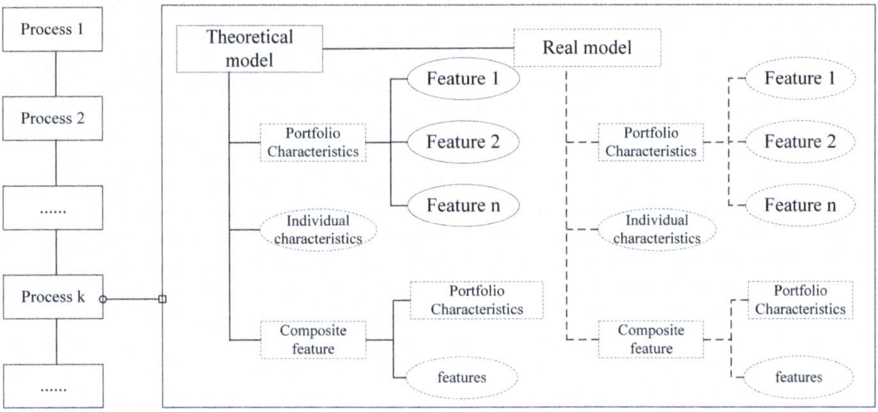

Fig. 11.25 Geometric data model structure diagram

which makes it easy to trace the product. Therefore, the twin model and data driving the digital twin system are mainly the geometric information of the product model, the behavioral information of the product, and the manufacturing process information, and the digital twin manufacturing needs to integrate the above information for the above processing to build a complete digital twin model.

Therefore, the expression for the digital twin model DTM is defined as follows:

$$DTM = 3D\_Model \cup Behaviour \cup Context$$

where 3D_Model-Product digital twin driver model, consists of three-dimensional geometric model 3D_Model, product behavior information Behaviour, processing process information Context together (Fig. 11.25)

$$3D\_Model = \{Material, Structure, Surface, Mechanical\_property\}$$

The 3D geometry of the 3D_Model-MBD twin model is composed of 3D geometric elements, which mainly include model geometric elements and auxiliary geometric elements (Fig. 11.26).

$$Behaviour = \{Material, Structure, Surface, Mechanical\_property\}$$

Behaviour—The behavioral information set responds to the functional properties of the machined workpiece, such as information on material hardness, structure, surface organization, and mechanical properties.

$$Context = Plan\_process \cup Machining\_process$$
$$Plan\_process = \{3D\_Model, Dimension, Tolerance, datum, process\_info\}$$

Fig. 11.26 Product behavior information structure

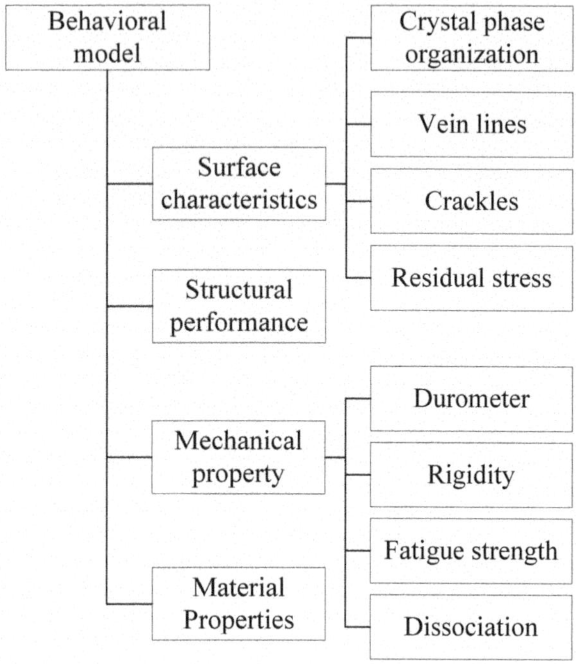

$$Machining\_process = \{Actual\_process,\ machine\ tool\_status,\ detection,\ tool\_status\}$$

Context—Contextual information includes planned process information and machining process information, design information including such information as tolerances, process information, benchmarks, roughness, technical requirements and preset processes; machining process information including real-time information on the workpiece, real-time process part of the machining process for the machining of the parameter information, the process monitoring information including tools, lathes, workpiece information (Figs. 11.27 and 11.28).

11.3.2 Intelligent Methods in Machining Processes

(1) A Multi-scale Knowledge Modeling Approach for Processed Products

Digital twin technology breaks the barrier of unidirectional data transmission from the traditional virtual space to the physical space and realizes the two-way interaction between the physical space and the virtual space [28]. Digital twin can run through the whole life cycle of product design, manufacturing to operation and maintenance, for the massive multi-source heterogeneous data generated in the virtual and real

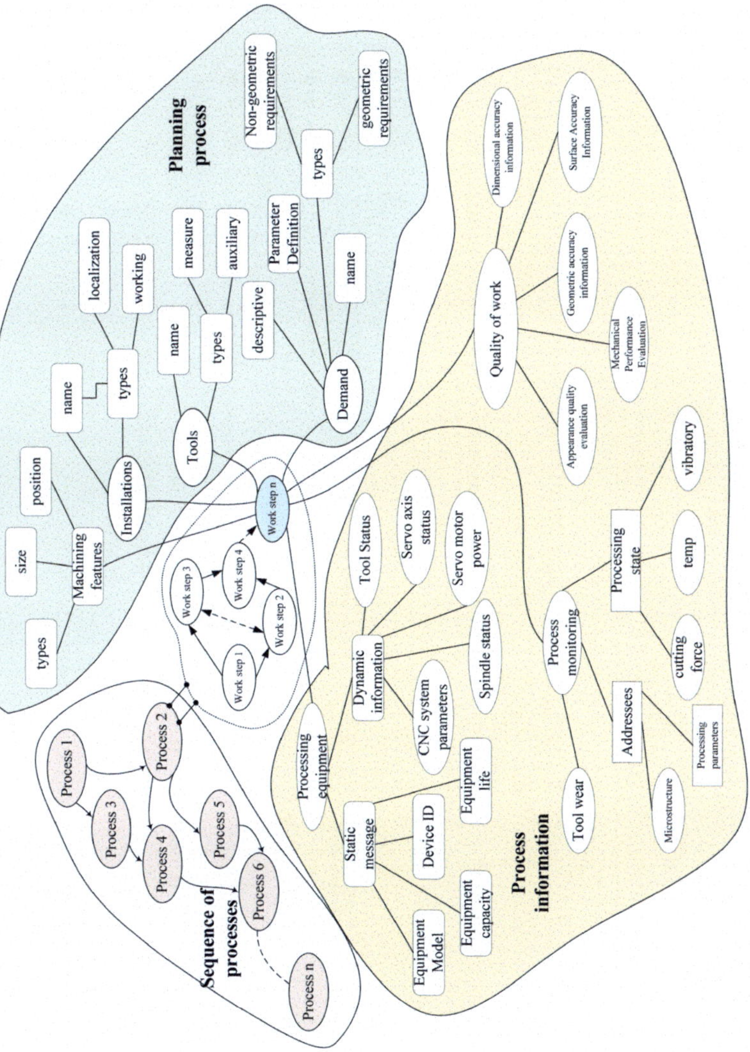

Fig. 11.27 Structure of the product digital twin drive model for CNC machine processing

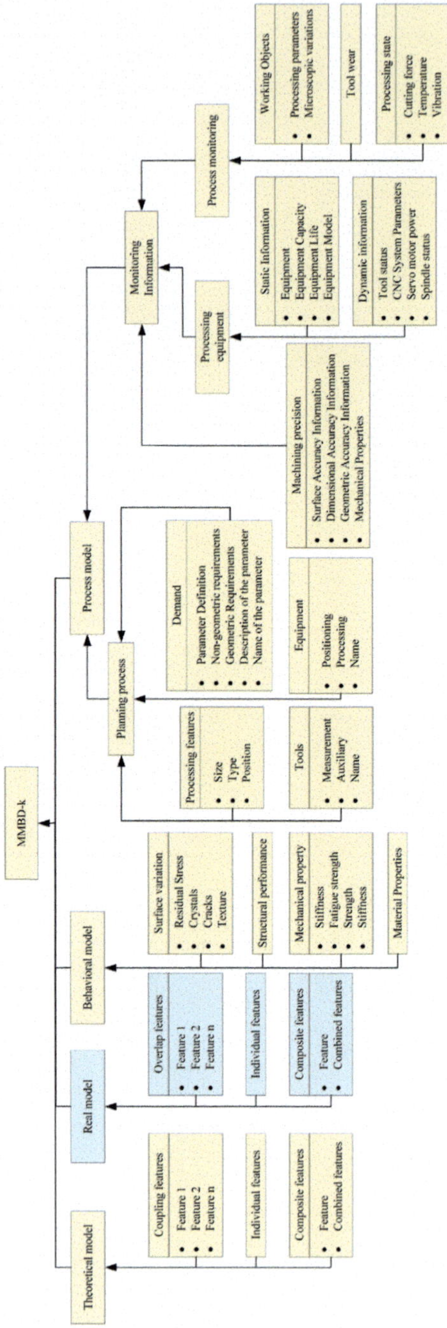

Fig. 11.28 Structure of the anamorphic digital twin model based on OWL modeling

space during product processing, data processing based on knowledge mapping. For the systematic decision-making of process parameters during product processing, the structured data provided by the knowledge graph is used as a dataset, and machine learning (neural networks, genetic algorithms, etc.) algorithms are used to realize the system according to the measurements, and at the same time, the knowledge is used to quickly retrieve the dynamic adjustment of the process (Fig. 11.29).

1. **Knowledge Graph-Based Data Processing Methods**

In the process of product manufacturing, the virtual space and physical space will produce a huge amount of data, and this type of data is characterized by multi-source heterogeneity, multi-scale and multi-dimensionality. As shown in Fig. 11.30, in the manufacturing process of products, data can be divided into two categories for storage: real-time database and knowledge base. The real-time database can store real-time collected data, such as equipment data, surface quality data, and residual stress data. The knowledge base can store historical machining process, machining quality data, equipment error data and intelligent algorithm models. Data are structured using data mining, machine learning and natural language processing, and a knowledge graph is formed for rapid knowledge retrieval. Finally, the processed data are delivered to the multi-scale visualization model module and the system control platform.

2. **Structured Processing of Data Formats**

In the processing and manufacturing of products, a large number of sensing devices are usually used to collect information, and these information types are of various kinds, which can be roughly classified into structured database information, semi-structured tables and lists, and unstructured data described by text and pictures, etc. These data exist in a dispersed and heterogeneous manner. How to effectively organize and utilize this information usually requires fusion of knowledge, so as to ensure the consistency of knowledge by fusing the collected data of different structures from different data sources into a unified knowledge graph, therefore, after completing the storage of the data, the work that needs to be done is to sift and extract the acquired data to obtain the ternary group of information and to fuse them, as shown in Fig. 11.31.

3. **Rapid Retrieval of Knowledge**

Knowledge graph mainly contains two layers of logical structure: schema layer and data layer. The order of completion of the data layer and schema layer is usually categorized into two construction methods: Up-Down and Bottom-Up. Top-down is to build the ontology structure of the schema layer of the knowledge graph first, and then extract the relevant entity information from other data sources and add it to the knowledge base; Bottom-up is to start from the construction of the data layer, with the help of some technical means, extract the entities from a large number of publicly

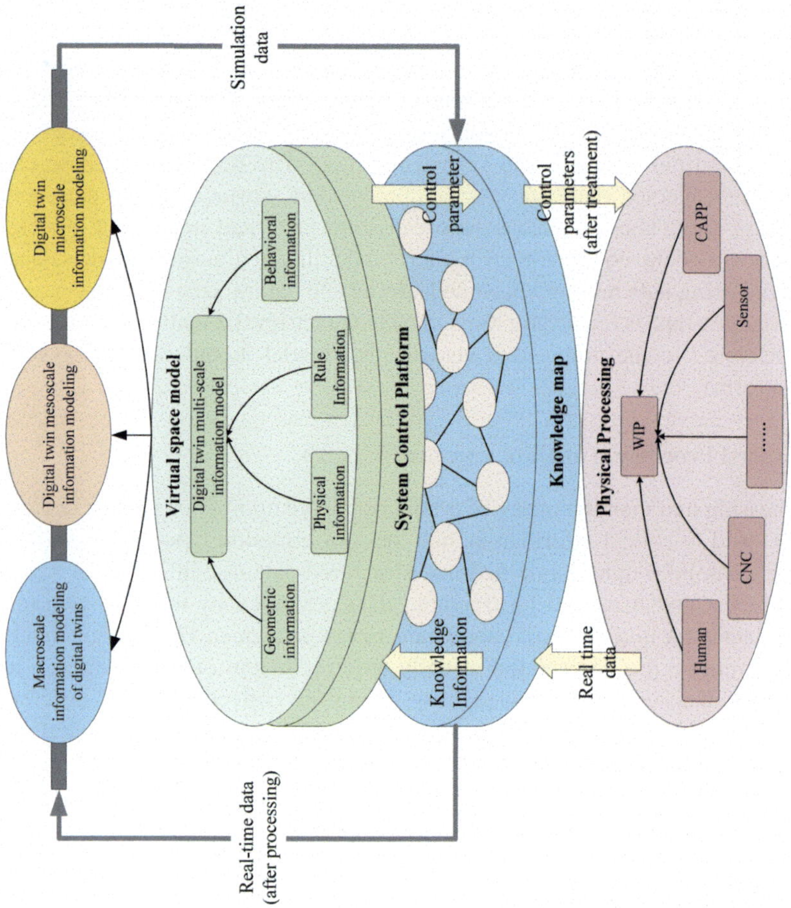

Fig. 11.29 Digital twin multiscale information model operational architecture

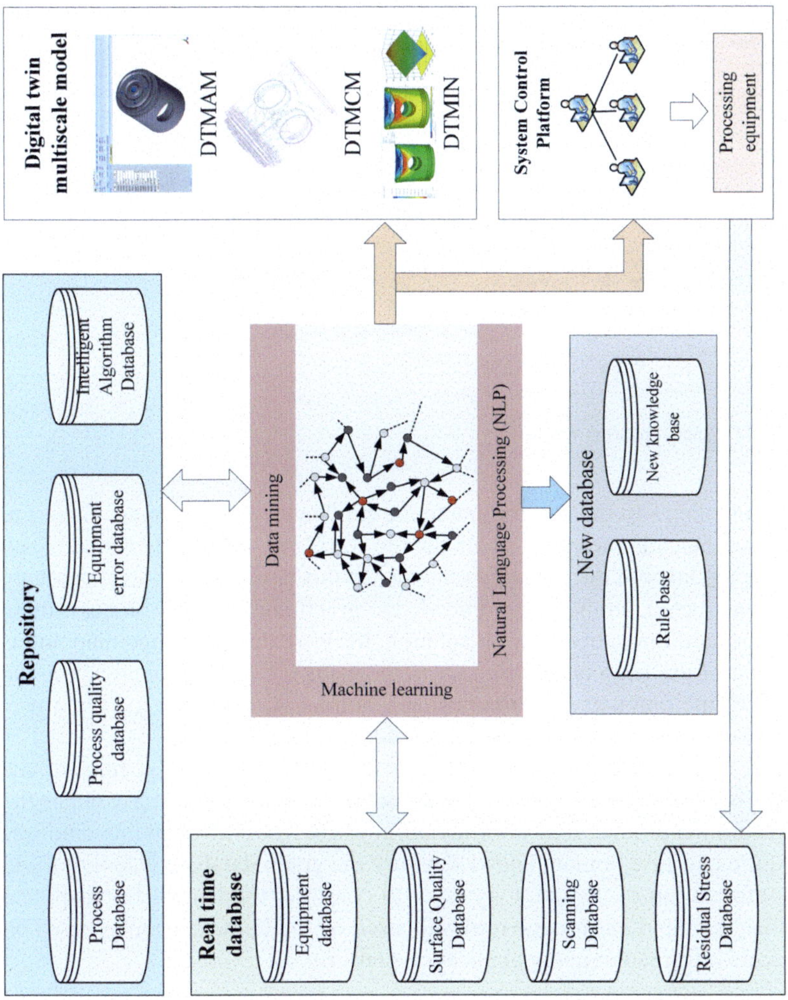

Fig. 11.30 Knowledge graph-based data preprocessing framework

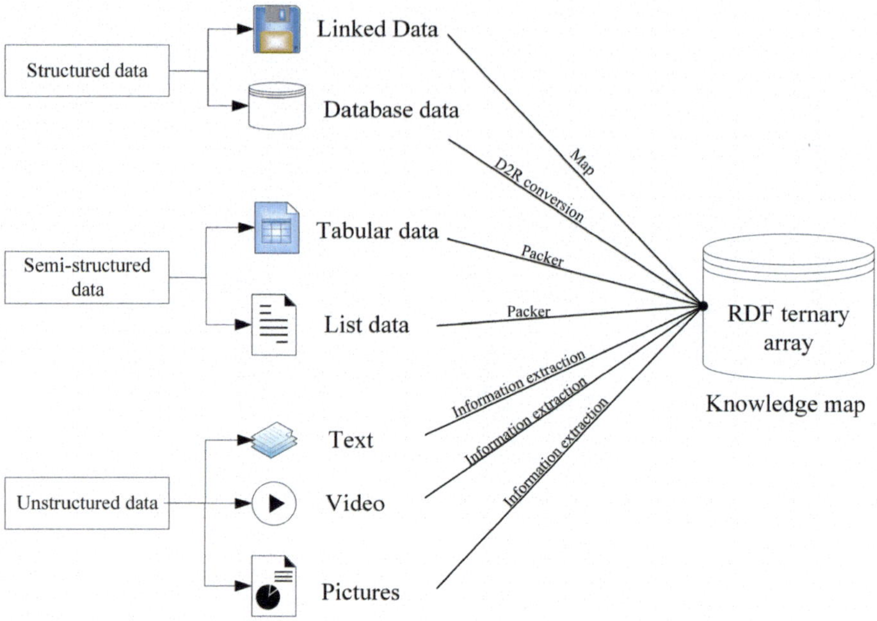

Fig. 11.31 RDF ternary information fusion

linked data, and then sort them according to the confidence level from the highest to the lowest, and then add the entities with the top sorting to the knowledge base, and then build the top layer of the ontology schema. On this basis, the top-level ontology schema is constructed. In the actual knowledge graph construction process, these two approaches are not carried out in isolation. Multi-scale knowledge mapping is the integration of the two construction methods, using a top-down approach in the construction of the ontology, and then using a bottom-up approach to obtain new knowledge to expand the existing knowledge map.

In the process of construction of multi-scale information knowledge map, firstly, various types of sensing devices are used in the actual production process to obtain the processing state and product state information, construct the corresponding ontology model, acquire and fuse domain knowledge, use the graph database Neo4j to fill in the entity information of the ontology layer to construct the knowledge map, and finally realize the rapid knowledge retrieval based on the knowledge map based on the multi-scale information query application platform.

(2) **Process Monitoring**

In order to realize the monitoring of the machining process driven by the manufacturing digital twin, a convenient and feasible monitoring system is required. The development of a tool health monitoring and prediction system based on MT Connect requires a combination of computer technology, network technology, sensor technology, etc., to perform multi-parameter online monitoring of CNC machine tools at the manufacturing site, and to store and centrally analyze the data and provide a user-friendly human–computer interface in the monitoring host computer [29].

In order to realize real-time and reliable tool health prediction, it is necessary to first collect multi-source information generated during the machining process, and the sources of these data are mainly the machine tool body, the tool part information database, the machine tool retrofitted sensors, the machining log, and manually-entered tool part quality inspection information.

As shown in Fig. 11.32, the digital twin-driven machine tool machining process monitoring is mainly composed of the Siemens 840D CNC data acquisition module based on the OPC UA communication protocol, i.e., the OPC UA client, the OPC UA-MT Connect adapter, the MT Connect agent, and the MTC client, which realizes the functions of visualization monitoring and tool health prediction. The first step is to establish a reliable connection with the CNC machine and the additional sensors to realize the data acquisition, and the following configurations need to be carried out before the connection is established.

1. Configure the MT Connect agent, mainly add the IP and port information of the OPC UA-MT Connect adapter connected to the acquisition device and the device name of the monitoring device, and then start the agent.
2. Start the OPC UA-MT Connect adapter, first of all, the adapter will open the OPC UA client's connection to maintain communication with the OPC UA server, and set the type of data subscription and sampling frequency to be collected.
3. The MTC client opens and sends a CURRENT request to the agent and waits for data changes to be passed up.

The above configurations allow connection to manufacturing physical entities. Figure 11.33 shows the CNC machine data acquisition process. In addition to sensor and CNC system data collected using this subscription method, for processing information read from the database or manually entered, the agent will immediately perform XML structured processing on the read/entered information and send the data to the agent.

Take the real-time prediction of tool health condition as an example, in mass production, frequent tool replacement not only causes production interruption, but also increases the cost of tool use; while the tool replacement cycle is too long will increase the probability of tool failure, affecting the quality of workpiece machining or even machine tool failure, which in turn increases the additional maintenance costs. The ultimate goal of real-time tool health prediction is to accurately grasp the timing of tool change.

In view of the different workpieces whose accuracy requirements can withstand the maximum tool wear value is different, so for different workpieces machining accuracy requirements in the tool severe wear before the tool change operation is not reasonable. Machining deformation and surface roughness are the two main parameters used to assess the machining quality of titanium alloy critical structural components. During high-speed milling of titanium alloy thin-walled parts, the oxidized wear, adhesive wear and diffusion wear of the tool are particularly serious, resulting

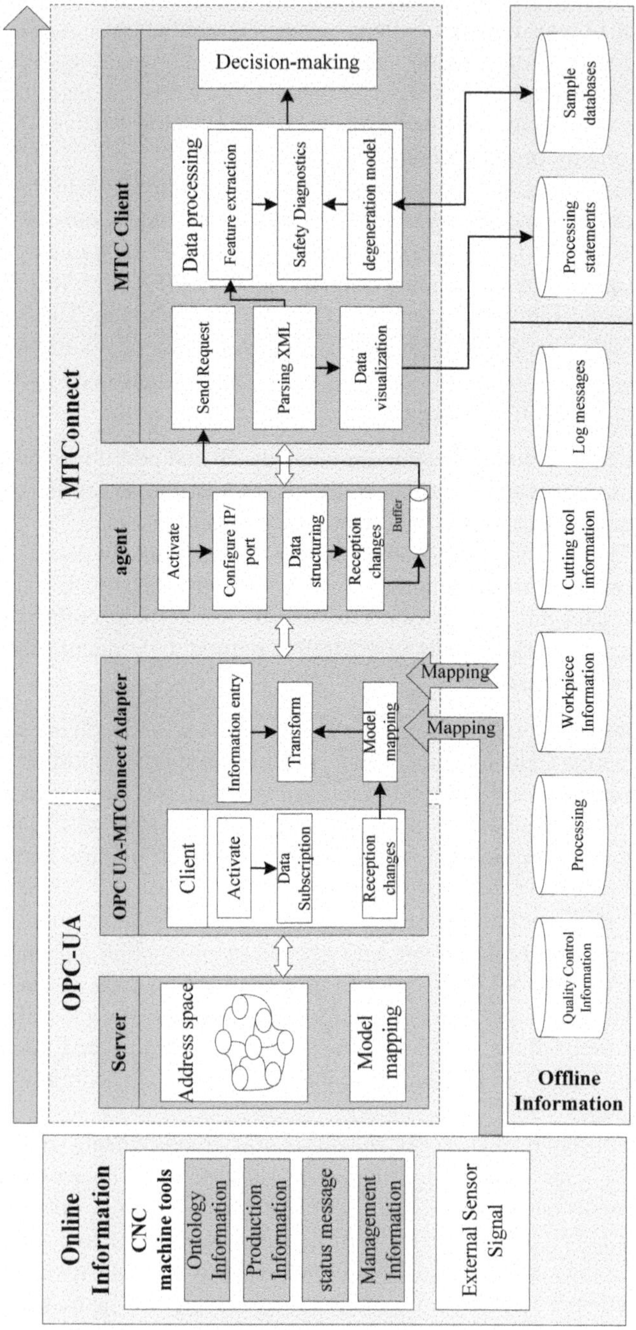

Fig. 11.32 Digital twin-driven machine tool machining process monitoring

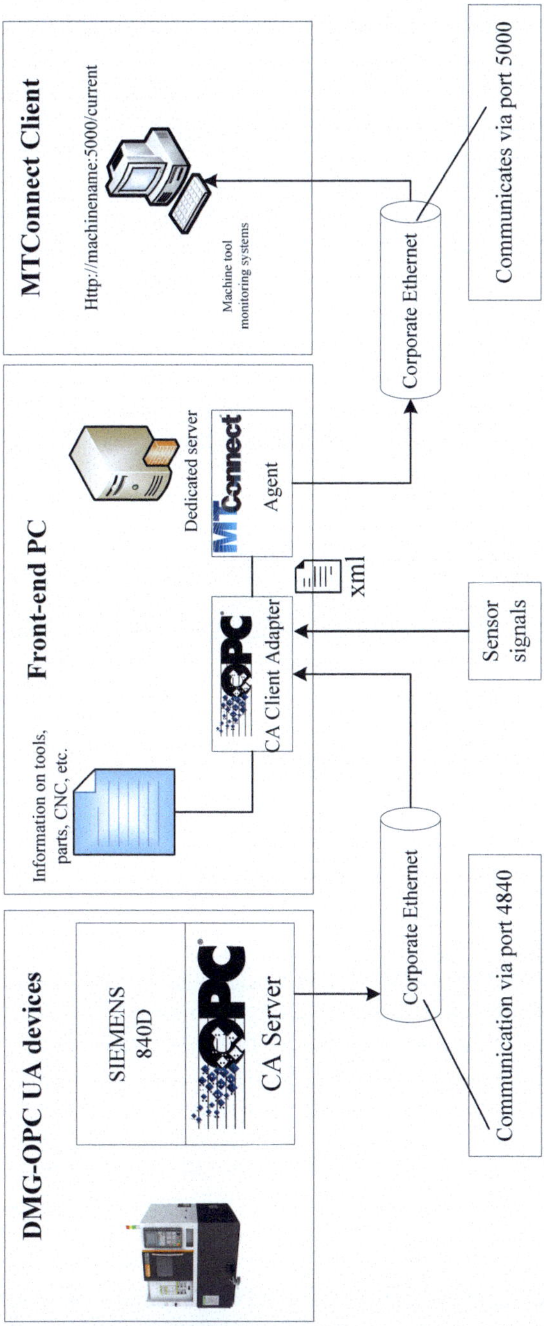

Fig. 11.33 Data acquisition flow for CNC machine tools

in a drastic reduction of the tool life, serious machining deformation of the workpiece, deterioration of the surface finish, and large surface roughness. Therefore, two indicators, tool wear value VB and surface roughness Ra, are used as the basis for tool change decision-making to ensure the safety of the process system.

As shown in Fig. 11.34, the "machine tool—tool—work-in-progress composite mapping model realized by the deep learning method in a data-driven way, predicts the health of the tool and the surface quality of the workpiece, and ultimately makes the decision to change the tool according to the prediction results. During the metal cutting test, the machine tool and process information parameters are extracted, in which the sensor parameters in the machine tool are processed by feature extraction to obtain a feature set plus the process information of its processing, i.e., cutting dosage information, as input to the classifier. The tool and work-in-process quality parameters, i.e., VB and workpiece surface roughness Ra, are measured offline during the test, and both are used as predicted outputs of the classifier, respectively. The two diagnostic results, VB and Ra, reflecting the health status of the tool, are used as the basis for decision making on whether to change the tool or not.

11.3.3 Digital Twin Processing Example

NIST gives a case where the machining of a feature can be simulated in 15 s, as shown in Fig. 11.35, and all feature simulations are completed in 80 s.

By acquiring actual machining data, based on the manufacturing digital twin model, machining features can be analyzed online to detect the actual impact of specific machining processes on machining features.

11.4 Digital Twin Intelligent Assessment

In recent years, with the rapid development of computer, modeling and digital twin technology, digital twin technology has been widely used in military, industrial, economic, environmental and social fields. With the continuous development of industrial informatization and digitization, the application scope and demand for processing digital twins are expanding, and the virtual models constructed are becoming more and more fine and complex strong. At the same time, with the deep and extensive application of digital twins in different fields, it prompts people to gradually pay attention to the question of whether the virtual model constructed based on digital twin technology can truly reflect the state of the physical world. If the constructed digital twin model is inaccurate and has low fidelity, it may mislead the decision makers and even cause irreparable military, political, and economic losses. Therefore, intelligent evaluation of digital twin models plays a crucial role in clarifying and improving the fidelity and applicability of digital twin models.

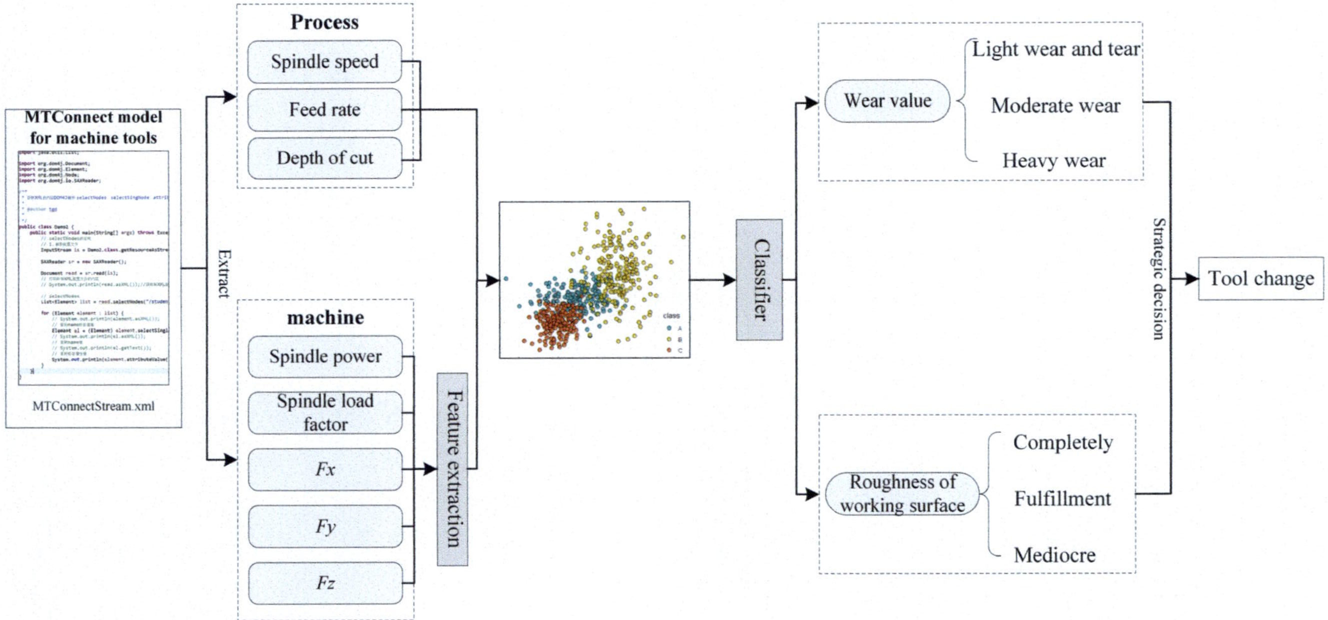

Fig. 11.34 Composite mapping model for manufacturing digital twins

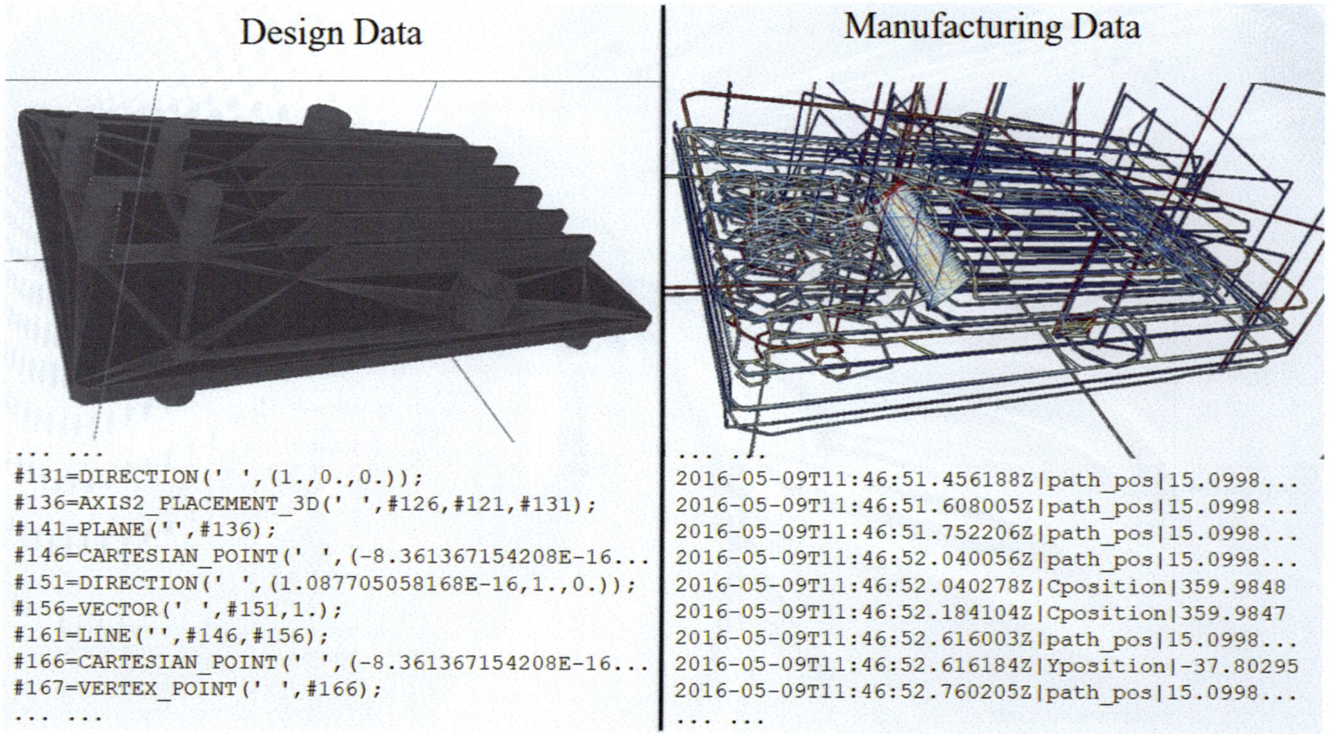

Fig. 11.35 Manufacturing digital twin example

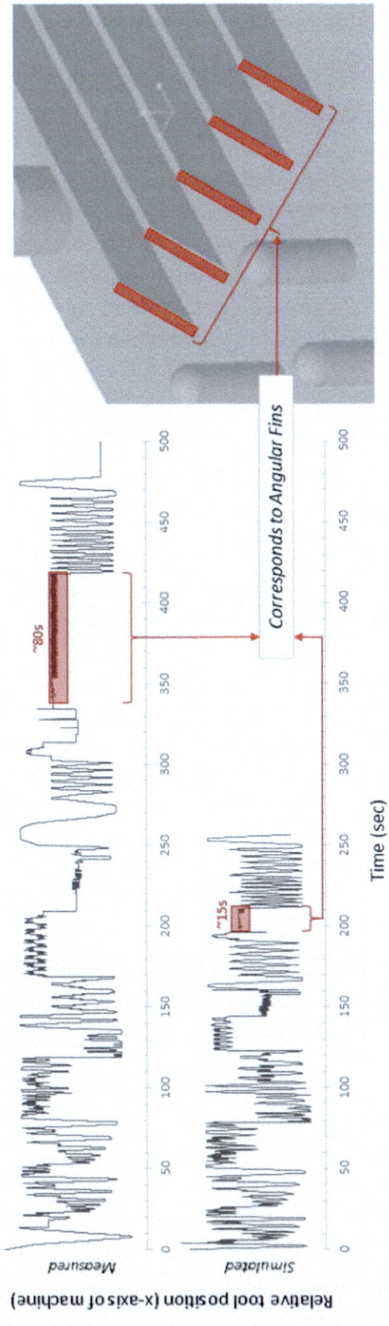

Fig. 11.35 (continued)

Intelligent assessment of digital twin models faces many challenges because the structure of simulation models is becoming more and more complex and the output characteristics are becoming more and more diverse as the functional requirements of simulation models continue to increase and the range of applications continues to expand. Therefore, the intelligent assessment of digital twin models involves the measurement and evaluation of a large number of qualitative and quantitative metrics, often relying on the empirical knowledge of domain experts, and requires the synthesis of different methods and metrics. Intelligent assessment of digital twin models is a complex multi-metric integrated assessment problem. In view of this, many researchers have proposed comprehensive assessment methods based on hierarchical analysis, fuzzy comprehensive evaluation, similarity theory, Bayesian nets, entropy weighting method, etc., which can obtain more robust and reliable comprehensive assessment results [30]. The main adjustments they face are as follows [31] (Table 11.2).

11.4.1 Model Intelligent Evaluation Ideas

(1) **Basic Concepts**

Intelligent assessment of digital twin models is a measure of the fidelity of the model. Fidelity is the degree to which a model reproduces the structure, function, and behavior of a real-world physical entity, and it measures the overall absolute difference between the digital twin virtual model and the physical entity. The digital twin model consists of several sub-models, and it is difficult to fully satisfy the need for fidelity assessment by only analyzing the results of data consistency analysis or qualitative assessment of a single behavior or a single application of the model. In addition, the processing digital twin model evolves continuously with the processing state of the product, and the assessment of the fidelity of the digital twin model should not only focus on the static state or performance of the model at a specific stage or condition, but also on the state and performance of the model during the use and management process. Therefore, for the static state, it should be evaluated from the three levels of model similarity, confidence, and accuracy; for the dynamic state, it should be evaluated from the level of model maturity. The specific definitions are as follows.

1. **Definition Based on Similarity**

Digital twin model similarity is determined by the similarity between the virtual model and the physical entity, being the degree of similarity between the two in terms of geometry, processing behavior, and functionality.

Table 11.2 Challenges to digital twin model evaluation

Challenge	Fundamental meaning
Combinatorial explosion of input variables and uncertainty	There are multiple inputs in the digital twin model, each of which may have multiple values, which are used to analyze and study the system response under different situations. Then, the combination of multiple input variables will lead to a combinatorial explosion, and it is not possible to fully test and validate each combination of inputs under the constraints of limited time and resources
Diversity and coupling of functional outputs	Digital twin models have multiple functional outputs under the same input conditions, and the multiple functional outputs have certain correlations and coupling characteristics, so it is difficult to accurately and completely define and analyze the output domain
Assessing multi-source heterogeneity of data	There are various sources and types of data for intelligent evaluation of digital twin models, including deterministic data, stochastic data, non-periodic time series, periodic time series, and descriptive functional properties, etc. Different types of data require different analytical validation methods
Difficulty in aggregating results from different types of credibility assessments	Intelligent assessment of digital twin models involves a large number of parameters and indicators, some of which can be measured to obtain accurate values, some of which can be expressed by probability statistics or interval estimation, and some of which do not have exact data and require empirical estimation or qualitative assessment by domain experts. Making full use of various types of data and studying effective methods of aggregating multiple types of data are key to improving the rationality and reliability of assessment results
Credibility assesses the complexity of indicator correlations	Intelligent assessment of digital twin models is a multi-indicator integrated assessment and decision-making problem, which often requires the establishment of an assessment indicator system. However, in practice, it is difficult to realize the complete independence of the assessment indicators from each other, and there may be complex relationships such as partial redundancy, weak correlation, etc.

2. **Definition Based on Confidence**

Digital twin model confidence is the probability that the output result of the virtual model is within the confidence interval.

3. **Definition Based on Accuracy**

Digital twin model accuracy is the degree of correctness of the model output results.

4. **Definition Based on Maturity**

Digital twin model maturity is the degree to which the virtual model meets the expected results and application requirements as the processing state of the physical

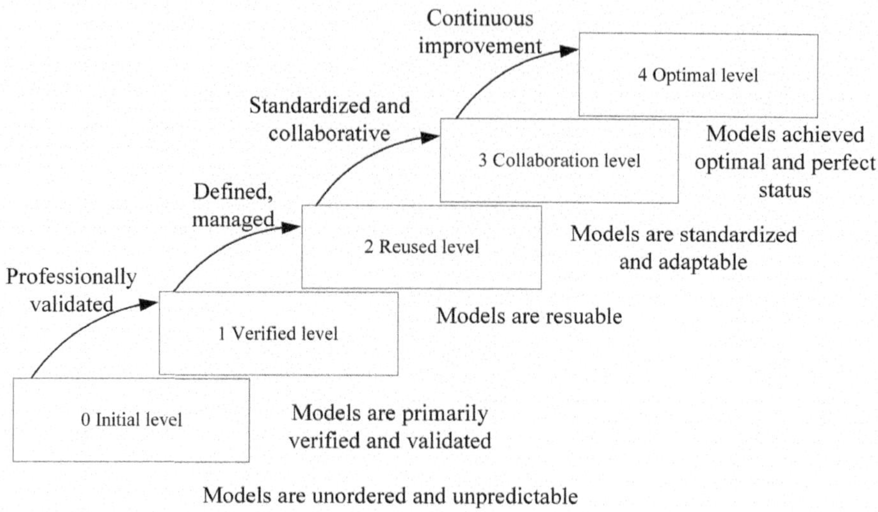

Fig. 11.36 Five levels of digital twin model fidelity

entity continues to change, which describes the evolution of the virtual model with the object being modeled in physical space.

(2) 5 Levels of Model Fidelity

Referring to the fidelity level classification methods used in other domains, model fidelity can be categorized into five levels, as shown in Fig. 11.36. These five levels are described as follows.

1. Initial Level

The modeling process at the initial level is unstandardized with no or little model documentation. The model is unvalidated and unverified, there are many errors, and there are many uncertainties in the use of the model.

2. Verification Level

The modeling process at the verification level is poorly standardized and the model documentation is incomplete and non-standard. The model has undergone initial validation, and although the model has very little error, there are still many problems.

3. Reusable Level

The reusable level has a well-organized modeling process with complete and standard model documents. The model has been professionally validated and approved with acceptable credibility, and there are no errors or problems that could lead to failure or malfunction.

4. **Collaboration Level**

The modeling process at the collaborative level has a high degree of standardization and the model has a high degree of credibility. The model is adaptable to heterogeneous environments and has few problems in use.

5. **Optimal Level**

The modeling process of the optimal level has a high degree of standardization and the model has a high degree of credibility, confidence and accuracy. The model has high adaptability to heterogeneous environments and has no problems in use.

(3) **Main Ideas for Qualitative Quantitative Analysis of Model Fidelity**

After constructing the model fidelity evaluation index system, the qualitative quantitative analysis of model fidelity should be carried out.

The basic idea of the model is to determine the weight of each indicator hierarchically. First of all, by defining an "abstract indicator" for each factor and stage (the abstract indicator is a virtual indicator, whose value does not come directly from the actual data) to "quantify" the factors of the first level and the stage of the second level (the abstract indicator is a virtual indicator, whose value does not come directly from the actual data), and then to "quantify" the fidelity of the model by defining an "abstract indicator" for the first level and the second level (the abstract indicator is a virtual indicator, whose value does not come directly from the actual data). (the abstract indicator is a virtual indicator, its value does not come directly from the actual data), and then for each level using different methods to determine the weight of each abstract indicator, and from the next level to obtain the value of each level of the abstract indicator, the value of the indicator from the bottom of the actual comparison, is the result of the model data and the actual processing of the data compared to the value of the results.

The impact of different indicator factors on model fidelity is different, therefore, considering all indicators as equal cannot truly reflect the actual impact of each indicator on model fidelity, and it is necessary to determine the impact weight of each indicator factor. Usually, the weights of the first layer of evaluation metrics can be set by expert experience, because for a specific type of model, domain experts have a certain consensus on the impact of model fidelity. The weights of the evaluation metrics in the middle layer can be determined by a combination of qualitative and quantitative methods, such as the hierarchical analysis method or the fuzzy analysis-hierarchy process. The weights of the bottom layer evaluation indexes are determined by different quantitative methods according to the actual data needs, such as entropy weighting method and neural network method.

This hierarchical evaluation method takes into account the influence of stages and factors at different levels, while reducing the impact of uncertainty in the combination of the underlying indicators due to incomplete or inaccurate data on the results of the overall fidelity evaluation, and is suitable for the evaluation of the fidelity of models with a complex indicator system and a variety of levels.

(4) **Principles**

When constructing the digital twin model fidelity index system, the characteristics of the simulation system should be considered in addition to the general principles of the evaluation index system. As many domain-independent indexes as possible should be selected to objectively evaluate the performance of the model in terms of cross-domain reuse and scalability. More importantly, indexes that change over time and model usage will play a key role in the indexing system, thus highlighting the evolution process of the model. In conclusion, the construction of the model fidelity evaluation index system should follow the following principles.

1. Basic principles: including completeness, computability, non-bindingness, and objectivity.
2. Usage- and time-dependent principle: choose indexes that change with usage and time to reflect the state of the modeling process being used and managed.
3. Domain-independent principle: Selecting domain-independent indexes as far as possible ensures the versatility and extensibility of the indexing system.

(5) **Model Fidelity Assessment Methods**

For the assessment of the fidelity of the digital twin model, it is usually necessary to establish a fidelity assessment index system, comprehensively use a variety of qualitative and quantitative assessment methods, and select appropriate verification and assessment methods according to the characteristics of each assessment index [32]. These methods can be divided into three categories: qualitative methods, quantitative methods and comprehensive methods, as shown in Table 11.3.

(6) **Model Fidelity Evaluation Index System Framework**

Based on the analysis of the composition and working principle of the digital twin model, the model fidelity evaluation index system shown in Fig. 11.37 is proposed [33]. The index system has multiple layers, the first layer is the evaluation indexes of the geometric model, mechanism model, and data model, the middle layer is the index factors affecting the model fidelity in the corresponding stages, and the bottom layer is the indexes corresponding to each index factor. In some cases, there may be more than one intermediate layer. In addition, the framework of the indicator system is open and can be adapted to the characteristics of the models in a particular application area.

In the index system, F is the comprehensive evaluation index of digital twin model, and its value is closer to 1 indicating higher fidelity and vice versa. Geometric model fidelity F1, mechanism model F2 and data model fidelity F3 are the initially established first-level evaluation indicators. These three independent modules are respectively refined and decomposed to obtain a multilevel indicator factor set. In this case, the lower level indicators are evaluated against the higher level indicators using a weighted average.

Table 11.3 Digital twin model fidelity assessment methods

Method classification			Method name
Qualitative methods			Turing test, graphical method, surface verification, curve method
Quantitative methods	Static		Point estimation, interval estimation, Bayesian estimation, tour test, confidence interval method, regression analysis
	Dynamic	Time domain Frequency domain Time–frequency domain	Gray correlation analysis, error analysis, feature matching, similarity coefficient method Fourier spectrum analysis, window spectrum analysis, cross spectrum analysis, instantaneous spectrum analysis Time–frequency domain discrete wavelet transform, discrete short-time Fourier transform Hierarchical analysis, fuzzy comprehensive judgment, neural network, support vector machine
	Integrated approach		Hierarchical analysis, fuzzy comprehensive judgment, neural networks, support vector machines

(7) **Model Fidelity Evaluation Process**

The following figure shows the general process of evaluating the fidelity degree of a digital twin model, and a specific description of the evaluation process is shown in Fig. 11.38:

1. Form a team. Model fidelity can be assessed by digital twin model users or third-party organizations. The team usually consists of assessment engineers in the field of smart manufacturing and industry experts in the field to which the digital twin belongs.
2. Develop a plan. Based on the complexity and scale of the model being evaluated, the capabilities and constraints of the evaluation team, develop a detailed evaluation plan.

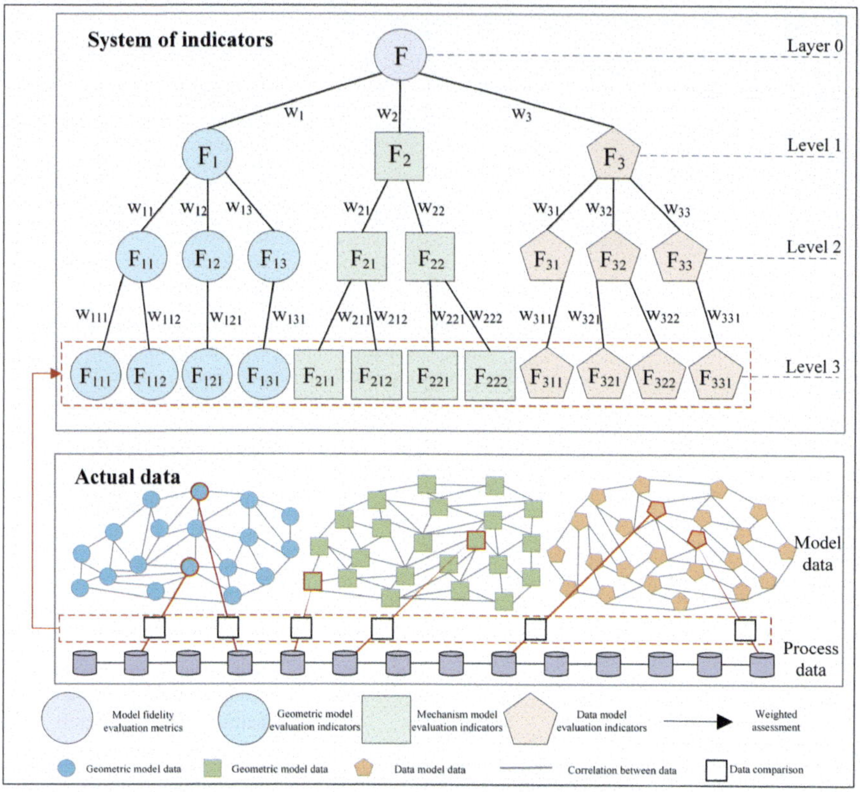

Fig. 11.37 Qualitative quantitative assessment system for high-fidelity models

3. Design an indicator system. According to the nature of the model and the feasibility of the evaluation, add new indicator factors to the previously mentioned indicator system.
4. Calculate the underlying indicators. According to the definition of each indicator, actually compare the model data with the actual processing data.
5. Determine the weights of the indicators. Use a combination of qualitative and quantitative methods to determine the weight values of each layer. The weights of the upper layers rely more on the experience of experts, and the weights of the lower layers rely more on data-based calculations.
6. Calculate the maturity of the model.

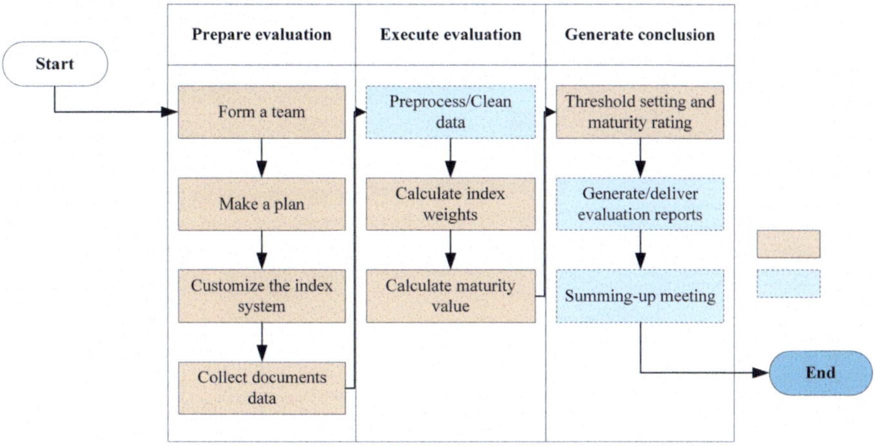

Fig. 11.38 General process for model fidelity evaluation

11.4.2 Model Fidelity Assessment Methods

(1) **Digital Twin High Fidelity Modeling**

The premise of constructing digital twin high-fidelity model is to create high-fidelity digital information model [34]. Figure 11.39 shows the modeling architecture of the processing digital twin high-fidelity model, firstly, the digital information model is constructed from the aspects of geometric structure, deformation mechanism, processing behavior, etc.; the ontology [35] technology is used to provide a unified digital description of the models, and the models are associated, combined and integrated through semantic matching and semantic mapping; then the ontology is instantiated to form the relational network, i.e., the knowledge graph [36], so that the knowledge can be acquired; this relational network will continuously change the "relationship" and "relationship" with the processing process. Knowledge is acquired; this relational network changes the "relationships" and "entities" continuously with the processing, and there is no need to re-import data when adding/modifying/deleting fields. In addition, as the product is processed, the physical processing data is continuously compared with the digital twin model data to validate the constructed digital twin model.

(2) **High-Fidelity Geometric Modeling**

The digital twin geometric model is essentially a three-dimensional solid model containing the real geometrical shape, which is iteratively updated with the evolutionary process of machined parts. In order to accurately reflect the true geometric state of the physical space, a complete description of the geometric model is provided by the technique based on Model Based Definition (MBD) [35, 36]. The MBD technique adds labeling information on the 3D model, which realizes the

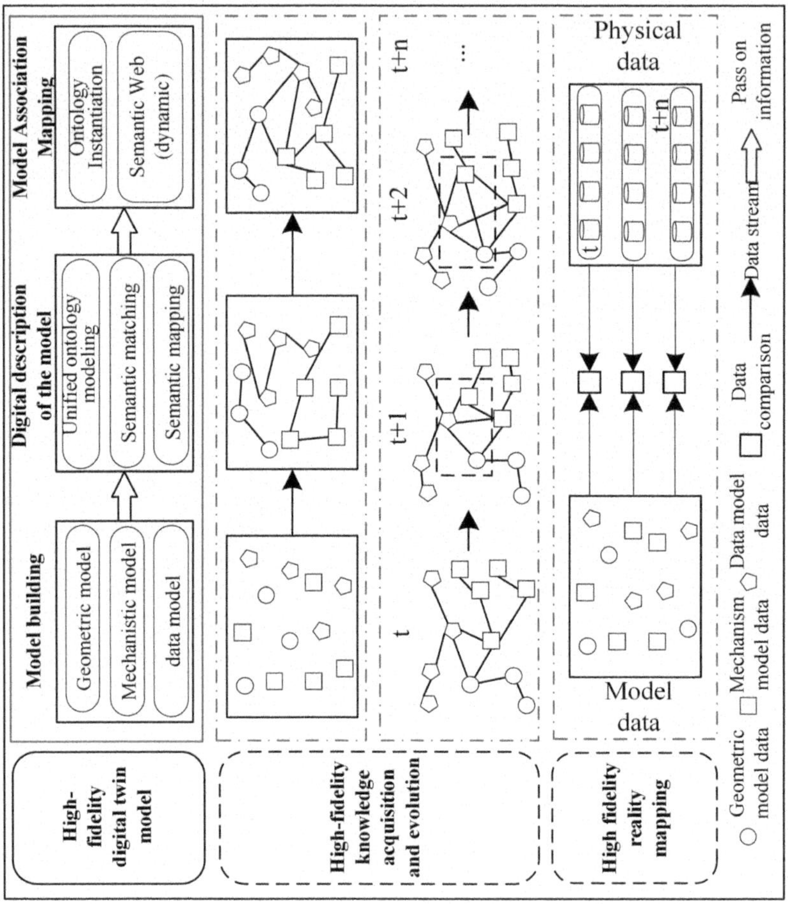

Fig. 11.39 High-fidelity digital twin modeling architecture for rotational molding processing

transition between Computer Aided Design (CAD) and Computer Aided Design (CAD) to Computer Aided Manufacturing (CAM) integration. The increased annotation information, including geometric information and non-geometric information, geometric information contains three-dimensional space solid model processing features, dimensions, tolerance information; non-geometric information for the engineering annotation information. The expression form of the high-fidelity geometric model for thin-walled parts is as follows:

$$Geo = \{Pr\_Feat,\ Mo\_Enti,\ Di\_Tole,\ En\_Note,\ Ot\_Info\} \tag{11.1}$$

$$Mo\_Enti = \{Mo\_Geom,\ Ai\_Geom\} \tag{11.2}$$

$$En\_Note = \{St\_Note,\ Pa\_Note,\ Ma\_Note\} \tag{11.3}$$

where: Pr_Feat for machining features, Mo_Enti for solid model, Di_Tole for dimensions and tolerances, En_Note for engineering notes, Ot_Info for other information, Mo_Geom for model geometry information, Ai_Geom for auxiliary geometry information, St_Note for standard notes, Pa_Note for part notes, Ma_Note is the material note.

(3) **High-Fidelity Mechanism Modeling**

The digital twin mechanism model includes physical model, data model and simulation. Before simulation, it is necessary to carry out a theoretical analysis of the mechanism of the deformation phenomenon of thin-walled parts, and establish a simplified physical model of the deformation state on the basis of the theoretical analysis, according to which a mathematical model representing the main parameters affecting the deformation phenomenon is deduced. In order to accurately reflect the state and behavior of the physical entity in the real environment, the simulation software is used to carry out spinning kinematics simulation, debugging each parameter in the mathematical model, and verifying the mutual influence relationship between the main parameters. Through efficient simulation, molding defects are controlled during the machining process.

(4) **High-Fidelity Data Modeling**

The digital twin data model is based on the real-time/historical data collected during the processing of thin-walled parts, and uses intelligent algorithms, such as machine learning and deep learning, to analyze and predict the unknown physical laws of rotational molding processing. In order to realize the real and complete mirroring from physical space to virtual space, factors such as behavior, rules and constraints of the spinning and forming process need to be taken into account when constructing the data model. Among them, the behavioral model covers the behavioral model of the spinning process, the competitive and collaborative behavioral model with other entities in the manufacturing space, and the failure behavior analysis model; the rule model covers all kinds of reasoning, correlation, and decision-making rule models in the thin-walled part forming process; and the constraint model covers all kinds of constraint models in the manufacturing environment.

All of the above models can be modeled uniformly in virtual space using a formal modeling language, which is represented as follows:

$$TP_i,\ SE = gMHf \propto mMHf \propto dMHf \tag{11.4}$$

where: TP_i and SE denote the digital information models of thin-walled machined parts and spinning machining equipment, respectively, i denotes the number of thin-walled machined parts, $gMHf$ denotes its geometrical model set, $mMHf$ denotes its mechanistic model set, $dMHf$ denotes its data model set, and $\propto$ is a natural connective.

(5) **High Fidelity Knowledge Acquisition and Evolution**

1. **Digital Description of High-Fidelity Models**

Before acquiring knowledge, it is first necessary to solve the problems such as poor knowledge generation caused by the lack of uniformity in data format among heterogeneous models. The high-fidelity models constructed with the help of ontology concepts and technical sections are modeled with unified descriptions to provide support for subsequent knowledge acquisition and knowledge evolution. The ontology can clearly describe the model concepts and the semantic relationship between them, and based on the rule-based reasoning of the domain ontology, it eliminates the semantic isomorphism between the terms and realizes the semantic matching between the concepts. The ontology description of high-fidelity models mainly includes geometric description (product design information, definition information), attribute description (basic metadata information, function information, parameter information) and behavioral description (static structure, dynamic behavior information).

2. **High-Fidelity Model Semantic Matching and Mapping**

There is an association relationship between high-fidelity models, which combines the ontology with geometric, mechanistic, and data meta-models to construct semantic matching and mapping between each meta-model element and the ontology concepts to realize information sharing and interaction between heterogeneous models [37]. Figure 11.40 shows the high-fidelity model semantic matching and mapping mechanism, in which the meta-model G, M, D is a unified digital description of the object model, a modeling language, which expresses the modeling specifications of the model GM, model MM, and model DM, respectively; the meta-ontology and domain ontology are the references for the mapping between different (meta-) models, which give semantic information to the elements within the meta-model and

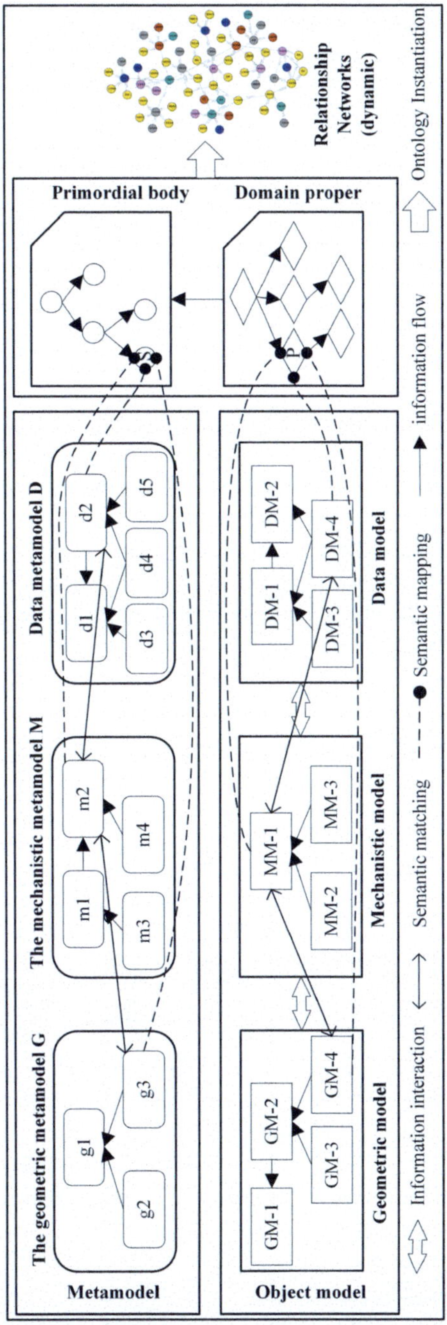

Fig. 11.40 High-fidelity model semantic matching and mapping mechanism

the domain model; and the relational network, which essentially is semantic network, which is an ontology-based knowledge base.

The realization of information sharing between heterogeneous models needs to consider the semantic mapping problem at both the meta-model and object model levels, and the semantic mapping between object models is based on the semantic mapping between meta-models.

1. Semantic Matching and Mapping of Metamodels. Assuming that g3 and m2 are common elements of meta-models G and M, then g3 and m2 have similar "type semantics" and there exists a semantic mapping between g3 and m2 and concepts S in the meta-ontology, so that the semantics of g3 and m2 are equivalent, denoted by: g3, m2 $\in$ a′, S $\in$ C′A. If mapping (g3 $\rightarrow$ S and m2 $\rightarrow$ S), then g3 = M(m2). where a′ = denotes the intersection of all metamodels, C'A denotes the syntactic rules and relational statutes between metamodels, and M denotes semantic equivalence.
2. Semantic Matching and Mapping of Object Models. Similar to the relationship between meta-models and meta-ontologies, there exists a semantic mapping between *GM-4* in the geometric model, *MM-1* in the mechanistic model and the concept P in the domain ontology, so the semantic equivalence between GM-4 and MM-1 is expressed as *GM-4* $\in$ geometric model $\wedge$ *MM-1* $\in$ mechanistic model, $Syn_{GM\text{-}4}$ = transformation ($Syn_{MM\text{-}1}$), if mapping (*GM-4* $\rightarrow$ *P* and *MM-1* $\rightarrow$ *P*), then *GM-4* = *M*(*MM-1*). where Syn denotes the syntactic rules and relational statutes between object models.
3. High-fidelity model knowledge evolution. The ontology models are instantiated to form a network of spinning relationships, and the knowledge is acquired by analyzing the relationships between the "entities". The acquired knowledge can be used to reduce the probability of errors in the operation of the digital twin virtual environment and improve the credibility of the twin model and twin data. In order to comprehensively acquire knowledge, it is necessary to build a dynamic relational network, add the time dimension to the cross-section data, and analyze the changes and trends of the graph structure over time by using temporal analysis techniques and graph similarity computation techniques, so as to expand the knowledge and grasp the key information.

(6) **Evaluation Process Under Virtual-Reality Mapping**

Digital twin high-fidelity virtual mapping is the basis for realizing self-feedback and self-improvement of high-fidelity models. Virtual model through the sensor, CNC spinning machine, measuring instruments and other equipment collected by the spinning processing information (blank information, tooling information, equipment operation information, thin-walled parts geometric information) synchronous update to sense its state, predict the situation that will occur, and then based on the history of the data and the analysis of the current data to make a decision, to achieve the dynamic virtual reality mapping.

Figure 11.41 shows a thin-walled parts spin molding processing high-fidelity virtual reality mapping process, thin-walled parts from the initial state (billet)

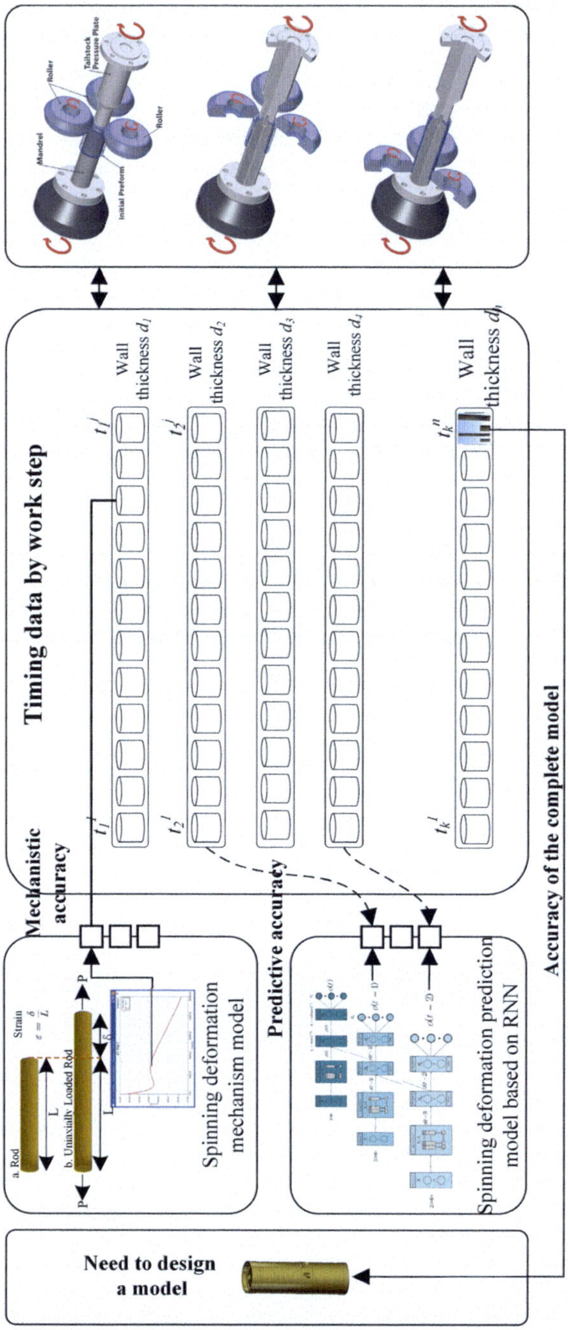

Fig. 11.41 High-fidelity reality-virtual mapping process for rotational molding processing

converted to the final state (semi-finished products) of the molding process is a time-dependent process, resulting in the processing data is the use of digital measuring instruments according to the step of the actual measurement obtained by the time series data. The machining process of a thin-walled part is divided into k time periods, denoted as {t1, t2, ..., tk}, and a time period is further divided into i moments, denoted as {t11, t12, ..., t1i}. During the t1 time period, the simulation data are compared with the processing data in real time to determine the model mechanism accuracy; At the same time, the prediction model predicts the values of deformation, such as wall thickness, at time t2 based on historical data; at time t2, in addition to determining the accuracy of the mechanism and predicting the values of deformation at time t3, it also compares the actual values of deformation with the predicted values of deformation at time t1 to determine the accuracy of the model's prediction; and so on until the completion of the machining process; and then it compares the semifinished parts with the design model to determine the accuracy of the complete model; and finally, the results of the comparison are used to generate a decision to guide the actual machining process. Finally, the results are compared to generate a decision to guide the actual machining process. See Algorithm 1 pseudo-code for the specific data mapping process. Where Mc represents the result of the mechanism accuracy comparison, Pc represents the result of the prediction accuracy comparison, and Dc represents the result of the complete model accuracy comparison.

Algorithm 1: Virtual Reality Mapping for Spin Molding of Thin-Walled Parts

Inputs: thin-walled part machining time period $T = \{t_1, t_2, \ldots, t_k\}$, time period corresponding to the set of moments $t = \left\{t_j^1, t_j^2, \ldots, t_j^i\right\}, j = 1, 2, \ldots, k$, spinning process dataset $X = X(t)$, simulation dataset $S = S(t)$, prediction dataset $P = P(t)$, design nominal dataset $D = D(t)$. data set $P = P(t)$, design nominal data set $D = D(t)$

Output: Comparison results

(1) for each t_k in T do
(2) $Mc = 0$;
/* Comparison of simulation data and process data */
(3) for each ${t_k}^i$ in t do
(4) $Mc = \text{abs}(X(t) - S(t))$
(5) $P'(t) = P(t)$;
(6) end
(7) $Pc = 0$;
/* Comparison of forecast and process data */
(8) for each ${t_{k+1}}^i$ in t do
(9) $Pc = abs(X(t) - P'(t))$;
(10) End
/* Comparison of design nominal data and finished product completion data*/
(11) $Dc = \text{abs}(X({t_k}^n) - D({t_k}^n))$
(12) return Mc, Pc, Dc

11.4.3 Evaluation Cases

An aerospace enterprise uses a strong spinning process to process the thin-walled shell parts of the combustion chamber, the processed parts undergo intense elastic–plastic deformation during the molding process, and there is a large work hardening phenomenon as well as rebound phenomenon, which makes it difficult to control the overall molding of the thin-walled parts, and therefore digital twinning technology is needed to improve the spinning process and enhance the quality of the spinning parts processing.

Figure 11.42 shows the MBD model of an aerospace thin-walled part, including 3D solid geometry, tolerance dimension, surface material and other attribute information. In the figure, the outer diameter of the workpiece is $D = 195.2$ mm, the inner diameter is $d = 174.6$ mm, and the three rotating wheels are uniformly distributed around the core mold axis at 120°, and the amount of thinning decreases gradually, accounting for 42%, 34%, and 24% of the total amount of thinning, respectively.

After the MBD model of thin-walled parts is constructed, Protégé is used as an ontology modeling tool to establish its ontology model based on the actual spinning process of thin-walled parts, which describes the design geometry of thin-walled shell parts, micro-organizational structure, spinning and forming machining process, dimensional inspection, and other information, and provides a framework basis for the establishment of the relational network as shown in Fig. 11.43.

After the ontology model is constructed, the Web Ontology Language (OWL) is used to describe the domain ontology, as shown in Fig. 11.44. According to the basic concepts in the domain ontology < Geometric Model, Mechanism Model, Data Model, Wall Thickness >, and the relationship Relationship {Wall Thickness|Geometric Model → Mechanism Model, Mechanism Model → Data Model}

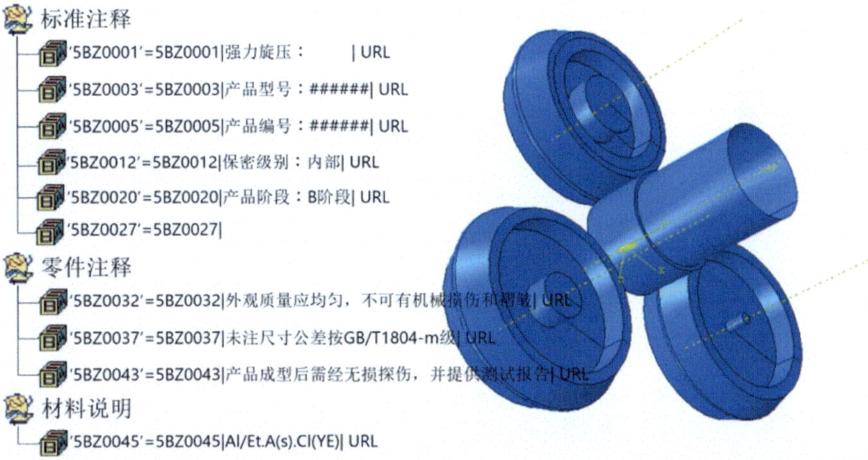

Fig. 11.42 MBD model of an aerospace thin-walled part

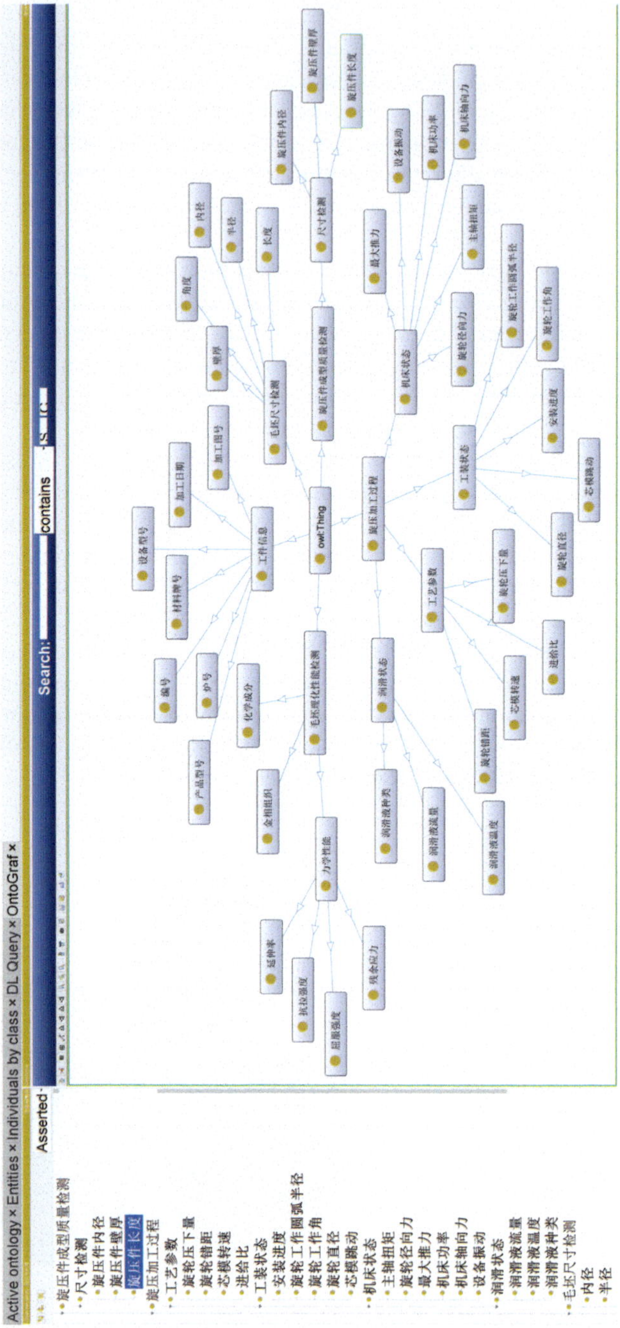

Fig. 11.43 Digital twin ontology model for a thin-walled part spinning

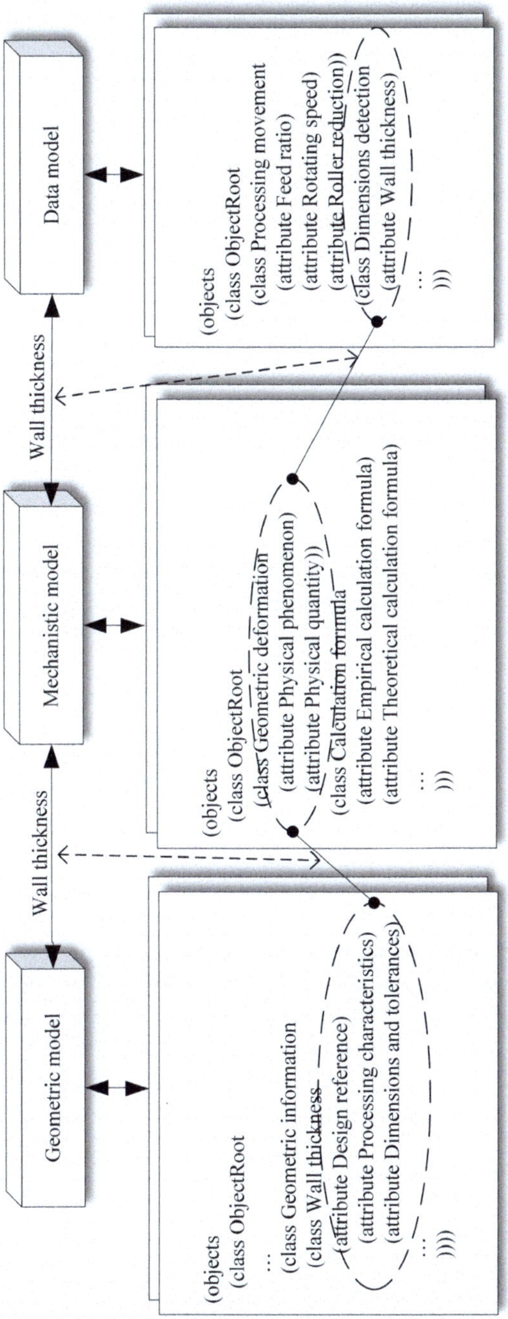

Fig. 11.44 Ontology-based heterogeneous information sharing

among them, the information interaction relationship among the three heterogeneous models sharing the object model can be reasoned out.

After sharing information between heterogeneous models, a javascript library (Data-Driven Documents, D3.js) based on data manipulation documents is used to construct a spinning processing relationship network, as shown in Fig. 11.45. The relational network abstracts the information in the ontology model into various entities, with nodes representing the entities, and the lines between the nodes representing the relationships between the entities, and describes their related attributes through "node-relationship-node". For example, "Dimension matching exceeds the tolerance-result-stop processing", "Spinning parts crack-reason-first feed ratio is too large".

After constructing the spinning digital twin model, the fidelity of the model needs to be measured and evaluated. The lowest metric of the metric system is the value that is directly quantified by data comparison. For example, Fig. 11.46 shows a thin-walled part model after meshing in finite element simulation, and the wall thickness value of the corresponding cross section can be calculated according to the extracted resultant coordinates.

Three combinations of spinning process parameters shown in Table 11.4 were selected for simulation, and strong spinning test was carried out to verify. The actual wall thickness was detected by ultrasonic thickness gauge to calculate the rebound amount according to the wall thickness value, and the results are shown in Table 11.4.

By comparing the simulation value and the test value to the thin-walled parts wall thickness simulation output similarity evaluation index value, the results show that the similarity is basically more than 95%, indicating that there is a good fidelity (Table 11.5).

In the data model, the multiple linear regression model was selected to predict the wall thickness value and the mean square error method was used for evaluation. Fifty sets of spinning processing data of a thin-walled part were collected, which were divided into training data and test data, and the training data included seven fields, which were working angle, spinning wheel diameter, spinning wheel working arc, feed ratio, spinning wheel misalignment distance, mandrel rotational speed, and wall thickness; and the test data included six fields, excluding the wall thickness value. Figure 11.47 shows the comparison of wall thickness prediction value and target value, the two folds in the figure have basically the same trend, indicating that the constructed model can predict the wall thickness value more accurately; in addition, the mean square error value of the model output is $MSE = 0.0043$, $MSE < 0.005$, indicating that it has a good degree of fidelity.

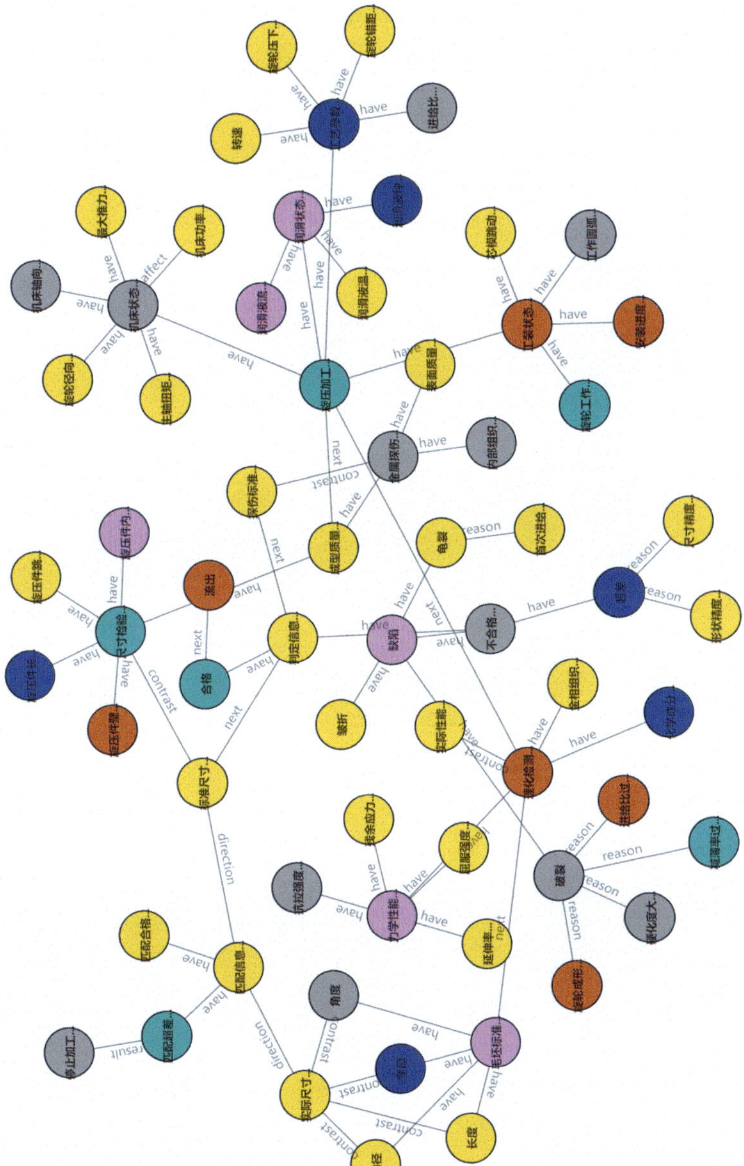

Fig. 11.45 Spinning processing relationship network

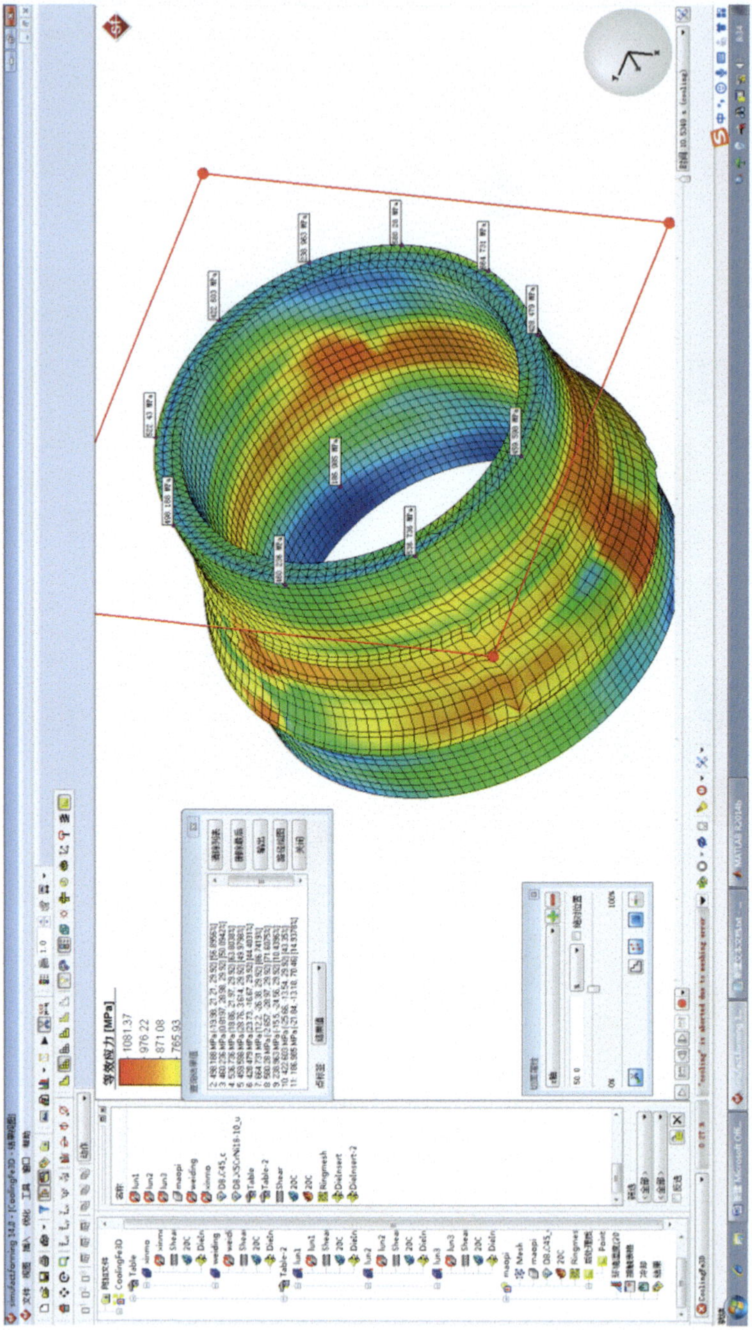

Fig. 11.46 Thin-walled part model after mesh division

Table 11.4 A thin-walled parts spinning process parameter combinations

Test number	Working angle (α/°)	Rotor diameter (ϕ/mm)	Working radius of spinning wheel (ρ/mm s^{-1})	Feed ratio (f/mm)	Rotor misalignment (c/mm)	Mandrel speed (r/r min^{-1})
1	20	125	2	0.5	2	60
2	22.5	145	7	1.25	6	60
3	25	165	12	0.75	5	60

Table 11.5 Simulated and tested values of spinning of a thin-walled part

Test number	Simulation value	Experimental value	Similarity (%)
1	0.424	0.437	97.1
2	0.381	0.402	94.9
3	0.439	0.451	97.4

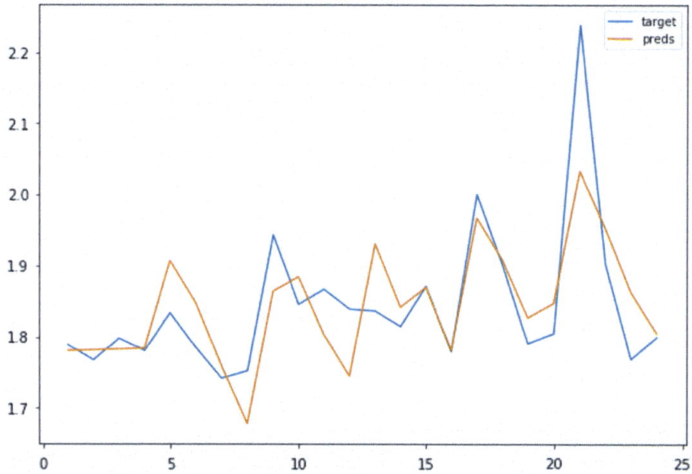

Fig. 11.47 Comparison of predicted and target wall thicknesses for a thin-walled part

References

1. Rosen R, Von WG, LO G et al (2015) About the importance of autonomy and digital twins for the future of manufacturing. IFAC Pap Online 48(3):567–572
2. Grieves M (2005) Product lifecycle management: the new paradigm for enterprises. Int J Prod Dev 2(1/2):71–84
3. Tuegel EJ, Ingraffea AR, Eason TG et al (2011) Reengineering aircraft structural life prediction using a digital twin. Int J Aerosp Eng. https://doi.org/10.1155/2011/154798

4. Tuegel EJ (1812) The airframe digital twin: some challenges to realization. In: Proceedings of the 53rd AIAA/ASME/ASCE/AHS/ASC structures, structural dynamics and materials conference 20th AIAA/ASME/AHS adaptive structures conference 14th AIAA, 2012
5. Glaessgen E, Stargel D (2012) The digital twin paradigm for future NASA and US air force vehicles. In: Proceedings of the 53rd structures, structural dynamics and materials conference. Special session: digital twin. Honolulu, Hi, USA, pp 1–14
6. Shaoftom M, Conroy, Doyle R et al (2012) Modeling, simulation, information technology & processing roadmap. NASA, Washington, DC, USA
7. Warwick, G (2015) GE advances analytical maintenance with digital twins. In: Aviation week & space technology, 19 Nov 2015
8. Tao F, Zhang M, Chen JF et al (2017) Digital twin workshop: a new paradigm for future workshop. Comput Integr Manuf Syst 23(1):1–9
9. Li H, Tao F, Song W et al (2019) Integration framework and key technologies of complex product design-manufacturing based on digital twin. Comput Integr Manuf Syst 25(6):1320–1336
10. Liu SM, Bao JS, Lu YQ et al (2020) Digital twin modeling method based on biomimicry for machining aerospace components. J Manuf Syst 58:180–195
11. Zhuang CB, Liu JH, Xiong H (2018) Digital twin-based smart production management and control framework for the complex product assembly shop-floor. Int J Adv Manufact Technol 96(1–4):1149–1163
12. Meraghni S, Terrissa L, Yue M et al (2021) A data-driven digital-twin prognostics method for proton exchange membrane fuel cell remaining useful life prediction. Int J Hydrogen Energy 46(2):2555–2564
13. GE Digital Twin (2016) Analytic engine for the digital power plant
14. Siemens. Digital twin. https://www.plm.automation.siemens.com/global/en/our-story/glossary/digital-twin/
15. Deloitte. Digital twin. https://www.deloitte.com
16. Boeing. Digital twin
17. SAP. Digital twin. https://www.sap.com
18. PTC. Digital twin
19. Ansys. Digital twin framework. www.ansys.com
20. Microsoft. Digital twin
21. Oracle. Digital twin. www.oracle.com
22. Fraunhofer. Digital twin. https://www.fraunhofer.de
23. Fei T, Weiran L, Qianhua L et al (2018) Exploration of digital twin and its application. Comput Integr Manuf Syst 24(01):1–18
24. Tao F, Liu WR, Liu JH et al (2018) Digital twin and its potential application exploration. Comput Integr Manuf Syst 24(01):1–18
25. Tao F, Cheng JF, Qi QL et al (2018) Digital twin-driven product design, manufacturing and service with big data. Int J Adv Manuf Technol 94(9–12):3563–3576
26. Zhuang C, Liu J, Xiong H et al (2017) Connotation, architecture and trends of product digital twin. Comput Integr Manuf Syst 23(4):753–768
27. Hedberg TD Jr, Hartman NW, Rosche P et al (2017) Identified research directions for using manufacturing knowledge earlier in the product life cycle. Int J Prod Res 55(3):819–827
28. Liu S, Bao J, Lu Y et al (2021) Digital twin modeling method based on biomimicry for machining aerospace components. J Manufact Syst
29. Liu S, Lu Y, Li J et al (2021) Multi-scale evolution mechanism and knowledge construction of a digital twin mimic model. Robot Comput Integr Manufact 71(5):102123
30. Yuan G, Bao J, Zheng X et al (2018) Tool life prediction for high-speed milling of titanium alloy based on CNC real-time monitoring data-driven method. China Mech Eng 29(004):457–462, 470
31. Zhai S, Gu J, Zhan K (2011) Research on the index system of simulation model credibility assessment. J Syst Simul 023(B07):26–29

32. Xu Z, Yang X, Tang J et al (2020) Credibility assessment of complex simulation models: challenges and methods. Sci Technol Eng 17
33. Li W, Lin S, Zhou Y et al (2018) Research progress on credibility assessment of complex simulation systems. Sci China Inf Sci 048(007):767–782
34. Hu F, Yang Y, Liu S, Zheng X, Lu X, Bao J, A high-fidelity modeling method for spin molding digital twins of aerospace thin-walled parts. Comput Integr Manufact Syst 1–17
35. Yang L, Chen S, Wang X et al (2019) Digital twins and parallel systems: development status, comparison and outlook. J Automat 45(11):2001–2031
36. Zhu J, Zhang H (2020) An ontology-based knowledge representation model for simulation thought space. J Syst Simul 6(11):1–10
37. Yu Y, Gu L, Yin P et al (2017) Research and application of MBD model ontology modeling and retrieval technology. J Beijing Univ Aeronaut Astronaut 43(02):260–269

Chapter 12
AI + AR-Assisted Manufacturing

This chapter first provides an overview of AR. Secondly, it introduces the AR-oriented multi-view construction and interaction method of machining process information and the model virtual-reality fusion method based on multiple information fusion. Finally, a machining process prototype system based on AR is developed and a case study is presented.

12.1 AR Overview

This subsection firstly explains the definition and development of AR, and secondly summarizes the key technologies involved in AR.

12.1.1 AR Definition and Development

A report by the Battelle Institute in the United States indicates that augmented reality (AR) and its research applications are set to be a strategically significant trend in the coming years. This technology is expected to have a long-term impact on global social development. Since Thomas and his colleagues introduced the concept of augmented reality in 1992, it has become a focal point for commercial entities, experts, scholars, and research units both domestically and internationally. At present, the academic community has yet to reach a consensus on an authoritative definition of augmented reality. Wikipedia defines augmented reality (AR) as a real-time, direct or indirect view of the physical environment in which computer-generated sensory information—such as sound, video, images, or data—is integrated. This integration enhances the physical world by adding digital elements to its components. Baidu Encyclopedia describes augmented reality as a technology that overlays simulated

T. Liu et al., *Industrial Intelligence: Methods and Applications*, Springer Series in Advanced Manufacturing, https://doi.org/10.1007/978-3-031-81477-8_12

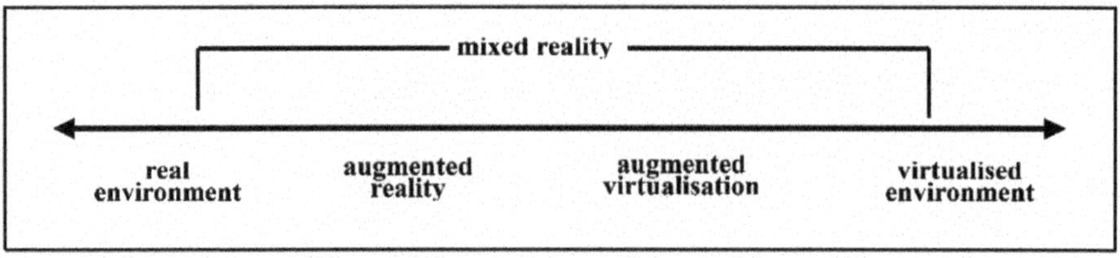

Fig. 12.1 Reality-virtual continuum model

sensory information (visual, auditory, olfactory, tactile, etc.) onto the real world, providing an experience that surpasses reality. Despite slight variations, the essence of the definitions from Wikipedia and Baidu Encyclopedia is fundamentally identical. This paper synthesizes existing definitions and relevant academic discussions to characterize augmented reality (AR) as a technology that integrates information technology, various sensing technologies, computer vision, and multimedia. AR precisely "overlays" computer-generated two-dimensional or three-dimensional virtual information—including 3D objects, animations, images, graphics, sounds, and text—onto the user's real environment. The goal is to enhance users' perception of the real environment by seamlessly integrating virtual digital content with the real world. In 1994, researchers at the University of Toronto studying augmented reality introduced the "Reality-Virtuality Continuum" model. This model illustrates the transition process between real and virtual environments, as shown in Fig. 12.1.

12.1.2 AR Key Technology

During the operational process, the augmented reality system typically undergoes a series of fundamental procedures. Firstly, the system captures genuine visual footage via the camera, subsequently analyzing and processing the imagery based on the acquired real-world scenes. Through the conversion calculations involving the world coordinate system, camera coordinate system, imaging plane coordinate system, and image coordinate system, the system derives the camera's external parameter data. This data allows for the determination of the camera's relative position and orientation within the real environment. Consequently, it facilitates the real-time generation and integration of virtual three-dimensional models, animations, images, sounds, text, and other digital content into the actual setting. The amalgamation of the virtual and real elements occurs through a process of fusion, ultimately conveyed to the viewer via a display device. Figure 12.2 exemplifies the fundamental workflow of a typical augmented reality system.

American laboratory scientists have delineated three fundamental characteristics: virtual-real fusion, real-time interaction, and three-dimensional registration. The ultimate objective of the system is to seamlessly overlay virtual digital content onto the

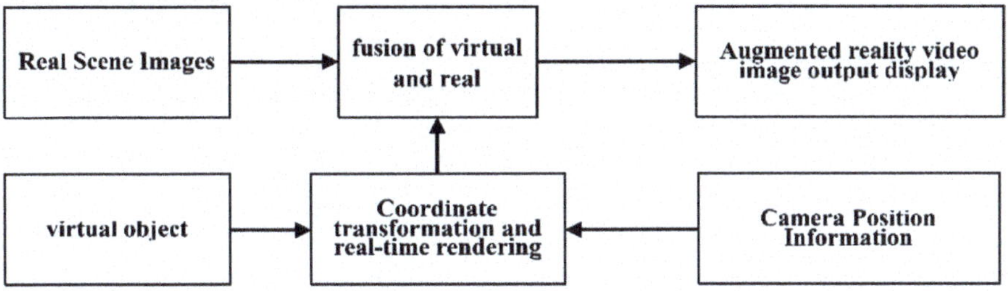

Fig. 12.2 Typical basic flow of an augmented reality system

real-world environment captured by the camera in real time. This allows for instantaneous display of the augmented information through a display device, facilitating user interaction. In this intricate process, 3D tracking registration serves as the pivotal mechanism for the convergence of virtual and real elements. Three-dimensional tracking registration involves the system's real-time detection and identification of image features within the physical environment. It determines the camera's position and orientation in three-dimensional space relative to the actual surroundings based on the video image frames acquired. This precision allows for the accurate placement of virtual objects within the real scene.

To attain the flawless amalgamation of virtual objects and real-world scenes, 3D tracking registration technology stands as the core, indispensable technology of the system.

In augmented reality systems, a multitude of artificial markers are utilized for achieving registration and tracking. These artificial markers feature high-contrast patterns with regular shapes, resulting in efficient and stable recognition. Due to its low computational cost, this approach maintains excellent performance even in challenging lighting conditions [1, 2]. Given the diverse challenges associated with manual tagging in complex environments, markerless registration tracking has progressively emerged as the focal point of recent research endeavors. Markerless recognition presents a formidable challenge, particularly when the identified object lacks surface texture. Certain scholars are endeavoring to address this issue through computer vision techniques. For example, the Histogram of Oriented Gradients (HOG) algorithm [3] computes the gradient direction of local image areas and shows promising results in recognizing images with significant gradient variations. Similarly, the Dominance Direction Template (DOT) algorithm [4] is a high-performance object detection method based on binary template representations of local dominant gradient directions. It is particularly effective for detecting untextured objects and estimating their poses. The LINE-MOD algorithm [5] is a real-time 3D object detection method that features rapid model training and the ability to detect textureless objects using 2D gradients and 3D surface normals. Empirical studies demonstrate the algorithm's speed and robustness, even in challenging environments. Park et al. [6] developed a texture-free object tracking method that uses online training with depth cameras. The DOT algorithm estimates the object's pose, which is then refined with object contours and depth information. Nguyen et al. [7] utilized point clouds for 3D object pose estimation, improving performance in matching and voting processes through peer-to-peer functionality. Recently, Newcombe et al. [8] introduced the DynamicFusion system, an innovative tracking method that uses an RGB-D sensor to simultaneously track the sensor's pose and map the scene geometry with a fine-grained iterative closest point (ICP) algorithm. Mur-Artal et al. [9] proposed the ORB-SLAM keyframe-based SLAM system for camera tracking and mapping, enhancing map reusability and supporting real-time repositioning and closed-loop operations with significant viewpoint invariance. Engel et al. [10] introduced the LSD-SLAM algorithm, a featureless monocular SLAM method that provides precise pose estimation through direct image alignment and real-time 3D environment reconstruction. Finally, Nuernberger et al. [11] presented a system that automatically aligns virtual objects with real-world constraints in real time by extracting 3D edge and planar constraints to facilitate virtual-real alignment.

In summary, manually marked methods are difficult to apply directly in workpiece processing environments. Additionally, object detection methods based on feature points are unsuitable for textureless workpieces. However, the LINE-MOD template matching method can address initialization registration issues in virtual-real fusion. An attitude estimation method that integrates multiple sensors can handle scenarios with significant viewing angle changes. By leveraging the strengths of each approach, a multi-stage virtual-real fusion method can be developed to meet the needs of virtual-real integration during processing.

12.2 AR-Oriented Multi-view Construction and Interaction of Process Information

Augmented reality (AR) offers an interface that enables users to interact with the digital world of a smart factory [12]. When operators rely on real-time information, AR can visually display this data in the field, immersing them in the digital landscape of Industry 4.0 solutions [13]. The use of multiple views enhances the systematic presentation of information.

With IoT technology, the system can gather extensive real-time data from workpieces, tools, and various equipment. Although this data is collected and organized, it is not directly accessible to workers for monitoring processing status. The primary issue is that data in the database is stored in a format optimized for computer storage and compression but not for human readability. Therefore, it is crucial to convert the information into an intuitive format, such as three-dimensional contour diagrams showing surface deformation to indicate machining quality, or line graphs representing current spindle speeds or feed rates.

Moreover, while large volumes of data facilitate process management, they also introduce the challenge of data redundancy. It is essential to extract the most relevant data for the current processing status. To enable efficient interaction with augmented reality systems, effective human–computer interaction methods must also be provided.

12.2.1 Description and Definition of AR-Based Processing Information Integration Model

For the machining process, the main sources of information are tool information, machine tool information and workpiece information. This information can be divided into static data and real-time dynamic data according to whether it changes with the processing, and in order to implement finer control of the processing process, it should also contain a reasonable estimate of this information, that is, inference data. Therefore, according to the ontology theory, an AR-oriented machining process information integration model can be constructed, including static data PSD (data that the workpiece remains unchanged during machining), real-time dynamic data RTD, and inference data RD, namely:

$$\text{DT-MPM} = (PSD, RTD, RD) \tag{12.1}$$

where

$$PSD = (GDD, PDD, TP, MP) \tag{12.2}$$

where GDD represents the geometric model data of the workpiece; PDD stands for process design data; TP stands for tool attribute data; MP stands for machine attribute data

$$RTD = (PQD, RTS, RMS) \tag{12.3}$$

where *PQD* represents product quality data; *RTS* represents the real-time status data of the tool; *RMS* stands the real-time status data of the machine tool.

$$RD = (PQP, MSP, TLP, CPR) \tag{12.4}$$

Among them: *PQP* represents the predicted value of product quality; *MSP* stands for machine condition prediction; *TLP* stands for tool life prediction; *CPR* stands for Current Machining Recommendation.

12.2.2 A Multi-view Method for Building Process Information for AR

The AR-based processing information integration model utilizes AR visualization technology to visually present data. The data within the processing system is complex and multifaceted, requiring the selection of various datasets to observe the processing procedure at different stages. Traditional AR interface views are pre-configured with static layouts, limiting the ability to independently select data of interest and thus failing to meet the needs of complex processing systems. The issue is addressed by dynamically generating and integrating multiple views, enabling a multi-view approach to model information.

The AR view object (AR-VO) in each AR multi-view is not fixed; its selection depends on the specific machining process. The system dynamically chooses AR view objects to create an AR multi-view, allowing users to adjust the content of the generated multi-view as needed.

$$ARDMV \subseteq \left(ARVO_{fixed}, ARVO_{select}, ARVO_{recommend}\right) \tag{12.5}$$

Equation (12.5) indicates that the augmented reality dynamic multi-view ARDMV consists of the inherent view $ARVO_{fixed}$, the selective view $ARVO_{select}$, and the system-recommended view $ARVO_{recommend}$.

Among them, $ARVO_{recommend}$ are related to the view object VO, the system context state and the process.

$$R\colon VO \times Context \times Process \rightarrow Rating \tag{12.6}$$

When the influence of Context on recommendation is not considered, view recommendation can be performed through a collaborative filtering algorithm, assuming that in process P, the score of a particular view v is $r_{v,p}$. The view v score is compared to the score of the other calculated operation view P_v and then the weighted average of the other operation scores is used as the prediction. If the view P_v is a set of views that have been rated in process P, the average rating of view v can be defined by the following formula, $\overline{r}_v$.

$$\overline{r}_v = \frac{1}{|P_v|} \sum_{p \in P_v} r_{v,p} \tag{12.7}$$

The collaborative filtering algorithm predicts the score of a particular operation on view v based on the score of a similar process on view v. The correlation between the processes is calculated using the Pearson correlation coefficient, and the similarity $w(p, a)$ can be determined by Eq. (12.8) based on the correlation of the rating vectors of process P and other process a.

$$w(p,a)=\frac{\sum_{i\in I_p\cap I_a}\left(r_{p,i}-\overline{r}_p\right)\left(r_{a,i}-\overline{r}_p\right)}{\sqrt{\sum_{i\in I_p\cap I_a}\left(r_{p,i}-\overline{r}_p\right)^2\cdot\sum_{i\in I_p\cap I_a}\left(r_{a,i}-\overline{r}_p\right)^2}} \tag{12.8}$$

where I_p denotes the view set of operation P, and I_a. denotes the view set of operation a.

The value of $w(p, a)$ is used to measure the similarity between the scoring vectors of the two processes, with a high absolute value indicating high similarity and $w(p, a) \in [-1, +1]$. The recommended score of view v in a process P can be obtained as follows, $p^{collab}(v, i)$, factor k is the normalized weight.

$$p^{collab}(p,v)=\overline{r}_p+k\sum_{a\in P_i}w(p,a)\left(r_{a,i}-\overline{r}_p\right) \tag{12.9}$$

$$k=\frac{1}{\sum_{a\in P_i}w(p,a)} \tag{12.10}$$

In information fusion computing, the system context information is defined as follows (12.11):

$$\text{Context}=(\mathrm{C}_1,\mathrm{C}_2,\ldots,\mathrm{C}_\mathrm{k}) \tag{12.11}$$

where, $C_i \subset C(i \in [1, k])$ Contextual information that represents a dimension in any type. Including factors such as machine status information, tool status information, workpiece machining geometry state, workpiece machining physical state and so on. The gray correlation algorithm is used to find the values of the context relationship and the view correlation matrix in the system

$$r_{m,v}=\begin{pmatrix} r_{11} & \cdots & r_{1n}\\ \vdots & \ddots & \vdots\\ r_{m1} & \cdots & r_{mn}\end{pmatrix} \tag{12.12}$$

Finally, the choice of the view object in $VO_{recommend}$ is determined by Eqs. (12.13) and (12.14)

$$p^{\text{collab}}(p,v)=\overline{nr}_p+k\left\{\left(nr_{m,i}-\overline{nr}_m\right)+\sum_{a\in V_i}w(m,a)\left(r_{a,i}-\overline{r}_a\right)\right\} \tag{12.13}$$

$$k=\frac{1}{\sum_{a\in V_i}w(m,a)} \tag{12.14}$$

In this method, you can select the top N views (N is determined by the user) as needed, and form AR multi-views together with fixed views and user-selected views.

12.2.3 AR-Oriented Multi-view Interaction Method for Machining Process Information

Human–computer interaction is a crucial element of augmented reality systems, with its efficiency and rationality being key indicators of the system's success. In augmented reality systems that rely on traditional displays, interaction is limited to mouse operations, making the display and interaction methods relatively outdated. In contrast, previous research on glasses-based augmented reality has shown a simplistic interaction mode, primarily based on gesture operations, and lacks a systematic interaction process, thus not fully utilizing the advantages of augmented reality. To develop a simple and intuitive interaction model for augmented reality-based processing systems, the human–computer interaction model shown in Fig. 12.3 was created. This model integrates three interaction modes: eye tracking, gesture recognition, and voice recognition, and includes three main components: interactive input, intent recognition, and application.

Combined with the existing technology and actual requirements, this paper mainly implements the interaction method. There are two main parts.

(1) **Visual-Based Interactions**

The Hololens is furnished with a pair of infrared cameras serving as eye-tracking sensors, enabling the acquisition of the user's present line of sight. As depicted in Fig. 12.4, the user's gaze constitutes captivating data, wherein green hues denote momentary fixation, while expansive regions of red signify prolonged fixation at a particular location.

The application scenarios pertaining to gaze tracking encompass the following:

1. Serving as a complement to gesture input and voice input, furnishing contextual information support. By fixating upon a target, the system can aid in discerning the user's input through voice or gestures.
2. Operating implicitly, diminishing the need for active interactions between users and systems. When users peruse lengthy textual messages or content replete with

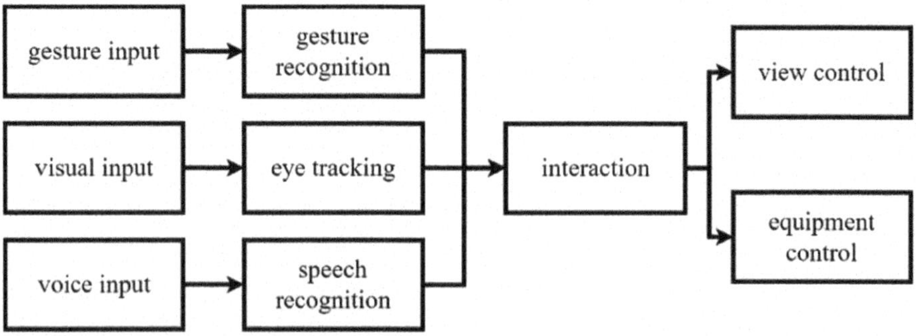

Fig. 12.3 Human–computer interaction framework

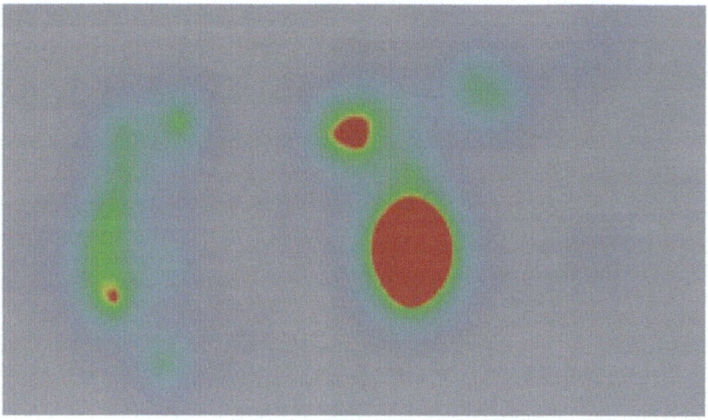

Fig. 12.4 Schematic diagram of gaze tracking

multi-tiered information, the system can automatically initiate text scrolling or display subsequent levels of information based on the gaze point.

3. Functioning as an input to evaluate the efficacy of an augmented reality (AR) view in the current operational state. Through the analysis of gaze landing point data, facilitated by the multi-view construction approach, the accuracy and effectiveness of view recommendations under similar conditions can be further enhanced.

(2) **Gesture-Based Interactions**

Gesture interaction has long been a primary method for engaging with augmented reality systems, as it closely aligns with the user's intuitive understanding. By directly interacting with virtual objects using hand gestures or rays emitted from the palm, users can perform actions such as pressing buttons, grabbing, picking up, moving, scaling, or rotating objects. These gestures mimic interactions on a touch screen and can be applied to manipulate textual content.

Using the RGB camera and hand joint recognition algorithm of the Hololens, a hand joint diagram can be obtained, as shown in Fig. 12.5. This diagram reveals a total of 25 key points in the palm joint model. By analyzing the spatial and temporal information of these key points, the system can accurately track palm movements and define various operations.

This article combines system and user experience requirements to design and define the actions shown here.

(3) **Voice-Based Interactions**

Speech is considered an important input method in augmented reality. In the Hololens system, you can set a specified keyword as an action command, and when the system recognizes the correct voice, it will execute the corresponding action command.

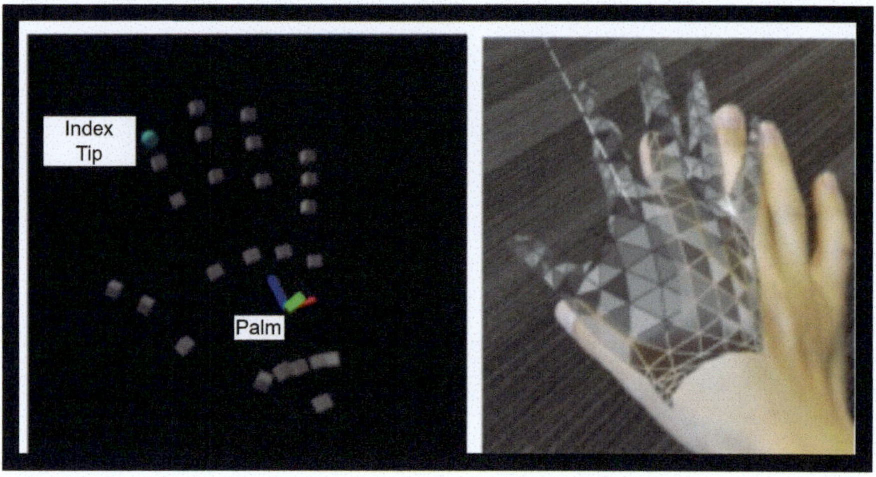

Fig. 12.5 Gesture recognition

You can also bind keywords to objects to replace gestures with voices, as shown in Fig. 12.6.

However, current developments in speech recognition technology are insufficient to support it as the primary input mode for the system. Instead, speech recognition serves more as an auxiliary means of interaction, such as controlling the system through a combination of gaze and voice when gesture-based input is impractical.

Fig. 12.6 Schematic diagram of keybinding keywords

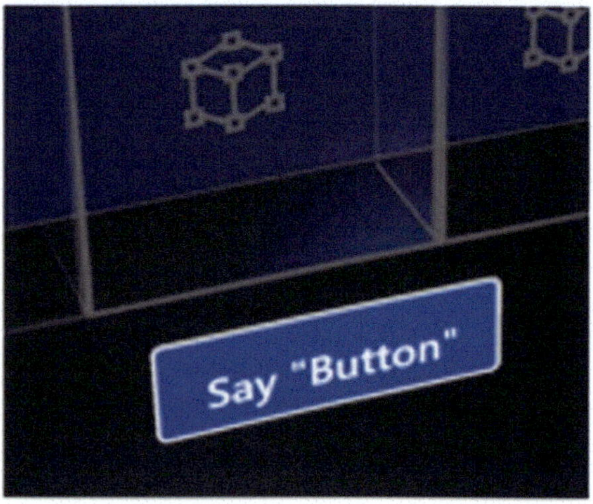

12.3 A Model Virtual and Real Fusion Method Based on Multivariate Information Fusion

The integration of virtual and real-world information within a scene is a critical factor in evaluating the usability of augmented reality systems. Researchers both domestically and internationally have conducted extensive studies and achieved notable results in this area. Traditional fusion methods rely heavily on artificial identification to merge machine tool information between virtual and real domains. However, these methods face challenges, particularly in fusing virtual and real workpieces. The generation of cutting materials and fluids during machining can interfere with artificial identification, leading to inaccurate registration. Additionally, variations in workpiece sizes and the difficulty of manual marking for certain parts can hinder the registration process.

Model-based recognition methods, which do not rely on artificial identification, have become prominent. Techniques such as model-based tracking [16], surface texture-based tracking [17], and edge-based tracking [18, 19] are commonly used. The surface texture-based approach utilizes 2D texture features for camera pose estimation but is prone to interference from cutting waste, especially on untextured objects. In contrast, the edge-based technique uses binary edge images obtained through contour extraction algorithms like Canny, but it is highly sensitive to changes in lighting, noise, and blur. Recent advancements in sensor technology have led to the adoption of multi-sensor fusion-based tracking methods [20–22].

Given the current processing environment and technological capabilities, employing a registration method based on the LINEMOD model during the registration stage, combined with visual and inertial sensor data for online tracking, has the potential to improve detection of textured objects and ensure stable tracking during processing.

12.3.1 The Basic Idea and Process of Virtual and Real Integration Are Integrated

Virtual-real fusion registration involves two stages: offline template training and online registration tracking. During the offline stage, multi-process multi-view image samples of the workpiece are generated. The gradient direction descriptor and camera pose are then calculated and saved as matching templates. In the online stage, the gradient direction descriptor is computed in real time based on images captured by the camera. The LINE-MOD template matching method is used to perform the matching. Once a match is achieved, the integrated representation model view is registered in the correct location. For tracking, camera and inertial sensor (IMU) data are fused and processed using a multi-state Kalman filter algorithm to calculate and estimate the device's attitude, thereby ensuring stable tracking.

12.3.2 Gradient Descriptors Are Learned Offline

(1) **Multi-process and Multi-view Part Image Acquisition**

In the offline stage, the template is generated from the 3D model of the workpiece in the system at different processes, as shown in Fig. 12.7.

It is more convenient and efficient compared to online learning methods. This is because there is no need to spend time and effort manually making a well-sampled training set. Besides, the greedy approach employed by the online learning method is difficult to strike a good balance between efficiency and robustness. In this article, in order to efficiently generate all templates, an automatic template generator has been developed using Unity3D. Using the Unity virtual camera, we can capture the template view of the workpiece during different processes and from various viewing angles. However, the method based on hemispherical sampling [14] has certain drawbacks. Specifically, it causes a problem where the sampling density at the hemispherical pole is too large, while the sampling density near the equator is too small. As shown in Fig. 12.8, points 1, 2, and 3 are noticeably denser compared to points 7, 8, and 9.

Therefore, in this paper, each face of the icosahedron is further divided on the surface of the icosahedron centered on the workpiece. The camera's axis is always centered, allowing the virtual camera to move and sample at the grid intersection. 162 templates are formed for each object at the current distance, and each sample point is about 15° apart. In addition to this, regular polyhedra of different radii are used in 15 cm steps, resulting in matching templates at different scales (Fig. 12.9).

Through the above sampling method, a multi-view and multi-scale template of the whole process of the workpiece can be obtained. However, due to the large number of symmetrical structures in the design process of the workpiece. For this special case, the recognition efficiency can be improved by reducing the sampling points of symmetrical workpieces and removing templates with images that are very close to each other. As shown in Fig. 12.10a, b, for this kind of center-symmetrical parts, under the premise of ensuring accuracy, only the workpiece sampling images need to be obtained in the range of 0°–45° in different longitudes and latitudes. As shown in Fig. 12.10c, d, for this kind of axisymmetric parts, only the sampling images need to be obtained in the range of 0°–90° at different longitudes and latitudes. Finally, extract the corresponding view template and the corresponding camera pose and save them to an XML file.

(2) **Gradient Vector Descriptor Extraction**

Before delving into the extraction of the gradient vector descriptor, let us briefly elucidate the rationale for employing the gradient vector as the template descriptor. Compared to other forms of template descriptors, vector gradients possess superior discriminative properties and exhibit robustness against variations in light direction, intensity, and noise. Furthermore, for workpieces devoid of intricate textures, image gradients typically represent the most dependable image feature. By considering

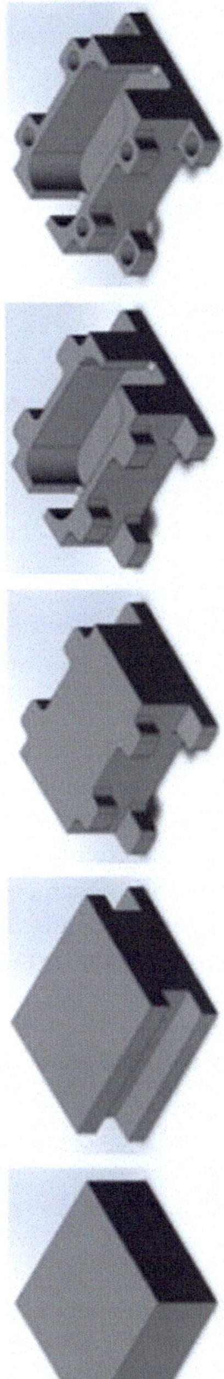

Fig. 12.7 Schematic diagram of different process models of the workpiece

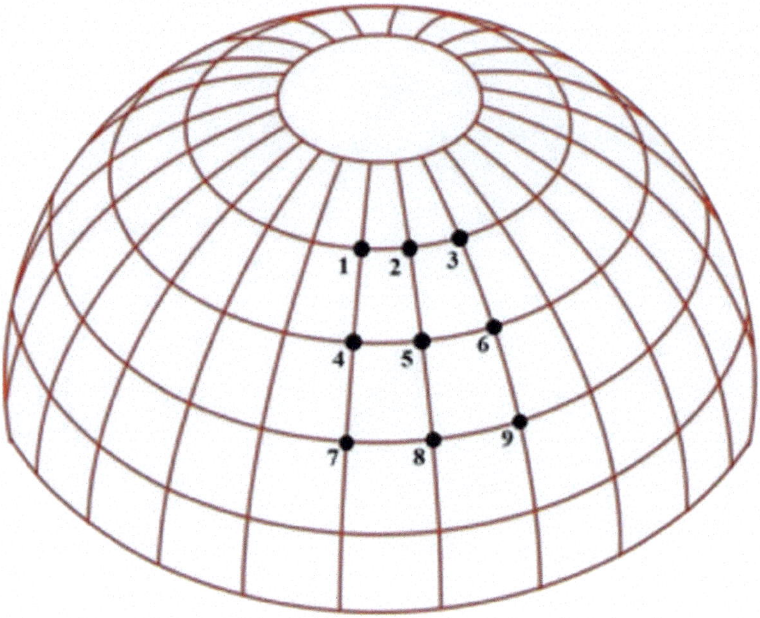

Fig. 12.8 Hemispheric sampling schematic

solely the gradient direction and disregarding its magnitude, it also becomes resilient to changes in contrast. Moreover, taking the absolute value of the cosine ensures that the occlusion boundary of the workpiece is correctly managed, irrespective of whether the workpiece resides in a dim or brightly lit machining environment. To enhance robustness, the gradient direction for each color channel (red, green, and blue) of the input image is calculated independently. For each location within the image, the channel with the highest absolute value is then designated as the gradient direction for that location:

$$I_g(x) = \mathrm{ori}\big(\hat{C}(x)\big) \tag{12.15}$$

where

$$\hat{C}(x) = \underset{C \in \{R,G,B\}}{\arg\max} \frac{\partial C}{\partial x} \tag{12.16}$$

where R, G, and B represent the three color channels, and $\hat{C}(x)$ represents the gradient vector at position x. As shown in Fig. 12.11, the left side is the view of the workpiece in a certain viewing angle, and the right side is the object outline gradient extracted by the above method.

In order to quantify the gradient pattern and improve the robustness of the gradient vector to the system noise. The collinear gradient direction vector is taken as the same

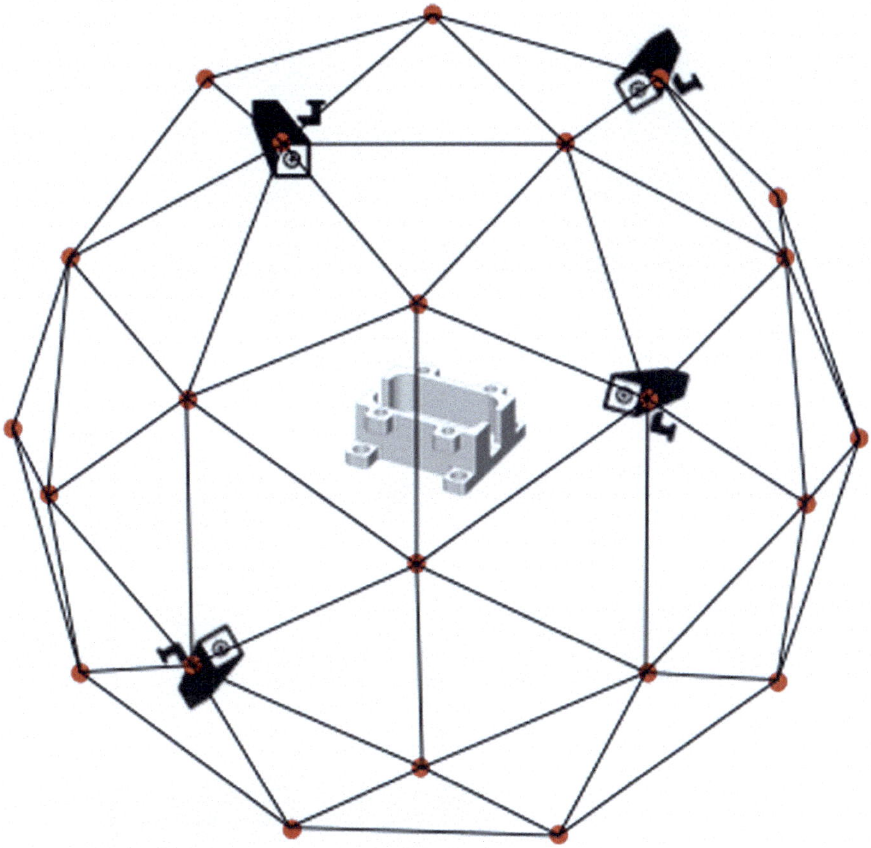

Fig. 12.9 Schematic diagram of virtual camera sampling

vector, so the gradient vector space is compressed into the range of 0°–180°. At the same time, the angular space is divided into 8 equally spaced angles. As shown in Fig. 12.12, the blue arrow indicates that the gradient vector projection is within the range of the second ballot box, so it belongs to the second ballot box. At the same time, to make the quantized gradient vector robust to noise, the gradient with the highest frequency in the 3 × 3 pixel neighborhood is considered. The gradient in the quantization direction at each position is then assigned to that position. While keeping those gradients where the specification is greater than a small threshold, are also retained.

(3) **Gradient Direction Spreading**

In the actual matching process, some distortion and displacement may occur in the acquired images. This can cause a difference between the gradient of the actual image and the gradient of the template image generated in the offline stage, leading

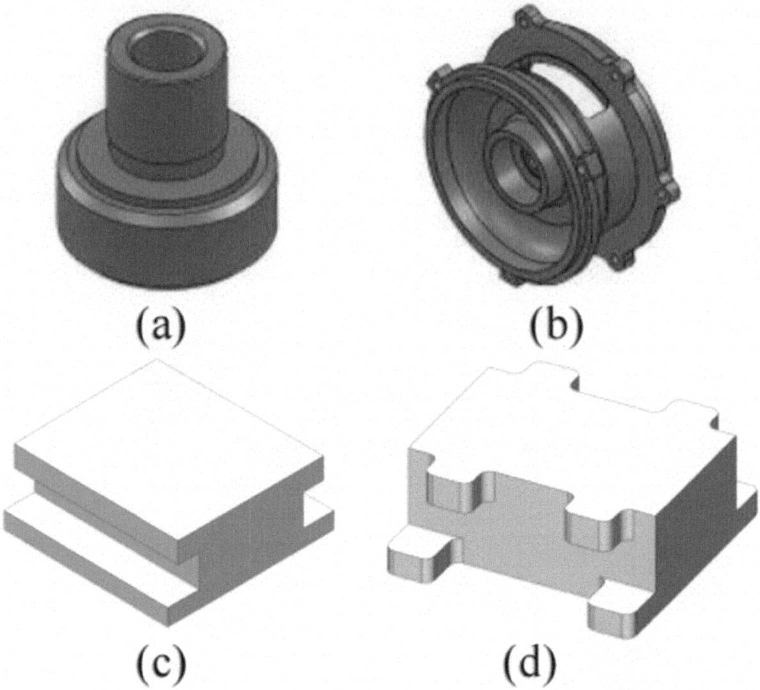

Fig. 12.10 Schematic diagram of the parts

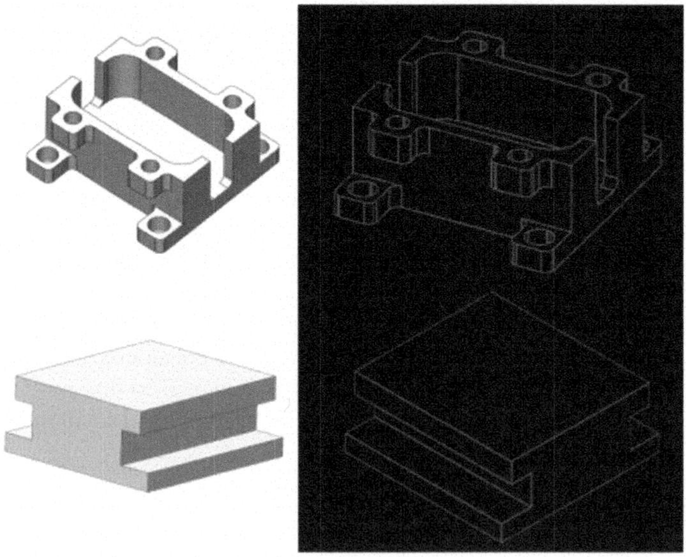

Fig. 12.11 Calculation of gradient directions for different models

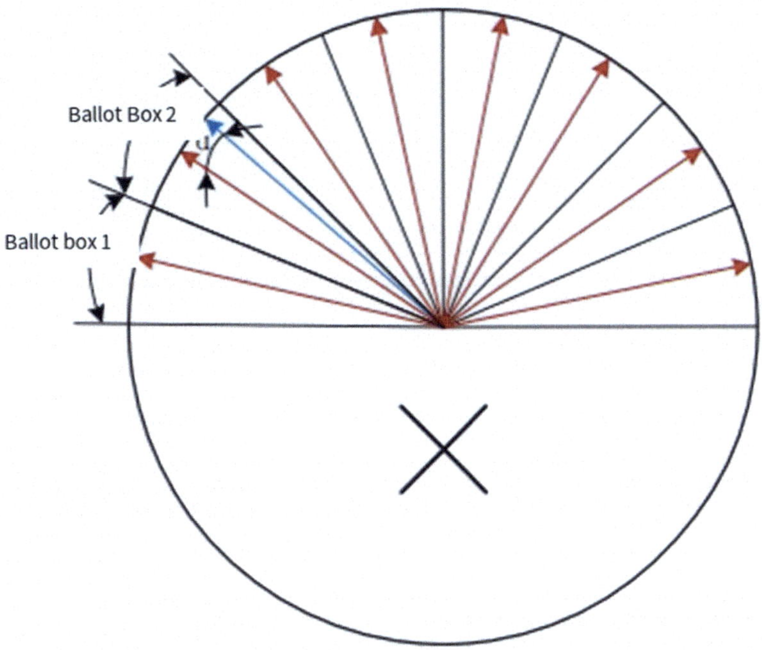

Fig. 12.12 Schematic diagram of the ballot box

to a failure in matching. In order to improve the robustness of online matching and reduce the matching time, the template image gradient can be diffused. The gradient diffusion method is to generate a new template gradient image by moving the original template gradient image in three dimensions: up, down, left and right, and superimposing the gradient images after these movements. Figure 12.13a shows the gradient image of the original template, and Fig. 12.13b shows the new gradient image obtained after diffusion. Where the solid arrow represents the original gradient descriptor, and the dashed arrow represents the gradient descriptor generated by the gradient spread.

Figure 12.14a shows the gradient direction of the same object at the same viewing angle extracted by the online process. As can be seen from the figure, due to perspective deformation or slight deviation of the viewing angle, the direction of the extracted contour gradient is inconsistent with that of the object extracted in the offline stage. However, due to the gradient direction divergence method, when the input image slides on the template image, it can still obtain a good matching effect with the template image, as shown in Fig. 12.14b, the red circle indicates the successfully matched gradient direction, in which there are both the original gradient and the gradient generated by diffusion. This improves the robustness of the method to small deformations and position shifts.

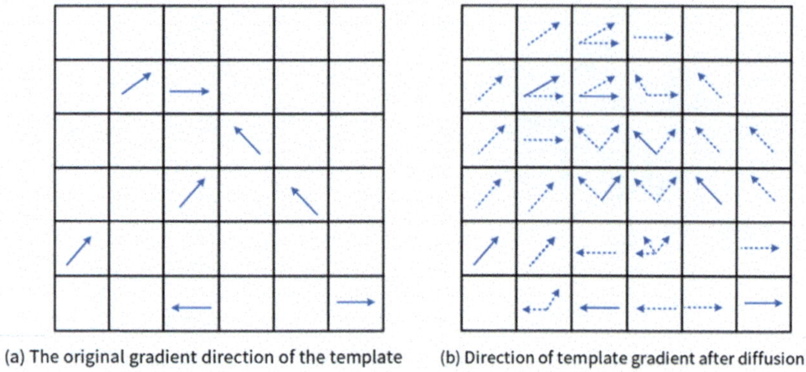

Fig. 12.13 Schematic diagram of the gradient of the offline training template

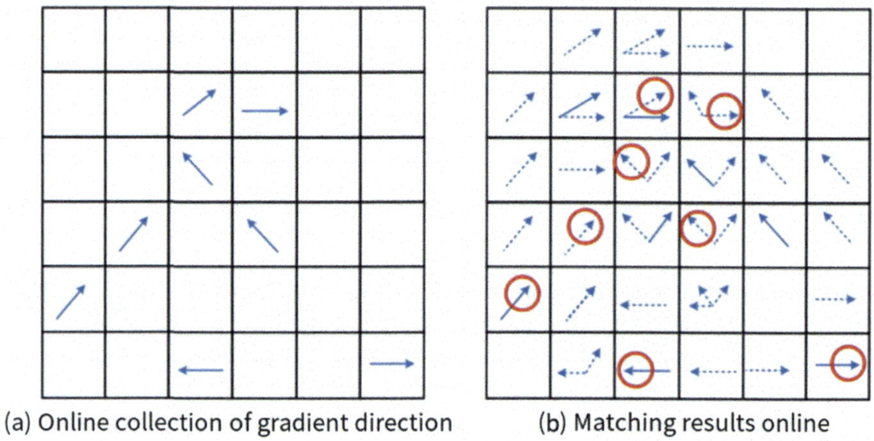

Fig. 12.14 Online template matching

12.3.3 Online Identification and Tracking of Multiple Information Fusion

After obtaining the template gradient descriptor, the real-time image can be registered, but in the process of processing, it is difficult to ensure that the registration target is always kept in the field of view. And a large amount of cutting waste will be generated during the cutting process of the workpiece, which causes great interference to the image-based registration. It is difficult to ensure continuous tracking by relying on the template registration method alone, so it is necessary to use an online tracking method that integrates other sensors after the registration is completed. Neither accelerometers nor angle sensors alone can construct accurate real-time attitude estimation. In order to give full play to the advantages of each sensor and

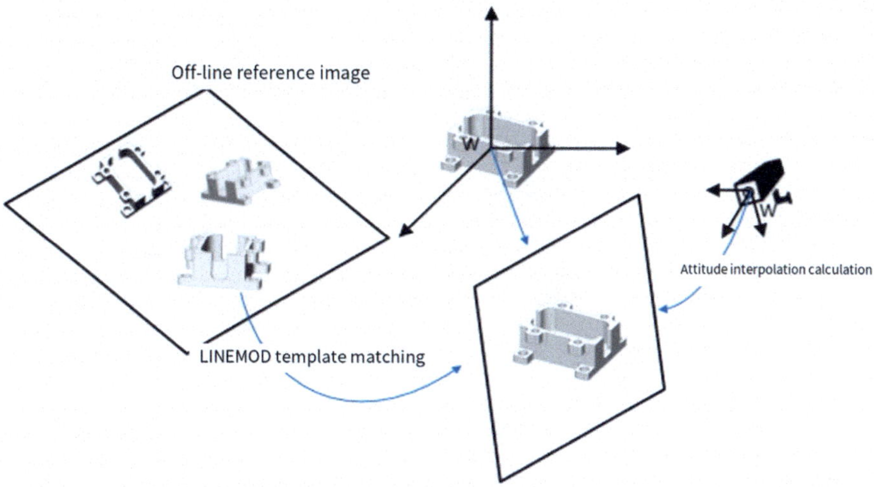

Fig. 12.15 Workpiece pose estimation process based on linemod

eliminate the limitations of the sensor, this paper adopts the online tracking method of vision and IMU sensor fusion.

(1) **Online Recognition Based on Visual Information**

The online workpiece pose estimation process, as shown in Fig. 12.15, is divided into two steps. Firstly, the real-time gradient is extracted based on the current camera image. This gradient is then matched with the template gradient vector of the training workpiece from the offline stage. As a result, the template view closest to the current angle of view and the corresponding camera space pose can be obtained.

Due to the limited sample template density, it is not possible to completely cover all possible viewpoints, so the camera viewpoint is likely to fall between multiple sampling points, as shown in Fig. 12.16. At this time, it is necessary to interpolate the calculation based on the camera spatial posture represented by the surrounding templates in order to obtain the precise spatial posture of the current camera.

In this, a similarity measurement method proposed by Hinterstoisser et al. [15] is used to compare the workpiece surface gradient vector obtained in the online stage and the template gradient vector obtained in the offline stage. The following equation is given (12.17).

$$\varepsilon(/, T, C) = \sum_{r \in p} \max_{t \in R(c+r)} |\cos(\mathrm{Ori}(O, r) - \mathrm{ori}(/, t))| \tag{12.17}$$

where ε the similarity between the component to be identified and the part training template obtained in the offline training stage; $\mathrm{Ori}(O, r)$ refers to the gradient direction at the point r at the O position of the reference image; $\mathrm{Ori}(/, t)$ input the direction of the gradient at the point t at the position I of the image.

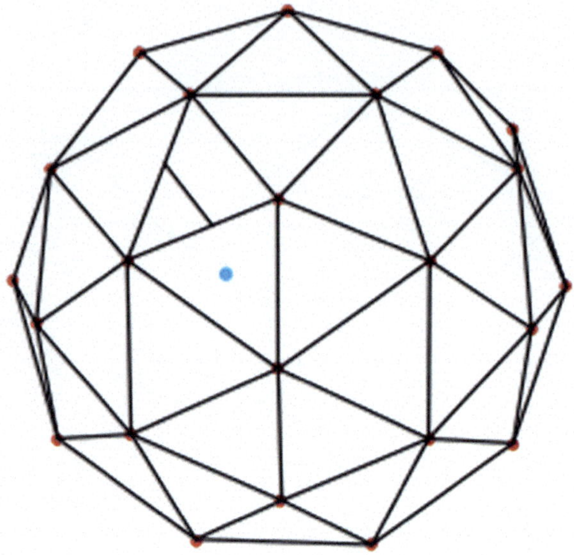

Fig. 12.16 Schematic diagram of the position relationship between the viewing angles

$$t \in R(c+r) = \left[c + r - \frac{\tau}{2}, c + r + \frac{\tau}{2}\right] \times \left[c + r - \frac{\tau}{2}, c + r + \frac{\tau}{2}\right] \quad (12.18)$$

It represents the area where c + r is the center and τ is the neighborhood; P is a list of positions r; T = (O, p) denotes the template of the object.

In online matching, the gradient descriptor D′ is first extracted from the real-time acquired images and then matched with all template gradient descriptors N(D, K) generated in the offline phase. As shown in Fig. 12.17, the blue dot represents the location of the current camera angle of view. The red dot indicates the position of the offline training angle that is most similar to the shape descriptor of the current perspective image. The other locations are adjacent nodes. The camera position target optimization function is as follows:

$$\varepsilon = \min \left\| D' - \omega_i D_i - \sum_{j=0}^{k} D_j w_{ij} \right\| \quad (12.19)$$

where: (k = 1, 2, 3, …, n) represents a set of gradient descriptors for template images; K represents the topological relationship between viewpoints; D_i denotes the gradient descriptor that is most similar to D'; D_j denotes the adjacent node gradient descriptor.

$\omega_i D_i$, w_{ij} denotes the central coordinate data of the gradient descriptor *Di* that is closest to D′ in N, and because the values of the weighting factors ωi and ωij can be solved by $\omega_i + \sum_{j=0}^{k} \omega_{ij} = 1$ we can use the values of ωi and ωij to complete the interpolation of the camera position, so Eq. (12.20) can be written

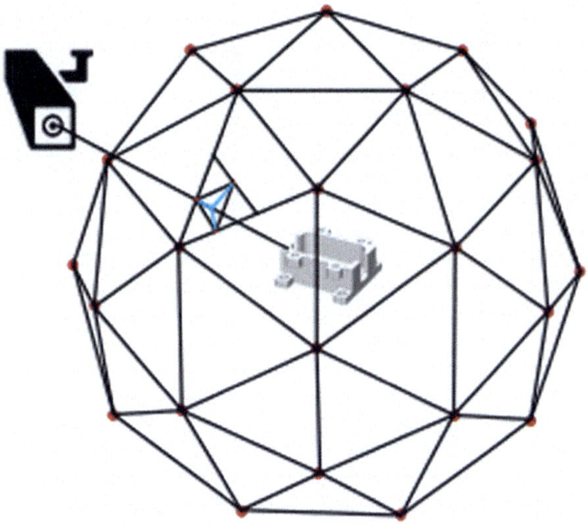

Fig. 12.17 Schematic diagram of pose interpolation

$$\varepsilon = \min\left((D' - D_i) - \sum_{j=0}^{k} (D_j - D_i)\omega_{ij}\right)^2 \tag{12.20}$$

where $\partial\varepsilon/\partial\omega_{ij} = 0$, so the weighting factors ωi and ω_{ij} can be solved. Finally, the current camera's position on the icosahedral surface pose P can be calculated by ωi and ωij.

(2) **Attitude Estimation Based on Inertial Sensor Data**

The purpose of sensor fusion is to improve the performance of individual sensors and to consolidate their information. IMUs alone do not provide accurate information, so vision is also required. Using Vision alone can not deal with occlusion, fast movement, and not all areas are covered due to the camera's field of view. Therefore, considering the shortcomings of each sensor, fusing IMU data with visual data can obtain better pose estimation results. The IMU gives the velocity, position, angular velocity, and direction, and the camera gives the position and direction. The multi-state constrained Kalman filter algorithm is used to realize the fusion of vision and IMU. The overall pose of the camera relative to the world coordinate system is then calculated. Before using the sensors of the camera and the IMU for fusion tracing, several reference coordinate systems need to be introduced. The four main coordinate systems are shown in Fig. 12.18.

Wherein: Global Coordinate System/World Coordinate System is used to describe the actual position of the object and determine the pose estimation relative to the IMU and camera frames. Object Coordinate System is used to describe the location of the binding and the virtual information on the object. IMU coordinate system represents the coordinate system of the IMU (accelerometer and gyroscope) itself. Camera

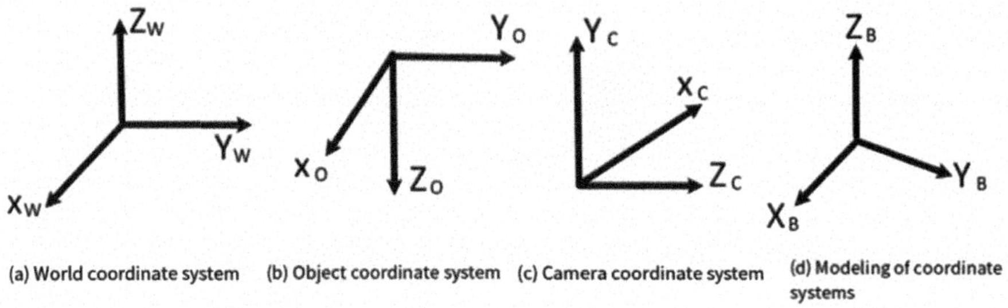

Fig. 12.18 Schematic diagram of the reference coordinate system

Coordinate System indicates the coordinate system of the camera itself. The IMU sensor provides the transformation method of the camera coordinate system relative to the world coordinate system, and the visual image provides the variation method of the object coordinate system relative to the camera coordinates.

(3) **Attitude Fusion Tracking Based on Kalman Filter Algorithm**

The multi-state constrained Kalman filter, while fundamentally an extended Kalman filter, diverges significantly from the traditional extended Kalman filter algorithm used for state estimation. The extended Kalman filter algorithm incorporates the three-dimensional coordinate information of landmarks in the environment along with the state of the IMU into the state vector simultaneously for estimation. As more feature points are observed, the dimensionality of the state vector increases exponentially. Conversely, the multi-state constrained Kalman filter algorithm excludes the feature point information from the state vector, instead adding the camera pose at various time instances to the state vector. Multiple cameras can concurrently observe identical feature points, creating geometric constraints between camera poses. Consequently, since the number of camera poses is substantially smaller than the number of feature points, the dimensionality of the state vector in the multi-state constrained Kalman filter is considerably reduced compared to that of the extended Kalman filter.

The multi-state constrained Kalman filter process consists of two parts: state prediction and state update. The state prediction process is done independently by the IMU. It integrates the output of the accelerometer and the output of the gyroscope. Equation (12.21) represents the state prediction vector of the IMU:

$$X_{IMU} = \left[{}^{I}_{G}q^{T} b_{g}^{T} G_{V_I}^{T} b_{a}^{T} G_{P_I}^{T} \right]^{T} \tag{12.21}$$

where: ${}^{I}_{G}$q is the quaternion rotation from the world coordinate system to the IMU coordinate system; b_g is the error value of the accelerometer; and G_{V1} is the position of the IMU in the world coordinate system. The part that follows the state prediction is the status update. Observing the model can explain the process of state updates.

The observation model of the multi-state constrained Kalman filter algorithm characterizes the geometric constraint relationship between different camera attitudes. The essence of this is to minimize the reprojection error in the beam adjustment. The projection coordinates of the object are solved by using the transformation matrix of the camera coordinate system and the world coordinate system, and the reprojection error refers to the difference between the calculated coordinates and the real coordinates. The observational model is represented by Eqs. (12.22) and (12.23):

$$r_i^{(j)} = z_i^{(j)} - \hat{z}_i^{(j)} \tag{12.22}$$

$$\hat{z}_i^{(j)} = \frac{1}{{}^{C_i}\hat{Z}_j}\begin{bmatrix} {}^{C_i}\hat{X}_j \\ {}^{C_i}\hat{Y}_j \end{bmatrix}, \quad \begin{bmatrix} {}^{C_i}\hat{X}_j \\ {}^{C_i}\hat{Y}_j \\ {}^{C_i}\hat{Z}_j \end{bmatrix} = C\left({}^{C_i}\hat{\bar{q}}\right)\left({}^{G}\hat{p}_{f_j} - \hat{p}_{C_j}\right) \tag{12.23}$$

In order to calculate the reprojection error, it is necessary to know the coordinates of the feature points, and obtain the 3D coordinates of the current image feature points by triangulation. The optical flow method was used to track the feature points. As shown in Fig. 12.19, the green dots in the image represent feature points in the environment, and the image on the right represents the optical flow field generated by motion.

Figure 12.20 shows the update mechanism of the multi-state Kalman filter based on the sliding window, where there are always two camera states. When a new camera state is entered, the oldest camera state will disappear from the sliding window. In this case, the feature points in the discarded camera state are used for observation and update. It is first triangulated, and then the deleted camera state is updated with the results of the triangulation. After performing the observed update, the status of the IMU will also be updated.

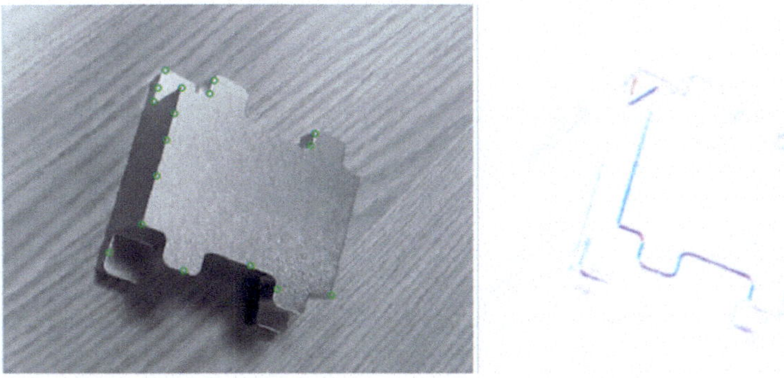

Fig. 12.19 Schematic diagram of feature point detection and optical flow calculation

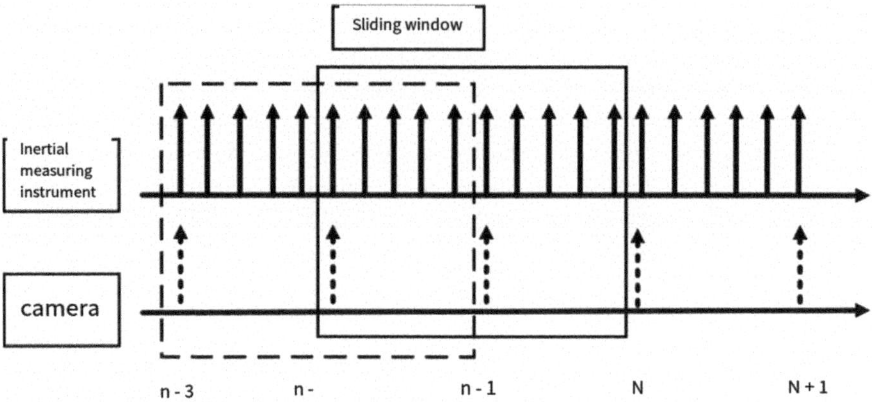

Fig. 12.20 Sliding window model

12.3.4 *Method Validation and Discussion*

To evaluate the efficacy and reliability of the algorithm delineated in this chapter, its performance must be rigorously tested. The hardware experiments employ an Intel Core i5-8400 quad-core eight-thread microprocessor, boasting a CPU frequency of 3.8 GHz, complemented by an NVIDIA GeForce GTX1060 graphics card. The sensor utilized is an Intel RealSense D415i binocular depth camera, with RGB and depth sensor resolutions of 1080p and 720p, respectively. Additionally, the hardware integrates IMU sensors, including accelerometers and gyroscopes, which facilitate data acquisition through APIs. The software system is powered by the Unity3D engine and developed within the .NET framework.

(1) **The Algorithm Performs Precision Analysis**

In order to evaluate the registration accuracy of the virtual-real fusion algorithm proposed in this chapter, multiple registration tracking experiments were carried out on the same object. Since it is not possible to conduct experiments in all directions and at all distances, the part registration distance is measured based on the size of the part itself. The position and orientation errors are measured within a distance range of 20–50 cm and an angle range of 0°–90°. After successful registration, the deviation between the model and the real object is measured and recorded for different registration poses.

The error relationship between the registered angle between the model and the actual object at different angles and distances. As can be seen from the figure, when the registration distance remains unchanged, the registration error of different registration angles varies between 2.0° and 3.0°, but there is no obvious relationship between the error size and the registration angle when the registration distance is different. It can be considered that there is no absolute advantage of the registration angle, which also shows that the matching template generation method based on the

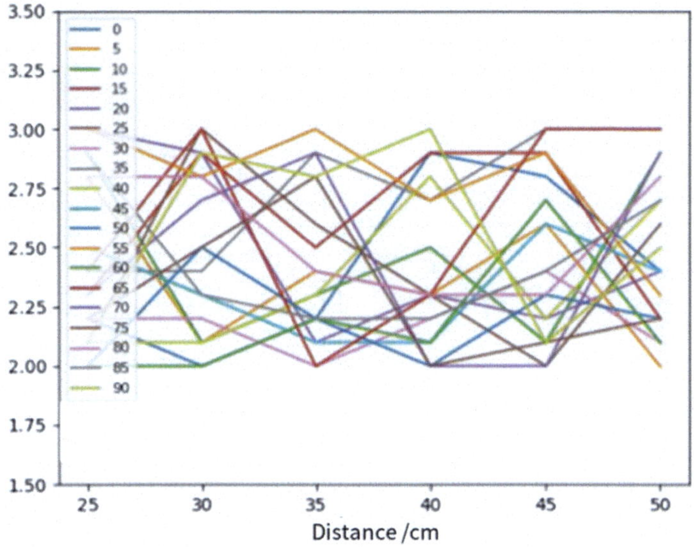

Fig. 12.21 Angular error

twin model can cover different registration situations. When the registration angle is kept constant, the error of the registration angle also tends to increase with the increase of the registration distance.

Figure 12.21 shows the relationship between the registered position error between the model and the actual object at different angles and distances. As can be seen from the figure, there is basically no difference in the registration distance error of different registration angles when the registration distance remains the same. When the registration angle is constant, with the increase of the registration distance, the registration distance error has a tendency to increase obviously, and it is approximately linear change (Fig. 12.22).

Based on the analysis of angle error and distance error, it can be found that under the premise of ensuring the success of registration, the closer the registration distance, the higher the registration accuracy.

(2) **Analysis of Algorithm Execution Efficiency**

In order to verify the registration efficiency and tracking stability of the algorithm proposed in this chapter, multiple registration tracking experiments were carried out on different recognition objects in the complex environment of simulated processing, as shown in Fig. 12.23.

Due to the impossibility of conducting experiments in all directions and distances, the registration distance of parts is related to the size of the parts themselves. Therefore, different process models of the same workpiece were selected to repeat the previous experiment. After successful registration, simulate the actions during the machining process on site and move freely for 1 min. Then, the registration

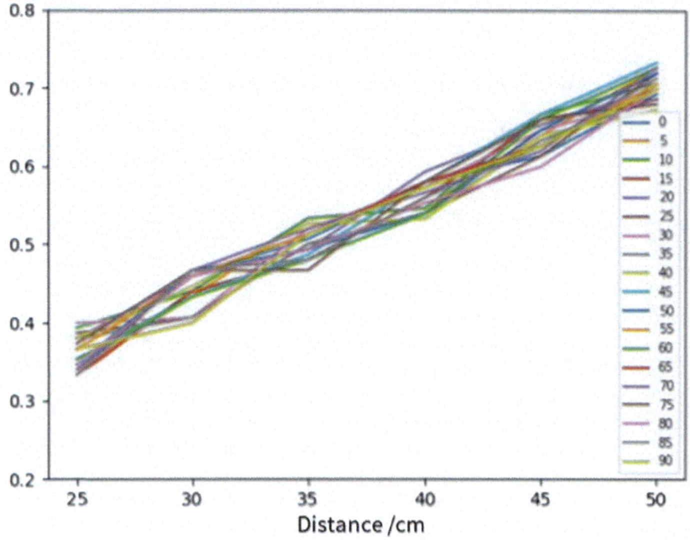

Fig. 12.22 Distance error

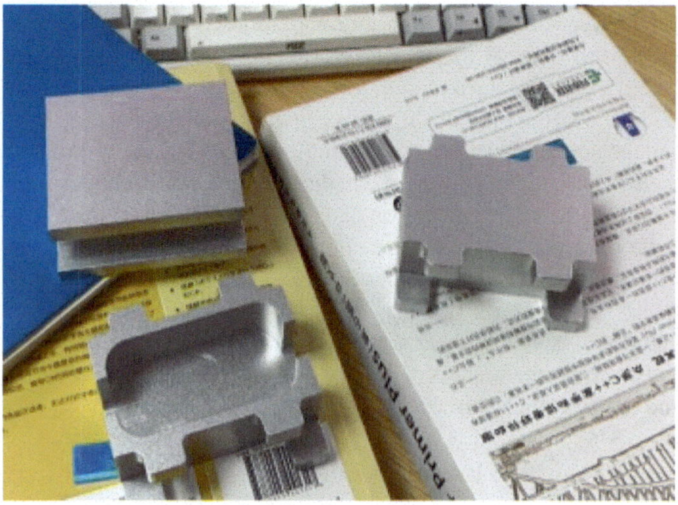

Fig. 12.23 Schematic diagram of the simulated environment

consumption time and whether the registration result is stable after the activity are counted.

Figure 12.24 shows the results of the registration time of different objects, and it can be found that the registration time of different targets at the same registration distance is basically the same. As the registration distance increases, the registration time also increases, which may be caused by the decrease in the effective gradient extracted from the image as the distance increases.

The results of the registration tracking of different models, as shown in Table 12.1, show that all models are able to maintain the correct registration state after simulated motion or even loss of target. It is proved that the online tracking method proposed in this paper can effectively estimate the camera pose and maintain the tracking stability.

(3) **LINEMOD Algorithm Comparison**

In order to compare the differences between the proposed method and the LINEMOD algorithm, different algorithms were used to register and track the target. As shown in Fig. 12.25, both methods have achieved good results in the registration stage.

However, when the model is blocked by hand and away, the LINEMOD algorithm has obvious registration errors, while the proposed algorithm remains stable (Fig. 12.26).

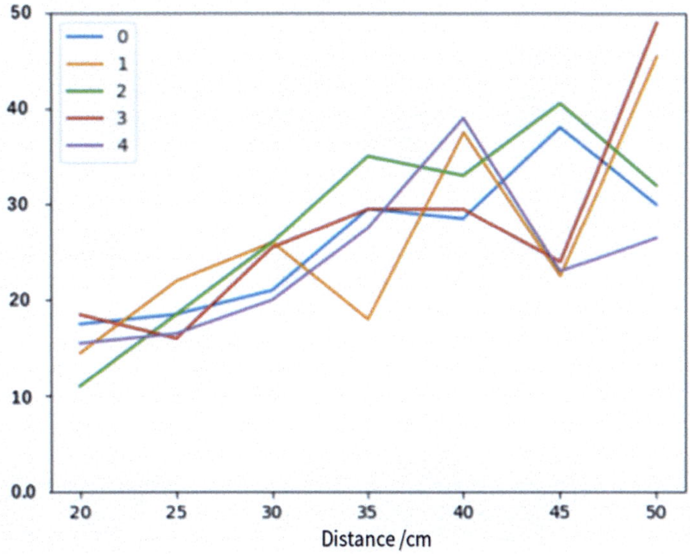

Fig. 12.24 Registration time

Table 12.1 Tracking results of different model registrations

Model	Recognition distance (cm)	Registration time (ms)	Tracking stability	Ability to track lost targets
	40	12.3	Stabilize	Can
	30	13.7	Stabilize	Can
	40	14.3	Stabilize	Can

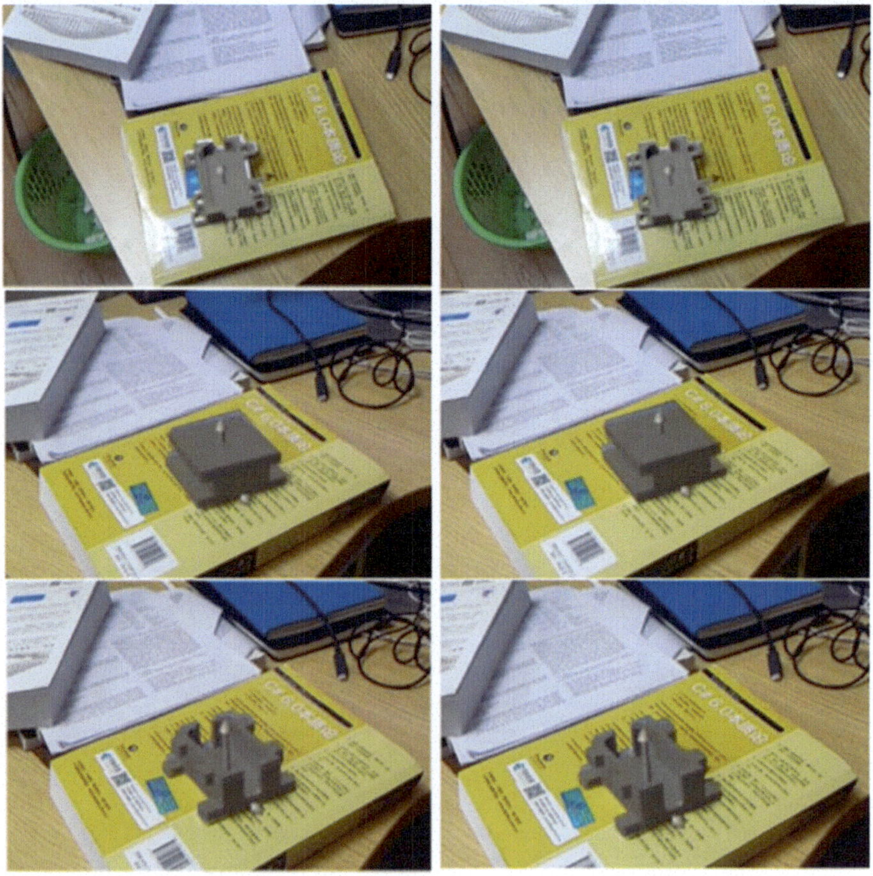

(a) The method of this article (b)LENMOD algorithm

Fig. 12.25 Compares the algorithms

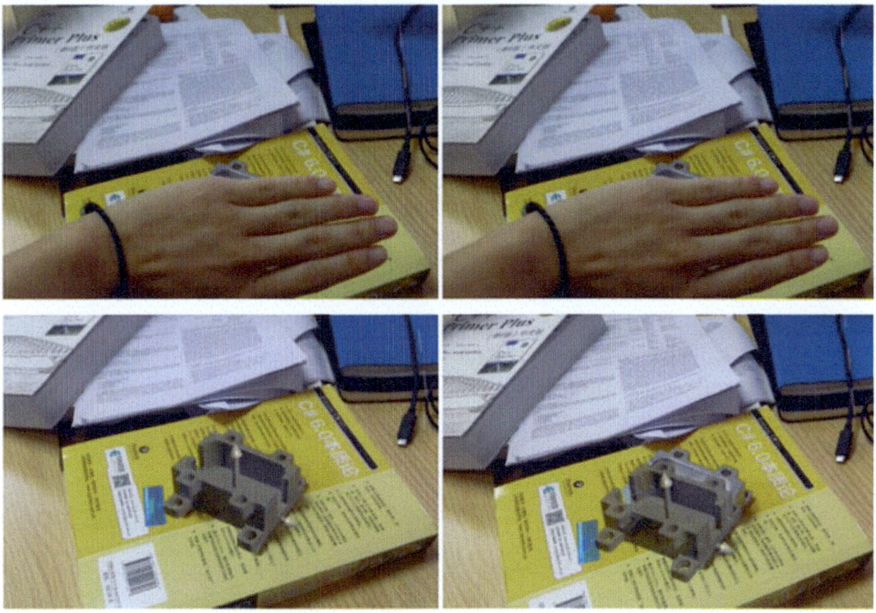

(a) The method of this article (b) LENMOD algorithm

Fig. 12.26 Comparison of simulated field of view loss

12.4 AR-Based Processing Prototype System Development and Case Analysis

This subsection begins with the development and implementation of a prototype system, followed by the validation and practical application of the system's functionality.

12.4.1 Prototype System Development and Implementation

1. **AR-Based Processing System Architecture**

The software architecture of the AR oriented process system consists of two parts: data monitoring and control. The monitoring system collects and processes data through various devices and sensors. This mainly includes the geometric shape, physical and equipment status information of the workpiece, and generates machining status predictions and machining suggestions based on this information. Finally, these contents are converted into AR views and displayed in real-time. The control section allows users to interact with the physical and virtual parts of the processing system. After obtaining intuitive and comprehensive process information from the

AR view of the processing system, users can use this information to make decisions and directly control digital and physical spaces through the AR system. Users can control virtual components through gestures and voice commands to influence the physical world.

(2) **Data Acquisition**

The data collection of the processing system is shown in Fig. 12.27. During system operation, read static data directly from the database as needed. There are two types of dynamic data: machine status data and external sensor data installed on the machine. Machine status data can be directly read from the machine controller through the OPC-UA protocol. External sensor data is read to the terminal through the corresponding communication protocol, and then the data is cleaned and sorted to eliminate zero and outlier values, and finally integrated into the information integration model (Fig. 12.28).

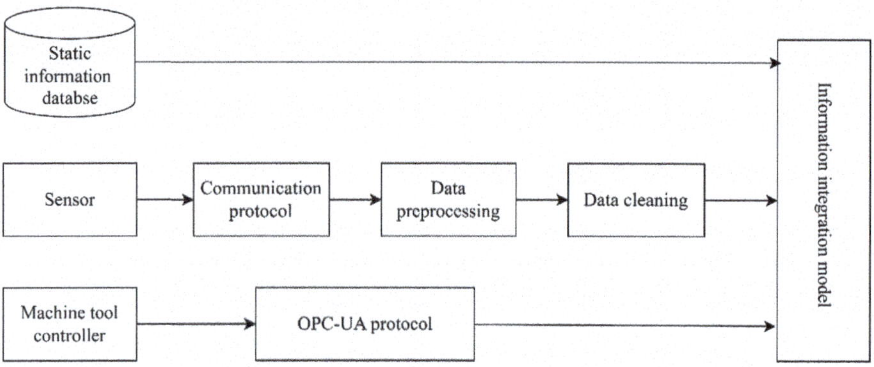

Fig. 12.27 Data collection method

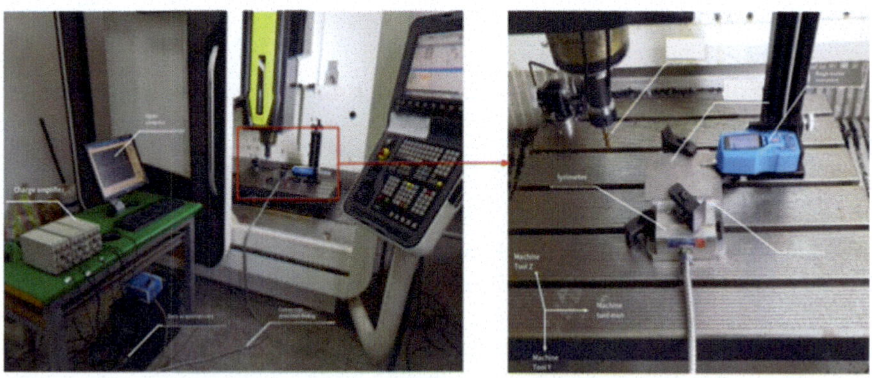

Fig. 12.28 Schematic diagram of the data collection environment

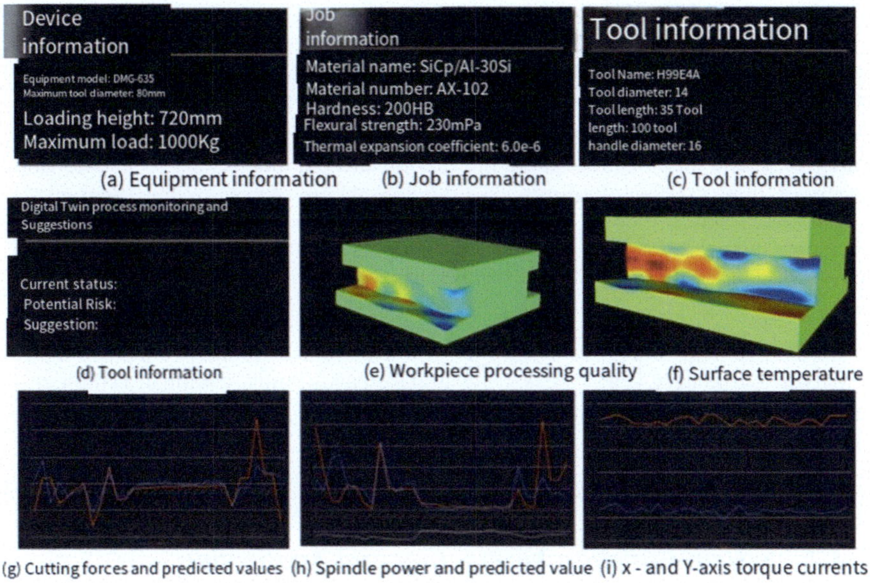

Fig. 12.29 System creation and release

(3) AR System Release

The augmented reality system mainly includes four steps: configuring the Hololens development environment, setting up the screen, importing the software development toolkit, compiling and uploading the project, as shown in Fig. 12.29.

1. Opt for the Hololens Development Environment

To craft a Hololens application, it is imperative to utilize the UWP platform. Select the UWP platform as the primary development platform in the system settings view and opt for the latest Windows 10 SDK. The target architecture should be x64, with the Hololens designated as the target device.

2. Configuration Interface

For a comprehensive AR encounter, incorporation of the Windows Mixed Reality package into Unity settings is essential. Additionally, select a 16-bit depth render to mitigate rendering overhead. Despite being the latest advanced display development device, Hololens still exhibits a discernible performance disparity when compared to computers, necessitating meticulous oversight of performance overhead.

3. **Integration of MRTK Software Package**

The MRTK Software Development Kit, an array of application development tools furnished by Microsoft explicitly for augmented reality systems, offers comprehensive support for Hololens development. This package encompasses a cross-platform input system and modules for spatial interaction and UI, functioning as an extensible framework that significantly diminishes the complexity of system development.

4. **System Deployment and Debugging**

Handpick the corresponding scene in Unity3D, package it, and generate .sln files. Subsequently, utilize VisualStudio 2019 to compile the project file by opening the project file. Ultimately, select the ARM64 and Release options, establish a connection with Hololens, execute system deployment, and conduct system testing within the Hololens environment.

12.4.2 System Function Verification and Application Examples

Taking the processing of a typical aerospace structural part with a carrier box as an example, the functions and parameters of the system are experimentally verified. The workpiece is made of high-volume fraction aluminium silicon carbide, which has the advantages of high strength, low density and excellent dimensional stability. During machining, the surface quality and machining accuracy of the workpiece are difficult to ensure due to the problem of fast tool wear caused by the low melting point of the material. Figure 12.30 shows the physical and digital model of the workpiece $\left(SiC_p/Al\right)$.

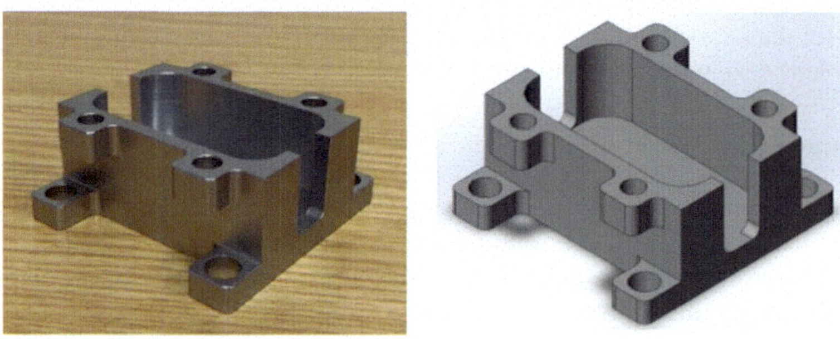

Fig. 12.30 With carrier cassette and digital model

(1) Dynamic View Building and Multi-view Generation Function Verification

Based on the types and characteristics of the data collected in Sect. 12.4.1, different dynamic AR view objects were constructed, as shown in Fig. 12.31. It mainly includes AR view objects such as equipment information, workpiece processing quality information, cutting force and its predicted values.

The method proposed in Sect. 12.2.2 is used to evaluate all dynamic AR views in the current process state, as shown in Table 12.2 for the scores of AR views.

Based on the score results, the system selects the top six views as recommended views. Then, by adding the fixed view (control option view and operation process view), an AR multi-view of the machining process as shown in Fig. 12.32 is generated.

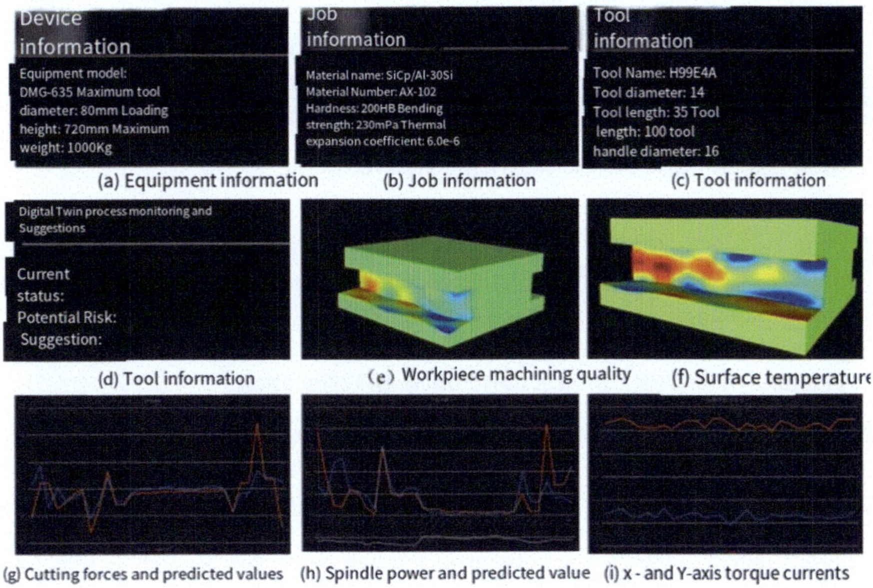

Fig. 12.31 Schematic diagram of the dynamic AR view object

Table 12.2 Partial AR view scores

View	Score	View	Score
View of workpiece machining quality	0.9236	Machining parameter view	0.6113
Machining suggestion view	0.8626	Workpiece surface temperature view	0.4936
Tool temperature view	0.8232	Job information view	0.4800
Spindle speed view	0.8210	Tool information view	0.4390
Program	0.7670	Device information view	0.4270

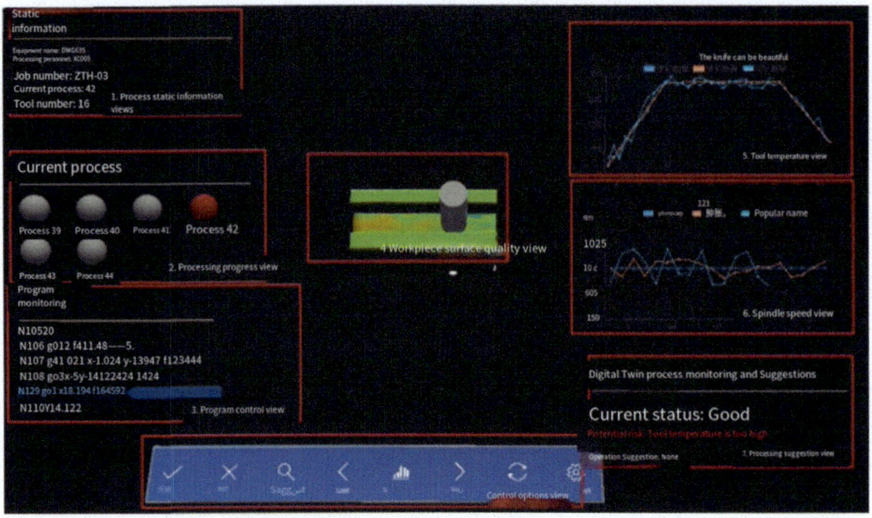

Fig. 12.32 AR multi-view diagram of the processing system

It includes machining static information view, machining progress view, program control view, workpiece surface quality view, tool temperature view, spindle speed view, machining suggestion view and control view.

(2) Human–Computer Interaction Function Verification

Users can interact and control the system in depth through gestures, voice and other ways. As shown in Fig. 12.33, most of the views in the system can be interacted with through gestures and voices, including but not limited to, adjusting the scale, position, and angle of the view object. Select different view objects from the View Gallery; Control of machining equipment through view objects, such as selecting the current operation, adjusting the NC code, changing tools, and so on. At the same time, when the operator gazes at an element in the view (the system can capture the user's gaze), the system will automatically pop up the specific value of that element. This enables more granular situational awareness.

(3) System Application and Analysis

Finally, the AR-oriented processing system was applied to the actual processing process with carrier box. Figure 12.34 shows the user's perspective of the system during machining, i.e., what the user actually sees and what details of the view. As you can see from the diagram, Hololens glasses provide an intuitive and comprehensive view of the data information in the manufacturing environment. At the same time, the operator can select the process and control the machine tool through the system.

The existing process information representation method and the proposed method. Compared with traditional methods, existing methods combined with augmented

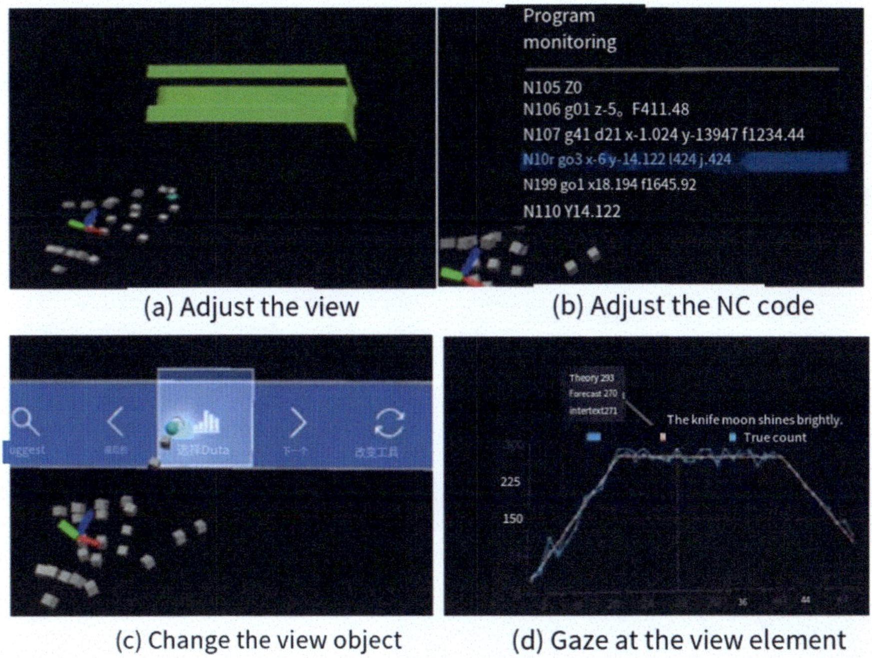

Fig. 12.33 Schematic diagram of real-time interaction

reality can perceive and predict the geometric and physical changes of products in real time, and recommend relevant decision-making methods. However, compared to the methods proposed in this article, they lack efficient and intuitive information visualization and interaction.

In order to verify the effectiveness of the system in practical use, 20 volunteers were invited to rate the system from eight dimensions. It includes the richness of information during the processing, satisfaction with view recommendations, smoothness of view interaction, comfort of system use, speed of virtual reality registration, accuracy of virtual reality tracking, and so on. The scoring criteria are 0–10 points, with higher scores indicating higher system satisfaction.

From the assessment findings, it is discernible that with regards to the visual system, the populace expresses contentment with the recommendations for views and the methods of interaction. However, the overall comfort of the system is somewhat lacking, primarily due to the requirement of wearing Hololens on the head during usage. This concentration of weight solely on the head poses certain inconveniences for prolonged periods of use. In the realm of virtual-real fusion tracking, the evaluators acknowledge the commendable performance of system registration and tracking. Nevertheless, there remains some room for enhancement in terms of view clarity. This limitation predominantly stems from the current augmented reality eyewear, which is substantially influenced by lighting conditions, thereby affecting display brightness and clarity.

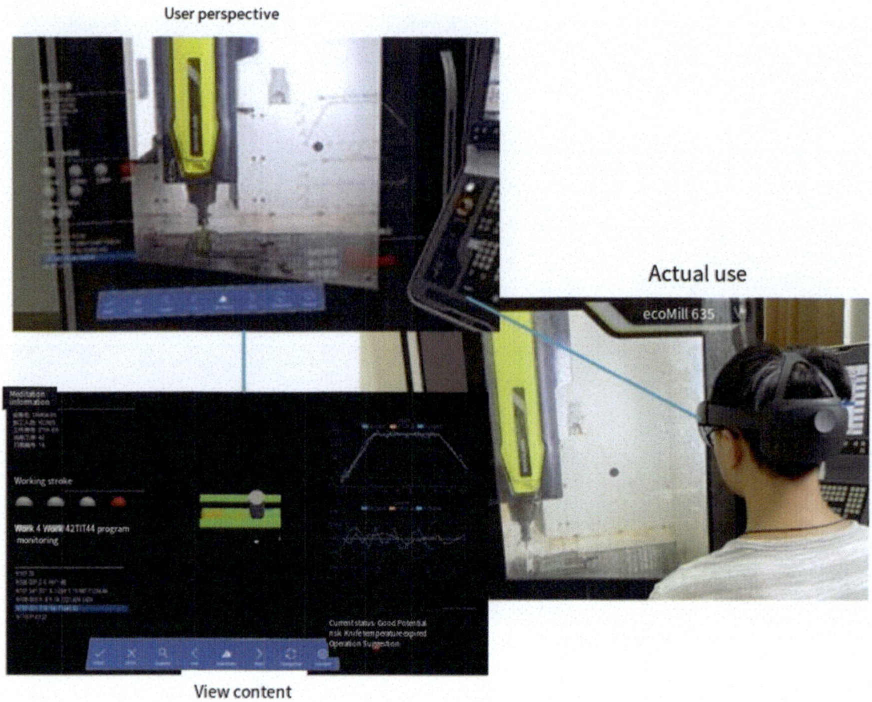

Fig. 12.34 Actual usage scenarios

References

1. Chen CJ, Hong J, Wang SF (2015) Automated positioning of 3D virtual scene in AR-based assembly and disassembly guiding system. Int J Adv Manuf Technol 76(5–8):753–764
2. Fang HC, Ong SK, Nee AYC (2013) Orientation planning of robot end-effector using augmented reality. Int J Adv Manuf Technol 67(9–12):2033–2049
3. Dalal N, Triggs B (2005) Histograms of oriented gradients for human detection. In: 2005 IEEE computer society conference on computer vision and pattern recognition (CVPR'05), San Diego, CA, USA, pp 886–893
4. Hinterstoisser S, Lepetit V, Ilic S et al (2010) Dominant orientation templates for real-time detection of texture-less objects. In: IEEE conference on computer vision and pattern recognition. IEEE, pp 2257–2264
5. Hinterstoisser S, Cagniart C, Ilic S et al (2012) Gradient response maps for real-time detection of textureless objects. IEEE Trans Pattern Anal Mach Intell 34(5):876–888
6. Park Y, Lepetit V, Woo W (2011) Texture-less object tracking with online training using an RGB-D camera. In: IEEE international symposium on mixed & augmented reality
7. Nguyen DD, Ko JP, Jeon JW (2015) Determination of 3D object pose in point cloud with CAD model. In: 2015 21st Korea-Japan joint workshop on frontiers of computer vision (FCV), p 7103725
8. Newcombe RA, Fox D, Seitz SM (2015) DynamicFusion: reconstruction and tracking of non-rigid scenes in real-time. In: IEEE conference on computer vision and pattern recognition, pp 343–352

9. Mur-Artal R, Montiel JMM, Tardos JD (2015) ORB-SLAM: a versatile and accurate monocular SLAM system. IEEE Trans Rob 31(5):1147–1163
10. Engel J, Schöps T, Cremers D (2014) LSD-SLAM: large-scale direct monocular SLAM. In: Lecture notes in computer science. Springer International Publishing, Cham, pp 834–849
11. Nuernberger B, Ofek E, Benko H et al (2016) SnapToReality: aligning augmented reality to the real world. In: Proceedings of the 2016 CHI conference on human factors in computing systems (CHI'16), NY, USA, 2016, pp 1233–1244
12. Egger J, Masood T (2020) Augmented reality in support of intelligent manufacturing—a systematic literature review. Comput Ind Eng 140:106195
13. Masood T, Egger J (2019) Augmented reality in support of Industry 4.0—implementation challenges and success factors. Robot Comput Integr Manufact 58:181–195
14. Wang Y, Zhang S, Yang S et al (2017) A LINE-MOD-based markerless tracking approach for AR applications. Int J Adv Manufact Technol 89(5–8):1699–1707
15. Hinterstoisser S, Cagniart C, Ilic S et al (2012) Gradient response maps for real-time detection of texture-less objects. IEEE Trans Pattern Anal Mach Intell 34(5):876–888
16. Thomas PC, David WM (1992) Augmented reality: an application of heads-up display technology to manual manufacturing processes. In: Proceedings of the twenty-fifth hawaii international conference on system sciences, Kauai, HI, USA, pp 659–669
17. Wu Y, He H, Zhang F et al (2015) Interactive operation method for augmented reality workshop layout design. Comput Integr Manufact Syst 21(05):1187–1192
18. Wohlhart P, Lepetit V (2015) Learning descriptors for object recognition and 3D pose estimation. In: 2015 IEEE conference on computer vision and pattern recognition (CVPR), pp 7–12
19. Wagner D, Reitmayr G, Mulloni A et al (2010) Real-time detection and tracking for augmented reality on mobile phones. IEEE Trans Vis Comput Graph 16(3):355–368
20. Nguyen DD, Ko JP, Jeon JW (2015) Determination of 3D object pose in point cloud with CAD model. In: 2015 21st Korea-Japan joint workshop on frontiers of computer vision (FCV). IEEE, 2015, pp 1–6
21. Guo J (2017) Attitude tracking based on monocular vision and IMU on mobile devices. Huazhong University of Science and Technology
22. Xiao B (2014) Research on mobile augmented reality technology based on natural features and sensing information. Xiamen University

The manufacturer's authorised representative in the EU is Springer Nature Customer Service Centre GmbH, Europaplatz 3, 69115 Heidelberg, Germany. If you have any concerns regarding our products, please contact ProductSafety@springernature.com

Printed and bound by CPI Group (UK) Ltd, Croydon, CR0 4YY
23/09/2026
02221471-0007